This is a textbook for the senior undergraduate or graduate student beginning a serious study of X-ray crystallography. It will be of interest both to those intending to become professional crystallographers and to those physicists, chemists, biologists, geologists, metallurgists and others who will use it as a tool in their research. All major aspects of crystallography are covered – the geometry of crystals and their symmetry, theoretical and practical aspects of diffracting X-rays by crystals and how the data may be analysed to find the symmetry of the crystal and its structure. Recent advances are fully covered, including the synchrotron as a source of X-rays, methods of solving structures from powder data and the full range of techniques for solving structures from single-crystal data. A suite of computer programs is provided for carrying out many operations of data-processing and solving crystal structures – including by direct methods. While these are limited to two dimensions they fully illustrate the characteristics of three-dimensional work. These programs are required for many of the problems given at the end of each chapter but may also be used to create new problems by which students can test themselves or each other.

T0213396

# An introduction to X-ray crystallography

# An introduction to
# X-ray crystallography

## SECOND EDITION

## M. M. WOOLFSON

*Emeritus Professor of Theoretical Physics*
*University of York*

PUBLISHED BY THE PRESS SYNDICATE OF THE UNIVERSITY OF CAMBRIDGE
The Pitt Building, Trumpington Street, Cambridge CB2 1RP, United Kingdom

CAMBRIDGE UNIVERSITY PRESS
The Edinburgh Building, Cambridge CB2 2RU, United Kingdom
40 West 20th Street, New York, NY 10011-4211, USA
10 Stamford Road, Oakleigh, Melbourne 3166, Australia

First published 1970
Second edition 1997

Typeset in Times 10/12 pt

*A catalogue record for this book is available from the British Library*

*Library of Congress cataloguing in publication data*

Woolfson, M. M.
An introduction to X-ray crystallography / M.M. Woolfson. – 2nd ed.
  p.  cm.
Includes bibliographical references and index.
ISBN 0 521 41271 4 (hardcover). – ISBN 0 521 42359 7 (pbk.)
1. X-ray crystallography.   I. Title.
QD945.W58   1997
548′.83–dc20      96–5700 CIP

ISBN 0 521 41271 4 hardback
ISBN 0 521 42359 7 paperback

Transferred to digital printing 2003

# Contents

# Preface to the First Edition

In 1912 von Laue proposed that X-rays could be diffracted by crystals and shortly afterwards the experiment which confirmed this brilliant prediction was carried out. At that time the full consequences of this discovery could not have been fully appreciated. From the solution of simple crystal structures, described in terms of two or three parameters, there has been steady progress to the point where now several complex biological structures have been solved and the solution of the structures of some crystalline viruses is a distinct possibility.

X-ray crystallography is sometimes regarded as a science in its own right and, indeed, there are many professional crystallographers who devote all their efforts to the development and practice of the subject. On the other hand, to many other scientists it is only a tool and, as such, it is a meeting point of many disciplines – mathematics, physics, chemistry, biology, medicine, geology, metallurgy, fibre technology and several others. However, for the crystallographer, the conventional boundaries between scientific subjects often seem rather nebulous.

In writing this book the aim has been to provide an elementary text which will serve either the undergraduate student or the postgraduate student beginning seriously to study the subject for the first time. There has been no attempt to compete in depth with specialized textbooks, some of which are listed in the Bibliography. Indeed, it has also been found desirable to restrict the breadth of treatment, and closely associated topics which fall outside the scope of the title – for example diffraction from semi- and non-crystalline materials, electron- and neutron diffraction – have been excluded. For those who wish to go no further it is hoped that the book gives a rounded, broad treatment, complete in itself, which explains the principles involved and adequately describes the present state of the subject. For those who wish to go further it should be regarded as a foundation for further study.

It has now become clear that there is wide acceptance of the SI system of units and by-and-large they are used in this book. However the ångstrom unit has been retained as a unit of length for X-ray wavelengths and unit-cell dimensions etc., since a great deal of the basic literature uses this unit. A brief explanation of the SI system and some important constants and equations are included in the section *Physical constants and tables* on pp. 325–326.

I am deeply indebted to Dr M. Bown and Dr S.G. Fleet of the Department of Mineralogy, University of Cambridge and to my colleague, Dr P. Main, for reading the manuscript and for their helpful criticism which included suggestions for many improvements of treatment.

My thanks are also due to Professor C. A. Taylor of the University of Cardiff for providing the material for figs. 8.9 and 8.10 and also to Mr W. Spellman and Mr B. Cooper of the University of York for help with some of the illustrations.

M.M.W.

# Preface to the Second Edition

Since the first edition of this book was published in 1970 there have been tremendous advances in X-ray crystallography. Much of this has been due to technological developments – for example new and powerful synchrotron sources of X-rays, improved detectors and increase in the power of computers by many orders of magnitude. Alongside these developments, and sometimes prompted by them, there have also been theoretical advances, in particular in methods of solution of crystal structures. In this second edition these new aspects of the subject have been included and described at a level which is appropriate to the nature of the book, which is still an introductory text.

A new feature of this edition is that advantage has been taken of the ready availability of powerful table-top computers to illustrate the procedures of X-ray crystallography with FORTRAN® computer programs. These are listed in the appendices and available on the WorldWideWeb*. While they are restricted to two-dimensional applications they apply to all the two-dimensional space groups and fully illustrate the principles of the more complicated three-dimensional programs that are available. The Problems at the end of each chapter include some in which the reader can use these programs and go through simulations of structure solutions – simulations in that the known structure is used to generate what is equivalent to observed data. More realistic exercises can be produced if readers will work in pairs, one providing the other with a data file containing simulated observed data for a synthetic structure of his own invention, while the other has to find the solution. It can be great fun as well as being very educational!

I am particularly grateful to Professor J. R. Helliwell for providing material on the new Laue method and on image-plate methods.

M. M. Woolfson
York 1996

*http://www.cup.cam.ac.uk/onlinepubs/412714/412714top.html

# 1 The geometry of the crystalline state

## 1.1 The general features of crystals

Materials in the crystalline state are commonplace and they play an important part in everyday life. The household chemicals salt, sugar and washing soda; the industrial materials, corundum and germanium; and the precious stones, diamonds and emeralds, are all examples of such materials.

A superficial examination of crystals reveals many of their interesting characteristics. The most obvious feature is the presence of facets and well-formed crystals are found to be completely bounded by flat surfaces – flat to a degree of precision capable of giving high-quality plane-mirror images. Planarity of this perfection is not common in nature. It may be seen in the surface of a still liquid but we could scarcely envisage that gravitation is instrumental in moulding flat crystal faces simultaneously in a variety of directions.

It can easily be verified that the significance of planar surfaces is not confined to the exterior morphology but is also inherent in the interior structure of a crystal. Crystals frequently cleave along preferred directions and, even when a crystal is crudely fractured, it can be seen through a microscope that the apparently rough, broken region is actually a myriad of small plane surfaces.

Another feature which may be readily observed is that the crystals of a given material tend to be alike – all needles or all plates for example – which implies that the chemical nature of the material plays an important role in determining the crystal habit. This suggests strongly that the macroscopic form of a crystal depends on structural arrangements at the atomic or molecular level and that the underlying factor controlling crystal formation is the way in which atoms and molecules can pack together. The flatness of crystal surfaces can then be attributed to the presence of regular layers of atoms in the structure and cleavage would correspond to the breaking of weaker links between particular layers of atoms.

## 1.2 The external symmetry of crystals

Many crystals are very regular in shape and clearly exhibit a great deal of symmetry. In fig. 1.1(a) there is shown a well-formed crystal of alum which has the shape of a perfect octahedron; the quartz crystal illustrated in fig. 1.1(b) has a cross-section which is a regular hexagon. However with many other crystals such symmetry is not evident and it might be thought that crystals with symmetry were an exception rather than a rule.

Although the crystals of a particular chemical species usually appear to

Fig. 1.1.
(*a*) Alum crystal.
(*b*) Quartz crystal.

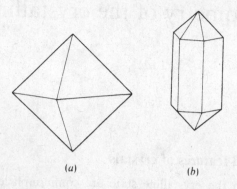

(*a*)                    (*b*)

have the same general habit a detailed examination reveals considerable variation in size and shape. In particular one may find a selection of platy crystals looking somewhat like those shown in fig. 1.2(*a*). The shapes of these seem to be quite unrelated but, if they are rearranged as in fig. 1.2(*b*), a rather striking relationship may be noted. Although the relative sizes of the sides of the crystal cross-sections are very different the normals to the sides (in the plane of the figure) form an identical set from crystal to crystal. Furthermore the set of normals is just that which would be obtained from a regular hexagonal cross-section although none of the crystals in fig. 1.2 displays the characteristics of a regular polygon. While this illustration is essentially two-dimensional the same general observations can be made in three dimensions. Although the crystals of a given species vary greatly in the shapes and sizes of corresponding faces, and may appear to lack symmetry altogether, the set of normals to the faces will be identical from crystal to crystal (although a crystal may occasionally lack a particular face completely) and will usually show symmetry that the crystals themselves lack. For example, fig. 1.3(*a*) shows the set of normals for an octahedron. These normals are drawn radiating from a single point and are of equal length. This set may well have been derived from a solid such as that shown in fig. 1.3(*b*) but the symmetry of the normals reveals that this solid has faces whose relative orientations have the same relationship as those of the octahedron.

The presentation of a three-dimensional distribution of normals as done in fig. 1.3 makes difficulties both for the illustrator and also for the viewer. The normals have a common origin and are of equal length so that their termini lie on the surface of a sphere. It is possible to represent a spherical distribution of points by a perspective projection on to a plane and the stereographic projection is the one most commonly used by the crystallographer. The projection procedure can be followed in fig. 1.4(*a*). Points on the surface of the sphere are projected on to a diametral plane with projection point either $O$ or $O'$, where $OO'$ is the diameter normal to the projection plane. Each point is projected from whichever of $O$ or $O'$ is on the opposite side of the plane and in this way all the projected points are contained within the diametral circle. The projected points may be conventionally represented as above or below the projection plane by full or open circles. Thus the points $A$, $B$, $C$ and $D$ project as $A'$, $B'$, $C'$ and $D'$ and, when viewed along $OO'$, the projection plane appears as in fig. 1.4(*b*).

## 1.2  The external symmetry of crystals

Fig. 1.2.
(a) Set of apparently
irregular plate-like
crystals.
(b) Crystals rearranged
to show parallelism of
faces and underlying
hexagonal symmetry.

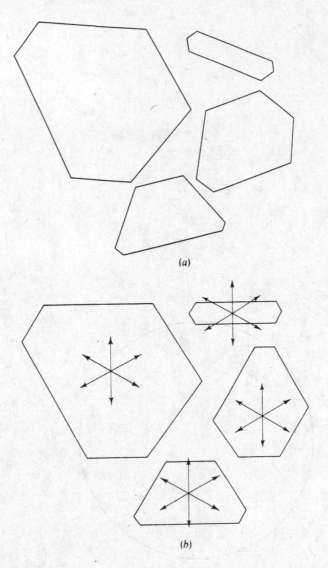

(a)

(b)

Fig. 1.3.
(a) Set of normals to the
faces of an octahedron.
(b) Solid whose faces
have same set of normals
as does an octahedron.

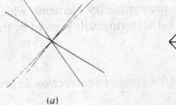

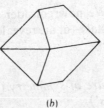

(a)

(b)

Fig. 1.4.
(a) The stereographic
projection of points
from the surface of a
sphere on to a
diametral plane.
(b) The final
stereographic
projection.

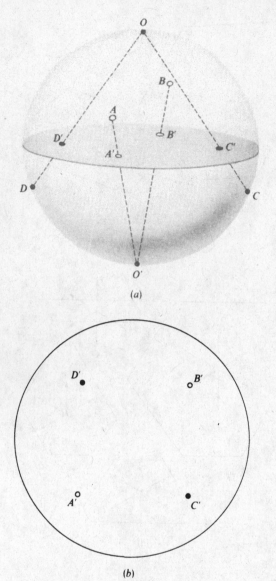

(a)

(b)

We now consider the symmetry elements which may be present in crystals – or are revealed as intrinsically present by the set of normals to the faces.

## Centre of symmetry (for symbol see section below entitled 'Inversion axes')

A crystal has a centre of symmetry if, for a point within it, faces occur in parallel pairs of equal dimensions on opposite sides of the point and equidistant from it. A selection of centrosymmetric crystals is shown in fig. 1.5(a). However even when the crystal itself does not have a centre of symmetry the intrinsic presence of a centre is shown when normals occur in collinear pairs. The way in which this shows up on a stereographic projection is illustrated in fig. 1.5(*b*).

### Mirror plane (written symbol m; graphical symbol —)

This is a plane in the crystal such that the halves on opposide sides of the plane are mirror images of each other. Some crystal forms possessing mirror planes are shown in fig. 1.6(*a*). Mirror planes show up clearly in a stereographic projection when the projecting plane is either parallel to or perpendicular to the mirror plane. The stereographic projections for each of the cases is shown in fig. 1.6(*b*).

Fig. 1.5.
(*a*) A selection of centrosymmetric crystals.
(*b*) The stereographic projection of a pair of centrosymmetrically related faces.

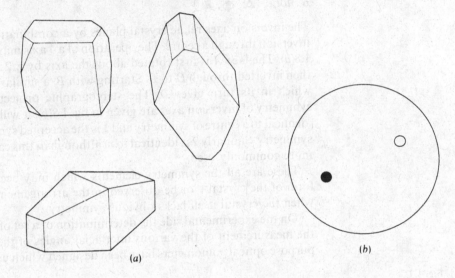

(*a*)

(*b*)

Fig. 1.6.
(*a*) Crystals with mirror planes.
(*b*) The stereographic projections of a pair of faces related by a mirror plane when the mirror plane is (i) in the plane of projection; (ii) perpendicular to the plane of projection.

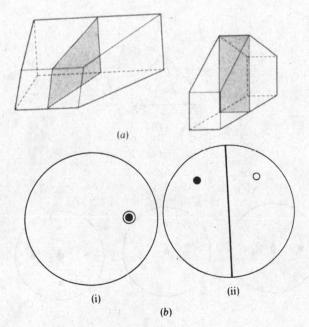

(*a*)

(i)                    (ii)

(*b*)

## Rotation axes (written symbols 2, 3, 4, 6; graphical symbols ◆, ▲, ◆, ⬢)

An *n*-fold rotation axis is one for which rotation through $2\pi/n$ leaves the appearance of the crystal unchanged. The values of *n* which may occur (apart from the trivial case $n = 1$) are 2, 3, 4 and 6 and examples of twofold (diad), threefold (triad), fourfold (tetrad) and sixfold (hexad) axes are illustrated in fig. 1.7 together with the stereographic projections on planes perpendicular to the symmetry axes.

## Inversion axes (written symbols $\bar{1}, \bar{2}, \bar{3}, \bar{4}, \bar{6}$; graphical symbols o, none, △, ◈, ⬡ )

The inversion axes relate crystal planes by a combination of rotation and inversion through a centre. The operation of a $\bar{4}$ axis may be followed in fig. 1.8(*a*). The face *A* is first rotated about the axis by $\pi/2$ to position *A'* and then inverted through *O* to *B*. Starting with *B*, a similar operation gives *C* which in its turn gives *D*. The stereographic projections showing the symmetry of inversion axes are given in fig. 1.8(*b*); it will be noted that $\bar{1}$ is identical to a centre of symmetry and $\bar{1}$ is the accepted symbol for a centre of symmetry. Similarly $\bar{2}$ is identical to *m* although in this case the symbol *m* is more commonly used.

These are all the symmetry elements which may occur in the external form of the crystal – or be observed in the arrangement of normals even when the crystal itself lacks obvious symmetry.

On the experimental side the determination of a set of normals involves the measurement of the various interfacial angles of the crystal. For this purpose optical goniometers have been designed which use the reflection of

Fig. 1.7.
(*a*) Perspective views and views down the axis for crystals possessing diad, triad, tetrad and hexad axes.
(*b*) The corresponding stereographic projections.

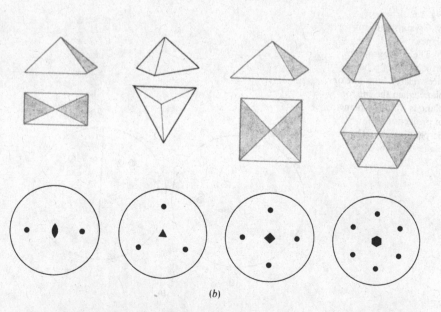

(*b*)

Fig. 1.8.
(*a*) A perspective view
of the operation of an
inverse tetrad axis.
(*b*) Stereographic
projections for $\bar{1}, \bar{2}, \bar{3}, \bar{4}$
and $\bar{6}$.

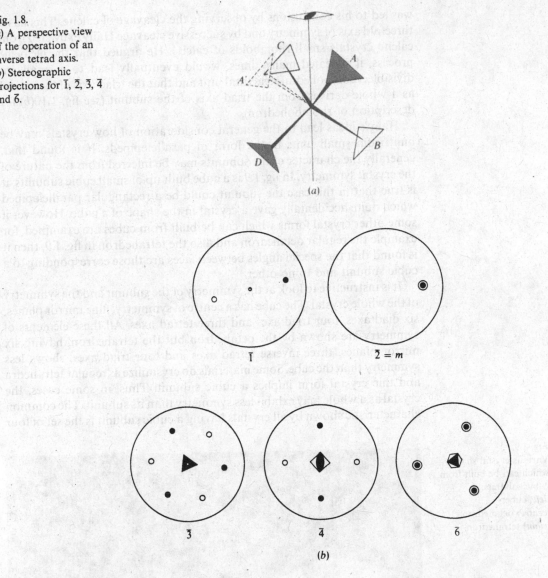

light from the mirror-like facets of the crystal to define their relative
orientations.

## 1.3 The seven crystal systems

Even from a limited observation of crystals it would be reasonable to
surmise that the symmetry of the crystal as a whole is somehow connected
with the symmetry of some smaller subunit within it. If a crystal is fractured
then the small plane surfaces exposed by the break, no matter in what part
of the body of the crystal they originate, show the same angular relationships
to the faces of the whole crystal and, indeed, are often parallel to the crystal
faces.

The idea of a structural subunit was first advanced in 1784 by Haüy who was led to his conclusions by observing the cleavage of calcite. This has a threefold axis of symmetry and by successive cleavage Haüy extracted from calcite crystals small rhomboids of calcite. He argued that the cleavage process, if repeated many times, would eventually lead to a small, indivisible, rhombohedral structural unit and that the triad axis of the crystal as a whole derives from the triad axis of the subunit (*see* fig. 1.10(*b*) for description of rhombohedron).

Haüy's ideas lead to the general consideration of how crystals may be built from small units in the form of parallelepipeds. It is found that, generally the character of the subunits may be inferred from the nature of the crystal symmetry. In fig. 1.9 is a cube built up of small cubic subunits; it is true that in this case the subunit could be a rectangular parallelepiped which quite accidentally gave a crystal in the shape of a cube. However if some other crystal forms which can be built from cubes are examined, for example the regular octahedron and also the tetrahedron in fig. 1.9, then it is found that the special angles between faces are those corresponding to a cubic subunit and to no other.

It is instructive to look at the symmetry of the subunit and the symmetry of the whole crystal. The cube has a centre of symmetry, nine mirror planes, six diad axes, four triad axes and three tetrad axes. All these elements of symmetry are shown by the octahedron but the tetrahedron, having six mirror planes, three inverse tetrad axes and four triad axes, shows less symmetry than the cube. Some materials do crystallize as regular tetrahedra and this crystal form implies a cubic subunit. Thus, in some cases, the crystal as a whole may exhibit less symmetry than its subunit. The common characteristic shown by all crystals having a cubic subunit is the set of four

Fig. 1.9.
Various crystal shapes which can be built from cubic subunits:
(*left*) cube;
(*centre*) octahedron;
(*right*) tetrahedron.

triad axes – and conversely all crystals having a set of four triad axes are cubic.

Similar considerations lead to the conclusion that there are seven distinct types of subunit and we associate these with seven *crystal systems*. The subunits are all parallelepipeds whose shapes are completely defined by the lengths of the three sides *a, b, c* (or the ratios of these lengths) and the values of the three angles $\alpha, \beta, \gamma$ (fig. 1.10(*a*)). The main characteristics of the seven crystal systems and their subunits are given in table 1.1.

## 1.4 The thirty-two crystal classes

In table 1.1 there is given the essential symmmetry for the seven crystal systems but, for each system, different symmetry arrangements are possible. A crystal in the triclinic system, for example, may or may not have a centre of symmetry and this leads us to refer to the two *crystal classes* $\bar{1}$ and 1 within the triclinic system. As has been previously noted $\bar{1}$ is the symbol for a centre of symmetry and the symbol 1, representing a onefold axis, corresponds to no symmetry at all. These two crystal classes may be shown conveniently in terms of stereographic projections as in fig. 1.11(*a*) and (*b*). The projections show the set of planes generated from a general crystal face by the complete group of symmetry elements.

The possible arrangements for the monoclinic system are now considered.

Fig. 1.10.
(*a*) A general parallelepiped subunit.
(*b*) A rhombohedron showing the triad axis.
(*c*) The basic hexagonal subunits which are packed as shown to give hexagonal symmetry.

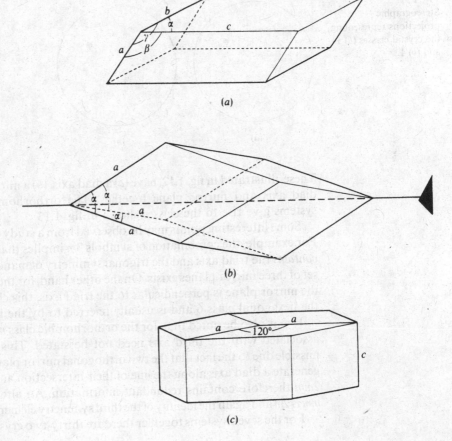

(*a*)

(*b*)

(*c*)

Table 1.1. *The seven crystal systems*

| System | Subunit | Essential symmetry of crystal |
|---|---|---|
| Triclinic | No special relationships | None |
| Monoclinic | $a \neq b \neq c$ | Diad axis or mirror plane |
| | $\beta \neq \alpha = \gamma = 90°$ | (inverse diad axis) |
| Orthorhombic | $a \neq b \neq c$ | Three orthogonal diad or inverse |
| | $\alpha = \beta = \gamma = 90°$ | diad axes |
| Tetragonal | $a = b \neq c$ | One tetrad or inverse tetrad |
| | $\alpha = \beta = \gamma = 90°$ | axis |
| Trigonal | $a = b = c$ | One triad or inverse triad |
| | $\alpha = \beta = \gamma \neq 90°$ | axis |
| | (see fig. 1.10(*b*)) | |
| | or as hexagonal | |
| Hexagonal | $a = b \neq c$ | One hexad or inverse hexad |
| | $\alpha = \beta = 90°, \gamma = 120°$ | axis |
| | (see fig. 1.10(*c*)) | |
| Cubic | $a = b = c$ | Four triad axes |
| | $\alpha = \beta = \gamma = 90°$ | |

Fig. 1.11.
Stereographic
projections representing
the crystal classes (*a*) 1
and (*b*) $\bar{1}$.

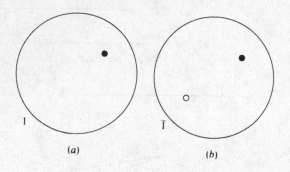

(*a*)          (*b*)

These, illustrated in fig. 1.12, have (*a*) a diad axis, (*b*) a mirror plane and (*c*) a diad axis and mirror plane together. The orthorhombic and trigonal systems give rise to the classes shown in fig. 1.13.

Some interesting points may be observed from a study of these diagrams. For example, the combination of symbols 3*m* implies that the mirror plane *contains* the triad axis and the trigonal symmetry demands therefore that a set of three mirror planes exists. On the other hand, for the crystal class 3/*m*, the mirror plane is perpendicular to the triad axis; this class is identical to the hexagonal class $\bar{6}$ and is usually referred to by the latter name.

It may also be noted that, for the orthorhombic class *mm*, the symmetry associated with the third axis need not be stated. This omission is permissible due to the fact that the two orthogonal mirror planes automatically generate a diad axis along the line of their intersection and a name such as 2*mm* therefore contains redundant information. An alternative name for *mm* is 2*m* and again the identity of the third symmetry element may be inferred.

For the seven systems together there are thirty-two crystal classes and all

Fig. 1.12.
Stereographic projections
representing the three
crystal classes in the
monoclinic system (a) 2,
(b) m and (c) 2/m.

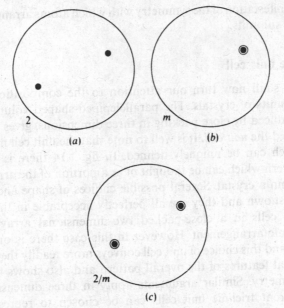

2

(a)

m

(b)

2/m

(c)

Fig. 1.13.
Stereographic
projections representing
the three crystal classes
in the orthorhombic
system and the six
classes in the trigonal
system.

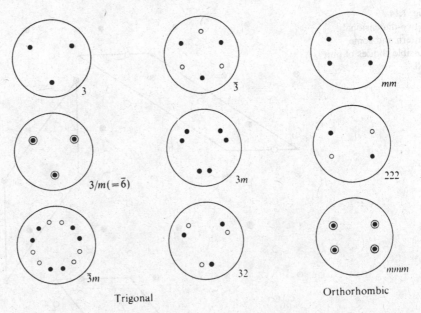

3

$\bar{3}$

mm

$3/m(=\bar{6})$

3m

222

$\bar{3}m$

32

mmm

Trigonal

Orthorhombic

crystals may be assigned to one or other of these classes. While the general
nature of the basic subunit determines the crystal system, for each system
there can be different elements of symmetry associated with the crystal. If a
material, satisfying some minimization-of-potential-energy criterion, crystallizes with some element of symmetry, it strongly implies that there is
some corresponding symmetry within the subunit itself. The collection of
symmetry elements which characterizes the crystal class, and which must
also be considered to be associated with the basic subunit, is called a *point
group*. It will be seen later that the point group is a macroscopic

manifestation of the symmetry with which atoms arrange themselves within the subunits.

## 1.5 The unit cell

We shall now turn our attention to the composition of the structural subunits of crystals. The parallelepiped-shaped volume which, when reproduced by close packing in three dimensions, gives the whole crystal is called the *unit cell*. It is well to note that the unit cell may not be an entity which can be uniquely defined. In fig. 1.14 there is a two-dimensional pattern which can be thought of as a portion of the arrangement of atoms within a crystal. Several possible choices of shape and origin of unit cell are shown and they are all perfectly acceptable in that reproducing the unit cells in a close-packed two-dimensional array gives the correct atomic arrangement. However in this case there is one rectangular unit cell and this choice of unit cell conveys more readily the special rectangular repeat features of the overall pattern and also shows the mirror plane of symmetry. Similar arguments apply in three dimensions in that many different triclinic unit cells can be chosen to represent the structural

Fig. 1.14.
A two-dimensional pattern and some possible choices of unit cell.

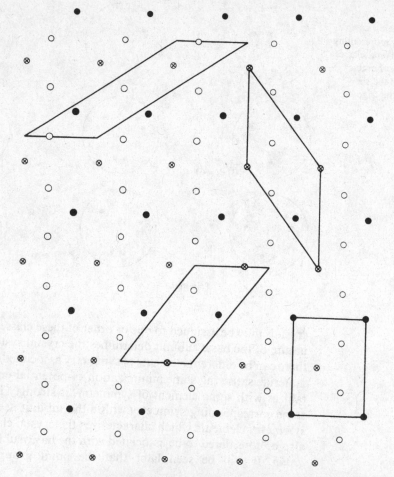

arrangement. One customarily chooses the unit cell which displays the highest possible symmetry, for this indicates far more clearly the symmetry of the underlying structure.

In §§ 1.3 and 1.4 the ideas were advanced that the symmetry of the crystal was linked with the symmetry of the unit cell and that the disposition of crystal faces depends on the shape of the unit cell. We shall now explore this idea in a little more detail and it helps, in the first instance, to restrict attention to a two-dimensional model. A crystal made of square unit cells is shown in fig. 1.15. The crystal is apparently irregular in shape but, when the set of normals to the faces is examined we have no doubt that the unit cell has a tetrad axis of symmetry. The reason why a square unit cell with a tetrad axis gives fourfold symmetry in the bulk crystal can also be seen. If the formation of the faces $AB$ and $BC$ is favoured because of the low potential energy associated with the atomic arrangement at these boundaries then $CD$, $DE$ and the other faces related by tetrad symmetry are also favoured because they lead to the same condition at the crystal boundary.

For the two-dimensional crystal in fig. 1.16 the set of normals reveals a mirror line of symmetry and from this we know that the unit cell is rectangular. It is required to determine the ratio of the sides of the rectangle from measurements of the angles between the faces. The mirror line can be located (we take the normal to it as the $b$ direction) and the angles made to this line by the faces can be found. In fig. 1.17 the face $AB$ is formed by points which are separated by $2a$ in one direction and $b$ in the other. The angle $\theta$, which the normal $AN$ makes with the $b$ direction, is clearly given by

$$\tan \theta = b/2a. \tag{1.1}$$

Fig. 1.15.
A two-dimensional crystal made up of unit cells with a tetrad axis of symmetry.

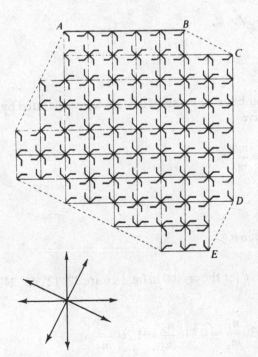

Fig. 1.16.
A two-dimensional
crystal built of
rectangular units.

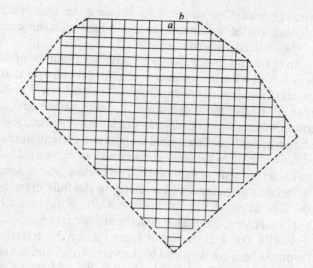

Fig. 1.17.
The relationship between
the crystal face AB and
the unit cell.

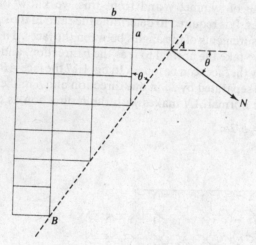

If the neighbouring points of the face were separated by $na$ and $mb$ then one
would have

$$\tan \theta = \frac{mb}{na}$$

or

$$\frac{b}{a} = \frac{n}{m} \tan \theta. \tag{1.2}$$

The angles $\theta$ for the crystal in fig. 1.16 are $32° \, 12'$, $43° \, 24'$ and $51° \, 33'$ so that
we have

$$\frac{b}{a} = 0.630 \frac{n_1}{m_1} = 0.946 \frac{n_2}{m_2} = 1.260 \frac{n_3}{m_3}. \tag{1.3}$$

We now look for the simplest sets of integers $n$ and $m$ which will satisfy equation (1.3) and these are found to give

$$\frac{b}{a} = 0.630 \times \frac{2}{1} = 0.946 \times \frac{4}{3} = 1.260 \times \frac{1}{1}.$$

From this we deduce the ratio $b:a = 1.260:1$.

This example is only illustrative and it is intended to demonstrate how measurements on the bulk crystal can give precise information about the substructure. For a real crystal, where one is dealing with a three-dimensional problem, the task of deducing axial ratios can be far more complicated.

Another type of two-dimensional crystal is one based on a general oblique cell as illustrated in fig. 1.18. The crystal symmetry shown here is a diad axis (although not essential for this system) and one must deduce from the interfacial angles not only the axial ratio but also the interaxial angle. Many choices of unit cell are possible for the oblique system.

The only unconsidered type of two-dimensional crystal is that based on a hexagonal cell where the interaxial angle and axial ratio are fixed.

All the above ideas can be carried over into three dimensions. Goniometric measurements enable one to determine the crystal systems, crystal class, axial ratios and interaxial angles.

## 1.6 Miller indices

In fig. 1.19 is shown the development of two faces $AB$ and $CD$ of a two-dimensional crystal. Face $AB$ is generated by steps of $2a, b$ and $CD$ by steps of $3a, 2b$. Now it is possible to draw lines parallel to the faces such that their intercepts on the unit-cell edges are $a/h, b/k$ where $h$ and $k$ are two integers.

The line $A'B'$ parallel to $AB$, for example, has intercepts $OA'$ and $OB'$ of the form $a/1$ and $b/2$; similarly $C'D'$ parallel to $CD$ has intercepts $a/2$ and $b/3$. The integers $h$ and $k$ may be chosen in other ways – the line with

Fig. 1.18.
A two-dimensional crystal based on an oblique unit cell.

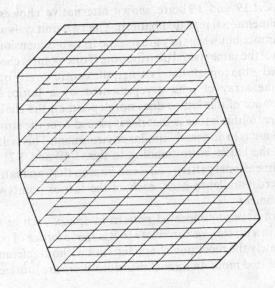

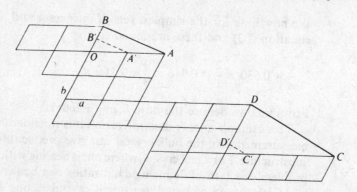

intercepts $a/2$ and $b/4$ is also parallel to $AB$. However, we are here concerned with the smallest possible integers and these are referred to as the *Miller indices* of the face.

In three dimensions a plane may always be found, parallel to a crystal face, which makes intercepts $a/h$, $b/k$ and $c/l$ on the unit-cell edges. The crystal face in fig. 1.20 is based on the unit cell shown with $OA = 3a$, $OB = 4b$ and $OC = 2c$. The plane $A'B'C'$ is parallel to $ABC$ and has intercepts $OA'$, $OB'$ and $OC'$ given by $a/4$, $b/3$ and $c/6$ (note that the condition for parallel planes $OA/OA' = OB/OB' = OC/OC'$ is satisfied). This face may be referred to by its Miller indices and $ABC$ is the face (436).

The Miller indices are related to a particular unit cell and are therefore not uniquely defined for a given crystal face. Returning to our two-dimensional example, the unit cell in fig. 1.21 is an alternative to that shown in fig. 1.19. The face $AB$ which was the $(1, 2)$ face for the cell in fig. 1.19 is the $(1, 1)$ face for the cell in fig. 1.21. However, no matter which unit cell is chosen, one can find a triplet of integers (generally small) to represent the Miller indices of the face.

## 1.7 Space lattices

In figs. 1.19 and 1.21 are shown alternative choices of unit cell for a two-dimensional repeated pattern. The two unit cells are quite different in appearance but when they are packed in two-dimensional arrays they each produce the same spatial distribution. If one point is chosen to represent the unit cell – the top left-hand corner, the centre or any other defined point – then the array of cells is represented by a lattice of points and the appearance of this lattice does not depend on the choice of unit cell. One property of this lattice is that if it is placed over the structural pattern then each point is in an exactly similar environment. This is illustrated in fig. 1.22 where the lattice corresponding to figs. 1.19 and 1.21 is placed over the two-dimensional pattern and it can be seen that, no matter how the lattice is displaced parallel to itself, each of the lattice points will have a similar environment.

If we have any repeated pattern in space, such as the distribution of atoms in a crystal, we can relate to it a *space lattice* of points which defines completely the repetition characteristics without reference to the details of the repeated motif. In three dimensions there are fourteen distinctive space

Fig. 1.20.
The plane $A'B'C'$ is
parallel to the crystal
face $ABC$ and makes
intercepts on the cell
edges of the form $a/h$,
$b/k$ and $c/l$ where $h$, $k$
and $l$ are integers.

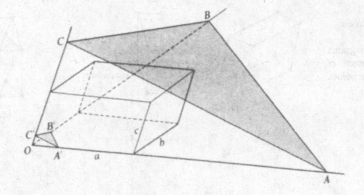

Fig. 1.21.
An alternative unit cell
to that shown in fig.
1.19. The faces $AB$ and
$CD$ now have different
Miller indices.

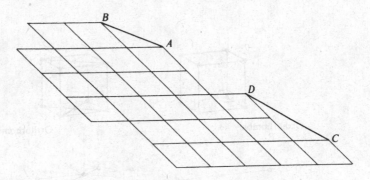

Fig. 1.22.
The lattice (small dark
circles) represents the
translational repeat
nature of the pattern
shown.

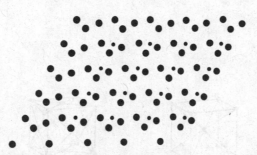

lattices known as *Bravais lattices*. The unit of each lattice is illustrated in fig.
1.23; lines connect the points to clarify the relationships between them.
Firstly there are seven simple lattices based on the unit-cell shapes
appropriate to the seven crystal systems. Six of these are indicated by the
symbol $P$ which means 'primitive', i.e. there is one point associated with
each unit cell of the structure; the primitive rhombohedral lattice is usually
denoted by $R$. But other space lattices can also occur. Consider the space
lattice corresponding to the two-dimensional pattern given in fig. 1.24. This
could be considered a primitive lattice corresponding to the unit cell shown
dashed in outline but such a choice would obscure the rectangular repeat
relationship in the pattern. It is appropriate in this case to take the unit cell
as the full line rectangle and to say that the cell is *centred* so that points
separated by $\frac{1}{2}a, \frac{1}{2}b$ are similar. Such a lattice is non-primitive. The three
possible types of non-primitive lattice are:

Fig. 1.23.
The fourteen Bravais
lattices. The
accompanying diagrams
show the environment of
each of the lattice points.

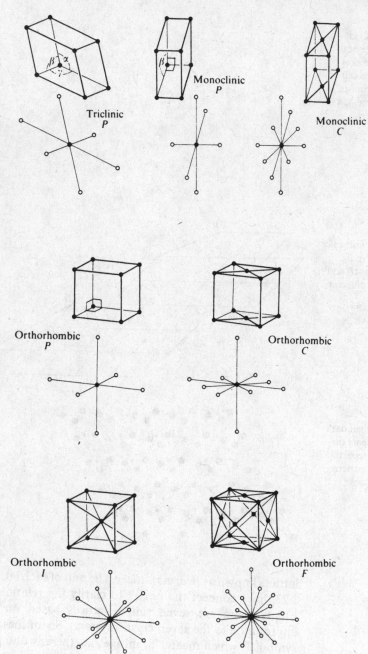

Triclinic
*P*

Monoclinic
*P*

Monoclinic
*C*

Orthorhombic
*P*

Orthorhombic
*C*

Orthorhombic
*I*

Orthorhombic
*F*

(1st Part)

## 1.7 Space lattices

Fig. 1.23. (cont.)

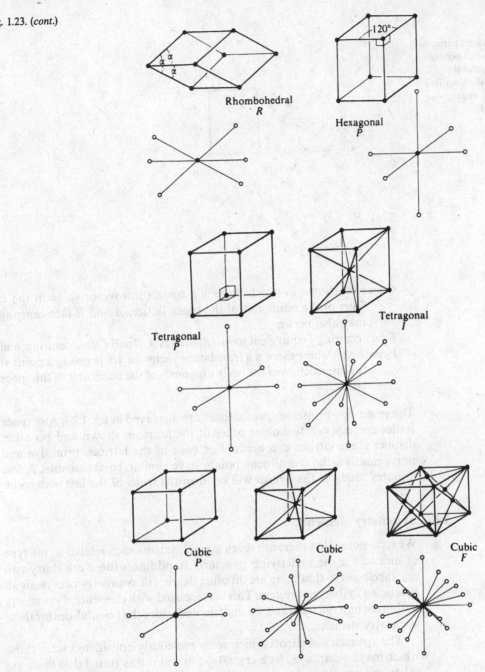

Fig. 1.23. (2nd Part)

Fig. 1.24.
Two-dimensionl pattern
showing two choices of
unit cell – general
oblique (dashed outline)
and centred rectangular
(full outline).

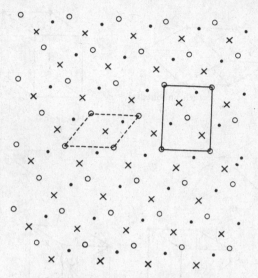

*C*-face centring – in which there is a translation vector $\frac{1}{2}a, \frac{1}{2}b$ in the *C* faces of the basic unit of the space lattice. *A* and *B*-face centring may also occur;

*F*-face centring – equivalent to simultaneous *A*, *B* and *C*-face centring; and
*I*-centring – where there is a translation vector $\frac{1}{2}a, \frac{1}{2}b, \frac{1}{2}c$ giving a point at the intersection of the body diagonals of the basic unit of the space lattice.

The seven non-primitive space lattices are displayed in fig. 1.23. Any space lattice corresponds to one or other of the fourteen shown and no other distinct space lattices can occur. For each of the lattices, primitive and non-primitive, the constituent points have similar environments. A few minutes' study of the figures will confirm the truth of the last statement.

## 1.8 Symmetry elements

We have noted that there are seven crystal systems each related to the type of unit cell of the underlying structure. In addition there are thirty-two crystal classes so that there are differing degrees of symmetry of crystals all belonging to the same system. This is associated with elements of symmetry within the unit cell itself and we shall now consider the possibilities for these symmetry elements.

The symmetry elements which were previously considered were those which may be displayed by a crystal (§ 1.2) and it was stated that there are thirty-two possible arrangements of symmetry elements or point groups. A crystal is a single unrepeated object and an arrangement of symmetry elements all associated with one point can represent the relationships of a crystal face to all symmetry-related faces.

The situation is different when we consider the symmetry within the unit cell, for the periodic repeat pattern of the atomic arrangement gives new possibilities for symmetry elements. A list of symmetry elements which can be associated with the atomic arrangement in a unit cell is now given.

### Centre of symmetry ($\bar{1}$)

This is a point in the unit cell such that if there is an atom at vector position **r** there is an equivalent atom located at $-\mathbf{r}$. The unit cell in fig. 1.25(a) has centres of symmetry at its corners. Since all the corners are equivalent points the pairs of atoms $A$ and $A'$ related by the centre of symmetry at $O$ are repeated at each of the other corners. This gives rise to other centres of symmetry which bisect the edges of the cell and lie also at the face and body

Fig. 1.25.
(a) A centrosymmetric unit cell showing the complete family of eight distinct centres of symmetry.
(b) A unit cell showing mirror planes.
(c) A unit cell showing glide planes.
(d) A view down a tetrad axis of symmetry showing the other symmetry axes which arise.
(e) The operation of a $2_1$ axis.
(f) The operation of $3_1$ and $3_2$ axes.

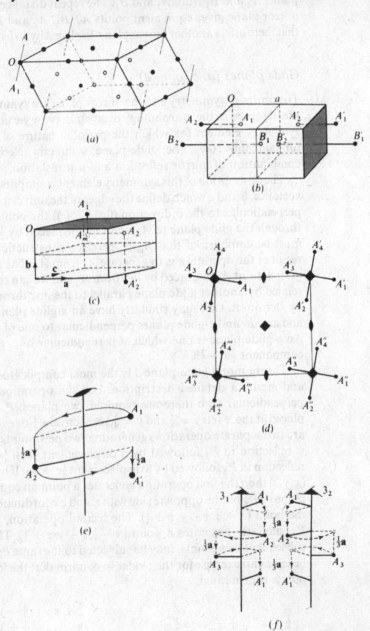

centres. While these extra points are also centres of symmetry they are not equivalent to those at the corners since they have different environments.

### Mirror plane (m)

In fig. 1.25(*b*) there is shown a unit cell with mirror planes across two opposite (equivalent) faces. The plane passing through $O$ generates the points $A_2$ and $B_2$ from $A_1$ and $B_1$. The repeat distance perpendicular to the mirror plane gives equivalent points $A'_1$, $B'_1$, $A'_2$ and $B'_2$ and it can be seen that there arises another mirror plane displaced by $\frac{1}{2}a$ from the one through $O$.

### Glide planes (a, b, c, n, d)

The centre of symmetry and the mirror plane are symmetry elements which are observed in the morphology of crystals. Now we are going to consider a symmetry element for which the periodic nature of the pattern plays a fundamental role. The glide-plane symmetry element operates by a combination of mirror reflection and a translation.

The description of this symmetry element is simplified by reference to the vectors $\mathbf{a}$, $\mathbf{b}$ and $\mathbf{c}$ which define the edges of the unit cell. For an $a$-glide plane perpendicular to the $b$ direction (fig. 1.25(*c*)) the point $A_1$ is first reflected through the glide plane to $A_m$ and then displaced by $\frac{1}{2}\mathbf{a}$ to the point $A_2$. It must be emphasized that $A_m$ is merely a construction point and the net result of the operation is to generate $A_2$ from $A_1$. The repeat of the pattern gives a point $A'_2$ displaced by $-\mathbf{b}$ from $A_2$ and we can see that $A'_2$ and $A_1$ are related by another glide plane parallel to the one through $O$ and displaced by $\frac{1}{2}\mathbf{b}$ from it. One may similarly have an $a$-glide plane perpendicular to $c$ and also $b$- and $c$-glide planes perpendicular to one of the other directions. An $n$-glide plane is one which, if perpendicular to $c$, gives a displacement component $\frac{1}{2}\mathbf{a} + \frac{1}{2}\mathbf{b}$.

The diamond glide-plane $d$ is the most complicated symmetry element and merits a detailed description. For the operation of a $d$-glide plane perpendicular to $\mathbf{b}$ there are required two planes, $P_1$ and $P_2$, which are placed at the levels $y = \frac{1}{8}$ and $y = \frac{3}{8}$, respectively. For each initial point there are two separate operations generating two new points. The first operation is reflection in $P_1$ followed by a displacement $\frac{1}{4}\mathbf{c} + \frac{1}{4}\mathbf{a}$, and the second a reflection in $P_2$ followed by a displacement $\frac{1}{4}\mathbf{c} + \frac{1}{4}\mathbf{a}$. If we begin with a point $(x, y, z)$ then the first operation generates a point an equal distance from the plane $y = \frac{1}{8}$ on the opposite side with $x$ and $z$ coordinates increased by $\frac{1}{4}$, i.e. the point $(x + \frac{1}{4}, \frac{3}{4} - y, z + \frac{1}{4})$. The second operation, involving the plane $P_2$, similarly generates a point $(x - \frac{1}{4}, \frac{3}{4} - y, z + \frac{1}{4})$. These points, and all subsequent new points, may be subjected to the same operations and it will be left as an exercise for the reader to confirm that the following set of eight points is generated:

| $x$ | $y$ | $z$ | $\frac{3}{4} + x$ | $\frac{3}{4} - y$ | $\frac{1}{4} + z$ |
|---|---|---|---|---|---|
| $\frac{1}{4} + x$ | $\frac{1}{4} - y$ | $\frac{1}{4} + z$ | $x$ | $\frac{1}{2} + y$ | $\frac{1}{2} + z$ |
| $\frac{1}{2} + x$ | $y$ | $\frac{1}{2} + z$ | $\frac{1}{4} + x$ | $\frac{3}{4} - y$ | $\frac{3}{4} + z$ |
| $\frac{3}{4} + x$ | $\frac{1}{4} - y$ | $\frac{3}{4} + z$ | $\frac{1}{2} + x$ | $\frac{1}{2} + y$ | $z.$ |

These coordinates show that there are two other glide planes at $y = \frac{5}{8}$ and $y = \frac{7}{8}$ associated with displacements $\frac{1}{4}\mathbf{c} + \frac{1}{4}\mathbf{a}$ and $\frac{1}{4}\mathbf{c} - \frac{1}{4}\mathbf{a}$, respectively.

### Rotation axes (2, 3, 4, 6)

The modes of operation of rotation axes are shown in fig. 1.7; the new feature which arises for a repeated pattern is the generation of subsidiary axes of symmetry other than those put in initially. This may be seen in fig. 1.25(*d*) which shows a projected view of a tetragonal unit cell down the tetrad axis. The point $A_1$ is operated on by the tetrad axis through $O$ to give $A_2$, $A_3$ and $A_4$ and this pattern is repeated about every equivalent tetrad axis. It is clear that $A_1$, $A_2'$, $A_3''$ and $A_4'''$ are related by a tetrad axis, non-equivalent to the one through $O$, through the centre of the cell. A system of diad axes also occurs and is indicated in the figure.

### Screw axes ($2_1$; $3_1$, $3_2$; $4_1$, $4_2$, $4_3$; $6_1$, $6_2$, $6_3$, $6_4$, $6_5$)

These symmetry elements, like glide planes, play no part in the macroscopic structure of crystals since they depend on the existence of a repeat distance. The behaviour of a $2_1$ axis parallel to $a$ is shown in fig. 1.25(*e*). The point $A_1$ is first rotated by an angle $\pi$ round the axis and then displaced by $\frac{1}{2}\mathbf{a}$ to give $A_2$. The same operation repeated on $A_2$ gives $A_1'$ which is the equivalent point to $A_1$ in the next cell. Thus the operation of the symmetry element $2_1$ is entirely consistent with the repeat nature of the structural pattern.

The actions of the symmetry elements $3_1$ and $3_2$ are illustrated in fig. 1.25(*f*). The point $A_1$ is rotated by $2\pi/3$ about the axis and then displaced by $\frac{1}{3}\mathbf{a}$ to give $A_2$. Two further operations give $A_3$ and $A_1'$, the latter point being displaced by $\mathbf{a}$ from $A_1$. The difference between $3_1$ and $3_2$ can either be considered as due to different directions of rotation or, alternatively, as due to having the same rotation sense but displacements of $\frac{1}{3}\mathbf{a}$ and $\frac{2}{3}\mathbf{a}$, respectively. The two arrangements produced by these symmetry elements are enantiomorphic (i.e. in mirror-image relationship).

In general, the symmetry element $R_D$ along the $a$ direction involves a rotation $2\pi/R$ followed by a displacement $(D/R)\mathbf{a}$.

### Inversion axes ($\bar{3}$, $\bar{4}$, $\bar{6}$)

The action of the inversion axis $\bar{R}$ is to rotate the point about the axis by an angle $2\pi/R$ and then invert through a point contained in the axis. Since $\bar{1}$ and $\bar{2}$ are equivalent to a centre of symmetry and mirror plane, respectively, they are not included here as inversion axes.

There is given in table 1.2 a list of symmetry elements and the graphical symbols used to represent them.

## 1.9  Space groups

Symmetry elements can be combined in groups and it can be shown that 230 distinctive arrangements are possible. Each of these arrangements is called a *space group* and they are all listed and described in volume A of the *International Tables for Crystallography*. Before describing a few of the 230

Table 1.2.

| Type of symmetry element | Written symbol | Graphical symbol | |
|---|---|---|---|
| Centre of symmetry | $\bar{1}$ | o | |
| | | Perpendicular to paper | In plane of paper |
| Mirror plane | $m$ | | |
| Glide planes | $a\ b\ c$ | | |
| | | glide in plane of paper | arrow shows glide direction |
| | | glide out of plane of paper | |
| | $n$ | | |
| Rotation axes | 2 | 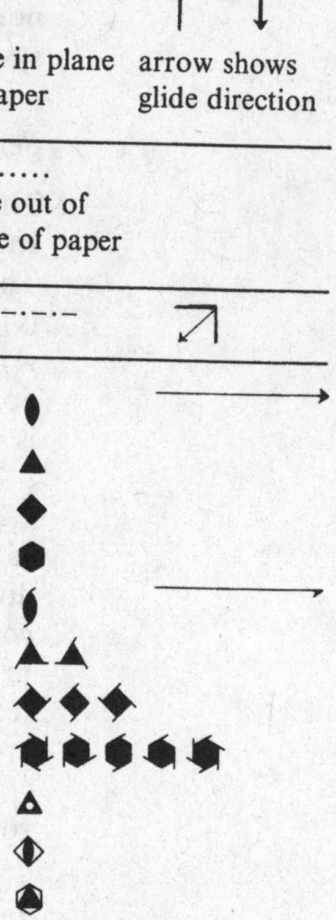 | |
| | 3 | | |
| | 4 | | |
| | 6 | | |
| Screw axes | $2_1$ | | |
| | $3_1, 3_2$ | | |
| | $4_1, 4_2, 4_3$ | | |
| | $6_1, 6_2, 6_3, 6_4, 6_5$ | | |
| Inversion axes | $\bar{3}$ | | |
| | $\bar{4}$ | | |
| | $\bar{6}$ | | |

space groups we shall look at two-dimensional space groups (sometimes called *plane groups*) which are the possible arrangements of symmetry elements in two dimensions. There are only 17 of these, reflecting the smaller number of possible systems, lattices and symmetry elements. Thus there are:

four crystal systems – oblique, rectangular, square and hexagonal;
two types of lattice – primitive ($p$) and centred ($c$); and
symmetry elements –

| | |
|---|---|
| rotation axes | 2, 3, 4 and 6 |
| mirror line | $m$ |
| glide line | $g$. |

We shall now look at four two-dimensional space groups which illustrate all possible features.

### Oblique p2

This is illustrated in fig. 1.26 in the form given in the *International Tables*. The twofold axis is at the origin of the cell and it will reproduce one of the structural units, represented by an open circle, in the way shown. The right-hand diagram shows the symmetry elements; the twofold axis manifests itself in two dimensions as a centre of symmetry. It will be seen that three other centres of symmetry are generated – at the points $(x, y) = (\frac{1}{2}, 0)$, $(0, \frac{1}{2})$ and $(\frac{1}{2}, \frac{1}{2})$. The four centres of symmetry are all different in that the structural arrangement is different as seen from each of them.

Fig. 1.26.
The two-dimensional space group *p2* as it appears in *International Tables for X-ray Crystallography*.

$p\,2$    No. 2    $p\,2\,1\,1$    2    Oblique

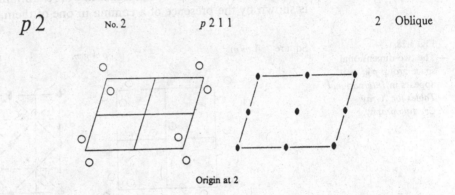

Origin at 2

### Rectangular cm

This rectangular space group is based on a centred cell and has a mirror line perpendicular to the y axis. In fig. 1.27 the centring of the cell is seen in that for each structural unit with coordinate $(x, y)$ there is another at $(\frac{1}{2} + x, \frac{1}{2} + y)$. In addition, the mirror line is shown relating empty open circles to those with commas within them. The significance of the comma is that it indicates a structural unit which is an enantiomorph of the one without a comma.

The right-hand diagram in fig. 1.27 shows the symmetry elements in the unit cell and mirror lines are indicated at $y = 0$ and $y = \frac{1}{2}$. What is also

Fig. 1.27.
The two-dimensional space group cm as it appears in *International Tables for X-ray Crystallography*.

$c\,m$    No. 5    $c\,1\,m\,1$    m    Rectangular

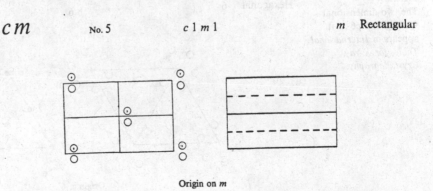

Origin on *m*

apparent, although it was not a part of the description of the two-dimensional space group, is the existence of a set of glide lines interleaving the mirror lines. The operation of a glide line involves reflection in a line followed by a translation $\frac{1}{2}a$. Because of the reflection part of the operation, the related structural units are enantiomorphs.

### Square p4g

This two-dimensional space group is illustrated in fig. 1.28 and shows the fourfold axes, two sets of glide lines at an angle $\pi/4$ to each other and a set of mirror lines at $\pi/4$ to the edges of the cell. Starting with a single structural unit there are generated seven others; the resultant eight structural units are the contents of the square cell. Wherever a pair of structural units are related by either a mirror line or a glide line the enantiomorphic relationship is shown by the presence of a comma in one of them.

Fig. 1.28.
The two-dimensional space group *p4g* as it appears in *International Tables for X-ray Crystallography.*

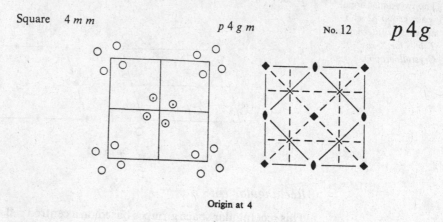

Square  4 *m m*       *p* 4 *g m*       No. 12       *p*4*g*

Origin at 4

### Hexagonal p6

As the name of this two-dimensional space group suggests it is based on a hexagonal cell, which is a rhombus with an angle $2\pi/3$ between the axes. As can be seen in fig. 1.29 the sixfold axis generates six structural units about each origin of the cell. A pair of threefold axes within the cell is also

Fig. 1.29.
The two-dimensional space group *p6* as it appears in *International Tables for X-ray Crystallography.*

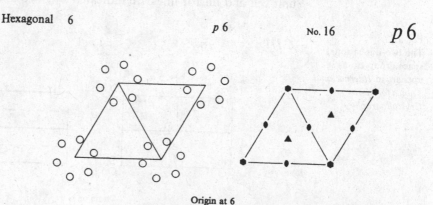

Hexagonal  6       *p* 6       No. 16       *p*6

Origin at 6

generated. The complete arrangement of symmetry elements is shown in the right-hand diagram of fig. 1.29.

Having established some general characteristics of space groups by the study of some relatively simple two-dimensional examples we shall now look at five three-dimensional space groups where the third dimension introduces complications not found in two dimensions.

### Triclinic P$\bar{1}$

This space group is based on a triclinic primitive cell which has a centre of symmetry. The representation of this space group, as given in the *International Tables*, is reproduced in fig. 1.30. The cell is shown in projection and the third coordinate (out of the plane of the paper), with respect to an origin at a centre of symmetry, is indicated by the signs associated with the open-circle symbols. This convention is interpreted as meaning that if one coordinate is $+t$ then the other is $-t$. The comma within the open-circle symbol indicates that if the symmetry operation is carried out on a group of objects and not just on a point, then the groups represented by $\bigcirc$ and $\odot$ are enantiomorphically related. The diagram on the right-hand side shows the distribution of symmetry elements.

The information which heads figs. 1.30–1.35 is taken from the *International Tables* and, reading across the page, is (i) the crystal system, (ii) the point group, (iii) symmetry associated with *a, b* and *c* axes (where appropriate), (iv) an assigned space-group number and (v) the space-group name according to the Hermann–Mauguin notation with, underneath, the older and somewhat outmoded Schoenflies notation.

Fig. 1.30.
The operation of the space group P$\bar{1}$ as shown in the *International Tables for X-ray Crystallography.*

Triclinic $\bar{1}$        P$\bar{1}$        No. 2        P$\bar{1}$
$C_i^1$

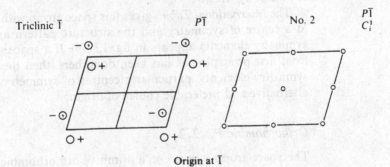

Origin at $\bar{1}$

### Monoclinic Cm

This space group is based on a monoclinic *C*-face centred cell with the mirror plane perpendicular to the unique axis. The unique axis for monoclinic space groups is the one perpendicular to the other two and, by convention, this is taken as the *b* axis. The letters shown in fig. 1.31 do not appear in the *International Tables* but they assist in a description of the generation of the complete pattern starting with a single unit.

We start with the structural unit $A_1$ and generate $A_2$ from it by the operation of the *C*-face centring. The mirror plane gives $A_3$ from $A_2$ and the

Fig. 1.31.
The operation of the
space group *Cm* as
shown in the
*International Tables for
X-ray Crystallography.*
The letters have been
added.

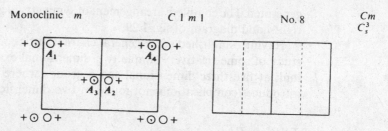

Monoclinic   *m*                                   *C* 1 *m* 1                              No. 8                          *Cm*
$C_s^3$

Origin on plane *m*; unique axis *b*

centring gives $A_4$ from $A_3$. This constitutes the entire pattern. The + signs against each symbol tell us that the units are all at the same level and the commas within the open circles indicate the enantiomorphic relationships.

It may be seen that this combination of *C*-face centring and mirror planes produces a set of *a*-glide planes.

## Monoclinic $P2_1/c$

This space group is based on a primitive, monoclinic unit cell with a $2_1$ axis along *b* and a *c*-glide plane perpendicular to it. In fig. 1.32(*a*) these basic symmetry elements are shown together with the general structural pattern produced by them. It can be found by inspection that other symmetry elements arise; $A_1$ is related to $A_4$ and $A_2$ to $A_3'$ by glide planes which interleave the original set. The pairs of units $A_4$, $A_2$ and $A_1$, $A_3$ are related by a centre of symmetry at a distance $\frac{1}{4}c$ out of the plane of the paper and a whole set of centres of symmetry may be found which are related as those shown in fig. 1.25(*a*).

The *International Tables* gives this space group with the unit-cell origin at a centre of symmetry and the structure pattern and complete set of symmetry elements appears in fig. 1.32(*b*). If a space group is developed from first principles, as has been done here, then the emergence of new symmetry elements, particularly centres of symmetry, often suggests an alternative and preferable choice of origin.

## Orthorhombic $P2_12_12_1$

This space group is based on a primitive orthorhombic cell and has screw axes along the three cell-edge directions. The name does not appear to define completely the disposition of the symmetry elements as it seems that there may be a number of ways of arranging the screw axes with respect to each other.

As was noted in § 1.4 in some point groups certain symmetry elements appear automatically due to the combination of two others. If this occurs in the point group it must also be so for any space group based on the point group. If we start with two sets of intersecting screws axes and generate the structural pattern from first principles we end up with the arrangement shown in fig. 1.33(*a*) which corresponds to the space group $P2_12_12$. The other possible arrangement, where the original two sets of screw axes do not intersect, is found to give a third set not intersecting the original sets and

## 1.9 Space groups

Fig. 1.32.
(a) The development from first principles of the structural pattern for the space group $P2_1/c$.
(b) The description of $P2_1/c$ as given in the *International Tables for X-ray Crystallography*.

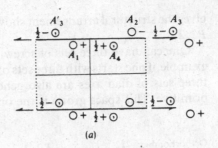

(a)

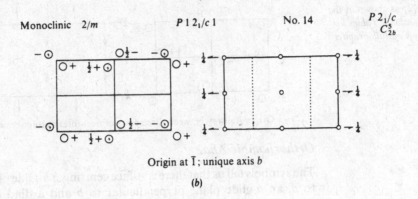

Monoclinic  2/m          $P 1 2_1/c 1$          No. 14          $P 2_1/c$
                                                                 $C_{2b}^5$

Origin at $\bar{1}$; unique axis $b$

(b)

Fig. 1.33.
(a) The development of the structural pattern for two sets of $2_1$ axes which intersect. This gives the space group $P2_12_12$.
(b) The development of the structural pattern for three sets of $2_1$ axes which intersect. This gives the space group $I222$. (*From International Tables for X-ray Crystallography*.)

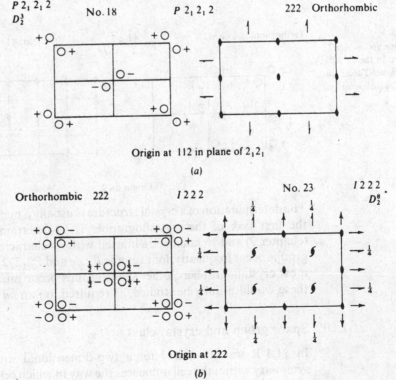

$P 2_1 2_1 2$          No. 18          $P 2_1 2_1 2$          222  Orthorhombic
$D_2^3$

Origin at  112 in plane of $2_1 2_1$

(a)

Orthorhombic  222          $I 2 2 2$          No. 23          $I 2 2 2$
                                                              $D_2^8$

Origin at 222

(b)

gives the structural arrangement shown in fig. 1.34 which is the space group $P2_12_12_1$.

One can have three sets of screw axes in a different arrangement. For example, if one starts with three sets of intersecting screw axes one finds that three sets of diad axes are also generated and that the unit cell is non-primitive. This space group is the one shown in fig. 1.33(*b*)–*I*222.

Fig. 1.34.
The space group
$P2_12_12_1$ as shown in the
*International Tables for*
*X-ray Crystallography.*

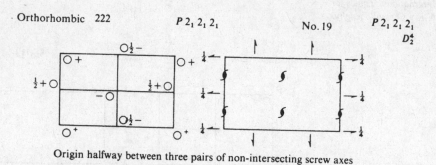

Orthorhombic 222          $P\,2_1\,2_1\,2_1$          No. 19          $P\,2_1\,2_1\,2_1$
$D_2^4$

Origin halfway between three pairs of non-intersecting screw axes

## Orthorhombic Aba2

The symbols tell us that there is *A*-face centring, a *b*-glide plane perpendicular to *a*, an *a*-glide plane perpendicular to *b* and a diad axis along *c*. The diagrammatic representation of this space group, as given in the *International Tables*, is shown in fig. 1.35. We should notice that the diad axis is automatically generated by the other two symmetry elements.

Fig. 1.35.
The space group *Aba2*
as shown in the
*International Tables for*
*X-ray Crystallography.*

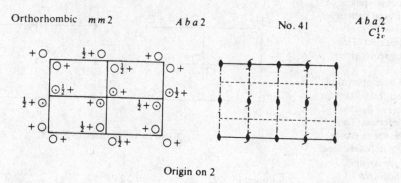

Orthorhombic   *m m* 2          *A b a* 2          No. 41          *A b a* 2
$C_{2v}^{17}$

Origin on 2

The determination of a crystal structure is usually a major undertaking and the first task of the crystallographer is to determine the space group (chapter 7) and to familiarize himself with its characteristics. Some space groups occur frequently, for example $P2_1/c$ and $P2_12_12_1$ are well known by most crystallographers; other space groups occur much more rarely and these would usually be studied, as required, on an *ad hoc* basis.

## 1.10   Space group and crystal class

In §1.4 it was illustrated for a two-dimensional square unit cell how symmetry within the cell influences the way in which cells associate to form the complete crystal. The formation of the faces of a crystal in the process of crystallization takes place in such a way that the crystal has a configuration of minimum potential energy. If the unit cell contains a diad axis then clearly, by symmetry, any association of cells giving a particular face will be matched by an associated face related by a crystal diad axis. However a screw axis in the unit cell will also, in the macroscopic aspect of the whole crystal, give rise to a crystal diad axis since the external appearance of the crystal will not be affected by atomic-scale displacements due to a screw axis. Similarly, mirror planes in the crystal are formed in response to both mirror planes and glide planes in the unit cell.

In the *International Tables* the point group is given for each of the listed space groups. The space groups described in this chapter, with the corresponding point groups, are:

| Space group | Point group |
|---|---|
| $P\bar{1}$ | $\bar{1}$ |
| $Cm$ | $m$ |
| $P2_1/c$ | $2/m$ |
| $P2_12_12_1$ | $222$ |
| $Aba2$ | $mm2$ |

A study of the crystal symmetry can be an important first step in the determination of a space group as for a particular point group a limited number of associated space groups are possible. However the art of examining crystals by optical goniometry is now largely ignored by the modern X-ray crystallographer who tends to use only diffraction information if possible.

## Problems to Chapter 1

1.1 A unit cell has the form of a cube. Find the angles between the normals to pairs of planes whose Miller indices are:
(a) (100) (010); (b) (100) (210); (c) (100) (111); (d) (121) (111).

1.2 The diagrams in fig. 1.36 show a set of equivalent positions in a unit cell. Find the crystal system and suggest a name for the space group.

1.3 Draw diagrams to show a set of equivalent positions and the set of symmetry elements for the following space groups:
(a) *pg*; (b) *p4mm*; (c) *Pm*; (d) *P2/m*; (e) *I4*.

Fig. 1.36.
Diagrams for Problem
1.2.

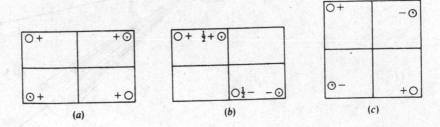
(a)                  (b)                  (c)

# 2 The scattering of X-rays

## 2.1 A general description of the scattering process

To a greater or lesser extent scattering occurs whenever electromagnetic radiation interacts with matter. Perhaps the best-known example is Rayleigh scattering the results of which are a matter of common everyday observation. The blue of the sky and the haloes which are seen to surround distant lights on a foggy evening are due to the Rayleigh scattering of visible light by molecules of gas or particles of dust in the atmosphere.

The type of scattering we are going to consider can be thought of as due to the absorption of incident radiation with subsequent re-emission. The absorbed incident radiation may be in the form of a parallel beam but the scattered radiation is re-emitted in all directions. The spatial distribution of energy in the scattered beam depends on the type of scattering process which is taking place but there are many general features common to all types of scattering.

In fig. 2.1 the point $O$ represents a scattering centre. The incident radiation is in the form of a parallel monochromatic beam and this is represented in the figure by the bundle of parallel rays. The intensity at a point within a beam of radiation is defined as the energy per unit time passing through unit cross-section perpendicular to the direction of propagation of the radiation. Thus for parallel incident radiation the intensity may be described as the power per unit cross-section of the beam. However, the scattered radiation emanates in all directions with some spatial distribution about the point $O$; this is shown in the figure by drawing a conical bundle of rays with apex at the point $O$ representing the rays scattered within a small solid angle in some particular direction. Clearly, in

Fig. 2.1.
Representation of the radiation incident on and scattered from a point scatterer.

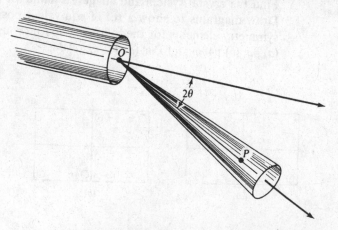

this case, the intensity of the scattered radiation will depend on the distance from $O$ and there will be an inverse-square law fall-off of intensity with distance. The intensity of the scattered radiation is thus usually described as the energy scattered per unit time per unit solid angle in a particular direction and is therefore a measure of what is happening at the scatterer itself.

If the incident radiation falling on $O$ is in the form of a simple monochromatic wave then the variation with time of the displacement $y$ of the incident wave can be described by the equation

$$y = A\cos(2\pi vt), \tag{2.1}$$

where $v$ is the frequency of the radiation and $A$ its amplitude.

The scattered wave will have a displacement at the point $P$ (fig. 2.1) which will depend on a number of factors:

(i) The distance $OP\,(=D)$ will introduce a phase shift with respect to the scattered wave at $O$ of $-2\pi D/\lambda$ where $\lambda$ is the wavelength of the radiation. This can also be expressed as $-2\pi Dv/c$ where $c$ is the velocity of propagation of the radiation.

(ii) The scattering process itself may introduce a phase shift so that the scattered wave at $O$ will be retarded with respect to the incident wave at $O$. This quantity $\alpha_s$ is called the *scattering phase shift*.

(iii) The inverse-square law of reduction of intensity with distance for the scattered radiation causes the fall-off of amplitude to be inversely proportional to distance $D$.

The displacement at $P$ can now be described by

$$y(2\theta, D, t) = f_{2\theta}\frac{A}{D}\cos[2\pi v(t - D/c) - \alpha_s]. \tag{2.2}$$

The influence of the factors (i), (ii) and (iii) may readily be seen in equation (2.2). The quantity $f_{2\theta}$ is a constant of proportionality with the dimension of length which is a function of the scattering angle and will be referred to as the scattering length. For a particular type of scatterer it will be a function of the scattering angle denoted by $2\theta$ in fig. 2.1. In X-ray diffraction theory the scattering angle is conventionally denoted by $2\theta$ (and not simply by $\theta$) as this leads to simplifications in notation in later developments of the theory and is also associated with the historical development of the subject (see § 3.6).

It is mathematically convenient to write the equation of a progressive wave in complex form as

$$\begin{aligned}Y &= Y_0 \exp[2\pi iv(t - x/c)]\\ &= Y_0\cos[2\pi v(t - x/c)] + iY_0\sin[2\pi v(t - x/c)].\end{aligned} \tag{2.3}$$

In equation (2.3) $Y_0$ is the amplitude of the wave, the real part of the expression is the displacement and the ratio (imaginary part/real part) is the tangent of the phase of the wave motion at $(x, t)$ with respect to that at the origin $(0, 0)$.

With this nomenclature one can express the time dependence of the disturbance at $P$ in fig. 2.1 as

$$y(2\theta D, t) = f_{2\theta} \frac{A}{D} \exp[2\pi i v(t - D/c) - i\alpha_s].$$ (2.4)

The amplitude of the disturbance at $P$ due to a single scatterer is then given by

$$\eta(2\theta, D) = f_{2\theta} \frac{A}{D}$$ (2.5)

and the phase lag of the disturbance at $P$ behind the incident wave at $O$ is

$$\alpha_{OP} = 2\pi v D/c + \alpha_s.$$ (2.6)

The intensity of the scattered beam in terms of power per unit solid angle is given by

$$\mathscr{I}_{2\theta} = K[\eta(2\theta, D)]^2 \times D^2 = f_{2\theta}^2 K A^2$$

or

$$\mathscr{I}_{2\theta} = f_{2\theta}^2 I_0,$$ (2.7)

where $K$ is the constant relating intensity to (amplitude)$^2$ and $I_0$ is the intensity of the incident beam on the scatterer. The use of the distinct symbols to represent the differently defined incident and scattered beam intensities should be noted.

## 2.2 Scattering from a pair of points

Consider the situation shown in fig. 2.2 where radiation is incident on two identical scattering centres $O_1$ and $O_2$. We shall find the resultant at $P$, a point at a distance $r$ from $O_1$ which is very large compared to the distance $O_1O_2$. Under this condition the scattered radiation which arrives at $P$ has been scattered from $O_1$ and $O_2$ through effectively the same angle $2\theta$. The planes defined by (i) $O_1$, $O_2$ and the incident beam direction and (ii) $O_1$, $O_2$ and $P$ are indicated in fig. 2.2 to emphasize the three-dimensional nature of the phenomenon we are considering.

Since the scatterers are identical the scattering phase shift $\alpha_s$ will be the same for each. Hence, for the radiation arriving at $P$ the phase difference of the radiation scattered at $O_2$ with respect to that scattered at $O_1$ is

Fig. 2.2.
Scattering from a pair of point scatterers.

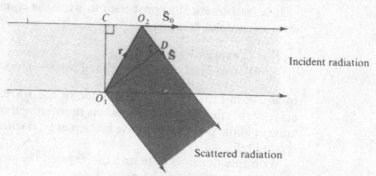

$$\alpha_{O_1O_2} = -\frac{2\pi}{\lambda}(CO_2 + O_2D). \tag{2.8}$$

Two unit vectors $\hat{S}_0$ and $\hat{S}$ are now defined which lie, respectively, along the directions of the incident and scattered beams. If the vector joining $O_1$ to $O_2$ is denoted by $\mathbf{r}$ then

$$CO_2 = \mathbf{r} \cdot \hat{S}_0, \ O_2D = -\mathbf{r} \cdot \hat{S}$$

and thus, from equation (2.8),

$$\alpha_{O_1O_2} = 2\pi\mathbf{r} \cdot \left(\frac{\hat{S} - \hat{S}_0}{\lambda}\right). \tag{2.9}$$

The bracketed quantity in equation (2.9) may be replaced by an equivalent vector

$$\mathbf{s} = \frac{\hat{S} - \hat{S}_0}{\lambda} \tag{2.10}$$

giving

$$\alpha_{O_1O_2} = 2\pi\mathbf{r} \cdot \mathbf{s}. \tag{2.11}$$

The vector $\mathbf{s}$ is highly significant in describing the scattering process and a geometrical interpretation of it is shown in fig. 2.3. The vectors $\hat{S}_0/\lambda$ and $\hat{S}/\lambda$ in the incident and scattered directions have equal magnitudes $1/\lambda$. It can be seen from simple geometry that $\mathbf{s}$ is perpendicular to the bisector of the angle between $\hat{S}_0$ and $\hat{S}$ and that its magnitude is given by

$$s = (2\sin\theta)/\lambda. \tag{2.12}$$

If the displacement due to the incident radiation at $O_1$ is described by equation (2.1) then the resultant disturbance at $P$, a distance $D$ from $O_1$, will be given by

$$y(2\theta, D, t) = f_{2\theta}\frac{A}{D}\{\exp[2\pi i\nu(t - D/c) - i\alpha_s]$$

$$+ \exp[2\pi i\nu(t - D/c) - i\alpha_s + 2\pi i\mathbf{r} \cdot \mathbf{s}]\}$$

$$= f_{2\theta}\frac{A}{D}\exp[2\pi i\nu(t - D/c) - i\alpha_s][1 + \exp(2\pi i\mathbf{r} \cdot \mathbf{s})]. \tag{2.13}$$

The amplitude of this resultant is

$$\eta_2(2\theta, D) = f_{2\theta}\frac{A}{D}[1 + \exp(2\pi i\mathbf{r} \cdot \mathbf{s})]$$

Fig. 2.3.
The relationship of $\mathbf{s}$ to $\hat{S}_0$ and $\hat{S}$.

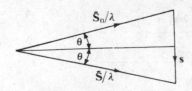

which, using equation (2.5), may be expressed in terms of the amplitude of scattering from a single unit as

$$\eta_2(2\theta, D) = \eta(2\theta, D)[1 + \exp(2\pi i\mathbf{r}\cdot\mathbf{s})]. \tag{2.14}$$

This equation is interpreted in terms of a phase-vector diagram in fig. 2.4(*a*). The amplitude of the disturbance at $P$ due to scattering at $O_1$ is represented by the vector **AB** and that due to scattering at $O_2$ by the vector **BC**. Both these vectors have the same magnitude, $\eta(2\theta, D)$, and the angle between them equals the difference of phase of the radiation scattered from $O_1$ and $O_2$, $2\pi\mathbf{r}\cdot\mathbf{s}$. The resultant **AC** has magnitude $\eta_2(2\theta, D)$ and differs in phase from the radiation scattered at $O_1$ by the angle $\phi$.

However in this description we have given a special role to one of the scatterers, $O_1$, with respect to which as origin all phases are quoted. The phase-vector diagram can be drawn with more generality if one measures phases with respect to radiation which would be scattered from some arbitrary point $O$ if in fact a scatterer was present there. Then, if the positions of $O_1$ and $O_2$ with respect to $O$ are given by the vectors $\mathbf{r}_1$ and $\mathbf{r}_2$, equation (2.14) appears as

$$\eta_2(2\theta, D) = \eta(2\theta, D)[\exp(2\pi i\mathbf{r}_1\cdot\mathbf{s}) + \exp(2\pi i\mathbf{r}_2\cdot\mathbf{r})] \tag{2.15}$$

and the phase-vector diagram appears as in fig. 2.4(*b*).

## 2.3 Scattering from a general distribution of point scatterers

Let us now examine the scattering from a system of identical point scatterers $O_1, O_2, \ldots, O_n$. We are interested in the amplitude of the disturbance in some direction corresponding to a scattering vector **s** at a distance which is large compared with the extent of the system of scatterers.

If the position of the scatterer at $O_m$ is denoted by its vector displacement

Fig. 2.4.
(*a*) A phase-vector diagram for a pair of point scatterers with one point as phase origin.
(*b*) A phase-vector diagram for a pair of point scatterers with a general point as phase origin.

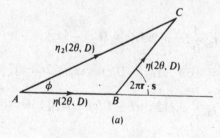
(a)

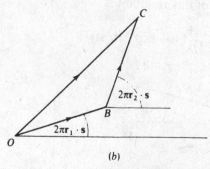
(b)

$\mathbf{r}_m$ from some origin point $O$ then, by an extension of the treatment which led to equation (2.15) we find

$$\eta_n(2\theta, D) = \eta(2\theta, D) \sum_{j=1}^{n} \exp(2\pi i \mathbf{r}_j \cdot \mathbf{s}). \qquad (2.16)$$

That this equation applies to identical scatterers is revealed by the factor $\eta(2\theta, D)$ appearing outside the summation. When the scatterers are non-equivalent the scattering amplitude must be written

$$\eta_n(2\theta, D) = \sum_{j=1}^{n} [\eta(2\theta, D)]_j \exp(2\pi i \mathbf{r}_j \cdot \mathbf{s})$$

$$= \frac{A}{D} \sum_{j=1}^{n} (f_{2\theta})_j \exp(2\pi i \mathbf{r}_j \cdot \mathbf{s}), \qquad (2.17)$$

where the scattering length for each of the scatterers now appears within the summation symbol. The phase-vector diagram for non-identical scatterers is shown in fig. 2.5 for the case $n = 6$. It will be evident that, although we refer to the scatterers as non-equivalent, we have assumed that they all have the same associated values of $\alpha_s$. This is the usual situation with X-ray diffraction. However it is sometimes possible to have the scatterers with differing phase shifts and, when this happens, useful information may be obtained (§8.5).

We shall find later that equation (2.17) is the basic equation for describing the phenomenon of X-ray diffraction and, when the symmetry of the atomic arrangements within crystals is taken into account, that it may appear in a number of modified forms.

## 2.4 Thomson scattering

We have discussed the results of scattering by distributions of scatterers without concerning ourselves with the nature of the scatterers or of the scattering process. It turns out that the scatterers of interest to us are electrons and the theory of the scattering of electromagnetic waves by free (i.e. unbound and unrestrained) electrons was first given by J. J. Thomson.

Fig. 2.5.
A phase-vector diagram for six non-identical scatterers.

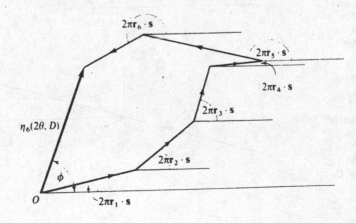

The basic mechanism of Thomson scattering is simple to understand. When an electromagnetic wave impinges on an electron the alternating electric-field vector imparts to the electron an alternating acceleration, and classical electromagnetic theory tells us that an accelerating charged particle emits electromagnetic waves. Thus the process may be envisaged as the absorption and re-emission of radiation and, although the incident radiation is unidirectional, the scattered radiation will be emitted in all directions. If we have the straightforward case where the incident radiation is a single, continuous and monochromatic wave then the acceleration of the electron will undergo a simple harmonic variation and the incident and emitted radiation will quite obviously have the same frequency.

If an electron at $O$, of charge $e$ and mass $m$, is undergoing an oscillation such that the acceleration is periodic with amplitude $a$ (fig. 2.6) then theory tells us that the scattered radiation at $P$, which is travelling in the direction $OP$, has an electric vector of amplitude

$$E = \frac{ea \sin \phi}{4\pi\varepsilon_0 rc^2} \tag{2.18}$$

which is perpendicular to $OP$ and in the plane defined by $OP$ and $a$. Here, $\varepsilon_0$ is the permittivity of free space.

In fig. 2.7 a parallel beam of electromagnetic radiation travelling along $OX$ falls upon an electron at $O$. We wish to determine the nature of the scattered wave at $P$. The amplitude of the electric vector $E$ of the incident wave is perpendicular to $OX$ and may be resolved into components $E_\perp$ and $E_\parallel$ perpendicular to and in the plane $OXP$. The electron will have corresponding components of acceleration of amplitudes

Fig. 2.6.
The relationship of the electric vector of scattered electromagnetic radiation at a point $P$ to the acceleration vector of an electron at $O$. The vectors are both in the plane of the diagram.

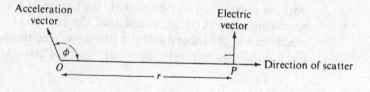

Fig. 2.7.
The relationship of the components of the electric vector of scattered electromagnetic radiation at $P$ to the components of the electric vector of the incident radiation at $O$.

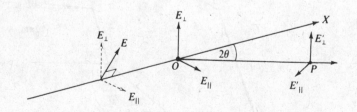

$$a_\perp = \frac{eE_\perp}{m}$$

and

$$a_\parallel = \frac{eE_\parallel}{m}. \tag{2.19}$$

Applying equation (2.18) we find the electric vector components of the scattered wave at $P$ as

$$E'_\perp = \frac{e^2}{4\pi\varepsilon_0 rc^2 m} E_\perp$$

and

$$E'_\parallel = \frac{e^2 \cos 2\theta}{4\pi\varepsilon_0 rc^2 m} E_\parallel. \tag{2.20}$$

The quantity $e^2/4\pi\varepsilon_0 c^2 m$, which has the dimensions of length and equals $2.82 \times 10^{-15}$ m, is referred to in classical electromagnetic theory as the radius of the electron.

Although we have been thinking about a simple, continuous, monochromatic, electromagnetic wave all the theory described above can be applied when the incident radiation is complex in form. A complicated incident wave may be analysed into simple components (see chapter 4) and the resultant electron acceleration and re-radiation may be found by adding together the effects of the simple components. Thus $E_\perp$ and $E_\parallel$ may be thought of as the components of the amplitude of any arbitrary electromagnetic radiation arriving at $O$.

If the intensity of the incident radiation is $I_0$ and if this radiation is unpolarized then

$$E_\perp^2 = E_\parallel^2 \propto \tfrac{1}{2} I_0$$
$$= CI_0, \text{ say.} \tag{2.21}$$

The intensity of the scattered radiation, defined as the power per unit solid angle scattered through an angle $2\theta$ is given by

$$\mathscr{I}_{2\theta} = \frac{1}{C} r^2 [(E')^2 + (E'_\parallel)^2]$$

$$= \frac{1}{2}\left(\frac{e^2}{4\pi\varepsilon_0 c^2 m}\right)^2 (1 + \cos^2 2\theta) I_0. \tag{2.22}$$

The factor $1/m^2$ in equation (2.22) shows why electrons are the only effective scatterers; the lightest nucleus, the proton, has the same magnitude of charge as the electron but 1837 times the mass.

Thomson scattering is coherent, that is to say there is a definite phase relationship between the incident and scattered radiation; in the case of a free electron the scattering phase shift is $\pi$. In all the processes concerned with the scattering of X-rays the electrons are bound into atoms and in §2.6 we shall investigate the form of the scattering from the total assemblage of electrons contained in an atom.

It is instructive to determine the proportion of the power of a beam incident on a material which will be scattered. First we calculate the total power scattered for each individual electron. In fig. 2.8 the point $O$ represents the electron and $OX$ the direction of the incident beam. The power $d\mathfrak{P}$ scattered into the solid angle $d\Omega$, defined by the region between the surfaces of the cones of semi-angles $\gamma$ and $\gamma + d\gamma$, is

$$d\mathfrak{P} = \mathscr{I}_\gamma \, d\Omega$$

and since $d\Omega = 2\pi \sin\gamma \, d\gamma$ and $\mathscr{I}_\gamma$ is given by equation (2.22) we have

$$d\mathfrak{P} = \pi \left( \frac{e^2}{4\pi\varepsilon_0 c^2 m} \right)^2 (1 + \cos^2\gamma) I_0 \sin\gamma \, d\gamma. \tag{2.23}$$

Hence the total power scattered by a single electron is

$$\mathfrak{P} = \pi \left( \frac{e^2}{4\pi\varepsilon_0 c^2 m} \right)^2 I_0 \int_{\gamma=0}^{\pi} (1 + \cos^2\gamma)\sin\gamma \, d\gamma$$

$$= \frac{8\pi}{3} \left( \frac{e^2}{4\pi\varepsilon_0 c^2 m} \right)^2 I_0. \tag{2.24}$$

For a material containing $n$ electrons per unit volume immersed in a parallel incident beam of cross-sectional area $\beta$ the power of the incident beam is $\beta I_0$ (since $I_0$, the intensity, is the power per unit area of the incident beam). The number of electrons traversed by the beam per unit length of path is $n\beta$ and hence the total power scattered per unit path is, by equation (2.24),

$$\mathfrak{P}_l = \frac{8\pi}{3} \left( \frac{e^2}{4\pi\varepsilon_0 c^2 m} \right)^2 \beta n I_0. \tag{2.25}$$

The ratio of $\mathfrak{P}_l$ to the power in the incident beam, $\beta I_0$, is called the *scattering power* of the material and is

Fig. 2.8.
The solid angle between the surfaces of the coaxial cones of semi-angles $\gamma$ and $\gamma + d\gamma$ is the area of the annular region on the surface of the unit sphere.

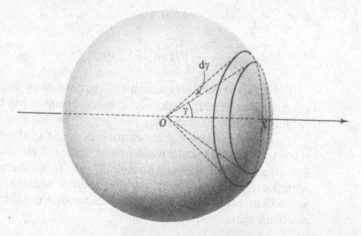

$$\sigma = \frac{\mathfrak{P}_l}{\beta I_0} = \frac{8\pi}{3}\left(\frac{e^2}{4\pi\varepsilon_0 c^2 m}\right)^2 n. \qquad (2.26)$$

The quantity $\sigma$ is the fraction of the incident radiation scattered per unit length of path (one metre in SI units).

If it is assumed that all the electrons in a material are free we can make an estimate of the fraction of the incident radiation which is scattered. For example a material of specific gravity 1.2 consisting of light atoms (say of atomic weight less than 30) would contain about $3 \times 10^{29}$ electrons $m^{-3}$. This value of $n$ substituted in equation (2.26) gives $\sigma \simeq 20$. Since the crystals used in X-ray diffraction usually have dimensions less than 1 mm it will be seen that only 2% or less of the incident X-ray beam is scattered.

However for the scattering which normally occurs when X-rays interact with matter the electrons are bound with various degrees of strength to the nuclei of atoms. The advent of the laser has made possible the direct measurement of Thomson scattering by the interaction of light from a ruby laser with an intense electron beam. In an experiment by Fiocco and Thompson (1963), illustrated in fig. 2.9, a 75 mA, 2 kV electron beam, magnetically focussed to give an electron density within the beam of $5 \times 10^{15}$ $m^{-3}$ was crossed by a beam from a ruby laser. This laser gave a burst of 20 joules of light in about 800 µs and the light measured by the photomultiplier detector had an intensity about $10^{-8}$ of that of the incident beam. The scattered light was of two types – some was scattered by the electron beam and there was also stray light reflected from the walls of the apparatus. The central feature of this experiment was that it was possible to separate the stray light from the Thomson-scattered component. Since the scattering was from rapidly moving electrons (speed about one-tenth of that of light) there was a Doppler shift in frequency. This shift was about 260 Å and with a suitable filter the stray radiation, whose wavelength was unshifted, could be prevented from reaching the detector.

This experiment barely detected the presence of Thomson scattering as there were very few (less than ten) scattered photons per pulse. Later work using higher-current electron beams and more powerful lasers has enabled a direct confirmation of equation (2.22) to be made (see Problem 2.3 at the end of this chapter).

Fig. 2.9.
The arrangement of
components of the
Fiocco and Thompson
(1963) experiment.

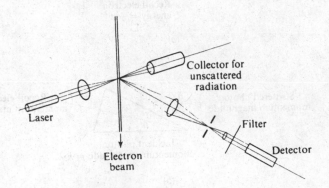

## 2.5   Compton scattering

Experimentally it is found that the radiation scattered by materials consists of two parts. The first part, that associated with Thomson scattering, has the same wavelength as the incident radiation; the second part has a wavelength longer than that of the incident radiation with the difference of wavelength depending on the angle of scatter. This latter component is due to what is known as Compton scattering and it is incoherent with the incident radiation. It is best described in terms of the elastic collision of a photon with an electron. In fig. 2.10(a) the incident photon moves along the direction $PO$ and, after collision with the electron, moves off along $OQ$ while the electron recoils along $OR$. From the conservation of energy in the elastic collision we find that

$$\frac{hc}{\lambda} = \frac{hc}{\lambda + d\lambda} + \tfrac{1}{2}mv^2$$

or, making usual approximations,

$$\frac{hc}{\lambda^2} d\lambda = \tfrac{1}{2}mv^2. \tag{2.27}$$

In addition to energy, momentum must also be conserved and in fig. 2.10(b) is shown the appropriate momentum vector diagram. It is a valid approximation to ignore the change in magnitude of the momentum of the scattered photon and thus we deduce from simple geometry that

$$\tfrac{1}{2}mv = \frac{h}{\lambda}\sin\theta. \tag{2.28}$$

Fig. 2.10.
(a) Velocity vector diagram for Compton scattering of a photon by an electron.
(b) Momentum vector diagram for Compton scattering.

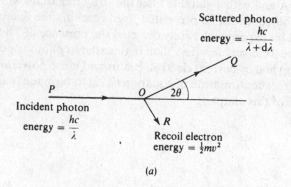

(a)

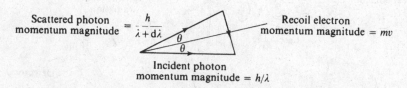

(b)

Eliminating $v$ from equations (2.27) and (2.28) we have

$$d\lambda = \frac{2h}{mc}\sin^2\theta$$

or

$$d\lambda = \frac{h}{mc}(1 - \cos 2\theta). \tag{2.29}$$

If the physical constants are now replaced by their magnitudes we find

$$d\lambda = 0.024(1 - \cos 2\theta) \text{ Å}. \tag{2.30}$$

The change of wavelength is seen to be independent of the wavelength of the incident radiation and depends only on the scattering angle. The maximum possible change of wavelength is for back scatter when $2\theta = \pi$ and this gives $d\lambda = 0.048$ Å. While this change of wavelength is small it is quite significant for X-rays with wavelengths of the order of 1 Å.

Thomson- and Compton scattering are examples of that dichotomy in physics summed up by the phrase 'wave–particle duality'. Classical theory regards light as a wave motion and the scattered light as a continuous outpouring of radiation from the scatterer simultaneously in all directions. When light is regarded in terms of photons, however, we are faced with the position that any change of direction of the photon on being scattered must be accompanied by an electron recoil to conserve momentum and hence a loss of energy of the photon to conserve energy. From the theoretical point of view there can be only Thomson scattering from a classical free electron and only Compton scattering from a wave-mechanical free electron. We shall see in the following section how these apparently diverse conclusions have been reconciled.

## 2.6 The scattering of X-rays by atoms

We are now going to consider how X-rays are scattered by electrons which are not free but are bound into definite energy states in atoms. Since the electron can only exist in discrete energy states then Thomson scattering must correspond to no change of energy of the electron and Compton scattering to a change of energy of some allowed magnitude. This latter change might be between one bound state and another or the electron might be ejected completely from the atom.

In general Thomson- and Compton scattering both occur but to determine the relative amounts of each type one must resort to a complete wave-mechanical treatment of the scattering process. Such a treatment shows that for a particular atomic electron the total intensity of scattering, both Thomson and Compton, equals the value given by Thomson's formula, equation (2.22). Furthermore, it shows that the coherently scattered component can be found from first principles by taking account of the fact that the electronic change is distributed and not located at a point. The solution of the wave equation for an atomic electron gives a wave function, $\Psi$, from which the distribution of electronic charge is found by

$$\rho = |\Psi|^2 \tag{2.31}$$

where $\rho$ represents the charge density in electron units per unit volume. In the special case when $|\Psi|$, and therefore $\rho$, is spherically symmetric we may represent the electron density by $\rho(r)$. If, for example, we express the positional parameters in spherical polar coordinates with respect to the centre of the atom as origin, then the charge associated with a small elemental volume is $\rho(r)r^2 \sin\psi \, dr \, d\psi \, d\phi$ (fig. 2.11). Thus, if the scattering vector is **s** and if the coordinate system is so arranged that **s** is parallel to the axis from which $\psi$ is measured, then the total amplitude of coherently scattered radiation may be found from equation (2.17) with integration replacing summation. We take the amplitude of the scattered wave from the small elemental volume as $C_s \times$ charge where $C_s$ is some constant dependent on the scattering vector **s** and we obtain the amplitude from the whole electron as

$$A_s = C_s \int_{r=0}^{\infty} \int_{\psi=0}^{\pi} \int_{\psi=0}^{2\pi} \rho(r)r^2 \exp(2\pi i r s \cos\psi)\sin\psi \, dr \, d\psi \, d\phi \qquad (2.32)$$

since $\mathbf{r}\cdot\mathbf{s} = rs\cos\psi$. Note particularly that the integration limits for $\psi$ and $\phi$ do cover the whole of space.

If the electron-density distribution is spherically symmetric then it is also centrosymmetric and this enables simplifications in equation (2.32) to be made. For every point $P$ with coordinates $(r, \psi, \phi)$ there is another point $P'$ with coordinates $(r, \pi - \psi, \pi + \phi)$ (see fig. 2.11) and the same electron density. The contribution of two elemental volumes round $P$ and $P'$ will give a resultant, the form of which can be appreciated by adding two terms such as

Fig. 2.11.
An elemental volume for spherical polar coordinates.

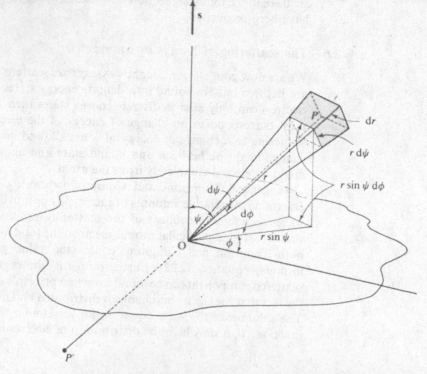

$$\exp(2\pi irs\cos\psi) + \exp[2\pi irs\cos(\pi - \psi)]$$
$$= \exp(2\pi irs\cos\psi) + \exp(-2\pi irs\cos\psi)$$
$$= 2\cos(2\pi rs\cos\psi).$$

It is clear from this that $A_s$ is a real quantity for a centrosymmetric distribution of electron density and that equation (2.32) can be rewritten as

$$A_s = C_s \int_{r=0}^{\infty} \int_{\psi=0}^{\pi} \int_{\psi=0}^{2\pi} \rho(r)r^2\cos(2\pi rs\cos\psi)\sin\psi \, dr \, d\psi \, d\phi \qquad (2.33)$$

The fact that $\rho(r)$ is independent of $\psi$ and $\phi$ enables one to integrate over these latter variables giving

$$A_s = 4\pi C_s \int_{r=0}^{\infty} \rho(r)r^2 \frac{\sin(2\pi rs)}{2\pi rs} \, dr. \qquad (2.34)$$

For a given value of s we shall now find the scattered amplitude from $\rho(r)$ as a fraction $p_s$ of that amplitude, say $(A_s)_0$, which would be given by a point electron at the origin. This may be found from equation (2.34) by noting that a point electron at the origin has an electron density $\rho(r) = \delta(r)$. The delta function $\delta(r)$ has the property

$$\delta(r) = 0, \quad r \neq 0$$
$$\delta(r) = \infty, \quad r = 0,$$

and $\int \delta(r) \, dv = 1$, where the integration can be over any finite volume of space around the origin.

For a delta-function electron density there is no contribution to equation (2.34) away from the origin and since

$$\left(\frac{\sin(2\pi rs)}{2\pi rs}\right)_{r\to 0} = 1$$

and, by definition

$$4\pi \int_{r=0}^{\infty} \delta(r)r^2 \, dr = 1$$

then it follows that

$$(A_s)_0 = C_s. \qquad (2.35)$$

From this we find

$$p_s = \frac{A_s}{(A_s)_0} = 4\pi \int_{r=0}^{\infty} \rho(r)r^2 \frac{\sin(2\pi rs)}{2\pi rs} \, dr \qquad (2.36)$$

where we recognize that $p$ depends only on the magnitude of s by dropping the vector notation in the subscript. If an atom contains $Z$ electrons then the total electron density, $\rho_a(r)$, will be the sum of the densities of the individual electrons, i.e.

$$\rho_a(r) = \sum_{j=1}^{Z} \rho_j(r). \tag{2.37}$$

The amplitude of coherent scattering from the total electron density will be obtained by adding the amplitudes for the electrons taken individually. We now define an *atomic scattering factor* $f_a$ as the ratio of the amplitude of the coherent scattered radiation from an atom to that from a single electron situated at the atomic centre. This is derived from equations (2.36) and (2.37) and is

$$f_a = 4\pi \int_{r=0}^{\infty} \rho_a(r) r^2 \frac{\sin(2\pi rs)}{2\pi rs} \, dr = \sum_{j=1}^{Z} (p_s)_j. \tag{2.38}$$

Atomic scattering factors are well tabulated in vol. III of the *International Tables for X-ray Crystallography*. Various models of atoms have been used to give electron-density distributions. For light atoms the Hartree self-consistent field method of computing wave functions is usually employed, while for heavy atoms the Thomas–Fermi approximation may be used.

We can now investigate the nature of Compton scattering from atoms. From equation (2.36) we can see that the intensity of coherent scattering from an atomic electron is $p_s^2 \mathscr{I}_{2\theta}$ and hence, since the *total* intensity of scattering as revealed by wave mechanics is $\mathscr{I}_{2\theta}$, the intensity of the Compton scattering must be $(1 - p_s^2)\mathscr{I}_{2\theta}$. However the Compton scattering from one atomic electron is incoherent with respect to that scattered by any other and hence the total intensity from all the electrons is obtained by adding the individual intensities from each of the electrons. Thus we have

$$\mathscr{I}_{\text{Compton}} = \sum_{j=1}^{Z} \{1 - (p_s)_j^2\} \times \mathscr{I}_{2\theta} \tag{2.39}$$

and

$$\mathscr{I}_{\text{Thomson}} = \left\{ \sum_{j=1}^{Z} (p_s)_j \right\}^2 \times \mathscr{I}_{2\theta} \tag{2.40}$$

It will be seen from equation (2.36) that since for $s = 0$ ($\theta = 0$) we have

$$\left( \frac{\sin(2\pi rs)}{2\pi rs} \right)_{s \to 1} = 1$$

then $p_0 = 1$. Hence, for radiation scattered in the incident-beam direction there is no incoherent component. As $\theta$ increases so $p$ decreases but the rate of decrease is less for those electrons which are most tightly bound in the atom. In fig. 2.12(*a*) there are given the radial electron-density distributions, $4\pi r^2 \rho(r)$, for the six electrons of the carbon atom calculated from Slater's approximate analytic wave functions. In fact there are two 1s electrons, two 2s electrons and two 2p electrons; when radial symmetry is assumed (and this is an assumption which, although usually made, is not really justified for 2p electrons) the 2s and 2p electrons give equivalent radial electron-density distributions.

## 2.6 The scattering of X-rays by atoms

Fig. 2.12.
(a) The radial electron-density functions of 1s and 2s electrons of a carbon atom as defined by Slater's analytic wave functions.
(b) The amplitude of scattering from the 1s and 2s electrons.
(c) The atomic scattering factor $f_a$ for the carbon atom and the coherent and incoherent scattered intensities from a single atom.

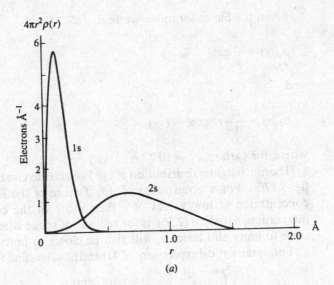

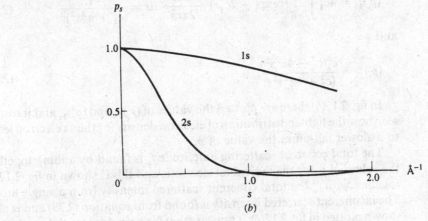

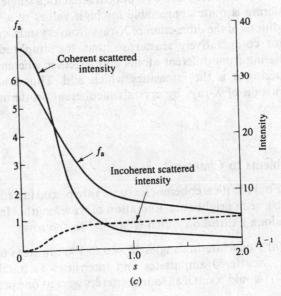

From the Slater formulae we find

$$\rho_{1s}(r) = \frac{c_1^3}{\pi} \exp(-2c_1 r) \tag{2.41}$$

and

$$\rho_{2s}(r) = \frac{c_2^5}{96\pi} r^2 \exp(-c_2 r) \tag{2.42}$$

where, for carbon, $c_1 = 10.77\,\text{Å}^{-1}$ and $c_2 = 6.15\,\text{Å}^{-1}$.

The much tighter distribution of the 1s electrons can be appreciated from fig. 2.12(a). For a given value of $s(\theta)$ if more of the electronic charge is concentrated at low values of $r$ then more of the contribution of the integration, equation (2.36), is for regions of space where $\sin(2\pi rs)/2\pi rs$ is close to unity and hence $p_s$ will also be closer to unity.

For the analytical expressions (2.41) and (2.42) we find from equation (2.36)

$$(p_s)_{1s} = 4c_1^3 \int_{r=0}^{\infty} r^2 \exp(-2c_1 r) \frac{\sin(2\pi rs)}{2\pi rs}\,dr = \frac{c_1^4}{(c_1^2 + \pi^2 s^2)^2} \tag{2.43}$$

and

$$(p_s)_{2s} = \frac{c_2^6(c_2^2 - 4\pi^2 s^2)}{(c_2^2 + 4\pi^2 s^2)^4}. \tag{2.44}$$

In fig. 2.12(b) there are plotted the values of $(p_s)_{1s}$ and $(p_s)_{2s}$ and it can be seen how the tighter distribution of electron density for the 1s electron leads to a slower fall-off in the value of $p_s$.

The total coherent scattering amplitude $f_a$ is found by adding together the values of $p_s$ for the individual electrons and this is shown in fig. 2.12(c) together with $f_a^2$ the total coherent scattered intensity from a single atom. The incoherent scattered intensity is found from equation (2.39) and is also shown plotted in fig. 2.12(c); it appears that for a single atom the incoherent scattering is quite appreciable for high values of $s$. However under the conditions of the diffraction of X-rays from crystals, very large numbers of atoms co-operatively scatter so that the amplitudes of the coherent scattering from different atoms add together whereas for the incoherent scattering it is the intensities which add. Thus when we consider the diffraction of X-rays by crystals incoherent scattering may generally be ignored.

## Problems to Chapter 2

2.1 Four identical coherent scatterers are placed in a row with a distance $3\lambda$ between neighbours. Radiation of wavelength $\lambda$ falls on the scatterers along a direction which is normal to the row.

(a) For scattering angles $2\theta = 0$ to $180°$, in steps of $20°$, determine the scattered amplitudes and intensities as fractions of that which would result if all four scatterers were at one point. Plot the results.

(b) For the same values of $2\theta$ find the phases of the resultant scattered radiation with respect to that from a scatterer at one end of the line.

2.2 Eight identical coherent scatterers are placed at the corners of a cube of side $\lambda$. A parallel beam of radiation of wavelength $\lambda$, moving along the direction of a body diagonal of the cube, falls on the scatterers. What is the ratio of the intensity of the scattering along the direction of one of the sides of the cube compared to that which would be obtained from a single scatterer for the same scattering angle?
(Note: there are two directions to be considered, one representing forward and the other backward scatter.)

2.3 In an experiment to investigate scattering from free electrons a field-emission discharge of 2 kV electrons of 5000 A is confined within a cylinder of 4 mm diameter. A Q-spoiled ruby laser produces a 10 joule pulse of duration shorter than that of the field emission and this passes in a fine pencil through the centre of, and perpendicular to, the electron beam. A detector is arranged to collect the light scattered through an angle of 45° from a solid angle of 0.01 steradian. Calculate, *approximately*, the number of photons entering the detector.

2.4 A lithium atom contains two 1s and one 2s electrons for which the Slater analytic wave functions are

$$\psi_{1s} = (c_1^3/\pi)^{\frac{1}{2}}\exp(-c_1 r)$$

and

$$\psi_{2s} = (c_2^5/96\pi)^{\frac{1}{2}}r\exp(-\tfrac{1}{2}c_2 r)$$

where

$$c_1 = 5.10\,\text{Å}^{-1} \text{ and } c_2 = 2.46\,\text{Å}^{-1}.$$

Show graphically the radial electron densities for a 1s and 2s electron and the amplitude of the coherent scattering from each of them. Plot also the intensity of (a) the coherent; and (b) the incoherent scattering from a single Li atom.
(Note: to calculate points for plotting the curves the recommended intervals are –

$$\rho_{1s}: r = 0\text{–}1.0\,\text{Å} \quad \text{by steps of } 0.1\,\text{Å};$$
$$\rho_{2s}: r = 0\text{–}2.0\,\text{Å} \quad \text{by steps of } 0.2\,\text{Å};$$

and

$$p: s = 0\text{–}2.0\,\text{Å}^{-1} \text{ by steps of } 0.2\,\text{Å}^{-1}.)$$

# 3 Diffraction from a crystal

## 3.1 Diffraction from a one-dimensional array of atoms

A crystal consists of a three-dimensional periodic arrangement of atoms each of which will scatter an incident X-ray beam. The scattered radiation from each of the atoms is coherent with respect to that from all others and we have seen in § 2.3 how to compute the resultant scattered amplitude. We are now going to examine in detail the influence of the periodic nature of the atomic arrangement in the diffraction of X-rays by a crystal. It is useful first of all to consider diffraction from a single row of $n$ atoms, each atom separated from its neighbour by a vector distance **a**.

Previously, in the reasoning which led to equation (2.33), it was found that the scattering amplitude from a centrosymmetric arrangement of scatterers is real if the centre of symmetry is used as origin. The analysis of the present case is simplified if we assume that $n$ is odd and if we take the centre atom, which is at the centre of symmetry, as the coordinate origin.

In a direction corresponding to a scattering vector **s**, and at some distance which is large compared with the extent of the lattice, the amplitude of the scattered radiation, expressed as a fraction of that from a point electron at the origin, is

$$A_n = f_a \sum_{q=-\frac{1}{2}(n-1)}^{\frac{1}{2}(n-1)} \cos(2\pi q \mathbf{a} \cdot \mathbf{s}), \tag{3.1}$$

where $f_a$ is the atomic scattering factor.

This expression can be simplified as follows:

$$A_n = \frac{f_a}{2\sin(\pi\mathbf{a}\cdot\mathbf{s})} \sum_{q=-\frac{1}{2}(n-1)}^{\frac{1}{2}(n-1)} 2\cos(2\pi q\mathbf{a}\cdot\mathbf{s})\sin(\pi\mathbf{a}\cdot\mathbf{s})$$

$$= \frac{f_a}{2\sin(\pi\mathbf{a}\cdot\mathbf{s})} \{\sin(\pi n\mathbf{a}\cdot\mathbf{s}) - \sin[\pi(n-2)\mathbf{a}\cdot\mathbf{s}]$$

$$+ \sin[\pi(n-2)\mathbf{a}\cdot\mathbf{s}] - \sin[\pi(n-4)\mathbf{a}\cdot\mathbf{s}]$$

$$+ \cdots + \sin[\pi(2-n)\mathbf{a}\cdot\mathbf{s}] - \sin[\pi(-n)\mathbf{a}\cdot\mathbf{s}]\}.$$

The terms on the right-hand side all cancel except the first and last, which are equal, and we have

$$A_n = f_a \frac{\sin(\pi n\mathbf{a}\cdot\mathbf{s})}{\sin(\pi\mathbf{a}\cdot\mathbf{s})}. \tag{3.2}$$

This equation can be found to hold also for $n$ even, if the centre of symmetry is used as origin, and equation (3.2) must give the magnitude of the scattered amplitude no matter what origin is chosen. The intensity of scattering is given by

$$(A_n)^2 = f_a^2 \frac{\sin^2(\pi na \cdot s)}{\sin^2(\pi a \cdot s)} = f_a^2 (K_n)^2, \text{ say,} \qquad (3.3)$$

and it will be instructive to examine the properties of $(K_n)^2$ as a function of $\mathbf{a} \cdot \mathbf{s}$ for various values of $n$. If we look at $(K_n)$ we see that for $\mathbf{a} \cdot \mathbf{s} = h$, where $h$ is an integer, both $\sin(\pi na \cdot s)$ and $\sin(\pi a \cdot s)$ are zero. However, we can determine the corresponding value of $(K_n)$ by noting that

$$\left[ \frac{\sin^2(nx)}{\sin^2 x} \right]_{x \to 0} = \frac{(nx)^2}{x^2} = n^2. \qquad (3.4)$$

In fig. 3.1 there are plotted the functions $\sin^2(\pi na \cdot s)$ and $\sin^2(\pi a \cdot s)$ for $n = 5$. It is evident that the ratio of these two functions will be periodic with a repeat distance of unity so that if $(K_n)^2 = n^2$ for $\mathbf{a} \cdot \mathbf{s} = 0$ it is also equal to $n^2$ for $\mathbf{a} \cdot \mathbf{s} = h$, where $h$ is any positive or negative integer.

In fig. 3.2, for the range $\mathbf{a} \cdot \mathbf{s} = 0$ to 1 the function $(K_n)^2$ is shown graphically for $n = 3$, 5 and 7. The vertical scale of these graphs has been adjusted to reveal more clearly the *relative* variation of $(K_n)^2$ as $\mathbf{a} \cdot \mathbf{s}$ changes. The following then are the general characteristics of the functions $(K_n)^2$.

(1) The main maxima of $(K_n)^2$ occur whenever $\mathbf{a} \cdot \mathbf{s}$ equals an integer.
(2) Between the main maxima there are $n - 1$ minima, where $(K_n)^2$ is zero, and $n - 2$ subsidiary maxima.
(3) The width of a main maximum is $2/n$ and the maxima therefore get narrower as $n$ increases.
(4) The main peak is twice as wide as the subsidiary ones.
(5) The ratio of the heights of the main to subsidiary maxima increases as $n$ increases.

The case $n = 4$ can be followed in the sequence of phase-vector diagrams shown in fig. 3.3. As $\mathbf{a} \cdot \mathbf{s}$ progresses from 0 to 1 we can see the development of the three minima and two subsidiary maxima. In fact the positions shown as subsidiary maxima in fig. 3.3 are close to but not actually at the maxima. The maxima of $(K_n)^2$ lie close to the maxima of $\sin^2(\pi na \cdot s)$, being closer

Fig. 3.1.
The functions $\sin^2(5\pi a \cdot s)$ and $\sin^2(\pi a \cdot s)$.

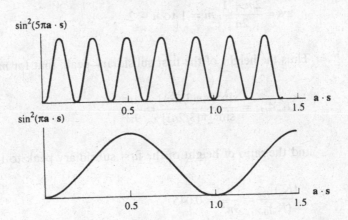

Fig. 3.2.
The intensity of
scattering from rows of
3, 5 and 7 scatterers as
functions of **a·s**.

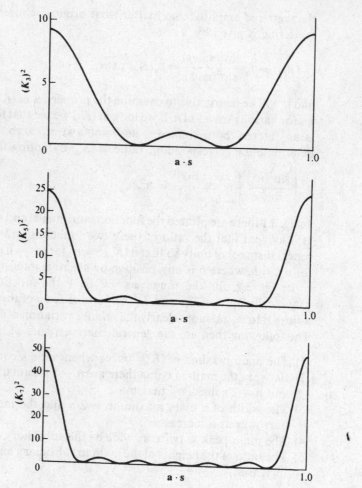

the larger the value of $n$ and the faster the variation of $\sin^2(\pi n\mathbf{a}\cdot\mathbf{s})$ compared to $\sin^2(\pi\mathbf{a}\cdot\mathbf{s})$. As $n$ tends to infinity, when the two sets of maxima tend to coincide, we may write the positions of the maxima of $(K_n)^2$ as

$$\mathbf{a}\cdot\mathbf{s} = \frac{2m+1}{2n}, \, m = 1 \text{ to } n-2.$$

Thus the height of the first subsidiary peak, that for $m = 1$, is given by

$$(K_n)^2_{3/2n} = \frac{\sin^2[\pi n(3/2n)]}{\sin^2[\pi(3/2n)]} \simeq \frac{4n^2}{9\pi^2}$$

and the ratio of height of the first subsidiary peak to the main peak is

$$\frac{(K_n)^2_{3/2n}}{(K_n)^2_0} = \frac{4}{9\pi^2} = 0.045. \tag{3.5a}$$

Fig. 3.3.
Phase-vector diagram
showing the formation
of maxima and minima
of intensity for a row of
four scatterers.

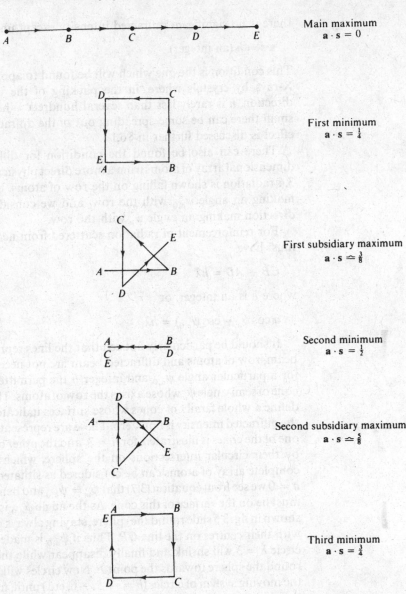

Main maximum
$\mathbf{a} \cdot \mathbf{s} = 0$

First minimum
$\mathbf{a} \cdot \mathbf{s} = \frac{1}{4}$

First subsidiary maximum
$\mathbf{a} \cdot \mathbf{s} \simeq \frac{3}{8}$

Second minimum
$\mathbf{a} \cdot \mathbf{s} = \frac{1}{2}$

Second subsidiary maximum
$\mathbf{a} \cdot \mathbf{s} \simeq \frac{5}{8}$

Third minimum
$\mathbf{a} \cdot \mathbf{s} = \frac{3}{4}$

Main maximum
$\mathbf{a} \cdot \mathbf{s} = 1$

Similarly for the *m*th peak, as long as $m \ll n$, we find

$$\frac{(K_n)^2_{(2m+1)/2n}}{(K_n)^2_0} = \frac{4}{(2m+1)^2\pi^2}. \tag{3.5b}$$

We may note that this ratio falls off rapidly with increasing *m* and that for
large *n* all the subsidiary peaks of appreciable magnitude are very close to
the main maximum. For very large *n* (theoretically $n = \infty$) we deduce that

there is no significant diffracted intensity except under the condition

$\mathbf{a \cdot s} = h$ (an integer).                                                      (3.6)

This condition is the one which will be found to apply in the diffraction of X-rays by crystals where, in the packing of the unit cells in any one direction, $n$ is rarely less than several hundred. Where the crystal is very small there can be some spreading out of the diffraction pattern and this effect is discussed further in §6.1.

There can also be found the condition for diffraction from a one-dimensional array of atoms from a more direct physical point of view. In fig. 3.4 radiation is shown falling on the row of atoms, the incident radiation making an angle $\psi_{a,0}$ with the row, and we consider the resultant in a direction making an angle $\psi_a$ with the row.

For reinforcement of radiation scattered from neighbouring atoms we must have

$$CB - AD = h\lambda$$

where $h$ is an integer, or

$$a(\cos\psi_a - \cos\psi_{a,0}) = h\lambda.$$                                             (3.7)

It should be particularly noticed that the lines representing the incident beam, row of atoms and diffracted beam are not necessarily coplanar and for a particular angle $\psi_{a,0}$ and integer $h$ the permitted angle $\psi_a$ defines a cone of semi-angle $\psi_a$ whose axis is the row of atoms. The various values of $h$ define a whole family of cones whose surfaces indicate directions in which the diffracted intensity is non-zero. These are represented in fig. 3.5($a$) where one of the cones is illustrated, for $h = 3$, and the other cones are represented by their circular intersections on the sphere, which is so large that the complete array of atoms can be considered as situated at the point $O$. For $h = 0$ we see from equation (3.7) that $\psi_a = \psi_{a,0}$ and hence the incident beam must lie on the surface of this cone. As the angle $\psi_{a,0}$ is varied so the circles shown in fig. 3.5 slide round the sphere, staying always perpendicular to and with their centres on the line $OP$. Thus if $\psi_{a,0}$ is made steadily smaller the circle $h = 3$ will shrink and finally disappear while the other circles move round the sphere towards the point $P$. New circles will appear at the rear of the moving system of circles ($h = -5, -6$, etc.) until, finally, when $\psi_{a,0} = 0$ the situation will appear as in fig. 3.5($b$).

Fig. 3.4.
The relationship of incident and diffracted X-rays to a row of atoms.

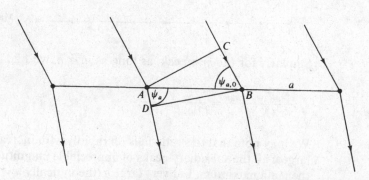

Fig. 3.5.
(a) The allowed cones of
diffraction from a
one-dimensional array of
atoms shown by their
intersections with a
sphere.
(b) The orders of
diffraction which occur
when the incident beam
is aligned with the row
of atoms.

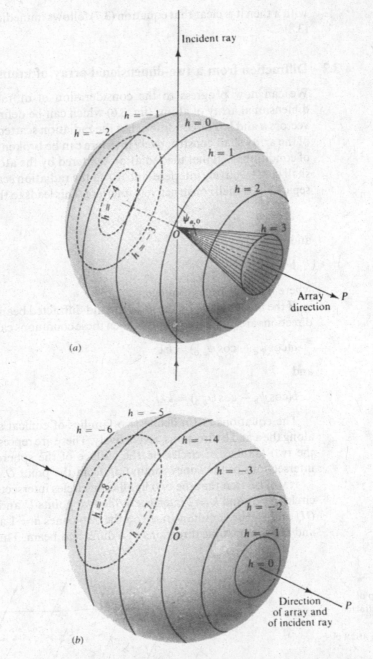

It should be noted that the two conditions for diffraction to occur,
equations (3.6) and (3.7), are equivalent. From equations (3.6) and (2.10) we
have

$$\frac{1}{\lambda}(\mathbf{a} \cdot \hat{\mathbf{S}} - \mathbf{a} \cdot \hat{\mathbf{S}}_0) = h. \tag{3.8}$$

Since $\mathbf{S}$ and $\hat{\mathbf{S}}_0$ are unit vectors making angles $\psi_a$ and $\psi_{a,0}$, respectively, with **a** then it is clear that equation (3.7) follows immediately from equation (3.8).

## 3.2  Diffraction from a two-dimensional array of atoms

We can now progress to the consideration of diffraction from a two-dimensional array of atoms (fig. 3.6) which can be defined in terms of two vectors **a** and **b**. The condition that the radiation scattered by all the atoms of the array shall constructively interfere can be broken down into the pair of conditions – (i) that the radiation scattered by the atoms of separation **a** shall constructively interfere, and (ii) all the radiation scattered by atoms of separation **b** shall constructively interfere. This leads to the pair of equations

$$\mathbf{a}\cdot\mathbf{s} = h$$

and

$$\mathbf{b}\cdot\mathbf{s} = k \tag{3.9}$$

where $h$ and $k$ are two integers.

If the angles made by the incident and diffracted beams with the **a** and **b** directions are $\psi_{a,0}, \psi_a, \psi_{b,0}$ and $\psi_b$ then these conditions can also be written as

$$a(\cos\psi_a - \cos\psi_{a,0}) = h\lambda$$

and

$$b(\cos\psi_b - \cos\psi_{b,0}) = k\lambda. \tag{3.10}$$

The equations (3.10) define two families of conical surfaces with axes along the **a** and **b** directions, respectively. These are represented in fig. 3.7 by the two families of circles on the surface of the sphere, these being the intersections of the cones emanating from the point $O$.

It will be seen that the two families of circles intersect – for example the circles $h = 1$ and $k = 2$ intersect in the two points $U$ and $V$. Thus the lines $OU$ and $OV$ lie simultaneously on the two cones $h = 1$ and $k = 2$ and $OU$ and $OV$ are possible directions of a diffracted beam. Thus for a particular

Fig. 3.6.
The relationship of incident and diffracted X-rays to a two-dimensional array of atoms.

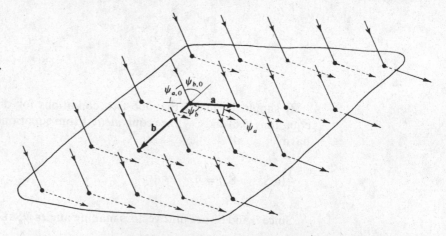

Fig. 3.7.
The cones of diffraction
corresponding to the **a**
and **b** directions of the
array are shown by
their intersections on a
sphere. Diffracted
beams are formed along
the lines joining *O* to
all the intersections of
circles.

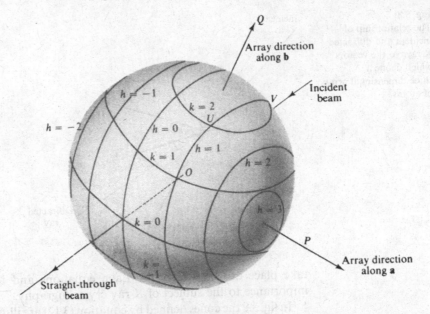

direction of incident radiation diffracted beams are produced in discrete directions as indicated by the intersection points in fig. 3.7 and to each diffracted beam a pair of integers (*hk*) can be assigned. We should note that, since the incident beam must lie on the surface of both the cones *h* = 0 and *k* = 0, it passes through the point *h* = 0, *k* = 0 on the surface of the sphere. Although the circle *h* = 3 appears in fig. 3.7 no diffracted beam can occur for *h* = 3 because it does not intersect any of the *k* circles. The integer pairs (*hk*) are equivalent to orders of diffraction from a two-dimensional grating; their significance will be further explained in the following section.

## 3.3  Diffraction from a three-dimensional array of atoms

The extension of the ideas of the previous two sections to diffraction from a three-dimensional array of atoms is quite straightforward. The condition that all the atoms should scatter in phase in some direction can be broken down into the three conditions that atoms separated by **a** or by **b** or by **c** should so scatter where the vectors **a**, **b** or **c** are the three vectors which define the array (fig. 3.8). The three conditions are

$$\mathbf{a \cdot s} = h$$
$$\mathbf{b \cdot s} = k$$
$$\mathbf{c \cdot s} = l$$

(3.11)

or

$$a(\cos \psi_a - \cos \psi_{a,0}) = h\lambda$$
$$b(\cos \psi_b - \cos \psi_{b,0}) = k\lambda$$
$$c(\cos \psi_c - \cos \psi_{c,0}) = l\lambda,$$

(3.12)

where the angles are as previously defined. These equations, in the form (3.11) or (3.12), which describe the conditions under which diffraction can

Fig. 3.8.
The relationship of
incident and diffracted
X-rays to the vectors
which define a
three-dimensional array
of atoms.

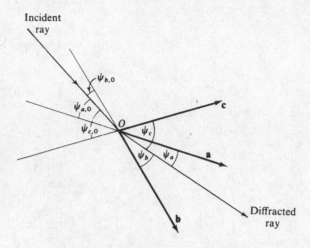

take place are known as the Laue equations and they are of central
importance to the subject of X-ray crystallography.

In fig. 3.9 the cones defined by equation (3.12) are illustrated for a typical
case by their intercepts with a sphere. We can see now that there is an
important new feature in this figure which has not appeared before. There is
no point on the surface of the sphere, except $h = 0$, $k = 0$, $l = 0$, through
which passes one member of each of the families of circles and therefore no
general diffracted beam can be produced. This will be the normal situation
when an incident beam has an arbitrary direction in relation to the array.

If we now wish to know what condition *does* give a diffraction beam we
can approach the problem in two ways. We can stipulate the triplet of
integers (*hkl*) and ask ourselves the question – 'How do we select a direction
of incident beam (i.e. $\psi_{a,0}, \psi_{b,0}, \psi_{c,0}$) such that the values of $\psi_a$, $\psi_b$ and $\psi_c$
given by equations (3.12) are possible directions of a diffracted beam?' These
latter three angles are to be 'possible' in the sense that for fixed vectors **a**, **b**
and **c** the angles are not independent and must therefore be related in the
correct way. Looked at in this way the problem seems a very formidable one
and this is not, in fact, the best way to approach the problem. Rather we
should ask ourselves whether we can find a vector **s** which satisfies
equations (3.11), since an incident and diffracted direction corresponding to
such a vector **s** would be what we are looking for. This approach also shows
us another new feature of the problem. The magnitude of the scattering
vector **s** gives the angle $\theta$ and the incident and diffracted beams must each
make an angle $\frac{1}{2}\pi - \theta$ with **s** (as in fig. 2.3) and contain **s** in the plane they
define. It is clear from fig. 3.10 that the vector $\hat{\mathbf{S}}_0$ which is along the incident
direction can lie anywhere along the surface of a cone of semi-angle $\frac{1}{2}\pi - \theta$
whose axis is the direction of **s**. We can see from this that if there is any
solution at all to the problem then there is a whole family of solutions. That
there *is* some solution can be appreciated by an examination of fig. 3.9. If the
crystal is rotated about some axis, say the *a* axis, then the family of cones
corresponding to various values of *h* does not move while the other two
families of cones vary. For a particular pair of indices *k* and *l* the point of
intersection on the sphere will move as the crystal rotates. Wherever it falls

*3.4   The reciprocal lattice*

Fig. 3.9.
The cones of diffraction corresponding to the **a**, **b** and **c** directions of the array are shown by their intersections on a sphere. A diffracted beam would be produced along a line joining $O$ to any trisection of circles. None occurs in the case shown (except $h = 0$, $k = 0$, $l = 0$).

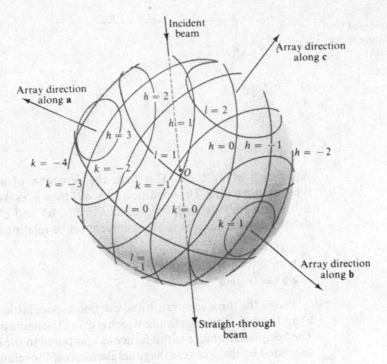

Fig. 3.10.
The conical surface is the locus of the incident and diffracted beams for a particular scattering vector **s**. The three vectors $\hat{\mathbf{S}}_0$, **s** and $\hat{\mathbf{S}}$ must always be coplanar.

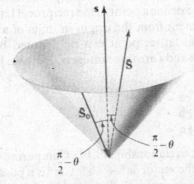

on one of the stationary $h$ circles a diffracted beam will be produced corresponding to that particular combination of $h$, $k$ and $l$.

## 3.4   The reciprocal lattice

We are now going to consider the Laue equations in the form (3.11) and examine the problem of finding scattering vectors **s** which will satisfy these relationships for a given set of integers $(hkl)$.

Consider the unit cell, defined by the vectors **a**, **b** and **c** and illustrated in fig. 3.11. We define a new vector **a\*** by the following relationships:

$$\mathbf{a^*}\cdot\mathbf{a} = 1;\ \mathbf{a^*}\cdot\mathbf{b} = \mathbf{a^*}\cdot\mathbf{c} = 0. \tag{3.13}$$

Since the scalar product of **a\*** with both **b** and **c** is zero then it must be perpendicular to each of them (or zero) and is hence perpendicular to the

Fig. 3.11.
The relationship of the
reciprocal-lattice vector
a* to a, b and c.

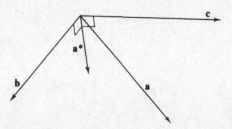

plane defined by them. This fixes the direction of $a^*$ and its angular relationship with $a$. The condition $a^* \cdot a = 1$ then fixes its magnitude and so it is completely defined. Two other vectors, $b^*$ and $c^*$, can be similarly defined and we now have the complete set of relationships

$$
\begin{array}{lll}
a \cdot a^* = 1 & a \cdot b^* = 0 & a \cdot c^* = 0 \\
b \cdot a^* = 0 & b \cdot b^* = 1 & b \cdot c^* = 0 \\
c \cdot a^* = 0 & c \cdot b^* = 0 & c \cdot c^* = 1.
\end{array}
\tag{3.14}
$$

Just as the three vectors $a$, $b$ and $c$ define a space lattice, so the vectors $a^*$, $b^*$ and $c^*$ define another lattice which is given the name of reciprocal lattice. The 'reciprocal' nature of this lattice as compared to the real space lattice is suggested by the non-zero diagonal elements of the relationships (3.14) and we shall see that the reciprocal relationship will also show itself in other ways. It is possible to define a point of the reciprocal lattice by three integers giving the displacement from the origin in units of $a^*$, $b^*$ and $c^*$.

Such a reciprocal lattice point can be denoted by the vector $ha^* + kb^* + lc^*$ where $h$, $k$ and $l$ are three integers. Now we look at the following equations:

$$
\begin{array}{l}
(ha^* + kb^* + lc^*) \cdot a = h \\
(ha^* + kb^* + lc^*) \cdot b = k \\
(ha^* + kb^* + lc^*) \cdot c = l
\end{array}
\tag{3.15}
$$

which follow from the relationships (3.14). Comparing equations (3.15) and (3.11) it is clear that the vector $ha^* + kb^* + lc^*$ is a possible value of $s$ which satisfies the Laue equations for the indices $(hkl)$ and, in fact, it turns out that this value is unique. We have now completely solved the problem, in principle at any rate, of how to set the crystal to give the $(hkl)$ reflection. First one finds the reciprocal lattice, which is fixed relative to the crystal, and this then gives the scattering vector

$$
s = ha^* + kb^* + lc^*
\tag{3.16}
$$

which, therefore, is also fixed relative to the crystal. The magnitude of $s$ gives the diffraction angle $2\theta$ from equation (2.12) and the crystal (and hence $s$) is positioned so that the incident beam makes an angle $\frac{1}{2}\pi - \theta$ with $s$. The diffracted beam will then be observed, also making an angle $\frac{1}{2}\pi - \theta$ with $s$, and in the plane defined by the incident beam and $s$. As we saw from fig. 3.10 this does not give a unique relationship between the incident beam and the crystal.

When space lattices were discussed in § 1.7 it was pointed out that, although for a particular atomic arrangement the space lattice was fixed, there are an infinity of possible choices of unit cell. It can be shown that similar considerations apply to the reciprocal lattice. The vectors $\mathbf{a}^*, \mathbf{b}^*$ and $\mathbf{c}^*$ by which it is defined are dependent on the non-unique description of the space lattice in terms of the vectors $\mathbf{a}, \mathbf{b}$ and $\mathbf{c}$. However it can be shown that for a particular space lattice the reciprocal space lattice is unique; if different vectors $\mathbf{a}, \mathbf{b}$ and $\mathbf{c}$ are chosen to define the real lattice then, although different vectors $\mathbf{a}^*, \mathbf{b}^*$ and $\mathbf{c}^*$ will define the reciprocal lattice, the reciprocal lattice will be unchanged. This is illustrated in fig. 3.12 for a two-dimensional example. The space lattice is shown by black dots and the reciprocal lattice by open circles. The vectors $\mathbf{a}_1$ and $\mathbf{b}_1$ can be used to define the space lattice and these lead to vectors $\mathbf{a}_1^*$ and $\mathbf{b}_1^*$ describing the reciprocal lattice. Alternatively the space lattice may be described in terms of $\mathbf{a}_2$ and $\mathbf{b}_2$ and the reciprocal lattice correspondingly in terms of $\mathbf{a}_2^*$ and $\mathbf{b}_2^*$ but the lattices themselves remain the same.

A unit cell of the crystal can alternatively be defined by the cell edges $a, b, c$ and interaxial angles $\alpha, \beta, \gamma$. Similarly there will be a reciprocal unit cell derived from this which can be defined by $a^*, b^*, c^*$ and $\alpha^*, \beta^*, \gamma^*$. We shall now derive some relationships involving these quantities.

First let us derive an expression for the volume of the unit cell. This is given by the triple-scalar product of the three sides as

$$V = \mathbf{a}\cdot\mathbf{b} \times \mathbf{c}. \tag{3.17}$$

To give the volume in terms of the scalar quantities $a, b, c, \alpha, \beta, \gamma$, it is advantageous to express $a, b$ and $c$ in terms of any orthogonal set of unit vectors, $\mathbf{i}, \mathbf{j}$ and $\mathbf{k}$ so that

$$\mathbf{a} = a_x\mathbf{i} + a_y\mathbf{j} + a_z\mathbf{k}$$
$$\mathbf{b} = b_x\mathbf{i} + b_y\mathbf{j} + b_z\mathbf{k}$$

and

$$\mathbf{c} = c_x\mathbf{i} + c_y\mathbf{j} + c_z\mathbf{k}. \tag{3.18}$$

Fig. 3.12.
A given space lattice and corresponding reciprocal-space lattice described in terms of different unit cells and related reciprocal cells.

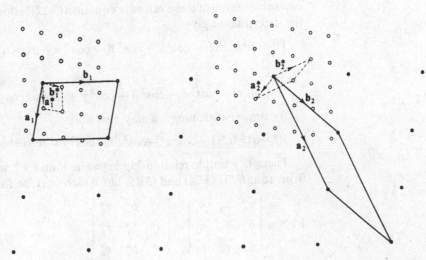

The volume of the unit cell can now be written as

$$V = \begin{vmatrix} a_x & a_y & a_z \\ b_x & b_y & b_z \\ c_x & c_y & c_z \end{vmatrix}. \tag{3.19}$$

The value of a determinant is unchanged if the rows and columns of the determinant are interchanged and it therefore follows that

$$V^2 = \begin{vmatrix} a_x & a_y & a_z \\ b_x & b_y & b_z \\ c_x & c_y & c_z \end{vmatrix} \times \begin{vmatrix} a_x & b_x & c_x \\ a_y & b_y & c_y \\ a_z & b_z & c_z \end{vmatrix}. \tag{3.20}$$

The rules for multiplying determinants are like those for multiplying matrices so that

$$V^2 = \begin{vmatrix} a_x a_x + a_y a_y + a_z a_z & b_x a_x + b_y a_y + b_z a_z & c_x a_x + c_y a_y + c_z a_z \\ a_x b_x + a_y b_y + a_z b_z & b_x b_x + b_y b_y + b_z b_z & c_x b_x + c_y b_y + c_z b_z \\ a_x c_x + a_y c_y + a_z c_z & b_x c_x + b_y c_y + b_z c_z & c_x c_x + c_y c_y + c_z c_z \end{vmatrix}. \tag{3.21}$$

It can be confirmed from the expressions (3.18) that this reduces to

$$V^2 = \begin{vmatrix} \mathbf{a \cdot a} & \mathbf{a \cdot b} & \mathbf{a \cdot c} \\ \mathbf{b \cdot a} & \mathbf{b \cdot b} & \mathbf{b \cdot c} \\ \mathbf{c \cdot a} & \mathbf{c \cdot b} & \mathbf{c \cdot c} \end{vmatrix}. \tag{3.22}$$

If the interaxial angles are $\alpha$, $\beta$ and $\gamma$ as defined in table 1.1 (p. 10) then we have

$$\mathbf{a \cdot a} = a^2 \quad \mathbf{b \cdot b} = b^2 \quad \mathbf{c \cdot c} = c^2$$
$$\mathbf{a \cdot b} = \mathbf{b \cdot a} = ab \cos \gamma$$
$$\mathbf{b \cdot c} = \mathbf{c \cdot b} = cb \cos \alpha$$

and

$$\mathbf{c \cdot a} = \mathbf{a \cdot c} = ac \cos \beta \tag{3.23}$$

and substitution of these values in equation (3.22) followed by evaluation of the determinant gives

$$V^2 = a^2 b^2 c^2 (1 - \cos^2 \alpha - \cos^2 \beta - \cos^2 \gamma + 2 \cos \alpha \cos \beta \cos \gamma)$$

or

$$V = abc(1 - \cos^2 \alpha - \cos^2 \beta - \cos^2 \gamma + 2 \cos \alpha \cos \beta \cos \gamma)^{\frac{1}{2}}. \tag{3.24}$$

By similar reasoning we may also find

$$V^* = a^* b^* c^* (1 - \cos^2 \alpha^* - \cos^2 \beta^* - \cos^2 \gamma^* + 2 \cos \alpha^* \cos \beta^* \cos \gamma^*)^{\frac{1}{2}}. \tag{3.25}$$

There is a simple relationship between $V$ and $V^*$ which is not obvious from equations (3.24) and (3.25) but which may be found as follows:

$$VV^* = \begin{vmatrix} a_x & a_y & a_z \\ b_x & b_y & b_z \\ c_x & c_y & c_z \end{vmatrix} \times \begin{vmatrix} a_x^* & b_x^* & c_x^* \\ a_y^* & b_y^* & c_y^* \\ a_z^* & b_z^* & c_z^* \end{vmatrix}$$

$$= \begin{vmatrix} a_x a_x^* + a_y a_y^* + a_z a_z^* & b_x a_x^* + b_y a_y^* + b_z a_z^* & c_x a_x^* + c_y a_y^* + c_z a_z^* \\ a_x b_x^* + a_y b_y^* + a_z b_z^* & b_x b_x^* + b_y b_y^* + b_z b_z^* & c_x b_x^* + c_y b_y^* + c_z b_z^* \\ a_x c_x^* + a_y c_y^* + a_z c_z^* & b_x c_x^* + b_y c_y^* + b_z c_z^* & c_x c_x^* + c_y c_y^* + c_z c_z^* \end{vmatrix}$$

$$= \begin{vmatrix} \mathbf{a \cdot a^*} & \mathbf{a \cdot b^*} & \mathbf{a \cdot c^*} \\ \mathbf{b \cdot a^*} & \mathbf{b \cdot b^*} & \mathbf{b \cdot c^*} \\ \mathbf{c \cdot a^*} & \mathbf{c \cdot b^*} & \mathbf{c \cdot c^*} \end{vmatrix}$$

which, from the relationships (3.14), gives

$$VV^* = \begin{vmatrix} 1 & 0 & 0 \\ 0 & 1 & 0 \\ 0 & 0 & 1 \end{vmatrix} = 1. \tag{3.26}$$

This shows once again the reciprocal relationship of the reciprocal lattice to the real space lattice. Any one of the parameters of the reciprocal lattice may be expressed in terms of the parameters of the real lattice and vice versa. For example we have

$$\mathbf{a \cdot a^*} = 1 \text{ from the relationships (3.14)}$$

and

$$\mathbf{a} \cdot \frac{\mathbf{b \times c}}{V} = 1 \text{ from the relationships (3.17)}$$

and since $\mathbf{a^*}$ is parallel to $\mathbf{b \times c}$ (both these vectors are perpendicular to the plane defined by $\mathbf{b}$ and $\mathbf{c}$) then we must have

$$\mathbf{a^*} = \frac{\mathbf{b \times c}}{V}. \tag{3.27}$$

The magnitude of these equal vectors must be equal and hence

$$a^* = \frac{bc \sin \alpha}{V}, \tag{3.28a}$$

where $V$ is given by equation (3.24). The real space parameter, $a$, is similarly related to the reciprocal-lattice parameters. By cyclic permutation of $a, b, c$ and $\alpha, \beta, \gamma$ and the corresponding reciprocal-lattice quantities in equation (3.28a) other relationships may be found. We can rewrite equation (3.28a) as

$$\sin \alpha = \frac{Va^*}{bc}$$

and since

$$V = \frac{1}{V^*}, \ b = \frac{c^* a^* \sin \beta^*}{V^*} \text{ and } c = \frac{a^* b^* \sin \gamma^*}{V^*}$$

we find

$$\sin \alpha = \frac{V^*}{a^* b^* c^* \sin \beta^* \sin \gamma^*}. \tag{3.28b}$$

Table 3.1. *Summary of relationships*

$$V = abc(1 - \cos^2\alpha - \cos^2\beta - \cos^2\gamma + 2\cos\alpha\cos\beta\cos\gamma)^{\frac{1}{2}}$$

$$V^* = a^*b^*c^*(1 - \cos^2\alpha^* - \cos^2\beta^* - \cos^2\gamma^* + 2\cos\alpha^*\cos\beta^*\cos\gamma^*)^{\frac{1}{2}}$$

$$VV^* = 1$$

$$a^* = \frac{bc\sin\alpha}{V}; \quad b^* = \frac{ac\sin\beta}{V}; \quad c^* = \frac{ab\sin\gamma}{V}$$

$$a = \frac{b^*c^*\sin\alpha^*}{V^*}; \quad b = \frac{a^*c^*\sin\beta^*}{V^*}; \quad c = \frac{a^*b^*\sin\gamma^*}{V^*}$$

$$\sin\alpha = \frac{V^*}{a^*b^*c^*\sin\beta^*\sin\gamma^*}; \quad \sin\alpha^* = \frac{V}{abc\sin\beta\sin\gamma}$$

$$\sin\beta = \frac{V^*}{a^*b^*c^*\sin\alpha^*\sin\gamma^*}; \quad \sin\beta^* = \frac{V}{abc\sin\alpha\sin\gamma}$$

$$\sin\gamma = \frac{V^*}{a^*b^*c^*\sin\alpha^*\sin\beta^*}; \quad \sin\gamma^* = \frac{V}{abc\sin\alpha\sin\beta}$$

Again by cyclic permutation of the symbols and by interchanging real and reciprocal parameters new equations can be found from equation (3.28b). A summary of all these relationships is given in table 3.1.

The magnitude of **s** can be found from equation (3.16), for

$$\begin{aligned}
s^2 = \mathbf{s}{\cdot}\mathbf{s} &= (h a^* + k b^* + l c^*){\cdot}(h a^* + k b^* + l c^*) \\
&= h^2 a^{*2} + k^2 b^{*2} + l^2 c^{*2} + 2hk a^* b^* \cos\gamma^* \\
&\quad + 2hl a^* c^* \cos\beta^* + 2kl b^* c^* \cos\alpha^*.
\end{aligned}$$

$$(3.29)$$

If we have dealt at some length with the reciprocal lattice and its properties it is because of its great importance in X-ray crystallography. It is useful to think of the crystal having rigidly attached to it a real space lattice and also a reciprocal lattice so that, as the crystal moves, so do the lattices. With this idea in mind many of the concepts of X-ray crystallography can be understood much more clearly.

## 3.5 Diffraction from a crystal – the structure factor

We must now look at the details of the arrangement in a crystal structure. This can be pictured as a close-packed three-dimensional array of unit cells within each of which there is a similar distribution of atoms. In fig. 3.13 eight unit cells are shown with four atoms in each unit cell; this can be thought of as a small part of a very large array which forms the complete crystal. If we consider the atoms labelled *A* alone then they form a three-dimensional array and from this array a non-zero diffracted beam can occur only if the Laue equations are satisfied. The same remarks apply to the array formed by atoms of type *B, C* or *D* alone. Thus the total crystal can be divided into four component intermeshed arrays and the diffraction from the whole

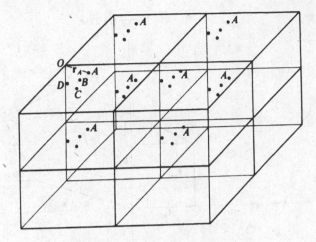

Fig. 3.13.
A group of eight unit cells containing four atoms per unit cell. Each atomic type forms a space lattice.

crystal will be the sum of the contributions from the four components. Hence there will be a diffracted beam from the whole crystal only when the Laue equations are satisfied and the Laue equations can thus be seen to be the controlling factor for diffraction from any three-dimensional periodic arrangement of atoms.

We shall refer the scattering from each component array of atoms to that which would be obtained from an array of point electrons at the origin points of the unit cell – marked $O$ in fig. 3.13. The scattering of any one atom $A$, with respect to an electron at the origin of the cell it occupies, will be $f_A \exp(2\pi i \mathbf{r}_A \cdot \mathbf{s})$ and the same ratio will apply for all the atoms $A$ with respect to electrons at all the origins. The total scattered amplitude from the crystal will thus be that from an array of electrons, one at each unit-cell origin, times a factor

$$f_A \exp(2\pi i \mathbf{r}_A \cdot \mathbf{s}) + f_B \exp(2\pi i \mathbf{r}_B \cdot \mathbf{s}) + f_C \exp(2\pi i \mathbf{r}_C \cdot \mathbf{s}) + f_D \exp(2\pi i \mathbf{r}_D \cdot \mathbf{s}). \quad (3.30)$$

The scattering vector $\mathbf{s}$ will correspond to a triplet of integers $(hkl)$ and, if there are $N$ atoms in the unit cell, we may, in general, write

$$F_{hkl} = \sum_{j=1}^{N} f_j \exp(2\pi i \mathbf{r}_j \cdot \mathbf{s}). \quad (3.31)$$

The quantity $F_{hkl}$, which we notice is a function of the contents of *one* unit cell, is called the *structure factor* and the integers $h$, $k$ and $l$ are referred to as the *indices of the reflection*. We shall see in § 3.6 the reason for referring to the diffracted beam as a reflection and at the same time we shall discover that the indices of the reflection are the same as the Miller indices defined in § 1.6.

The composition of the structure factor $F_{hkl}$ can be seen in a phase-vector diagram and is shown in fig. 3.14($a$). It can be divided into a real and imaginary part and written as

$$F_{hkl} = A_{hkl} + iB_{hkl} \quad (3.32)$$

where

$$A_{hkl} = \sum_{j=1}^{N} f_j \cos(2\pi \mathbf{r}_j \cdot \mathbf{s}) \quad (3.33a)$$

Fig. 3.14.
(a) Phase-vector
diagram for a structure
factor with
contributions from four
atoms.
(b) Phase-vector
diagram for a structure
factor corresponding to
a centrosymmetric
structure.

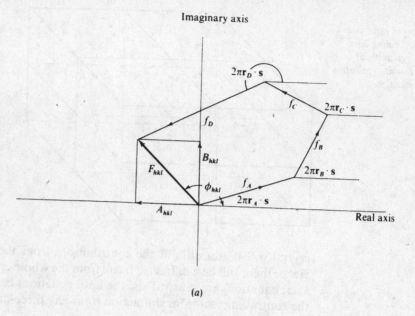

(a)

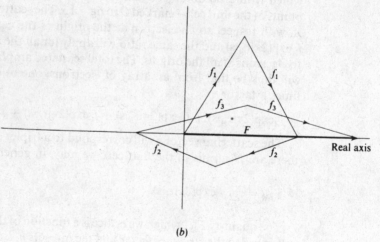

(b)

and

$$B_{hkl} = \sum_{j=1}^{N} f_j \sin(2\pi \mathbf{r}_j \cdot \mathbf{s}).$$

(3.33b)

There is clearly a phase angle $\phi_{hkl}$ associated with $F_{hkl}$ which is given by

$$\tan \phi_{hkl} = \frac{B_{hkl}}{A_{hkl}},$$

(3.34)

where $B_{hkl}$ and $A_{hkl}$ have the signs of $\sin \phi_{hkl}$ and $\cos \phi_{hkl}$, respectively.

The intensity of the diffracted beam from the total crystal, consisting as it does of large numbers of unit cells, will clearly be proportional to

$$\mathscr{I}_{hkl} = |F_{hkl}|^2 = A_{hkl}^2 + B_{hkl}^2$$

$$= \left\{ \sum_{j=1}^{N} f_j \cos(2\pi\mathbf{r}_j\cdot\mathbf{s}) \right\}^2 + \left\{ \sum_{j=1}^{N} f_j \sin(2\pi\mathbf{r}_j\cdot\mathbf{s}) \right\}^2$$

$$= \sum_{i=1}^{N} \sum_{j=1}^{N} f_i f_j [\cos(2\pi\mathbf{r}_i\cdot\mathbf{s})\cos(2\pi\mathbf{r}_j\cdot\mathbf{s}) + \sin(2\pi\mathbf{r}_i\cdot\mathbf{s})\sin(2\pi\mathbf{r}_j\cdot\mathbf{s})]$$

or

$$\mathscr{I}_{hkl} = \sum_{i=1}^{N} \sum_{j=1}^{N} f_i f_j \cos[2\pi(\mathbf{r}_i - \mathbf{r}_j)\cdot\mathbf{s}]. \tag{3.35}$$

The intensity is thus seen to depend only on the interatomic vectors and not on the actual atomic coordinates which depend on an arbitrary choice of origin. This only confirms what we know from physical considerations anyway, for we could hardly expect the intensity of the diffracted X-ray beam to depend on an arbitrary choice of an origin or an arbitrary unit cell.

An important case, which often occurs, is when the structure has a centre of symmetry. In § 2.6 we noted that for a centrosymmetric distribution of scattering material the amplitude of scattering is real. Thus for a centrosymmetric unit cell, with the centre of symmetry as origin, the expression for the structure factor is found by taking only the real part of equation (3.31) to give

$$F_{hkl} = \sum_{j=1}^{N} f_j(\cos 2\pi\mathbf{r}_j\cdot\mathbf{s}). \tag{3.36}$$

This explains why a centrosymmetric unit cell is always referred to a centre of symmetry as origin in the *International Tables* – a phenomenon we noticed with the space group $P2_1/c$ when we considered that space group in § 1.9.

A phase-vector diagram for the structure factor of a centrosymmetric structure shows very convincingly why the $F$ is real. In fig. 3.14(b) the scattering contribution is taken in turn from pairs of atoms which are centrosymmetrically related and after every pair of steps in the diagram one is always back on the real axis.

Other types of symmetry lead to other special forms of the structure-factor equation and this is dealt with in some detail in § 7.3.

## 3.6  Bragg's law

In 1912, von Laue first proposed that X-rays might be diffracted by crystals. At this time it had been deduced indirectly from the geometry of crystals that they were almost certainly in the form of a three-dimensional periodic array of atoms. From the known values of Avogadro's number, atomic weights and crystal densities the average distance apart of the atoms was found to be of the order of 1–2 Å and it could be presumed that the crystal repeat distances would be of the order of tens of ångström units. von Laue reasoned that an optical diffraction grating repeat distance was some few times the wavelength of visible light and the same relationship might obtain between the repeat distance in a crystal and X-ray wavelengths. This inspired prediction was confirmed by Friedrich and Knipping whose X-ray pictures, crude by modern standards as they were, caused a stir of excitement throughout scientific circles.

In 1913, W. L. Bragg gave the first mathematical explanation of the actual positions of the X-ray diffraction spots. We can follow the main lines of his explanation by considering diffraction by a two-dimensional array of scatterers as in fig. 3.15. A number of sets of equi-spaced parallel lines are shown which pass through all the scatterers and, according to the Bragg picture, these planes would act as partial reflectors for X-rays. When a parallel wave train falls on a specularly reflecting surface the conditions are such that reflection from any two points of the surface will give rays in phase in the reflected beam. In fig. 3.16(a) this can be clearly seen, as the phase difference for rays reflected from $A$ and $B$ is $CB - AD = 0$. However this condition would be true for *any* value of $\theta$ and we must now explain why it is only for special angles that the reflection can occur – as we know from experimental observations. To do this we must bring in the distance between the parallel reflecting planes. Since reflections from all points in one plane are in phase, it is only necessary to show that reflection from any point on one plane is in phase with that from any point on another plane to show that the total reflection from the two planes is in phase. The two points we shall consider are $A$ and $A'$ separated by the distance $d$ which is the interplane spacing [fig. 3.16(b)]. The condition for reinforcement is clearly given by

$$BA' + A'C = n\lambda$$

where $n$ is an integer. For reasons which will be clearer later we shall only

Fig. 3.15.
A selection of sets of parallel equally spaced lines on which lie all points of an array of scatterers.

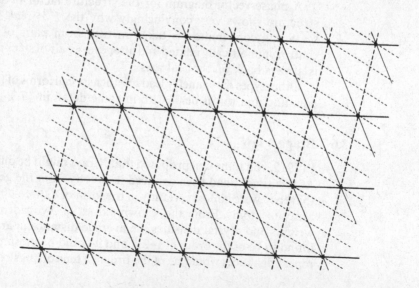

Fig. 3.16.
(*a*) Parallel rays
reflected from different
points of a plane are in
phase after reflection.
(*b*) Parallel rays
reflected from points on
neighbouring partially
reflecting planes are in
phase when Bragg's law
is obeyed.

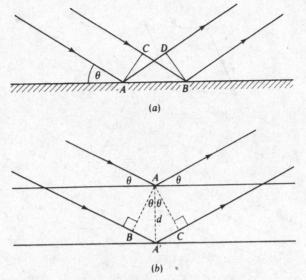

consider the case $n = 1$. From fig. 3.16(*b*) we can see that

$$BA' = A'C = d \sin \theta$$

giving

$$2d \sin \theta = \lambda. \tag{3.37}$$

This is the expression known as Bragg's law and it gives the permitted angles of reflection, $\theta$, in terms of $\lambda$ and the spacing of the reflecting planes, $d$. It can be seen now why we refer to the diffracted beams as 'reflections' and why the angle between the incident and diffracted beam is denoted by $2\theta$ and not simply by $\theta$.

If we look at the atom-rich lines in fig. 3.17(*a*) in terms of the two-dimensional unit cell which is outlined then it can be seen that they correspond to directions given by intercepts of $a/2$ and $b/1$ on the cell edges. In three dimensions one would have atom-rich planes parallel to the planes defined by three points with intercepts on the unit-cell axes of $a/h$, $b/k$, $c/l$ where $h, k$ and $l$ are three integers. These we shall refer to as the indices of the X-ray reflection and the spacing of the planes so defined is denoted by $d_{hkl}$. In fig. 3.17(*a*) the indices of the lines shown are (21) and in fig. 3.17(*b*) there are shown the lines with indices (42). Now the latter lines are not all atom-rich – in fact alternate ones are completely empty. In terms of Bragg's law the angle $\theta$ corresponding to such a spacing is given by

$$\sin \theta = \frac{\lambda}{2d_{42}}$$

but this can also be expressed as

$$\sin \theta = \frac{2\lambda}{2d_{21}}.$$

Seen in this way, a reflection from the line with indices $(2 \times 2 \quad 2 \times 1)$ can be regarded as a second-order reflection (difference of path $2\lambda$ by reflection

Fig. 3.17.
(*a*) The (21) reflecting
lines for a particular
two-dimensional unit
cell.
(*b*) The (42) reflecting
lines for the same unit
cell.

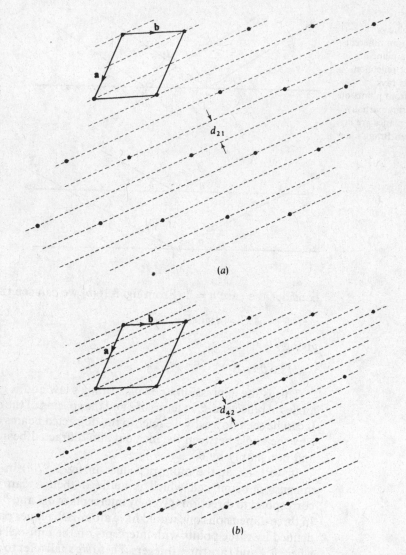

(*a*)

(*b*)

from successive lines) from the lines with spacing $d_{21}$. For three dimensions this generalizes to being able to consider reflection from planes with indices $(nh\,nk\,nl)$ as the $n$th-order reflection from planes with indices $(hkl)$. In fig. 3.18 there is shown part of a three-dimensional unit cell with planes drawn corresponding to the indices $(hkl)$. If $h$, $k$ and $l$ have no common factor then these planes would be those with Miller indices $(hkl)$ according to the definition given in §1.6.

Let us see if we can now relate Bragg's law to the Laue equations for, if they are describing the same phenomenon, they must be closely related. In fig. 3.18 $OD$ is perpendicular to $ABC$ and since $\overrightarrow{OA} = \mathbf{a}/h$ then

$$\overrightarrow{OA} \cdot \mathbf{d}_{hkl} = \frac{1}{h}\mathbf{a} \cdot \mathbf{d}_{hkl} = OA \times d_{hkl}\cos\chi = d_{hkl}^2.$$

Fig. 3.18.
Successive (hkl)
reflecting planes with
interplanar spacing $d_{hkl}$.

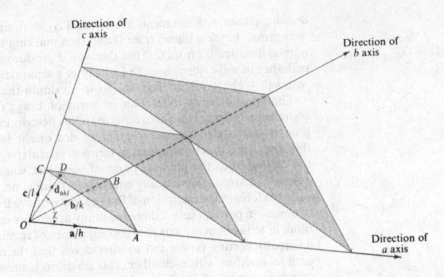

Thus

$$\mathbf{a} \cdot \frac{\mathbf{d}_{hkl}}{d_{hkl}^2} = h$$

and similarly

$$\mathbf{b} \cdot \frac{\mathbf{d}_{hkl}}{d_{hkl}^2} = k$$

and

$$\mathbf{c} \cdot \frac{\mathbf{d}_{hkl}}{d_{hkl}^2} = l. \tag{3.38}$$

Comparing this with equations (3.11) we see that, assuming a unique solution for **s**, we must have

$$\mathbf{s} = \mathbf{d}_{hkl}/d_{hkl}^2 \tag{3.39a}$$

that is to say that **s** is a vector in the direction of $\mathbf{d}_{hkl}$ but with a magnitude $1/d_{hkl}$. In terms of magnitudes

$$s = 1/d_{hkl} \tag{3.39b}$$

or, from equation (2.12),

$$\frac{2 \sin \theta}{\lambda} = \frac{1}{d_{hkl}}$$

which is a rearranged form of equation (3.37).

From equations (3.29) and (3.39b) the values of $d_{hkl}$ may be deduced for a particular unit cell and set of indices.

If we look again at fig. 2.3 and interpret $\hat{\mathbf{S}}_0$ and $\hat{\mathbf{S}}$ as vectors along the direction of the incident and reflected beams from a mirror reflection then clearly **s** is perpendicular to the plane of the mirror, as indeed is $d$ which we have seen is parallel to **s**. We could construct a reciprocal lattice by first drawing planes with intercepts $a/h$, $b/k$ and $c/l$, as in fig. 3.18, then finding the normals to these planes from $O$ and then marking the points along the normal distance $1/d$ from $O$. That this would produce a lattice of points is perhaps not self-evident but the identity of $s$ separately with $d_{hkl}/d_{hkl}^2$ and with $ha^* + kb^* + lc^*$ should leave us in no doubt that it would do so.

Thinking of X-ray diffraction in terms of Bragg's law is often very convenient and we shall have recourse to this description of the diffraction process from time to time. Whether or not one is formally justified in thinking of a diffraction process in terms of specular-type reflection is not too certain from the physical point of view but, since the two types of description are mathematically equivalent, there is no harm in doing so.

In the foregoing description of Bragg's law we have treated only the case of one atom per unit cell; if there are many atoms per cell then one should think in terms of many sets of reflecting planes of spacing $d_{hkl}$ each with a different reflecting power and so interleaved that the reflected rays from each set interfere with each other. This situation is shown in fig. 3.19 for a two-dimensional unit cell containing three atoms per unit cell.

## 3.7   The structure factor in terms of indices of reflection

It is convenient to express the positions of atoms in the unit cell with respect to the chosen origin in terms of fractional coordinates. Thus we say that an atom has coordinates $(x, y, z)$ where $x$, $y$ and $z$ are three fractional numbers in the range 0–1. If the cell edges are denoted by the vectors $\mathbf{a}$, $\mathbf{b}$ and $\mathbf{c}$ then the vector position of an atom would be

$$\mathbf{r} = x\mathbf{a} + y\mathbf{b} + z\mathbf{c} \tag{3.40}$$

as is shown for the point $P$ in fig. 3.20.

Previously, in equation (3.16), we have found an expression for $\mathbf{s}$ in terms of $h$, $k$ and $l$ and now we can rewrite equation (3.31) as

$$F_{hkl} = \sum_{j=1}^{N} f_j \exp(2\pi i \mathbf{r}_j \cdot \mathbf{s})$$

Fig. 3.19.
A two-dimensional
structure with three
different atoms showing
interleaved reflecting
planes.

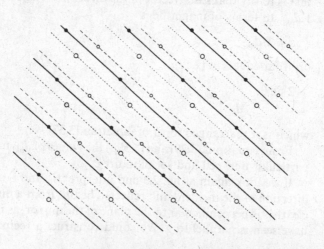

Fig. 3.20.
The relationship of the
vector $\overline{OP} = \mathbf{r}$ to the
fractional coordinates of
$P(x, y, z)$.

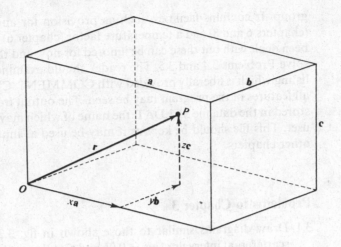

$$= \sum_{j=1}^{N} f_j \exp[2\pi i(x_j\mathbf{a} + y_j\mathbf{b} + z_j\mathbf{c})\cdot(h\mathbf{a}^* + k\mathbf{b}^* + l\mathbf{c}^*)].$$

From this and the relationships (3.14) we find

$$F_{hkl} = \sum_{j=1}^{N} f_j \exp[2\pi i(hx_j + ky_j + lz_j)]. \tag{3.41}$$

In the case of a centrosymmetric structure the structure-factor equation appears as

$$F_{hkl} = \sum_{j=1}^{N} f_j \cos[2\pi(hx_j + ky_j + lz_j)]. \tag{3.42}$$

and, using this nomenclature, equation (3.35) becomes

$$\mathscr{I}_{hkl} = \sum_{i=1}^{N} \sum_{j=1}^{N} f_i f_j \cos\{2\pi[h(x_i - x_j) + k(y_i - y_j) + l(z_i - z_j)]\}. \tag{3.43}$$

The experimentally observed intensity of a diffracted beam is proportional to $\mathscr{I}_{hkl}$ and it can be seen from equation (3.43) that

$$\mathscr{I}_{hkl} = \mathscr{I}_{\bar{h}\bar{k}\bar{l}}. \tag{3.44}$$

This equation holds whether or not the crystal structure has a centre of symmetry. There is a useful concept of a *weighted reciprocal lattice* where we associate with each reciprocal lattice point the corresponding $\mathscr{I}_{hkl}$. Equation (3.44) implies that the weighted reciprocal lattice always has a centre of symmetry. This important intensity relationship is known as Friedel's law.

The form of the structure-factor equation, (3.41), is the most useful one in practice, for normally, as we shall see, one wishes to calculate structure factors when the atomic coordinates $(x, y, z)$ are known for particular indices of reflection $(hkl)$.

In appendix I there is given a FORTRAN® program STRUCFAC which can be used to calculate structure factors for any two-dimensional space group. It contains facilities, such as provision for anomalous scattering (chapters 6 and 8) and a temperature factor (chapter 6) which have not yet been dealt with but these can be ignored for now and the program used to solve Problems 3.4 and 3.5. The reader should examine the FORTRAN® listing which is liberally provided with COMMENT 'C' statements so that all features of the program may be seen. The output from STRUCFAC is stored in the data file SF1.DAT, the name of which may be changed by the user. This file should be kept as it may be used as input for Problems to other chapters.

## Problems to Chapter 3

3.1 Draw diagrams similar to those shown in fig. 3.2 for a row of six scatterers at intervals of $\mathbf{a \cdot s} = 0.05$ and plot the resultant amplitude as a function of $\mathbf{a \cdot s}$.

3.2 A parallel beam of radiation of wavelength $\lambda$ falls on a row of scatterers of spacing $5\lambda$ so that its direction makes an angle of $30°$ with the row. Find what orders of diffraction will be produced and the angle the diffracted beams will make with the row.

3.3 A unit cell has the following parameters:

$$a = 5\,\text{Å} \ b = 10\,\text{Å}, \ c = 15\,\text{Å}, \ \alpha = \beta = 90°, \ \gamma = 120°.$$

  (i) Determine the parameters of the reciprocal cell.
  (ii) Find the volume of the real and reciprocal cells.
  (iii) Find the spacing of the (321) planes.
  (iv) If Cu $K\alpha$ radiation is used ($\lambda = 1.54\,\text{Å}$) what is the angle of diffraction ($2\theta$) for the (321) reflection?

3.4 The crystal and molecular structure of the phenylethylammonium salt of fosfomycin, $C_8H_{12}N^+.C_9H_6OPO_9^-.H_2O$, was solved by A. Perales and S. García-Blanco (*Acta Cryst.* (1978) **B34** 238–42). The non-centrosymmetric monoclinic space group is $P2_1$. The cell dimensions are:

$$a = 11.54\,\text{Å}, \ b = 6.15\text{Å}, \ c = 10.21\,\text{Å}; \ \beta = 102.5°,$$

and the cooordinates of the non-hydrogen atoms (the hydrogen atoms will make a small and negligible contribution to the structure factors) in one asymmetric unit are shown in table 3.2. The projection down the $b$ axis has the centrosymmetric two-dimensional space group $p2$. If the X-ray wavelength is $1.542\,\text{Å}$ (Cu $K\alpha$ radiation) then use the program STRUCFAC (appendix I) to calculate the complete set of observable structure factors for this projection.

3.5 The $c$-axis projection of the phenylethylammonium salt of fosfomycin has the non-centrosymmetric two-dimensional space group $pg$. Calculate the observable structure factors for this projection taking all necessary information from Problem 3.4.

3.6 A crystal of density very near $1.2 \times 10^3 \, \text{kg m}^{-3}$ is found to have a unit cell with $a = 5.601\,\text{Å}$, $b = 6.002\,\text{Å}$, $c = 8.994\,\text{Å}$, $\alpha = 100.7°$, $\beta = 95.2°$, $\gamma = 115.3°$.

Table 3.2. *Coordinates for Problem 3.4*

| Atom | $x$ | $y$ | $z$ |
| --- | --- | --- | --- |
| P | 0.1194 | 0.2500 | 0.2740 |
| O(1) | 0.0654 | 0.1303 | 0.1479 |
| O(2) | 0.1001 | 0.1560 | 0.4021 |
| O(3) | 0.0740 | 0.4928 | 0.2655 |
| O(4) | 0.3587 | 0.3229 | 0.3982 |
| O(water) | 0.1141 | 0.7220 | 0.0717 |
| N | 0.0697 | 0.3238 | 0.6410 |
| C(1) | 0.2755 | 0.2676 | 0.2737 |
| C(2) | 0.3677 | 0.1194 | 0.3322 |
| C(3) | 0.3482 | 0.9165 | 0.4017 |
| C(4) | 0.1748 | 0.4318 | 0.7318 |
| C(5) | 0.2156 | 0.6140 | 0.6519 |
| C(6) | 0.2670 | 0.2621 | 0.7892 |
| C(7) | 0.3650 | 0.2229 | 0.7347 |
| C(8) | 0.4479 | 0.0693 | 0.7909 |
| C(9) | 0.4363 | 0.9542 | 0.9017 |
| C(10) | 0.3394 | 0.9882 | 0.9551 |
| C(11) | 0.2564 | 0.1433 | 0.9011 |

(i) How many molecules of chemical formula $C_6H_{12}$ are there in the unit cell?

(ii) Give a better estimate of the density.

Note:
atomic weight of carbon (mean of normal isotopic mixture) = 12.0150 amu
atomic weight of hydrogen (mean isotopic mixture) = 1.0083 amu

1 amu = $1.6598 \times 10^{-27}$ kg

# 4 The Fourier transform

## 4.1 The Fourier series

Let us consider a function $f(X)$ which is single-valued, continuous (except for a finite number of finite discontinuities), with a finite number of maxima and minima and which is defined in the range $\frac{1}{2}a \geqslant X \geqslant -\frac{1}{2}a$. Within this range such a function may be represented by a series of sine and cosine terms of the form

$$f(X) = A_0 + 2 \sum_{h=1}^{\infty} A_h \cos\left(2\pi h \frac{X}{a}\right) + 2 \sum_{h=1}^{\infty} B_h \sin\left(2\pi h \frac{X}{a}\right). \tag{4.1}$$

This particular form of the series, known as the Fourier series, with factors of two outside the summation signs, we have adopted because it simplifies the forms of expressions we shall obtain by subsequent manipulation. We need not be too troubled by the various mathematical restrictions placed on the function $f(X)$; all the functions with which we shall find ourselves concerned are going to be very well behaved.

We might look at the form of some of the terms which are included in the summation. In fig. 4.1 there are shown in the range $\frac{1}{2}a \geqslant X \geqslant -\frac{1}{2}a$ the functions $\cos[2\pi(X/a)]$, $\cos[2\pi 2(X/a)]$, $\sin[2\pi(X/a)]$ and $\sin[2\pi 2(X/a)]$. It will be seen that the main characteristic of $\cos[2\pi h(X/a)]$ is that it is an even function of $X$ whereas $\sin[2\pi h(X/a)]$ is an odd function of $X$.

If, for a given $f(X)$, we wish to determine the coefficients in equation (4.1) we make use of the following relationships:

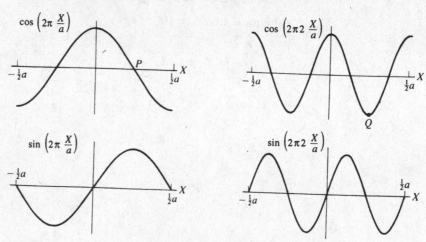

Fig. 4.1.
The functions $\cos\{2\pi(X/a)\}$, $\cos\{2\pi 2(X/a)\}$, $\sin\{2\pi(X/a)\}$ and $\sin\{2\pi 2(X/a)\}$.

$$\int_{-a/2}^{a/2} \cos\left(2\pi n\frac{X}{a}\right) dX = 0 \tag{4.2}$$

for $n \neq 0$, where $n$ is an integer;

$$\int_{-a/2}^{a/2} \cos\left(2\pi n\frac{X}{a}\right) dX = a \tag{4.3}$$

for $n = 0$ (since then $\cos[2\pi n(X/a)] = 1$ for all $X$); and

$$\int_{-a/2}^{a/2} \sin\left(2\pi n\frac{X}{a}\right) dX = 0 \tag{4.4}$$

for all integers $n$.

The results (4.2) and (4.4) can be seen in fig. 4.1 from the fact that there are equivalent matching positive and negative regions in the cosine and sine functions.

Other results follow from equations (4.2), (4.3) and (4.4). For example,

$$\int_{-a/2}^{a/2} \cos\left(2\pi n\frac{X}{a}\right) \cos\left(2\pi m\frac{X}{a}\right) dX$$

$$= \frac{1}{2}\int_{-a/2}^{a/2} \cos\left[2\pi(n+m)\frac{X}{a}\right] dX + \frac{1}{2}\int_{-a/2}^{a/2} \cos\left[2\pi(n-m)\frac{X}{a}\right] dX = 0 \tag{4.5}$$

unless $n = \pm m$ when

$$\int_{-a/2}^{a/2} \cos^2\left(2\pi n\frac{X}{a}\right) dX = \int_{-a/2}^{a/2} \left[\frac{1}{2} + \frac{1}{2}\cos\left(2\pi 2n\frac{X}{a}\right)\right] dX = \tfrac{1}{2}a. \tag{4.6}$$

We can also find

$$\int_{-a/2}^{a/2} \sin\left(2\pi n\frac{X}{a}\right) \sin\left(2\pi m\frac{X}{a}\right) dX$$

$$= \frac{1}{2}\int_{-a/2}^{a/2} \cos\left[2\pi(n-m)\frac{X}{a}\right] dX - \frac{1}{2}\int_{-a/2}^{a/2} \cos\left[2\pi(n+m)\frac{X}{a}\right] dX = 0 \tag{4.7}$$

unless $n = \pm m$ when

$$\int_{-a/2}^{a/2} \pm\sin^2\left(2\pi n\frac{X}{a}\right) dX = \int_{-a/2}^{a/2} \pm\left[\frac{1}{2} - \frac{1}{2}\cos\left(2\pi 2n\frac{X}{a}\right)\right] dX = \pm\tfrac{1}{2}a. \tag{4.8}$$

The final result we shall require is

$$\int_{-a/2}^{a/2} \sin\left(2\pi n\frac{X}{a}\right) \cos\left(2\pi m\frac{X}{a}\right) dX$$

$$= \frac{1}{2}\int_{-a/2}^{a/2} \sin\left[2\pi(n+m)\frac{X}{a}\right] dX + \frac{1}{2}\int_{-a/2}^{a/2} \sin\left[2\pi(n-m)\frac{X}{a}\right] dX = 0 \tag{4.9}$$

for all $n$ and $m$.

We now note that, for an integer $H$,

$$\int_{-a/2}^{a/2} f(X)\cos\left(2\pi H\frac{X}{a}\right)dX = A_0 \int_{-a/2}^{a/2}\cos\left(2\pi H\frac{X}{a}\right)dX$$

$$+ 2\sum_{h=1}^{\infty} A_h \int_{-a/2}^{a/2}\cos\left(2\pi h\frac{X}{a}\right)\cos\left(2\pi H\frac{X}{a}\right)dX$$

$$+ 2\sum_{h=1}^{\infty} B_h \int_{-a/2}^{a/2}\sin\left(2\pi h\frac{X}{a}\right)\cos\left(2\pi H\frac{X}{a}\right)dX \qquad (4.10)$$

and, from equations (4.5), (4.6) and (4.9), we find that all the components on the right-hand side are zero except for

$$2A_H \int_{-a/2}^{a/2}\cos^2\left(2\pi H\frac{X}{a}\right)dX = A_H a. \qquad (4.11)$$

Thus we have

$$A_H = \frac{1}{a}\int_{-a/2}^{a/2} f(X)\cos\left(2\pi H\frac{X}{a}\right)dX \qquad (4.12)$$

and similarly we find

$$B_H = \frac{1}{a}\int_{-a/2}^{a/2} f(X)\sin\left(2\pi H\frac{X}{a}\right)dX. \qquad (4.13)$$

The expression (4.1) can be rewritten as

$$f(X) = \sum_{h=-\infty}^{\infty} A_h \cos\left(2\pi h\frac{X}{a}\right) + \sum_{h=-\infty}^{\infty} B_h \sin\left(2\pi h\frac{X}{a}\right), \qquad (4.14)$$

where $A_{-h} = A_h$, $B_{-h} = -B_h$ and $B_0 = 0$.

What we are doing here is to extend the range of indices to all integers, both positive and negative. The relationships between the Fourier coefficient which are given in equation (4.14) are consistent with what we would obtain from equations (4.12) and (4.13) since $\cos\theta = \cos(-\theta)$, $\sin\theta = -\sin(-\theta)$ and $\sin 0 = 0$.

If we now write

$$C_h = A_h + iB_h \qquad (4.15)$$

then

$$C_h \exp\left(-2\pi h\frac{X}{a}\right) + C_{-h}\exp\left(2\pi h\frac{X}{a}\right)$$

$$= (A_h + iB_h)\left[\cos\left(2\pi h\frac{X}{a}\right) - i\sin\left(2\pi h\frac{X}{a}\right)\right]$$

$$+ (A_h - iB_h)\left[\cos\left(2\pi h\frac{X}{a}\right) + i\sin\left(2\pi h\frac{X}{a}\right)\right]$$

$$= 2A_h \cos\left(2\pi h\frac{X}{a}\right) + 2B_h \sin\left(2\pi h\frac{X}{a}\right)$$

$$= A_h \cos\left(2\pi h \frac{X}{a}\right) + A_{-h} \cos\left[2\pi(-h)\frac{X}{a}\right] + B_h \sin\left(2\pi h \frac{X}{a}\right)$$

$$+ B_{-h} \sin\left[2\pi(-h)\frac{X}{a}\right]. \tag{4.16}$$

If we examine equations (4.16) and (4.14) then it is clear that we may write

$$f(X) = \sum_{h=-\infty}^{\infty} C_h \exp\left(-2\pi i h \frac{X}{a}\right). \tag{4.17}$$

In addition, from equations (4.15), (4.12) and (4.13) we find

$$C_h = \frac{1}{a}\int_{-a/2}^{a/2} f(X)\cos\left(2\pi h \frac{X}{a}\right)dX + i\frac{1}{a}\int_{-a/2}^{a/2} f(X)\sin\left(2\pi h \frac{X}{a}\right)dX$$

$$= \frac{1}{a}\int_{-a/2}^{a/2} f(X)\exp\left(2\pi i h \frac{X}{a}\right)dX. \tag{4.18}$$

Relationships of the form (4.17) and (4.18) are not of much value for numerical work, involving as they do complex quantities, but they are a neat and concise form of expression for a Fourier series and the integrals which give the coefficients.

## 4.2 Numerical application of Fourier series

The best method of getting a 'feeling' for the properties of Fourier series is actually to do some numerical calculations. If we are dealing with one-dimensional functions then the equations we shall want to use are (4.1), (4.12) and (4.13).

To illustrate the application of these relationships we shall consider the function shown in fig. 4.2 which is

$$f(X) = 1 - 2\left|\frac{X}{a}\right|, \text{ for } -\frac{1}{4} \leqslant \frac{X}{a} \leqslant \frac{1}{4}$$

$$= \frac{1}{2}, \text{ for } -\frac{1}{2} \leqslant \frac{X}{a} \leqslant -\frac{1}{4} \text{ and } \frac{1}{2} \geqslant \frac{X}{a} \geqslant \frac{1}{4}. \tag{4.19}$$

Then from equation (4.12)

$$A_H = \frac{1}{a}\int_{-a/4}^{a/4}\left(1 - 2\left|\frac{X}{a}\right|\right)\cos\left(2\pi H \frac{X}{a}\right)dX$$

$$+ \frac{1}{a}\int_{a/4}^{a/2}\frac{1}{2}\cos\left(2\pi H \frac{X}{a}\right)dX + \frac{1}{a}\int_{-a/2}^{-a/4}\frac{1}{2}\cos\left(2\pi H \frac{X}{a}\right)dX$$

$$= \frac{1}{a}\int_{0}^{a/4}\left(1 - 2\frac{X}{a}\right)\cos\left(2\pi H \frac{X}{a}\right)dX$$

$$+ \frac{1}{a}\int_{-a/4}^{0}\left(1 + 2\frac{X}{a}\right)\cos\left(2\pi H \frac{X}{a}\right)dX + \frac{1}{a}\int_{a/4}^{a/2}\cos\left(2\pi H \frac{X}{a}\right)dX.$$

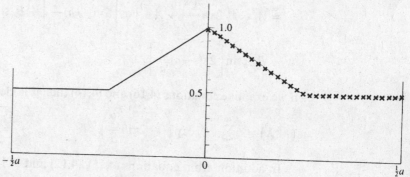

Fig. 4.2.

The function $f(X) = 1 - 2\left|\dfrac{X}{a}\right|$, for $-\tfrac{1}{4}a \leqslant X \leqslant \tfrac{1}{4}a$

$\qquad\qquad = \tfrac{1}{2}$, for $\tfrac{1}{4}a < |X| \leqslant \tfrac{1}{2}a$.

The crosses show the values of the reconstituted function computed from the Fourier series with $h_{\max} = 10$.

and these integrations give

$$A_H = \frac{1}{\pi^2 H^2}\left(1 - \cos\frac{\pi H}{2}\right) \text{ for } H \neq 0 \text{ and } A_0 = 0.625. \qquad (4.20)$$

From equation (4.12) it can be seen that $A_0 = (1/a) \times$ area under the curve and this can be confirmed for this simple function by an examination of fig. 4.2.

If one formally goes through the process of using equation (4.13) to determine the coefficients $B_H$ it would be found that they are all zero. This could be predicted from the fact that $f(X)$ is an even function of $X$ (i.e. $f(X) = f(-X)$) and the Fourier series which represents it must contain only even (i.e. cosine) terms.

Having obtained the values of the coefficients $A_h$ we are now in a position to see how well we can synthesize $f(X)$ from

$$f(X) = A_0 + 2\sum_{h=1}^{\infty} A_h \cos\left(2\pi h\frac{X}{a}\right). \qquad (4.21)$$

The coefficients, $A_h$, for $h = 0$ to 10 are:

| $h$ | 0 | 1 | 2 | 3 | 4 | 5 | 6 | 7 | 8 | 9 | 10 |
|-----|---|---|---|---|---|---|---|---|---|---|----|
| $A_h$ | 0.6250 | 0.1013 0.0506 | 0.0113 | 0 | 0.0041 | 0.0056 | | 0.0021 | 0 | 0.0013 | 0.0020 |

In practice one must always terminate the summation at some point and the similarity of the computed $f(X)$ to the true function will depend on the number of terms in the summation. In fig. 4.3 there are shown the results of terminating with $h_{\max} = 1$, 3 and 7 and in fig. 4.2 the result of having $h_{\max} = 10$. It can be seen how the summation gradually evolves towards the correct shape and the result with $h = 10$ is a very good representation of the

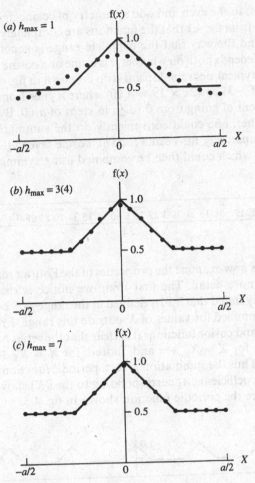

(a) $h_{max} = 1$

f(x)

1.0

0.5

$-a/2$   0   $a/2$   X

(b) $h_{max} = 3(4)$

f(x)

1.0

0.5

$-a/2$   0   $a/2$   X

(c) $h_{max} = 7$

f(x)

1.0

0.5

$-a/2$   0   $a/2$   X

Fig. 4.3.
Simulation of the function shown in fig. 4.2 using a Fourier series with: (a) $h_{max} = 1$; (b) $h_{max} = 3$ (since $A_4 = B_4 = 0$ this is equivalent to $h_{max} = 4$); and (c) $h_{max} = 7$. The computed values are shown by dots.

function; indeed, one could represent the function by the summation with any desired degree of precision by increasing the number of terms. Even $h_{max} = 3$ gives quite a good reproduction of $f(X)$ with the main discrepancy being in the region of the origin. The summations shown in fig. 4.3 were made with the computer program FOUR1 given in appendix II. It is a very simple program, written to be easy to understand; it exploits the symmetry of cosine and sine functions in that the former are even functions, so that $f(-X) = f(X)$ and the latter are odd, so that $f(-X) = -f(X)$.

In the early days of crystallography, before electronic computers were available, the task of computing a Fourier series – which can be in more than one dimension – was a formidable one. An early and important contribution to this problem was made by Beevers and Lipson (1934) who precalculated components of individual contributions to the series on strips of cardboard which were provided in convenient-to-use wooden boxes. In addition to the even and odd symmetry of cosine and sine functions it will be seen from fig. 4.1 that the functions are completely defined in the range 0 to $\frac{1}{4}a$ and the way that the complete range is generated from the partial range depends both on whether it is a sine or a cosine and also on the parity of $h$. A typical Beevers–Lipson strip is shown in fig. 4.4. The strip gives the value of $-31\cos(2\pi \times 19 \times n/60)$ where $n$ goes from 0 to 15, which is the equivalent of going from 0 to $\frac{1}{4}a$ in steps of $a/60$. By placing strips under each other, one could conveniently do the summations over the quarter range separately for cosine (odd $h$), cosine (even $h$), sine (odd $h$) and sine (even $h$) which could then be combined using symmetry to give the whole range.

Fig. 4.4.
A typical
Beevers–Lipson strip
giving
$-31\cos(2\pi \times 19 \times n/60)$
where $n$ is from 0 to 15.
CE indicates that it is a
cosine strip for even
$\frac{1}{120}$ths. The odd $\frac{1}{120}$ths
are on the reverse side.
As a check for the user
the last figures is the
sum of the 16 values.

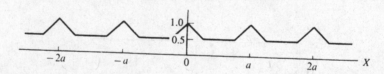

31 CE 19 $\quad$ $\overline{31}$ 13 21 $\overline{29}$ 3 27 $\overline{25}$ 6 30 $\overline{18}$ $\overline{15}$ 31 $\overline{10}$ $\overline{23}$ 28 0 $\quad$ (4)

Let us now examine the properties of the Fourier series, equation (4.1), in a little more detail. The first thing we notice is that although we have originally stated that $f(X)$ is defined in the range $-a/2 \leqslant X \leqslant a/2$ the series can be summed for values of $X$ outside this range. From the properties of the sine and cosine functions it is clear that the series has the same value for $X = X_1$, for $X = X_1 + a$ and, indeed, for $X = X_1 + ma$ where $m$ is any integer. Thus the summation gives a periodic function of spacing $a$ and the Fourier coefficients $A_h$ corresponding to the $f(X)$ shown in fig. 4.2 actually reproduce the periodic function shown in fig. 4.5.

Fig. 4.5.
The periodic function
reproduced by the
Fourier series which is
derived from the
function $f(X)$ shown in
fig. 4.2.

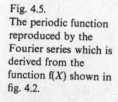

Often it is convenient to define a function in the range of $X$ from 0 to $a$ and it is clear that the Fourier series can be applied to this range. The coefficients $A_h$ which reproduced the $f(X)$ defined in the range $-a/2 \leqslant X \leqslant a/2$ in fig. 4.2 also reproduce the function $f(X)$ defined in the range $0 \leqslant X \leqslant a$ as shown in fig. 4.5. One may replace all the limits $a/2, -a/2$ by $a, 0$ in the relationships (4.2) to (4.13).

It is also convenient to change from the variable $X$ which has the units of $a$ to the variable $x = X/a$ so that $x$ is now a fractional coordinate defined in the range $\frac{1}{2} \geqslant x \geqslant -\frac{1}{2}$ or $1 \geqslant x \geqslant 0$.

Let us suppose that we have a periodic function $f(x)$ but that $f(x)$ is now not an analytical function but one which is known in graphical or tabular form.

Such a function, whose periodicity is unity is shown in fig. 4.6 both graphically and also tabulated at intervals of $\frac{1}{60}$ in $x$. We wish to carry out by a numerical method the integrals

$$A_h = \int_0^1 f(x)\cos(2\pi hx)\,\mathrm{d}x$$

(4.22a)

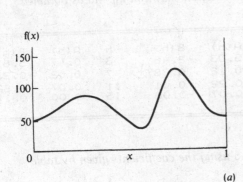

| 60x | f(x) | 60x | f(x) | 60x | f(x) | 60x | f(x) | 60x | f(x) | 60x | f(x) |
|---|---|---|---|---|---|---|---|---|---|---|---|
| 0 | 50 | 10 | 76 | 20 | 79 | 30 | 42 | 40 | 90 | 50 | 95 |
| 1 | 51 | 11 | 79 | 21 | 76 | 31 | 39 | 41 | 106 | 51 | 85 |
| 2 | 52 | 12 | 81 | 22 | 73 | 32 | 37 | 42 | 117 | 52 | 78 |
| 3 | 54 | 13 | 83 | 23 | 70 | 33 | 34 | 43 | 123 | 53 | 71 |
| 4 | 57 | 14 | 84 | 24 | 66 | 34 | 32 | 44 | 126 | 54 | 65 |
| 5 | 60 | 15 | 85 | 25 | 63 | 35 | 35 | 45 | 126 | 55 | 60 |
| 6 | 63 | 16 | 85 | 26 | 60 | 36 | 41 | 46 | 123 | 56 | 56 |
| 7 | 66 | 17 | 84 | 27 | 55 | 37 | 51 | 47 | 120 | 57 | 52 |
| 8 | 69 | 18 | 83 | 28 | 50 | 38 | 62 | 48 | 113 | 58 | 51 |
| 9 | 72 | 19 | 80 | 29 | 47 | 39 | 74 | 49 | 104 | 59 | 50 |
|  |  |  |  |  |  |  |  |  |  | 60 | 50 |

Fig. 4.6.
A non-analytical function f(x) shown graphically and by a table.

and

$$B_h = \int_0^1 f(x)\sin(2\pi hx)\mathrm{d}x \qquad (4.22b)$$

There are various methods available for numerical integration and in appendix III there is given a simple program, SIMP1, for applying *The Simpson's Rule* method. This replaces an integral

$$\int_a^b g(x)\mathrm{d}x \text{ by } \frac{h}{3}\sum_{n=0}^{N} w_n\, g(x_a + nh)$$

where the range of x values, from a to b is divided into an even number of intervals N each of width h and the weights $w_n$ follow the pattern 1 4 2 4 2 ... 2 4 2 4 1. Simpson's Rule is equivalent to adding the areas of parabolae fitted to three function values at points defining successive pairs of intervals. Using this method the integrals in equation (4.22) can be replaced by the summations

$$A_h = \sum_{n=0}^{60} w_n f\left(\frac{n}{60}\right)\cos\left[2\pi h\left(\frac{n}{60}\right)\right] \qquad (4.23a)$$

and

$$B_h = \sum_{n=0}^{60} w_n f\left(\frac{n}{60}\right)\sin\left[2\pi h\left(\frac{n}{60}\right)\right]. \qquad (4.23b)$$

The coefficients obtained from these summations are shown in table 4.1 up to $h = 15$ and these values of $A_h$ and $B_h$ are used with FOUR1 to reconstitute the function f(x) which is shown in table 4.2. These results may be compared with the tabulated values of the original function given in fig. 4.6. The small disagreements are due to rounding off the calculated values of $A_h$ and $B_h$ and the termination of the series.

## 4.3  Fourier series in two and three dimensions

The Fourier series can also be used to express two- and three-dimensional functions. For example if we have a two-dimensional, single-valued and

Table 4.1. *The coefficients found by equations (4.23) for the function shown in fig. 4.6 as obtained by the program SIMP1*

| h | A(h) | B(h) | h | A(h) | B(h) | h | A(h) | B(h) | h | A(h) | B(h) |
|---|------|------|---|------|------|---|------|------|---|------|------|
| 0 | 71.84 | 0.00 | 1 | 2.16 | -4.64 | 2 | -15.03 | -2.48 | 3 | 0.47 | 5.83 |
| 4 | 2.24 | -1.45 | 5 | -0.89 | -0.26 | 6 | 0.12 | 0.38 | 7 | 0.02 | -0.23 |
| 8 | -0.17 | 0.05 | 9 | 0.03 | 0.13 | 10 | 0.22 | 0.01 | 11 | 0.07 | -0.22 |
| 12 | -0.09 | 0.14 | 13 | 0.10 | 0.01 | 14 | -0.07 | -0.08 | 15 | 0.00 | 0.11 |

Table 4.2. *A reconstruction of the function shown in fig. 4.6 using the coefficients given by table 4.1 in the program FOUR1*

```
THE COORDINATES ARE X/A = n/N WHERE N =  60
```

| n | f(x) | n | f(x) | n | f(x) | n | f(x) |
|---|------|---|------|---|------|---|------|
| 0 | 50.20 | 1 | 51.01 | 2 | 52.10 | 3 | 53.93 |
| 4 | 56.83 | 5 | 60.14 | 6 | 63.09 | 7 | 65.77 |
| 8 | 68.80 | 9 | 72.27 | 10 | 75.73 | 11 | 78.82 |
| 12 | 81.29 | 13 | 82.97 | 14 | 84.04 | 15 | 84.82 |
| 16 | 85.09 | 17 | 84.21 | 18 | 82.32 | 19 | 80.32 |
| 20 | 78.52 | 21 | 76.21 | 22 | 73.05 | 23 | 69.68 |
| 24 | 66.52 | 25 | 63.16 | 26 | 59.24 | 27 | 55.08 |
| 28 | 50.89 | 29 | 46.51 | 30 | 42.36 | 31 | 39.16 |
| 32 | 36.60 | 33 | 33.97 | 34 | 32.43 | 35 | 34.70 |
| 36 | 41.58 | 37 | 51.02 | 38 | 61.51 | 39 | 74.25 |
| 40 | 89.98 | 41 | 105.75 | 42 | 117.22 | 43 | 123.17 |
| 44 | 125.50 | 45 | 125.82 | 46 | 123.99 | 47 | 119.60 |
| 48 | 112.90 | 49 | 104.33 | 50 | 94.61 | 51 | 85.21 |
| 52 | 77.42 | 53 | 71.10 | 54 | 65.31 | 55 | 59.88 |
| 56 | 55.48 | 57 | 52.44 | 58 | 50.61 | 59 | 49.91 |
| 60 | 50.20 | | | | | | |

continuous function $f(X, Y)$ defined in a parallelogram whose sides are the vectors **a** and **b**, where the coordinates $X$ and $Y$ are measured in the directions of **a** and **b**, then we may write

$$f(X, Y) = \sum_{h=-\infty}^{\infty} \sum_{k=-\infty}^{\infty} A_{hk} \cos\left[2\pi\left(h\frac{X}{a} + k\frac{Y}{b}\right)\right]$$

$$+ \sum_{h=-\infty}^{\infty} \sum_{k=-\infty}^{\infty} B_{hk} \sin\left[2\pi\left(h\frac{X}{a} + k\frac{Y}{b}\right)\right] \tag{4.24}$$

where $A_{hk} = A_{\bar{h}\bar{k}}$, $B_{hk} = -B_{\bar{h}\bar{k}}$ and $B_{oo} = 0$. We use the convention that $\bar{h} = -h$. By mathematical arguments similar to those used in §4.2 we can find

$$A_{hk} = \frac{1}{ab} \int_{-a/2}^{a/2} \int_{-b/2}^{b/2} f(X, Y) \cos\left[2\pi\left(h\frac{X}{a} + k\frac{Y}{b}\right)\right] dX\, dY \tag{4.25a}$$

and

$$B_{hk} = \frac{1}{ab} \int_{-a/2}^{a/2} \int_{-b/2}^{b/2} f(X, Y) \sin\left[2\pi\left(h\frac{X}{a} + k\frac{Y}{b}\right)\right] dX \, dY. \tag{4.25b}$$

If we write

$$C_{hk} = A_{hk} + iB_{hk} \tag{4.26}$$

then we have

$$f(X, Y) = \sum_{h=-\infty}^{\infty} \sum_{k=-\infty}^{\infty} C_{hk} \exp\left[-2\pi i\left(h\frac{X}{a} + k\frac{Y}{b}\right)\right] \tag{4.27}$$

and

$$C_{hk} = \frac{1}{ab} \int_{-a/2}^{a/2} \int_{-b/2}^{b/2} f(X, Y) \exp\left[2\pi i\left(h\frac{X}{a} + k\frac{Y}{b}\right)\right] dX \, dY. \tag{4.28}$$

There are corresponding equations for a three-dimensional Fourier series applied to a function $f(X, Y, Z)$ defined in the parallelepiped given by the vectors **a**, **b** and **c**.

The final equations corresponding to (4.27) and (4.28) are

$$f(X, Y, Z) = \sum_{h=-\infty}^{\infty} \sum_{k=-\infty}^{\infty} \sum_{l=-\infty}^{\infty} C_{hkl} \exp\left[-2\pi i\left(h\frac{X}{a} + k\frac{Y}{b} + l\frac{Z}{c}\right)\right] \tag{4.29}$$

and

$$C_{hkl} = \frac{1}{abc} \int_{-a/2}^{a/2} \int_{-b/2}^{b/2} \int_{-c/2}^{c/2} f(X, Y, Z)$$

$$\exp\left[2\pi i\left(h\frac{X}{a} + k\frac{Y}{b} + l\frac{Z}{c}\right)\right] dX \, dY \, dZ. \tag{4.30}$$

In fig. 4.1 are shown the forms of some of the terms included in a one-dimensional Fourier summation. It is not quite so straightforward to represent the terms in a two- or three-dimensional Fourier summation. The contours corresponding to $\cos\{2\pi[3(X/a) + 2(Y/b)]\} = 1, 0$ and $-1$ are shown in the $XY$ plane in fig. 4.7(a) and the planes corresponding to $\cos\{2\pi[3(X/a) + (Y/b) + 2(Z/c)]\} = 1, 0$ and $-1$ are shown in fig. 4.7(b) in relation to the volume defined by $a \geqslant X \geqslant 0$, $b \geqslant Y \geqslant 0$ and $c \geqslant Z \geqslant 0$.

## 4.4 The Fourier transform

We have seen that although the functions we have been representing by a Fourier series were defined in some one-, two- or three-dimensional region of space the series itself had a periodic nature. If in equations (4.17) and (4.18) we change our variable of summation from $h$ to $s = h/a$ and make $F(s) = aC_h$, then

$$f(X) = \frac{1}{a} \sum_{-\infty}^{\infty} F(s) \exp(-2\pi isX) \tag{4.31}$$

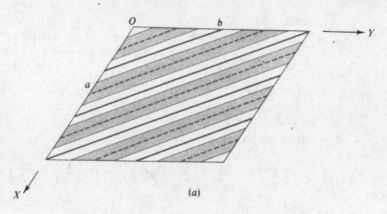

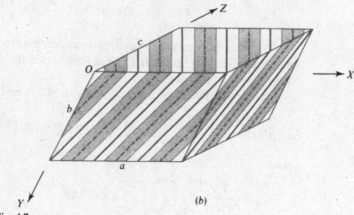

Fig. 4.7.

(*a*) Contours in the $XY$ plane of

$$\cos\left[2\pi\left(3\frac{X}{a}+2\frac{Y}{b}\right)\right] = 1 \text{ (thick full line)}$$

$$= 0 \text{ (thin full line)}$$
$$= -1 \text{ (thick broken line).}$$

Negative regions are shaded.

(*b*) Intersections of the faces of a unit cell with planes for which

$$\cos\left[2\pi\left(3\frac{X}{a}+\frac{Y}{b}+2\frac{Z}{c}\right)\right] = 1 \text{ (thick full line)}$$

$$= 0 \text{ (thin full line)}$$
$$= -1 \text{ (thick broken line).}$$

Negative regions are shaded.

and

$$F(s) = \int_{-a/2}^{a/2} f(X)\exp(2\pi i s X)\,dX. \tag{4.32}$$

Now let us consider what happens when $a$ becomes very large. The discrete values of $s$ ($= h/a$) become very close together and values of $F(s)$

determined from equation (4.32) for neighbouring values of $s$ should not be very different. The sum in equation (4.31) is clearly a real quantity (see equation (4.14)) and thus can be expressed as

$$f(X) = \frac{1}{a} \sum_{-\infty}^{\infty} R\{F(s) \exp(-2\pi i s X)\} \tag{4.33}$$

where $R(z)$ means the real part of $z$. In fig. 4.8 there is shown, for a particular value of $X$, a typical set of $R\{F(s)\exp(-2\pi i s X)\}$ for $s = 0, \pm 1/a, \pm 2/a$, etc. The rectangular blocks have areas $(1/a)R\{F(s)\exp(-2\pi i s X)\}$ and the total area of these blocks from $s = -\infty$ to $s = +\infty$ will equal $f(X)$. But if $a$ is very large, or as $a$ tends to infinity, so $F(s)$ tends to be a function of a continuously variable $s$ and one will then have

$$f(X) = \int_{-\infty}^{\infty} R\{F(s) \exp(-2\pi i s X)\} ds. \tag{4.34}$$

However we may now take out the 'real-part' symbol within the integration for the integral will still be real without this (see equation (4.16)). Thus we have as $a \to \infty$

$$f(X) = \int_{-\infty}^{\infty} F(s) \exp(-2\pi i s X) ds. \tag{4.35}$$

and

$$F(s) = \int_{-\infty}^{\infty} f(X) \exp(2\pi i s X) dX. \tag{4.36}$$

A function $f(X)$ related to $F(s)$ in the way shown is defined as the Fourier transform of $F(s)$ and similarly $F(s)$ is the Fourier transform of $f(X)$. Sometimes the integral in equation (4.36) is called the *inverse* Fourier transform to indicate the difference between the two forms. On this point it should be stressed that if the transformation from $f(X)$ to $F(s)$ involves a negative sign in the exponential term then the back transformation from

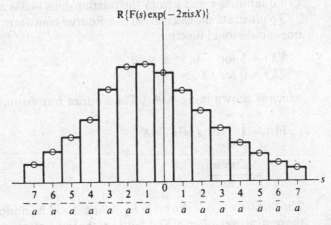

Fig. 4.8.
Values of $R\{F(s)\exp(-2\pi i s X)\}$ at $s = 0, \pm(1/a), \pm(2/a)$. The sum of the areas of the blocks gives $f(X)$.

$F(s)$ to $f(X)$ must have a corresponding positive sign. The two signs can both be reversed – this is only a matter of the way the transformation is defined in the first place in equation (4.15) where one could also have taken $C_h = A_h - iB_h$.

We shall now generalize our description of Fourier transforms to the three-dimensional function given in equations (4.29) and (4.30). The position of a point in the parallelepiped whose sides are the vectors **a**, **b** and **c** may be represented by

$$\mathbf{r} = X\hat{\mathbf{a}} + Y\hat{\mathbf{b}} + Z\hat{\mathbf{c}} \tag{4.37}$$

where $\hat{\mathbf{a}}$, $\hat{\mathbf{b}}$ and $\hat{\mathbf{c}}$ are the unit vectors in the directions of **a**, **b** and **c**. This may also be written as

$$\mathbf{r} = X\frac{\mathbf{a}}{a} + Y\frac{\mathbf{b}}{b} + Z\frac{\mathbf{c}}{c}. \tag{4.38}$$

Corresponding to **a**, **b** and **c** there will be reciprocal vectors **a***, **b*** and **c*** defined by the equations (3.14) and we may define a vector **s** by

$$\mathbf{s} = h\mathbf{a}^* + k\mathbf{b}^* + l\mathbf{c}^*. \tag{4.39}$$

If we now write $F(\mathbf{s}) = VC_{hkl}$ then by similar arguments which lead to equations (4.35) and (4.36) we find

$$f(\mathbf{r}) = \int_{v*} F(\mathbf{s}) \exp(-2\pi i\mathbf{s}\cdot\mathbf{r}) \, \mathrm{d}v^* \tag{4.40}$$

and

$$F(\mathbf{s}) = \int_{v} f(\mathbf{r}) \exp(2\pi i\mathbf{s}\cdot\mathbf{r}) \, \mathrm{d}v \tag{4.41}$$

In equations (4.40) and (4.41) $f(\mathbf{r})$ represents a function of position in some space represented by the vector **r** in which an element of volume is $\mathrm{d}v$. The Fourier transform of $f(\mathbf{r})$ is $F(\mathbf{s})$, a function in a reciprocal space in which an element of volume is $\mathrm{d}v^*$. We can see the mutually reciprocal nature of the two quantities **r** and **s** from the relationships (4.38) and (4.39).

To illustrate the calculation of a Fourier transform we consider first the one-dimensional function

$$f(X) = 1 \text{ for } -\tfrac{1}{2}a \leqslant X \leqslant \tfrac{1}{2}a$$
$$f(X) = 0 \text{ for } |X| > \tfrac{1}{2}a$$

which is shown in fig. 4.9(*a*). The Fourier transform is given by

$$F(s) = \int_{-\frac{1}{2}a}^{\frac{1}{2}a} \exp(2\pi isX)\mathrm{d}X$$

$$= \frac{\sin(\pi sa)}{\pi s}. \tag{4.42}$$

The function $F(s)$ is shown in fig. 4.9(*b*); it will be noticed that the width of the central maximum of $F(s)$, between the two points at which it falls to zero, is $2/a$ and is thus reciprocally related to the width of the function $f(X)$. This is a characteristic relationship between a function and its Fourier transform;

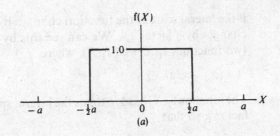

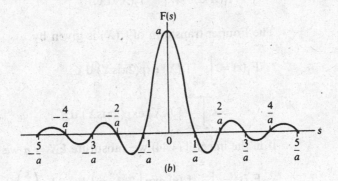

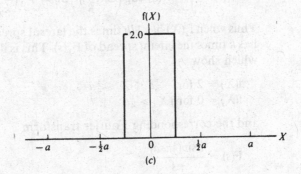

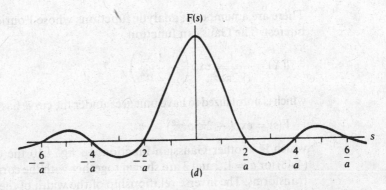

Fig. 4.9.

(a) The function f($X$) = 1, $-\frac{1}{2}a \leqslant X \leqslant \frac{1}{2}a$
$= 0, |X| > \frac{1}{2}a$.

(b) The Fourier transform of the function shown in (a)

$$F(s) = \frac{\sin(\pi s a)}{\pi s}.$$

(c) The function f($X$) = 2, $-\frac{1}{4}a \leqslant X \leqslant \frac{1}{4}a$
$= 0, |X| > \frac{1}{4}a$.

(d) The Fourier transform of the function shown in (c)

$$F(s) = \frac{2\sin(\frac{1}{2}\pi s a)}{\pi s}.$$

if the lateral scale of the function changes by a factor $k$ then the transform changes by a factor $1/k$. We can see this by considering the transforms of two functions $f_1(X)$ and $f_2(X)$, where

$$f_1(X) = kf_2(kX). \tag{4.43}$$

The function $f_2(X)$ has $1/k$ times the lateral spread of $f_1(X)$ and is scaled by a factor $k$ so that

$$\int_{-\infty}^{\infty} f_1(X) dX = \int_{-\infty}^{\infty} f_2(X) dX.$$

The Fourier transform of $f_1(X)$ is given by

$$F_1(s) = \int_{-\infty}^{\infty} f_1(X) \exp(2\pi i s X) \, dX$$

$$= k \int_{-\infty}^{\infty} f_2(kX) \exp(2\pi i s X) \, dX. \tag{4.44}$$

If in the integral (4.44) we substitute $kX = \eta$ we find

$$F_1(s) = \int_{-\infty}^{\infty} f_2(\eta) \exp\left(2\pi i \frac{s}{k}\eta\right) d\eta = F_2\left(\frac{s}{k}\right). \tag{4.45}$$

Thus when $f_2(X)$ has $1/k$ times the lateral spread of $f_1(X)$ we find that $F_2(s)$ has $k$ times the lateral spread of $F_1(s)$. This is illustrated in fig. 4.9(c) and (d) which show

$$f(X) = 2 \text{ for } -\tfrac{1}{4}a \leqslant X \leqslant \tfrac{1}{4}a$$
$$f(X) = 0 \text{ for } |X| > \tfrac{1}{4}a \tag{4.46}$$

and the corresponding Fourier transform

$$F(s) = \frac{2\sin(\tfrac{1}{2}\pi s a)}{\pi s}. \tag{4.47}$$

There are a number of analytic functions whose Fourier transforms are of interest. The Gaussian function

$$f(X) = \frac{1}{\sqrt{2\pi a^2}} \exp\left(-\frac{1}{2}\frac{X^2}{a^2}\right) \tag{4.48}$$

which is normalized to have unit area under the curve has a Fourier transform

$$F(s) = \exp(-2\pi^2 a^2 s^2) \tag{4.49}$$

which is another Gaussian function. In fig. 4.10 the Gaussian functions (4.48) for $a = 1, 2$ and $3$ are shown together with the corresponding Fourier transforms. The inverse relationship of the width of these functions can be readily seen.

Another function of great importance to the study of X-ray diffraction is the Dirac $\delta$-function. This function has the following properties:

$$\delta(X) = 0, \text{ for } X \neq 0$$
$$\delta(0) = \infty$$

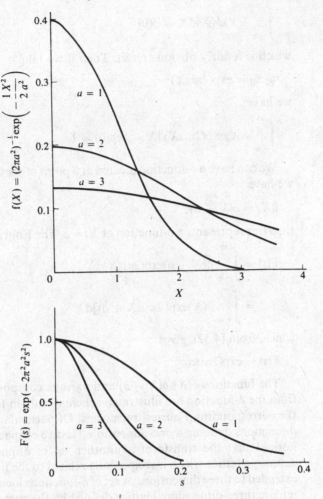

Fig. 4.10.
The function $f(X) = (2\pi a^2)^{-\frac{1}{2}}\exp(-\frac{1}{2}X^2/a^2)$ and its Fourier transforms $F(s) = \exp(-2\pi^2 a^2 s^2)$ for $a = 1, 2$ and 3.

and

$$\int_{-\infty}^{\infty} \delta(X)\mathrm{d}X = 1. \tag{4.50}$$

It is not the sort of function we can draw but we can imagine it like the functions shown in figs. 4.9 $(a)$, $(c)$ and 4.10 as the width of the curve tends to zero. In fact we can find the Fourier transform of the $\delta$-function by noting that if in equation (4.48) we make $a \to 0$ then $f(X) \to \delta(X)$ and, at the limit where $a = 0$, we have the Fourier transform of the $\delta$-function given as

$$F(s) = 1. \tag{4.51}$$

This result would also be found from the general expression

$$\int_{-\infty}^{\infty} \delta(X)\phi(X)\mathrm{d}X = \phi(0) \tag{4.52}$$

which is a fairly obvious result. Then if we take

$$\phi(X) = \exp(2\pi \mathrm{i} s X)$$

we have

$$\int_{-\infty}^{\infty} \delta(X)\exp(2\pi \mathrm{i} s X)\mathrm{d}X = \exp(0) = 1.$$

We can have a $\delta$-function located at a point other than the origin. Thus if we have

$$f(X) = \delta(X - a) \tag{4.53}$$

then this represents a $\delta$-function at $X = a$. The Fourier transform of this is

$$F(s) = \int_{-\infty}^{\infty} \delta(X - a)\exp(2\pi \mathrm{i} s X)\mathrm{d}X$$

$$= \int_{-\infty}^{\infty} \delta(X)\exp[2\pi \mathrm{i} s(X + a)]\mathrm{d}X$$

which, from (4.52), gives

$$F(s) = \exp(2\pi \mathrm{i} s a). \tag{4.54}$$

The functions $\delta(X), \delta(X - a)$ and various composite functions derived from the $\delta$-function are illustrated schematically in fig. 4.11 together with the corresponding Fourier transforms. Of particular interest is the set of $\delta$-functions defining a one-dimensional lattice of spacing $a$ where it will be noticed that the transform is another set of $\delta$-functions of weight $1/a$ defining a lattice which is reciprocally related to the first. This result can be extended to three dimensions. A set of $\delta$-functions located at the nodes of an infinite three-dimensional lattice defined by the vectors $\mathbf{a}$, $\mathbf{b}$ and $\mathbf{c}$ may be represented by

$$f(\mathbf{r}) = \sum_{n_1 = -\infty}^{\infty} \sum_{n_2 = -\infty}^{\infty} \sum_{n_3 = -\infty}^{\infty} \delta[\mathbf{r} - (n_1\mathbf{a} + n_2\mathbf{b} + n_3\mathbf{c})]. \tag{4.55}$$

The Fourier transform of equation (4.55) is

$$F(\mathbf{s}) = \frac{1}{V} \sum_{h = -\infty}^{\infty} \sum_{k = -\infty}^{\infty} \sum_{l = -\infty}^{\infty} \delta[\mathbf{s} - (h\mathbf{a}^* + k\mathbf{b}^* + l\mathbf{c}^*)] \tag{4.56}$$

a set of $\delta$-functions at the nodes of the lattice which is reciprocal to that defined by $\mathbf{a}$, $\mathbf{b}$ and $\mathbf{c}$ (as defined by the relationships (3.14)) and with weight $1/V$ where $V = \mathbf{a} \cdot \mathbf{b} \wedge \mathbf{c}$. The function $f(\mathbf{r})$ and $F(\mathbf{s})$ are related by the equations (4.40) and (4.41) which give the general relationship between any three-dimensional function and its Fourier transform.

## 4.5   Diffraction and the Fourier transform

In equation (2.17) the amplitude of scattering from a distribution of $n$ scatterers was found to be

Fig. 4.11.
Various composite
functions involving
δ-functions and their
Fourier transforms.

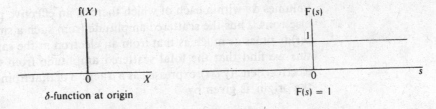

δ-function at origin

$F(s) = 1$

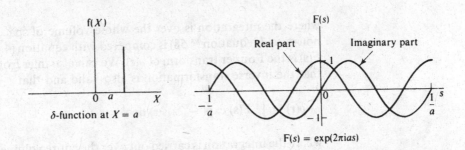

δ-function at $X = a$

$F(s) = \exp(2\pi i a s)$

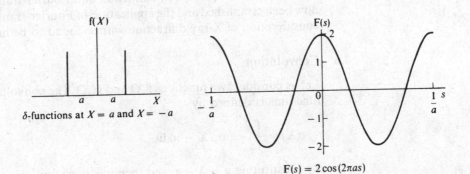

δ-functions at $X = a$ and $X = -a$

$F(s) = 2\cos(2\pi a s)$

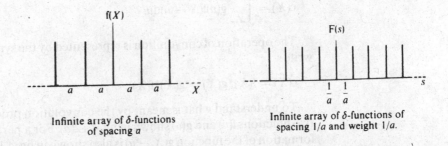

Infinite array of δ-functions
of spacing $a$

Infinite array of δ-functions of
spacing $1/a$ and weight $1/a$.

$$\eta_n(2\theta, D) = \sum_{j=1}^{n} [\eta(2\theta, D)]_j \exp(2\pi i \mathbf{s} \cdot \mathbf{r}_j) \qquad (4.57)$$

where $[\eta(2\theta, D)]_j$ is the amplitude of scattering at a distance $D$ from the $j$th
scatterer at an angle $2\theta$ with the incident radiation. If we have a continuous
distribution of electron density $\rho(\mathbf{r})$, expressed in electrons per unit volume,
then we can imagine that our space is divided into a number of small

volumes $dv$ within each of which there is an effective point charge $\rho(\mathbf{r})dv$ electrons. Thus the scattered amplitude from such a small volume will be $\rho(\mathbf{r})dv$ times as much as that from an electron at the same position. From this we find that the total scattered amplitude from the distribution of electron density $\rho(\mathbf{r})$, expressed as a fraction of that from a point electron at the origin, is given by

$$F(\mathbf{s}) = \int_{v} \rho(\mathbf{r}) \exp(2\pi i \mathbf{s} \cdot \mathbf{r})\, dv \tag{4.58}$$

where the integration is over the whole volume of space in which $\rho(\mathbf{r})$ is non-zero. If equation (4.58) is compared with equation (4.41) it is clear that $F(\mathbf{s})$ is the Fourier transform of $\rho(\mathbf{r})$. We can also infer from equation (4.40) that the inverse transformation is also valid and that

$$\rho(\mathbf{r}) = \int_{v*} F(\mathbf{s}) \exp(-2\pi i \mathbf{s} \cdot \mathbf{r})\, dv* \tag{4.59}$$

where the integration is carried out over the entire volume of the reciprocal space in which $\mathbf{s}$ is defined.

The relationship between diffraction and Fourier-transform theory has now been established and the application of Fourier-transform ideas to the consideration of X-ray diffraction will be found to be most useful.

## 4.6 Convolution

Let us consider two functions $f(X)$ and $g(X)$. The convolution of these two functions is defined by

$$c(X) = \int_{-\infty}^{\infty} f(u)g(X - u)du. \tag{4.60a}$$

By substituting $u = X - v$ and then $v = u$ we find also

$$c(X) = \int_{-\infty}^{\infty} g(u)f(X - u)du. \tag{4.60b}$$

The operation of convolution is represented by the symbol * and we may write

$$c(X) = f(X)*g(X) = g(X)*f(X). \tag{4.61}$$

To understand what is meant by the convolution process let us take the two functions $f(u)$ and $g(u)$ shown in fig. 4.12($a$). For a particular value $X$ the formation of the function $g(X - u)$ is also shown in fig. 4.12($b$); it will be seen that this function may be produced from $g(u)$ by (i) inversion about the point $u = X$ and (ii) translation of this point to the origin. The operation by which $g(X - u)$ is produced from $g(u)$ is referred to as *folding* $g(u)$ about $X$. In fig. 4.12($c$) there are shown superimposed the function $f(u)$, $g(X - u)$ and their product $f(u)\,g(X - u)$. Then the integral (4.60a) which defines $c(X)$ is given by the shaded area. The result of doing this for all values of $X$ completely defines the function $c(X)$ and, for the particular example we have considered, this is shown in fig. 4.12($d$).

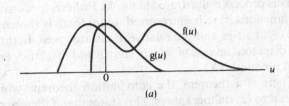

(a)

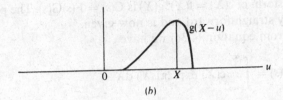

(b)

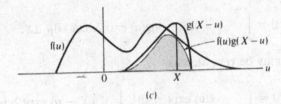

(c)

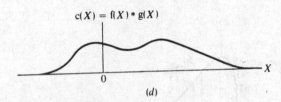

(d)

Fig. 4.12.
(a) Functions f(u) and g(u).
(b) The function g(X − u).
(c) The derivation of f(u)g(X − u). The shaded area gives c(X) = f(X)*g(X).
(d) The complete function c(X) = f(X)*g(X).

Another method, which can be used to determine the whole function c(X) at once, is to express c(X) as a summation by

$$c(X) = \int_{-\infty}^{\infty} f(u)g(X-u)\mathrm{d}u = \left\{ \sum_n f(u_n)g(X-u_n)\delta u \right\}_{\delta u \to 0}$$

where the summation is for the $n$ values of $u_n$ at the centres of the strips of width $\delta u$ which cover the non-zero ranges of the function f(u). The steps in the process are:

(i) Divide f(u) into strips of width $\delta u$.
(ii) Find the areas of the strips $\delta\mathfrak{A}_n = f(u_n)\delta u$.
(iii) Plot $g(u)\delta\mathfrak{A}_n$ with origin at centre of $n$th strip for all $n$.
(iv) Add the resultant curves to obtain c(X).

This process is illustrated in fig. 4.13 where, for convenience, the roles of the functions have been reversed. In (a) there is shown the division of $g(u)$ into eight strips and the relative areas are given. In (b) the eight weighted and displaced images of $f(u)$ are shown and in (c) they are added to give the total function $c(X)$.

There is a theorem, the convolution theorem, which is of particular interest to crystallographers. This states that if there are two functions $f(X)$ and $g(X)$ whose Fourier transforms are $F(s)$ and $G(s)$ then the Fourier transform of $c(X) = f(X)*g(X)$ is $C(s) = F(s)G(s)$. The proof of this result is fairly straightforward and is now given.

From equation (4.36) we have

$$C(s) = \int_{-\infty}^{\infty} c(X) \exp(2\pi i s X) \, dX \tag{4.62}$$

which, from equation (4.60b) gives

$$C(s) = \int_{-\infty}^{\infty} \int_{-\infty}^{\infty} f(X - u)g(u) \exp(2\pi i s X) \, du \, dX. \tag{4.63}$$

This may be rewritten as

$$C(s) = \int_{u=-\infty}^{\infty} g(u) \exp(2\pi s u) \left[ \int_{-\infty}^{\infty} f(X - u) \exp\{2\pi i s(X - u)\} dX \right] du. \tag{4.64}$$

Fig. 4.13.
(a) $g(u)$ divided into strips of width $\delta u$.
(b) The superposition of the functions $f(u - u_n)g(u_n)\delta u$.
(c) The sum of the superimposed functions rescaled to agree with fig. 4.12(d).

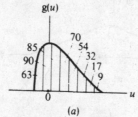

(a)

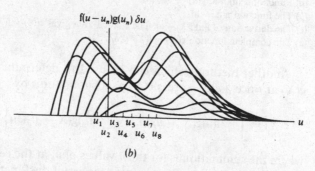

(b)

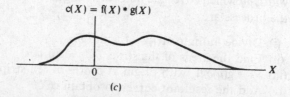

(c)

The integral in the square brackets is independent of $u$ (substitute $y = x - u$) and equals $F(s)$ and it follows that

$$C(s) = F(s)G(s). \tag{4.65}$$

It may be deduced from the reciprocal nature of Fourier transforms that $c(X)$ is the transform of $C(s)$ and is also the transform of the product $F(s)G(s)$. Thus we have the general result that the Fourier transform of a product of functions is the convolution of the transforms of the individual functions. We can express this as

$$\int_{-\infty}^{\infty} F(s)G(s) \exp(-2\pi i X s) \, ds = f(X)*g(X). \tag{4.66}$$

or

$$\int_{-\infty}^{\infty} f(X)g(X) \exp(2\pi i X s) \, dX = F(s)*G(s). \tag{4.67}$$

Let us look again at the process of convolution and in particular the case shown in fig. 4.14 where the function $g(X)$ is a spiky function of very small width. If the function $g(X)$ is normalized to have unit area under the curve then the convolution $c(X)$ differs very little from $f(X)$. If we take this process to the limit so that $g(X) = \delta(X)$ then we should find that $c(X) = f(X)$. This can also be found by applying the expression (4.52) since

$$c(X) = \int_{-\infty}^{\infty} g(u)f(X - u)du$$

$$= \int_{-\infty}^{\infty} \delta(u)f(X - u)du = f(X). \tag{4.68}$$

Fig. 4.14.
The convolution of $f(X)$
and $g(X)$, a spiky
function with unit area
under the curve, is little
different from $f(X)$.

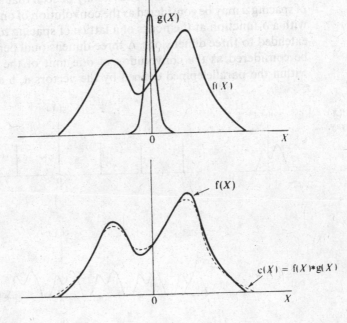

Extending the application of equation (4.52) we can determine the convolution of f($X$) with $\delta(X - a)$. This is

$$c(X) = \int_{-\infty}^{\infty} \delta(u - a)f(X - u)\,du$$

$$= f(X - a) \tag{4.69}$$

and this result is illustrated in fig. 4.15.

**Fig. 4.15.**
The convolution of f($X$) with a $\delta$-function not at the origin.

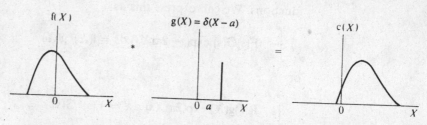

This process is taken another stage by considering the convolution of f($X$) with an infinite linear array of $\delta$ functions given by

$$g(X) = \sum_{n=-\infty}^{\infty} \delta(X - na). \tag{4.70}$$

The convolution is given by

$$c(X) = \sum_{n=-\infty}^{\infty} f(X - na) \tag{4.71}$$

and c($X$) is a periodic function of spacing $a$. What is more if f($X$) is completely defined in the range $0 \leqslant X \leqslant a$ (or any range of width $a$) then a unit of c($X$) is identical in appearance with f($X$) in the defined range. This point is illustrated in fig. 4.16. Thus it may be seen that any periodic pattern of spacing $a$ may be considered as the convolution of one unit of the pattern with a $\delta$-function at the nodes of a lattice of spacing $a$. The result may be extended to three dimensions. A three-dimensional periodic function may be considered as the convolution of one unit of the pattern, contained within the parallelepiped defined by the vectors **a**, **b** and **c**, with a set of

**Fig. 4.16.**
The convolution of f($X$) with a $\delta$-function array.

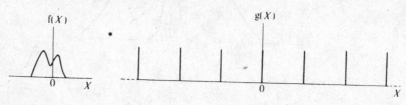

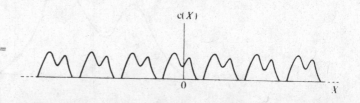

δ-functions at the nodes of a three-dimensional lattice defined by the vectors **a**, **b** and **c**. In particular, the electron density within a crystal may be considered as the convolution of the density within one unit cell with the three-dimensional real-space lattice of spacing **a**, **b** and **c**.

## 4.7 Diffraction by a periodic distribution

In § 4.5 the relationship between diffraction and the Fourier transform was established and this is exemplified by equations (4.58) and (4.59). The scattered amplitude, as a fraction of that scattered by a point electron, is the Fourier transform of the electron density. Thus, from equation (4.65) and the periodic nature of the electron density in the crystal, the scattered amplitude is seen to be the product of the transform of the electron density within one unit cell of the crystal with the transform of a set of δ-functions at the nodes of a unit-cell lattice which covers the volume of the crystal. The transform of this set of δ-functions will, for all practical purposes, be the same as the transform of an infinite set – that is another set of δ-functions at the nodes of the reciprocal lattice. Thus the diffracted amplitude for the whole crystal can be thought of as the Fourier transform of the electron density in one unit cell of the crystal, F(**s**), *sampled* at points of the reciprocal lattice, **s** = h**a*** + k**b*** + l**c***. Since the weight associated with each of the reciprocal-lattice δ-function points is $1/V$ (see equation (4.56)) the scattered amplitude corresponding to the point (hkl) of the reciprocal lattice is given by $(1/V)F_{hkl}$ where the reciprocal-lattice vector is now indicated by the triplet of integers (we write F(**s**) = $F_{hkl}$).

If the diffracted amplitudes are expressed as a fraction of that from an array of point electrons at the chosen origins then $F_{hkl}$ is the structure factor as defined in § 3.5. The structure factor is also given by equation (4.58) if the integration is carried out over the volume of the unit cell and if $\rho(\mathbf{r})$ is expressed in electron units per unit volume.

## 4.8 The electron-density equation

In equation (4.59) an expression has been given for the electron density in a crystal in terms of the scattering amplitudes F(**s**) expressed as a fraction of the scattering from a point electron. For the whole crystal, assumed to be infinite in extent, F(**s**) exists only at reciprocal lattice points and has weight at these points equal to $(1/V)F_{hkl}$.

Thus we may take equation (4.71),

$$\rho(\mathbf{r}) = \int_{v*} F(\mathbf{s}) \exp(-2\pi i \mathbf{s} \cdot \mathbf{r}) \, dv^*,$$

and replace it by

$$\rho(\mathbf{r}) = \frac{1}{V} \sum_{h=-\infty}^{\infty} \sum_{k=-\infty}^{\infty} \sum_{l=-\infty}^{\infty} F_{hkl} \exp(-2\pi i \mathbf{s} \cdot \mathbf{r}). \tag{4.72}$$

In this equation if $F_{hkl}$ is the normal structure factor then $\rho(\mathbf{r})$ will be the electron density expressed in electrons per unit volume. The electron density is completely defined in one unit cell; substituting

$$\mathbf{r} = x\mathbf{a} + y\mathbf{b} + z\mathbf{c}$$

and

$$\mathbf{s} = h\mathbf{a}^* + k\mathbf{b}^* + l\mathbf{c}^*$$

we find

$$\rho(xyz) = \frac{1}{V} \sum_{h=-\infty}^{\infty} \sum_{k=-\infty}^{\infty} \sum_{l=-\infty}^{\infty} F_{hkl} \exp\{-2\pi i(hx + ky + lz)\} \qquad (4.73)$$

where $x$, $y$ and $z$ are the fractional coordinates of a point in the unit cell.

The electron-density equation can take different forms depending on the space group of the crystal structure. For example with a centre of symmetry at the origin all the $F$'s are real and, since $F_{\bar{h}\bar{k}\bar{l}} = F_{hkl}$, we find

$$\rho(xyz) = \frac{1}{V} \sum_{h=-\infty}^{\infty} \sum_{k=-\infty}^{\infty} \sum_{l=-\infty}^{\infty} F_{hkl} \cos\{2\pi(hx + ky + lz)\}. \qquad (4.74)$$

The various forms of the structure-factor equation (3.41) and the electron-density equation are given for each of the 230 space groups in the *International Tables for X-ray Crystallography*. Some examples of this space-group dependence will be given later, in chapter 7, where it is shown that the distribution of the values of $|F|$ or $|F|^2$ can give space-group information.

The structure-factor equation (4.58), written in terms of indices of reflection ($hkl$) and fractional coordinates within the unit cell ($xyz$), is

$$F_{hkl} = V \int_{x=0}^{1} \int_{y=0}^{1} \int_{z=0}^{1} \rho(xyz) \exp\{2\pi i(hx + ky + lz)\} dx\,dy\,dz. \qquad (4.75)$$

For the class of structure factors for which $l = 0$

$$F_{hk0} = V \int_{x=0}^{1} \int_{y=0}^{1} \left[ \int_{z=0}^{1} \rho(xyz) dz \right] \exp\{2\pi i(hx + ky)\} dx\,dy. \qquad (4.76)$$

To attach a meaning to the integral in square brackets in equation (4.76) consider the total electron content in the fine filament parallel to the $z$ axis in the unit cell shown in fig. 4.17. The cross-section of this filament in the $xy$ plane is $\alpha$ and $\psi$ is the angle between the $z$ axis and the normal to the $xy$ plane. The electron content of the small shaded volume shown is $\rho(xyz)\alpha \cos \psi\, cdz$ and hence the total electron content in the filament is

$$\alpha c \cos \psi \int_{z=0}^{1} \rho(xyz)\, dz.$$

If we now imagine that we project all the electron content of the unit cell along $z$ on to the $xy$ plane then the electron density at the point ($xy$), expressed as electrons per unit area, will clearly be given by

$$\rho_p(xy) = c \cos \psi \int_{0}^{1} \rho(xyz)\, dz. \qquad (4.77)$$

From equations (4.77) and (4.76) we find

$$F_{hk0} = \frac{V}{c\cos\psi} \int_{x=0}^{1} \int_{y=0}^{1} \rho_p(xy)\exp\{2\pi i(hx + ky)\}dx\,dy$$

$$= \mathfrak{A} \int_{x=0}^{1} \int_{y=0}^{1} \rho_p(xy)\exp\{2\pi i(hx + ky)\}dx\,dy, \tag{4.78}$$

where $\mathfrak{A}$ is the area of the $xy$ plane of the unit cell.

It can be seen that $F_{hk0}$ is the Fourier transform of the two-dimensional function, the projected electron density $\rho_p(xy)$. Consequently, by analogy with our previous results, we may write

$$F_{hk0} = \int_{\mathfrak{A}} \rho_p(xy)\exp\{2\pi i(hx + ky)\}d\alpha \tag{4.79}$$

where $d\alpha$ is an element of area in the $xy$ plane and

$$\rho_p(xy) = \frac{1}{\mathfrak{A}} \sum_{h=-\infty}^{\infty} \sum_{k=-\infty}^{\infty} F_{hk0}\exp\{-2\pi i(hx + ky)\}. \tag{4.80}$$

The X-ray crystallographer frequently wishes to compute electron density since this gives a picture of the atomic arrangement in the crystal. At the resolution corresponding to the X-ray wavelengths usually used, the electron density has its highest value at the location of the atomic centre and the separate atoms of the structure are usually well resolved. The electron-density equation (4.73) contains imaginary quantities on the right-hand side but of course $\rho(xyz)$ is a real quantity. In §3.5 the structure factor was given as

$$F_{hkl} = A_{hkl} + iB_{hkl}$$

where

$$A_{hkl} = \sum_{j=1}^{N} f_j \cos\{2\pi(hx_j + ky_j + lz_j)\} \tag{4.81a}$$

and

Fig. 4.17.
Electron density
projected on to the $xy$
plane.

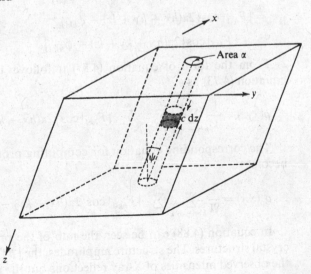

$$B_{hkl} = \sum_{j=1}^{N} f_j \sin\{2\pi(hx_j + ky_j + lz_j)\}. \tag{4.81b}$$

There is also a phase angle, $\phi_{hkl}$, associated with each structure factor given by

$$\tan \phi_{hkl} = \frac{B_{hkl}}{A_{hkl}}.$$

From the equations (4.81) it can be seen that

$$|F_{\bar{h}\bar{k}\bar{l}}| = |F_{hkl}| \tag{4.82}$$

and

$$\phi_{\bar{h}\bar{k}\bar{l}} = -\phi_{hkl}. \tag{4.83}$$

In addition from fig. 3.14($a$) we see that

$$A_{hkl} = |F_{hkl}| \cos \phi_{hkl} \tag{4.84}$$

and

$$B_{hkl} = |F_{hkl}| \sin \phi_{hkl} \tag{4.85}$$

so that from equation (4.81) we find

$$F_{hkl} = |F_{hkl}| \exp(i\phi_{hkl}). \tag{4.86}$$

Let us now consider the contributions to the summation on the right-hand side of equation (4.73) of two terms of indices ($hkl$) and ($\bar{h}\bar{k}\,\bar{l}$). This will be

$$F_{hkl} \exp\{-2\pi i(hx + ky + lz)\} + F_{\bar{h}\bar{k}\bar{l}} \exp\{2\pi i(hx + ky + lz)\}$$

$$= |F_{hkl}| \exp\{-2\pi i(hx + ky + lz) + i\phi_{hkl}\}$$

$$+ |F_{hkl}| \exp\{2\pi i(hx + ky + lz) + i\phi_{hkl}\}$$

$$= 2|F_{hkl}| \cos\{2\pi(hx + ky + lz) - \phi_{hkl}\}$$

$$= |F_{hkl}| \cos\{2\pi(hx + ky + lz) - \phi_{hkl}\}$$

$$+ |F_{\bar{h}\bar{k}\bar{l}}| \cos\{2\pi(\bar{h}x + \bar{k}y + \bar{l}z) - \phi_{\bar{h}\bar{k}\bar{l}}\}. \tag{4.87}$$

From the form of equation (4.87) it follows that one may re-write equation (4.73) as

$$\rho(xyz) = \frac{1}{V} \sum_{h=-\infty}^{\infty} \sum_{k=-\infty}^{\infty} \sum_{l=-\infty}^{\infty} |F_{hkl}| \cos\{2\pi(hx + ky + lz) - \phi_{hkl}\}. \tag{4.88}$$

The corresponding equation for computing projected electron density becomes

$$\rho_p(x,y) = \frac{1}{\mathfrak{A}} \sum_{h=-\infty}^{\infty} \sum_{k=-\infty}^{\infty} |F_{hk0}| \cos\{2\pi(hx + ky) - \phi_{hk0}\}. \tag{4.89}$$

In equation (4.88) can be seen the nub of the problem of determining crystal structures. The structure amplitudes, the $|F|$'s, can be derived from the observed intensities of X-ray reflections but the phase angles $\phi$ cannot

Fig. 4.18.
The z-axis projection of
D-xylose. The dashed
contour is at $3e\,\text{Å}^{-2}$ and
the others at intervals of
$2e\,\text{Å}^{-2}$. The positions of
carbon and oxygen
atoms are shown by the
line framework. (From
Woolfson, 1958.)

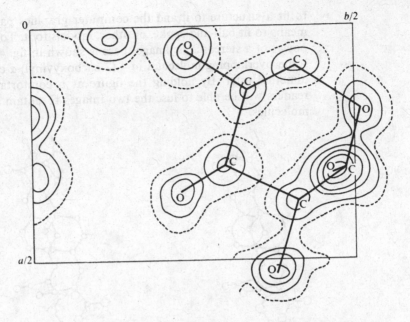

be directly determined. If the phases of the structure factors are known then
the crystal structure is known, for one can compute the electron density
from equation (4.88) and the atomic centres are located at peaks of electron
density. The electron density is usually computed at the points of a uniform
grid covering the unit cell and contours of constant electron density may
then be drawn. An example of such a contour map is given in fig. 4.18 which
shows a projection of the electron density of D-xylose. It will be noticed that
the carbon (c) and oxygen (o) atoms may all be clearly seen but no trace of
the hydrogen atoms is evident. The contribution of the single electron of the
rather diffuse hydrogen atom has been swamped by the contributions of the
heavier atoms. However, with careful experimental measurements and
special computing techniques hydrogen atoms can be detected and we shall
return later to this problem.

It is more difficult to present an electron-density map in three dimensions.
This can be done by computing two-dimensional sections keeping one
coordinate constant. These can be examined section-by-section or reproduced
on transparent material (glass or plastic) and stacked so as to get a direct
three-dimensional view of the electron density. The advent of modern
computer graphics has given the crystallographer very powerful ways of
examining and utilizing three-dimensional maps. On the computer screen
there can be projected stereoscopic pairs of images of surfaces of constant
electron density, reproduced by cage-like structures, and these are
accompanied by viewing glasses. In one system the stereo images are
projected on the screen alternately and glasses worn by the viewer are
equipped with opto–electronic devices synchronized with what is happening
on the screen so that the left and right eye see different images. If the
switching rate is high enough, greater than about 30 Hz, then there will be
no flicker and a constant three-dimensional image will be seen. With a
three-dimensional view of the electron density available it is then possible

to fit a structure to it and the computer graphics packages provide the means to fit ball-and-spoke, or other, models to it. To give an idea of the power of a stereoscopic image there is shown in fig. 4.19 a stereoview of some overlapping molecules of 4-(2-carboxyvinyl)-α-cyanocinnamic acid dimethyl ester. By holding the figure at a comfortable distance many readers will be able to fuse the two images to obtain a stereoview of the molecules.

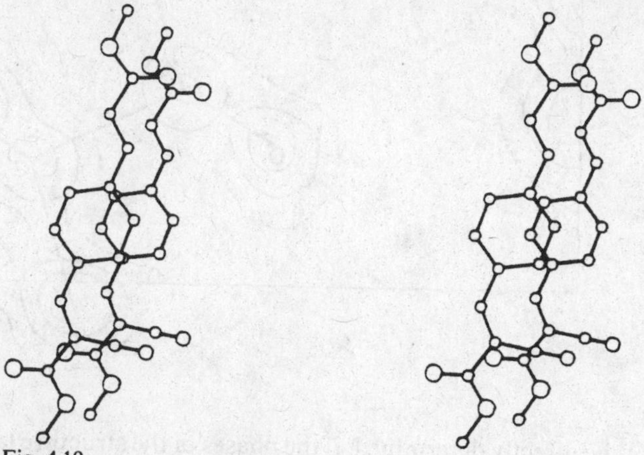

Fig. 4.19.
Stereopair of overlapping molecules of 4-(2-carboxyvinyl)-α-cyanocinnamic acid dimethyl ester (Nakanishi & Sarada, 1978).

In appendix IV there is provided a program, FOUR2, for computing two-dimensional electron density. This will accept the output of the program STRUCFAC and can be used to do some of the problems at the end of this chapter. Since STRUCFAC produces a whole semi-circle of data in reciprocal space the program FOUR2 is simplified by being able to treat the space group as $p1$. The most efficient programs for carrying out this kind of calculation employ the so-called Fast Fourier Transform algorithm. However, although FOUR2 is less sophisticated in its approach it does have some efficient features such as factorizing the calculation in the $x$ and $y$ directions and using some of the symmetry of cosine and sine functions. Transforming equation (4.89), using the results (4.84) and (4.85) applied to the two-dimensional structure factors we find

$$\rho_p(xy) = \frac{1}{\mathfrak{A}}\Bigg[\sum_k \{A_c(k,x) + B_s(k,x)\}\cos(2\pi ky)$$

$$+ \sum_k \{B_c(k,x) - A_s(k,x)\}\sin(2\pi ky)\Bigg] \qquad (4.90)$$

where

$$A_c(k,x) = \sum_h A_{hk0}\cos(2\pi hx) \qquad (4.91a)$$

$$A_s(k,x) = \sum_h A_{hk0}\sin(2\pi hx) \qquad (4.91b)$$

$$B_c(k, x) = \sum_h B_{hk0} \cos(2\pi hx) \qquad (4.91c)$$

$$B_s(k, x) = \sum_h B_{hk0} \sin(2\pi hx). \qquad (4.91d)$$

The summations over $h$ only need to be done for $x = 0$ to 0.5, the remainder of the range being found from the symmetry of the cosine and sine functions. Similarly, for each value of $x$, the two summations over $k$ need only be done for $y = 0$ to 0.5 and then combined for the whole range of $y$ using symmetry again.

One of the first decisions to be made when computing an electron-density map is the grid spacing to be used along each of the unit-cell axes. If the grid is too fine a great deal of needless computing is done whereas if the grid is too coarse then interpolation between grid points will be uncertain. The rule which gives the best interval is to take, along the $a$ axis for example, intervals of $\frac{1}{4}h_{max}$ where $h_{max}$ is the maximum $h$ index being included in the summation. Thus, if $h_{max} = 14$, intervals of $\frac{1}{56}$ are indicated although, in practice, one would take some slightly different and more convenient interval – for example $\frac{1}{50}$ or $\frac{1}{60}$.

The output of FOUR2 is scaled values of projected electron density at grid points. To visualize the actual projected density, contours of constant projected density can be drawn. An example of part of a map with density contours is shown in fig. 4.20 and should be used as a guide to the reader in drawing contours for Problem 4.7 which follows.

Fig. 4.20.
Part of the projected electron density produced as a solution of Problem 4.7 with contours drawn at the level of 20 and 30 in the arbitrary scaled units of projected density produced by the program FOUR2.

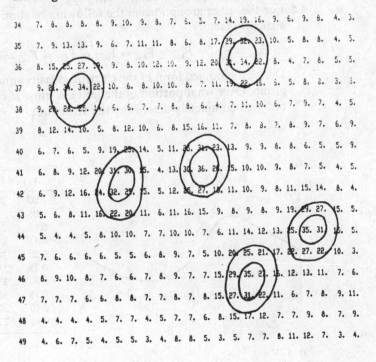

## Problems to Chapter 4

4.1 For the periodic function shown in fig. 4.21 both graphically and in tabulated form use the program SIMP1 to compute the Fourier coefficients $A_h$ and $B_h$ for $h = 0$ to 5.

4.2 With the coefficients found in Problem 4.1 (or given in the solution on page 374) reconstruct the function $f(x)$ by use of the program FOUR1 given in appendix II.

4.3 Find the Fourier transforms of the functions

(i) $f(x) = e^{-k|x|}, \quad a \geqslant x \geqslant -a$
$f(x) = 0, \qquad |x| > a.$

(ii) $f(x) = 1 - |x|, 1 \geqslant x \geqslant -1$
$\quad = 0, \qquad |x| > 1.$

4.4 Find by an analytical approach the convolution of

$$f(x) = \sin(2\pi x), \quad 0 \leqslant x \leqslant 1$$
$$= 0, \qquad\qquad x < 0 \text{ or } x > 1$$

and

$$g(x) = 1, \quad -\tfrac{1}{4} \leqslant x \leqslant \tfrac{1}{4}$$
$$= 0, \qquad |x| > \tfrac{1}{4}.$$

4.5 Determine the Fourier transform of $f(x) = (2\pi)^{-\frac{1}{2}}\exp[-\tfrac{1}{2}(x-2)^2]$. Compare the answer you get by the straightforward method with that found by considering $f(x)$ as the convolution of $g(x) = (2\pi)^{-\frac{1}{2}}\exp(-\tfrac{1}{2}x^2)$ with $h(x) = \delta(x - 2)$.

4.6 For the space group *Pmmm* the following relationships exist between structure factors:

$$F_{hkl} = F_{\bar{h}kl} = F_{h\bar{k}l} = F_{hk\bar{l}} = F_{h\bar{k}\bar{l}} = F_{\bar{h}k\bar{l}} = F_{\bar{h}\bar{k}l} = F_{\bar{h}\bar{k}\bar{l}}.$$

Derive a modified form of the electron-density equation for this space group. Over what portion of the unit cell must $\rho(xyz)$ be computed in order to completely determine the structure?

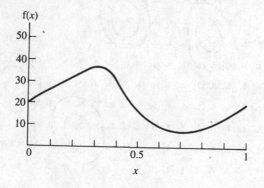

| x | f(x) | x | f(x) |
|------|------|------|------|
| 0.00 | 20.0 | 0.50 | 16.5 |
| 0.05 | 23.5 | 0.55 | 12.5 |
| 0.10 | 26.5 | 0.60 | 9.0 |
| 0.15 | 29.0 | 0.65 | 7.5 |
| 0.20 | 31.5 | 0.70 | 7.0 |
| 0.25 | 34.5 | 0.75 | 8.0 |
| 0.30 | 37.0 | 0.80 | 9.0 |
| 0.35 | 36.0 | 0.85 | 11.0 |
| 0.40 | 30.0 | 0.90 | 13.0 |
| 0.45 | 22.5 | 0.95 | 16.5 |

Fig. 4.21.
Periodic function for Problem 4.1.

4.7 Using the output file containing the structure factors generated in Problem 3.4 (p. 74) calculate a projected electron-density map using the program FOUR2 given in appendix IV. Remember that the name of the input data file for FOUR2 must be the same as that of the output file of STRUCFAC.

# 5 Experimental collection of diffraction data

## 5.1 The conditions for diffraction to occur

The basic conditions under which X-ray diffraction occurs from a crystal were given in § 3.3. We recall that there is associated with each set of indices a reciprocal-lattice vector $\mathbf{s} = h\mathbf{a}^* + k\mathbf{b}^* + l\mathbf{c}^*$; the magnitude of this vector is $(2\sin\theta)/\lambda$, where $\theta$ is the Bragg angle, and its direction is normal to the Bragg reflecting planes. If the incident X-rays make an angle $\frac{1}{2}\pi - \theta$ with $\mathbf{s}$ then a diffracted beam is produced coplanar with the incident beam and $\mathbf{s}$. Let us see how the crystal can be systematically moved to a position for a particular diffracted beam to be produced. In fig. 5.1 there is depicted a crystal, an incident beam of X-rays and the appropriate reciprocal-lattice vector $\mathbf{s}$. If, for example, the crystal is rotated about the axis $OA$ perpendicular to $IO$ then the whole reciprocal lattice rotates with it and the end of the vector $\mathbf{s}$ will move along the indicated circular path. As it does so the angle between the direction of the incident beam, $IO$, and $\mathbf{s}$ continuously changes. We denote the direction of the incident beam by the vector $\hat{\mathbf{S}}_0$ and, if ever the angle $\psi$ between $-\hat{\mathbf{S}}_0$ and $\mathbf{s}$ becomes $\frac{1}{2}\pi - \theta$, a diffracted beam results. The range of angles between $\mathbf{s}$ and $-\hat{\mathbf{S}}_0$ as the crystal rotates may be followed in fig. 5.2. Since the vector $-\mathbf{s}$ corresponds to the reflection with indices $\bar{h}\bar{k}\bar{l}$ which is, in general, equivalent to the $hkl$ reflection [see equation (3.44)], we also note its motion. In fig. 5.2($a$) is one extreme position where the vector $\mathbf{s}$ points in the same general direction as the incident beam and is in the plane defined by $IOA$. The angle between $\mathbf{s}$ and $-\hat{\mathbf{S}}_0$ is $\alpha$ and that between $-\mathbf{s}$ and $-\hat{\mathbf{S}}_0$ is $\pi - \alpha$. As the crystal rotates about $OA$ so does the vector $\mathbf{s}$ and the angle between $\mathbf{s}$ and $-\hat{\mathbf{S}}_0$ decreases until the

Fig. 5.1.
The motion of the scattering vector s as the crystal rotates.

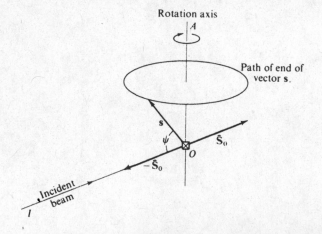

Fig. 5.2.
(*a*) One extreme position of the scattering vector where it makes the minimum angle with the direction of the incident beam.
(*b*) The scattering vector making an angle $\pi/2$ with the incident beam.

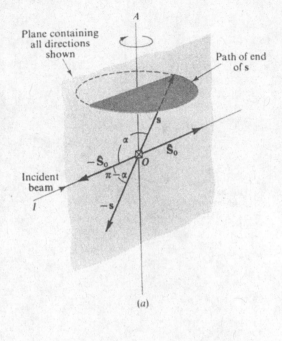

(*a*)

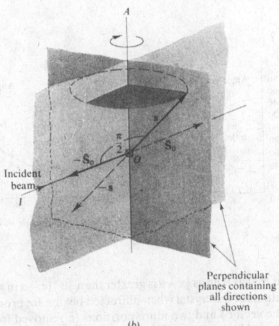

(*b*)

position shown in fig. 5.2(*b*) is reached when the angle equals $\frac{1}{2}\pi$. Further rotation of the crystal and of s reduces the angle between s and $-\hat{S}_0$ until finally s is again in the plane defined by *IOA* and the angle between s and $-\hat{S}_0$ is equal to $\pi - \alpha$ [fig. 5.2(*c*)]. The angle between $-s$ and $-\hat{S}_0$ is always the supplement of that made by s with $-\hat{S}_0$. The variation of these angles as the crystal is rotated through $2\pi$ is plotted in fig. 5.2(*d*) for $\alpha = 150°$. It is

Fig. 5.2. (*cont.*)
(*c*) Second extreme
position of the scattering
vector where it makes
the maximum angle with
the direction of the
incident beam.
(*d*) The variation of the
angle between **s** and
$-\hat{\mathbf{S}}_0$ and $-\mathbf{s}$ and $-\hat{\mathbf{S}}_0$
as the crystal rotates.

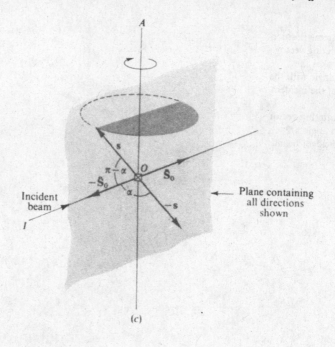

(*c*)

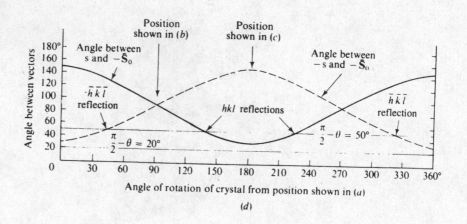

(*d*)

clear that as long as $\frac{1}{2}\pi - \theta$ is greater than 30 ° ($\pi - \alpha$ in general) there are two
positions of the crystal when diffracted beams are produced corresponding
to the vector **s** and two more positions ($\pi$ removed from the others) when
diffracted beams are produced corresponding to the vector $-\mathbf{s}$. If, however,
$\frac{1}{2}\pi - \theta < 30°$ then no diffracted beams will be produced corresponding to
either **s** or $-\mathbf{s}$. These points are illustrated by drawing the lines corresponding
to $\frac{1}{2}\pi - \theta$ equal to 50° and 20° in fig. 5.2(*d*). The situation we have considered
here, where the axis of rotation of the crystal is perpendicular to the
direction of the incident beam, is a particularly simple one to analyse. There
is an interesting geometrical way of looking at the diffraction condition
which can be very helpful in considering more general axes of rotation.

In fig. 5.3, $IO$ represents the direction of the incident beam and a sphere is drawn which passes through the crystal at $O$, has $IO$ as a diameter and is of radius $1/\lambda$. Let us assume that there is a reciprocal-lattice vector s whose end just touches the sphere at $P$. Since $IO$ is a diameter the angle $IPO$ is $\frac{1}{2}\pi$ and consequently $IOP\ (=\phi)$ is related to s by

$$s = \frac{2\cos\phi}{\lambda}. \tag{5.1}$$

However since $s = (2\sin\theta)/\lambda$ this gives

$$\phi = \frac{\pi}{2} - \theta. \tag{5.2}$$

But when the angle between the incident beam (actually with reversed direction) and the scattering vector s is $\frac{1}{2}\pi - \theta$ a diffracted beam is produced. The condition for a diffracted beam to be produced is now seen to be that the end of the scattering vector s, which is attached to the crystal at $O$, should just touch the surface of the sphere. The sphere of radius $1/\lambda$ which is defined above is known as the *sphere of reflection* or *Ewald sphere*. One can now follow the process of rotating the crystal about an axis not perpendicular to the incident beam by using the concept of the sphere of reflection. In fig. 5.4(*a*) the locus of the end of the vector s as the crystal is rotated is shown and this cuts the sphere at $P$ and $Q$. When the crystal is in such a position that the vector s is along $OP$ or $OQ$ a diffracted beam is produced with indices *hkl* appropriate to the vector s. The direction of the diffracted beam can also be found from this geometrical construction. In fig. 5.4(*b*) a cross-section of the sphere of reflection in the plane $IOP$ is shown. The diffracted beam must be along $OD$ which makes an angle $\frac{1}{2}\pi - \theta$ with $OP$. If $C$ is the centre of the circle then $CP$ equals $CO$ and therefore $\widehat{CPO}$ equals $\widehat{POC}$ and is also equal to $\widehat{POD}$. Hence $CP$ is parallel to $OD$ and gives the direction of the diffracted beam.

By rotation of the crystal about some axis any diffracted beam can be produced as long as the end of the scattering vector can be made to lie on the surface of the sphere. The condition for this is that

Fig. 5.3.
The sphere of reflection. When the tip of the vector s touches the sphere the corresponding reflection is produced.

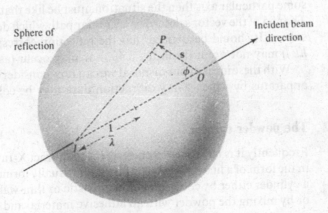

Fig. 5.4.
(a) As the crystal
rotates the end of the
scattering vector
touches the sphere of
reflection at $P$ and $Q$.
(b) $CP$ gives the
direction of the
diffracted beam.
(c) Neither $s$ nor $-s$
cuts the sphere of
reflection as the crystal
is rotated.
(d) $s$ does not but $-s$
does cut the sphere of
reflection as the crystal
is rotated.

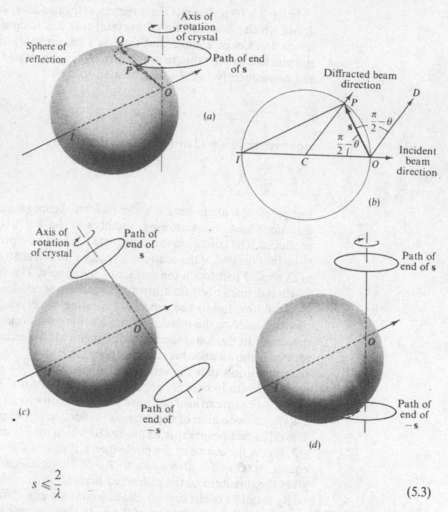

$$s \leqslant \frac{2}{\lambda} \tag{5.3}$$

where $2/\lambda$ is the diameter of the sphere of reflection. The sphere of radius $2/\lambda$ centred on $O$ within which a vector $s$ must lie in order to satisfy equation (5.3) is known as the *limiting sphere*.

If a possible reflection is not produced by rotation of the crystal about some particular axis then the situation must be like that shown in fig. 5.4(c). The end of the vector $s$ describes a circular path which does not intersect the sphere. It should be noticed that the reflection corresponding to $-s$ (i.e. $\bar{h}\bar{k}\bar{l}$) may not occur (as in fig. 5.4(c)) or may occur (as in fig. 5.4(d)).

With these ideas at our disposal we can now consider the various types of apparatus by which X-ray diffraction data may be collected.

## 5.2 The powder camera

Frequently it is necessary or convenient to diffract X-rays from a specimen in the form of a fine powder. The powder is usually formed into the shape of a cylinder either by containing it in a plastic or thin-walled glass container or by mixing the powder with an adhesive material and shaping it before it dries hard. Such a specimen will be a myriad of tiny crystallites arranged in random orientations. The volume of the individual crystallites in a typical powder might be about $(5 \mu m)^3$ so that in a specimen of total volume $(1 mm)^3$ the total number of crystallites will be of order $10^7$.

Let us examine what happens when a monochromatic X-ray beam falls on such a specimen. For a particular reflection there will be a large number of the crystallites in a diffracting position – that is to say, with their scattering vectors for this reflection making an angle $\frac{1}{2}\pi - \theta$ (or very near) with the reversed direction of the incident beam. It should be mentioned here, anticipating some of the results of the next chapter, that a crystal will reflect over a range of positions near to that corresponding to the Bragg angle so that the number of crystallites giving a particular reflection might be quite high. Any crystallite giving the reflection will produce a diffracted beam making an angle $2\theta$ with the incident beam and, as is shown in fig. 5.5, the locus of all such diffracted beams is a cone of semi-angle $2\theta$ with apex at the specimen.

Powder cameras vary somewhat in their design but the arrangement of components for a fairly typical type of camera is shown schematically in fig. 5.6(*a*). A beam of monochromatic X-rays is collimated by passing it through a fine tube (or through a slit system) and falls on the specimen. The film, in the form of a strip, is contained in a black paper envelope to protect it from the light and is wrapped round the inside of the cylindrical camera as shown. The direct beam, which is not diffracted, passes through a hole cut in the film and is absorbed by a beam trap. This is necessary as otherwise this radiation would be scattered by the film and fog it over a considerable area thereby obscuring some of the diffraction pattern at low Bragg angles. The diffraction pattern is recorded by the film as the intersection of cones similar to the one shown in fig. 5.5 with the film. When the film is straightened out it appears as in fig. 5.6(*b*); this film comes from a camera for which both the incident and straight-through beam pass through holes in the film.

If the powder is fairly coarse so that the crystallites are larger in size and fewer in number then the diffraction lines will not be uniform but tend to be rather spotty as in fig. 5.6(*c*). This can be overcome by rotating the specimen as then many more of the crystallites contribute to each of the diffraction lines. The films shown in fig. 5.6(*b*) and fig. 5.6(*c*) have been taken with and without rotation of the specimen, respectively; the improvement in appearance produced by rotation is evident.

Frequently the data required by the X-ray crystallographer are a

Fig. 5.5.
The locus of the diffracted beams for a given *hkl* from all the crystallites of a powder specimen.

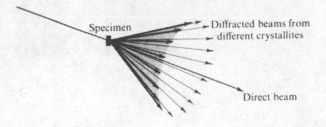

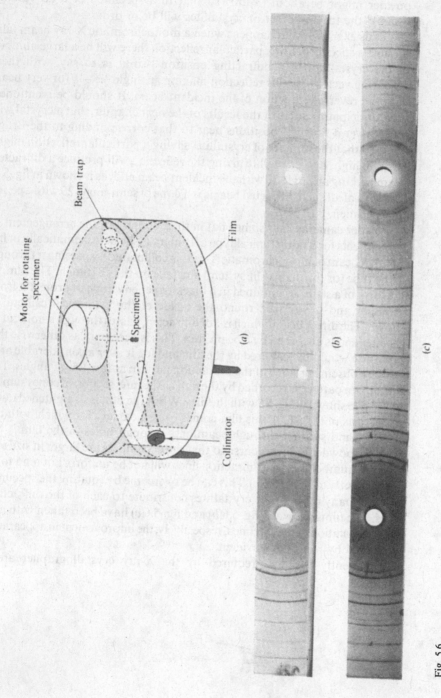

Fig. 5.6.
(a) A typical arrangement for a powder photograph.
(b) A powder photograph with rotation of the specimen.
(c) A powder photograph without rotation of the specimen.

complete set of diffracted intensities together with the corresponding indices of reflection *hkl*. From a powder photograph the only way of determining the indices of a diffraction line, assuming that one knows the unit-cell parameters, is by means of the diffraction angle $2\theta$. In fact many diffracted beams will have similar $2\theta$ values (similar magnitudes of **s**); not only does this give a problem of identification but, more important still, diffracted beams may overlap completely. If two or more beams fall on top of one another then it is impossible to ascertain the intensities of the individual components of the composite line and valuable information is lost. To get some idea of this problem let us consider how many reflections might be obtained for a crystal structure with a moderate-sized unit cell – say of volume $500\,\text{Å}^3$. The total volume within the limiting sphere is

$$V_\bigcirc = \tfrac{4}{3}\pi \left(\frac{2}{\lambda}\right)^3 \tag{5.4}$$

and the volume of the reciprocal unit cell will be, from equation (3.26),

$$V^* = (500\,\text{Å}^3)^{-1}. \tag{5.5}$$

Thus the number of reciprocal-lattice points within the limiting sphere will be

$$\frac{V_\bigcirc}{V^*} = V_\bigcirc V. \tag{5.6}$$

If the wavelength of the X-rays is that for Cu K$\alpha$ radiation, $1.54\,\text{Å}$, then the total number of reciprocal-lattice points equals

$$\tfrac{4}{3}\pi \left(\frac{2}{1.54}\right)^3 \times 500 \simeq 4000.$$

The number of independent reflections will be lower than this – for example reflections related by Friedel's law have the same Bragg angle and the same intensity and always overlap on the powder film. For space groups of higher symmetry there may be fourfold, sixfold or eightfold coincidences of equivalent reflections but nevertheless it can be seen that for all but the smallest unit cells the data consist of a large number of independent items. The overlap problem is so severe that powder photographs are normally not used for the purposes of collecting data for structure determination unless the alternative of obtaining data from a single crystal is, for some reason, impracticable.

   A combination of new needs, new technology and new theory has given the powder method a fresh lease of life. Many materials of great potential technological importance, e.g. high-temperature superconductors, may sometimes only be obtained in the form of fine powders. Their powder diffraction patterns may be collected with a powder diffractometer, a schematic representation of which is shown in fig. 5.7(*a*). The incident highly monochromatic X-ray beam is very well collimated which means that the angular spread of the powder lines is reduced as much as possible. The diffracted radiation is measured by some form of X-ray photon counter (see §5.7) with a very fine entrance slit which is slowly rotated about the central axis of the instrument. The resultant intensity profile, as shown in fig. 5.7(*b*),

Fig. 5.7.
(a) A schematic
representation of a
powder diffractometer.
(b) A typical
powder-diffractometer
trace.

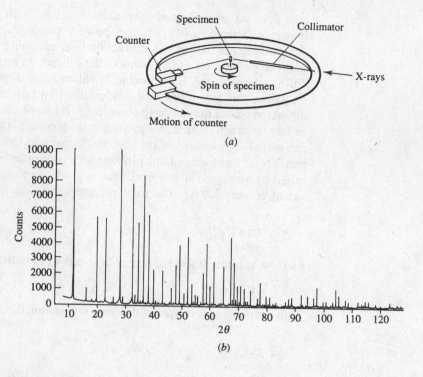

(a)

(b)

can be drawn by a pen recorder and also stored in digital form for subsequent computer analysis.

Although the incident X-ray beam is well collimated and highly monochromatic which, together with the fine entrance slit of the counter ensures high resolution in the diffraction pattern, there is nevertheless a considerable overlap of diffraction lines. This is because the small crystallites of which the powder specimen is formed will diffract over some small angular region around the Bragg angle. The theoretical basis of this line spread will be found in equation (3.3) which gives the shape of diffraction pattern from a finite array of scatterers and the theory of this is enlarged in § 6.1. However the cross-sections of the diffraction lines will all be of the same shape, albeit of different intensities and, if this shape is known, then it should be possible, at least in principle, to unravel the total intensity pattern as a sum of individual single-line contributions. The problem is one of deconvolution, the converse process to convolution as described in § 4.6 and illustrated in fig. 4.12. In fig. 5.8(a) there is represented an idealized powder diffraction pattern of $\delta$-function powder lines with different intensities and in fig. 5.8(b) the standard shape of each line. The convolution of the two gives the profile shown in fig. 5.8(c). The deconvolution problem is to find fig. 5.8(a) given fig. 5.8(c) and some model shape fig. 5.8(b). There have been a number of successful computer-based methods for carrying out the deconvolution of powder patterns but, on the whole, these are restricted to fairly simple compounds where the overlap is not too severe.

There are some deconvolution problems which cannot be solved by any form of analysis. For example for a cubic crystal with cell dimension $a$ it is found from equation (3.29) that

Fig. 5.8.
(*a*) An idealized
δ-function powder
pattern.
(*b*) The shape of an
isolated powder line.
(*c*) The actual powder
pattern – the
convolution of (*a*) and
(*b*).

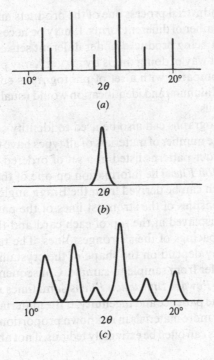

$$\frac{4\sin^2\theta}{\lambda^2} = \frac{1}{a^2}(h^2 + k^2 + l^2). \tag{5.7}$$

Thus any combinations of Miller indices giving the same values of $h^2 + k^2 + l^2$ will give exactly overlapping powder lines which no decon-volution process can disentangle. This will occur with pairs of indices [(550), (710)] and [(410), (322)] and the triple of indices [(621), (540), (443)]. For a cubic crystal, simple permutations of indices just give equivalent reflections which do not present a problem for powder diffraction analysis. However, in general the number of exactly overlapping powder lines is usually sufficiently small for the crystal structure analysis to be completed without using the information from them, although this information can be used subsequently in the process of structure refinement.

An important contribution to the use of powder patterns was made by H. M. Rietveld in 1967 who showed how it was possible to use the whole pattern, including the overlapped lines, to refine crystal structures. He described a mathematical least-squares procedure (see §9.4) for refining atomic positions, thermal vibration parameters (§6.5), lattice parameters and line-profile shapes which are all modified to give the best possible fit with the observed powder pattern. The number of parameters which could be accommodated by this method is of the order of 100, which means that it can only be applied to comparatively simple structures. Most commercial powder diffractometers will be provided with a range of software for deconvoluting the diffraction patterns and also for carrying out Rietveld refinement.

Powder cameras are often used for the identification of materials. For example in an industrial process one of the products may be alumina which can exist in a number of different forms. It may be necessary to ascertain the form of alumina being produced under different sets of conditions and the most convenient way of doing this is by taking X-ray powder photographs. These can be compared with a set of photographs covering the complete range of forms of alumina and identification would usually be straightforward and unambiguous.

Powder photographs can also be used to identify completely unknown materials. A large number of materials of all types have the major features of their X-ray powder patterns listed in a set of ordered cards known as the *Powder Diffraction File*. The information on one of the cards includes the *d*-spacings (which can be derived from the Bragg angle by equation (3.37)) and relative intensities of the strongest lines of the pattern. The strongest three lines are displayed at the top of each card and they are listed in the order of the *d*-spacings of their strongest lines. The relative intensities of powder lines may depend on the shape of the crystallites in the specimen and this may differ from sample to sample. Consequently identification of materials by the *Powder Diffraction File* is sometimes not straightforward and requires some patience and ingenuity. If the material to be identified is a mixture of two or more materials in unknown proportions then identification of the components can often be extremely tedious, if not absolutely impossible.

## 5.3   The oscillation camera

The simplest type of single-crystal camera in common use is the oscillation camera shown schematically in fig. 5.9. The crystal is glued to a suitable

Fig. 5.9.
A typical arrangement
for an oscillation camera.

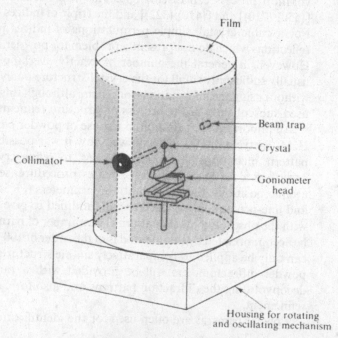

Film

Beam trap

Crystal

Collimator

Goniometer
head

Housing for rotating
and oscillating mechanism

support (e.g. a fine glass fibre) and this is attached to the rotatable goniometer head. The goniometer head consists of two perpendicular sets of arcs for rotating the crystal and two perpendicular slides for lateral motion of the crystal. Adjustments of the arcs and slides enable one to set a prominent axis of the crystal coincident with the axis of rotation of the goniometer head. This prominent axis may be one of the main edges of the crystal so that the axis can be set quite accurately by observation through the microscope which is usually attached to the camera. We shall assume for now that a prominent axis has been so set and that it corresponds to the direction of one of the edges of the unit cell, say the *a* direction.

The X-ray beam is collimated and is perpendicular to the axis of rotation of the crystal. As the crystal is rotated about the *a* axis then, from the Laue equations (3.12), we know that all the diffracted beams will lie along a series of cones coaxial with the *a* axis. For normal incidence $\psi_{a,0} = \frac{1}{2}\pi$ and the first of the Laue equations becomes

$$a \cos \psi_a = h\lambda. \tag{5.8}$$

The 'cone' corresponding to $h = 0$ is, for normal incidence, the plane perpendicular to the *a* axis containing the incident beam and the cones corresponding to other values of *h* are shown in fig. 5.10. The film, wrapped in black paper to protect it from the light, is mounted inside a metal cylinder which, in position in the camera, is coaxial with the rotation axis. As the crystal rotates so the X-ray reflections are produced one by one and are recorded by the film. When the film is developed and straightened out the diffraction spots are found to lie on a series of straight lines which are the locus of the intersection of the cones of diffraction with the film (see fig. 5.10). From the distances between these straight lines, called layer lines, we can find the length of the *a* axis of the unit cell. From fig. 5.10 it is clear that if the camera radius is *r* then the distance of the line with index *h* from that with $h = 0$ is given by

$$d_h = r \cot \psi_a. \tag{5.9}$$

By the use of equation (5.8) we may eliminate $\psi_a$ to obtain

$$a = h\lambda \frac{(r^2 + d_h^2)^{\frac{1}{2}}}{d_h}. \tag{5.10}$$

Fig. 5.11 shows, at two-thirds its natural size, a reproduction of a photograph taken on an oscillation camera. The X-rays used were unfiltered copper radiation and the layer lines for both $Cu\,K\alpha$ and $Cu\,K\beta$ radiation (see §5.8) of wavelengths 1.542 and 1.389 Å may be seen. The camera diameter $(2r)$ was 57.3 mm and the reader may, by direct measurements on the film, confirm that the unit-cell axis about which the crystal was rotated was of length 11.0 Å. From a rotating crystal photograph the unit-cell dimension may usually be found with an accuracy of about 1%. There are errors of measurement of distances on the film which are aggravated by the finite size of the spots due to the divergence of the X-ray beam, the size of the crystal and the spread in wavelengths of the X-rays.

Fig. 5.10.
The diffraction cones
with the incident beam
normal to the
oscillation axis which is
also a prominent cell
axis.

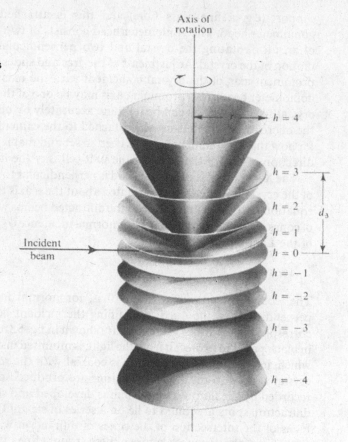

Fig. 5.11.
A typical oscillation
photograph (two-thirds
natural size).

Fig. 5.12.
(*a*) Location of the zero
layer when the crystal
axis is tilted from the
rotation axis in a
vertical plane
containing the incident
beam.
(*b*) Locus of zero layer
in relation to cylindrical
film.
(*c*) Appearance of zero
layer on the flattened
film when the upper
part of the crystal axis
is tilted towards the
incident beam.
(*d*) Appearance of zero
layer on the flattened
film when the lower
part of the crystal axis
is tilted towards the
incident beam.

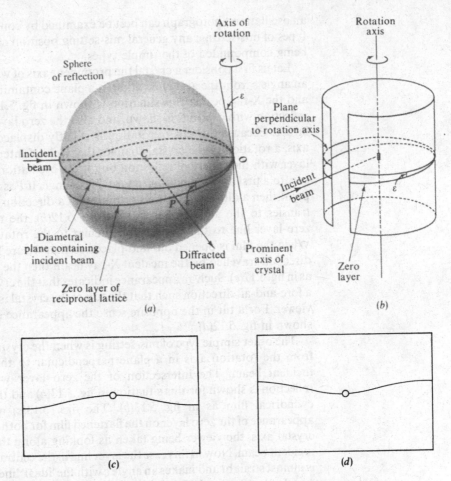

Changes in the physical dimensions of the films when they are processed are also important sources of error.

Although the separation of the diffraction data into layers makes the situation better than for powder photographs there is still nearly always some overlap of diffraction spots. In fact one rarely takes rotation photographs where the crystal rotates continuously. The cameras are provided with mechanisms operated by cams which enable one to oscillate the crystal to-and-fro through a few degrees – 5, 10 or 15° being the usual ranges. The cams are so designed that, apart from the inevitable discontinuities at the ends of the range, the angular speed is uniform. With a small angle of oscillation there are many fewer spots on the film and the chances of overlap are thereby much reduced. The photograph in fig. 5.11 is an oscillation photograph and it will be seen that the spots are all resolved.

Oscillation cameras are rarely used for the collection of intensity data but are sometimes used for the preliminary examination of crystals – the determination of unit-cell dimensions and crystal system (see §7.3). The crystal can also be accurately set about an axis on an oscillation camera and then transferred on its goniometer head to one of the more elaborate cameras which are described later. The effect of mis-setting of the crystal in an oscillation photograph can best be examined by considering two simple types of mis-setting; any general mis-setting position can be thought of as being compounded of the simple types.

Let us first consider a crystal the prominent axis of which is displaced by an angle $\varepsilon$ from the axis of rotation in a plane containing the rotation axis and the X-ray beam. This situation is shown in fig. 5.12(a). The sphere of reflection with centre $C$ is shown, and also the zero layer of the reciprocal lattice. Because the axis of rotation is slightly displaced from the crystal axis, a rotation of the crystal changes the circle of intersection of the zero layer with the sphere of reflection but, if the oscillation angle is small, we can, to a first approximation, ignore this change. If $P$ is a reciprocal-lattice point then a diffracted beam is produced in a direction given by $CP$. If we transfer to the actual camera, as in fig. 5.12(b), the relationship of the zero-layer line to the plane perpendicular to the rotation axis is shown. When the film is opened out and placed so that we are looking at it in the direction travelled by the incident X-ray beam then the zero layer appears as in fig. 5.12(c). Such an appearance indicates that the crystal axis is tilted in a fore-and-aft direction such that the top of the crystal is tilted towards the viewer. For a tilt in the opposite sense the appearance of the zero layer is shown in fig. 5.12(d).

The other simple type of mis-setting is when the crystal axis is displaced from the rotation axis in a plane perpendicular to the direction of the incident beam. The intersection of the zero layer with the sphere of reflection is shown for this situation in fig. 5.13(a) and this appears on the cylindrical film as in fig. 5.13(b). The figs. 5.13(c) and (d) show the appearance of the zero layer on the flattened film for both senses of tilt of the crystal axis, the viewer being taken as looking along the direction of the incident beam. Now in this case the layer line in the central region of the film is almost straight and makes an angle $\varepsilon$ with the 'ideal' line corresponding to a perfectly set crystal. This 'ideal' line does not actually appear on the film, although one can usually guess its position to within a degree or so, and therefore the angle $\varepsilon$ cannot be directly measured. However if the crystal is rotated through 180° and another oscillation picture is taken about this new position then one has gone from situation (c) to situation (d). If both layer lines are recorded on the same film and if one, say the first recorded, is made stronger than the other then the appearance of the film will be as shown in fig. 5.13(e). The angle between the two straight sections of the layer lines is $2\varepsilon$ which can be measured easily and the crystal is known to be in the position corresponding to the second (weaker) layer line.

In practice, in order to set a crystal, one would take the first oscillation picture with the two arcs along and perpendicular to the direction of the X-ray beam. If the mis-setting has the major characteristics shown in fig. 5.13 the two layer lines at 180° separation are recorded and the error in setting is corrected on the arc perpendicular to the beam. If the mis-setting has the major characteristics of fig. 5.12 then a rotation through $\frac{1}{2}\pi$ gives the fig. 5.13 situation and the correction can be made as before. By taking a succession of photographs and correcting the setting each time the crystal axis can be set along the rotation axis with a high degree of precision.

The indexing of oscillation photographs can best be done by a graphical

Fig. 5.13.
(*a*) Location of the zero layer when the crystal axis is tilted from the rotation axis in a plane perpendicular to the incident beam.
(*b*) Locus of zero layer in relation to cylindrical film.
(*c*) Appearance of zero layer on the flattened film when the crystal axis is tilted in an anticlockwise direction as seen looking along the direction of the incident beam.
(*d*) Appearance of zero layer on the flattened film when the crystal axis is tilted in a clockwise direction as seen looking along the direction of the incident beam.
(*e*) The double zero layer produced by taking two oscillation photographs with the crystal rotated 180° in between.

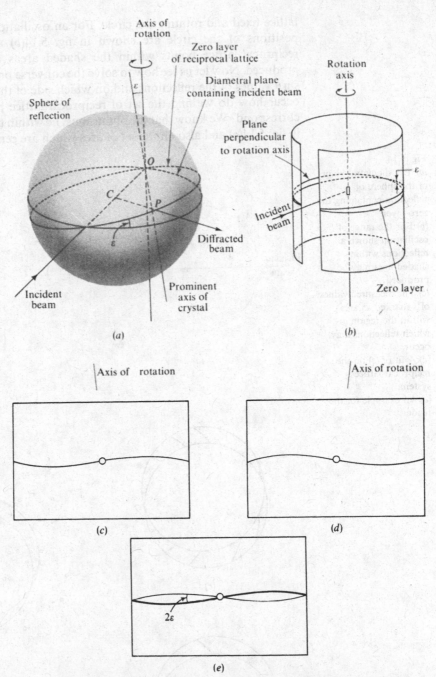

process which will be illustrated for the zero layer. In fig. 5.14(*a*) there is shown a section of the zero layer of the reciprocal lattice of a crystal situated at *O* and the diametral plane of the sphere of reflection (the circle of reflection) which contains the incident beam. As the crystal rotates so does the reciprocal lattice and reciprocal-lattice points pass through the circle of reflection giving diffracted beams as they do so. The relative motion of the circle and the reciprocal lattice can also be shown by keeping the reciprocal lattice fixed and rotating the circle. For an oscillation of 15° the extreme positions of the circle are shown in fig. 5.14(*b*) and for each of the reciprocal-lattice points within the shaded areas a reflection may be produced. Now let us see how to solve the converse problem – knowing the sin θ values of the reflections and on which side of the incident beam they occur how do we find the set of reciprocal-lattice points to which they correspond? We know that the points must lie within two areas arranged as in fig. 5.14(*c*) and also on the set of arcs which are centred at *O* and whose

Fig. 5.14.
(*a*) The diametral plane of the sphere of reflection containing the zero layer.
(*b*) For the range of oscillation shown all reflections within shaded area may be produced.
(*c*) The measured values of *s* marked as arcs within the region in which reflections may occur.
(*d*) A fit of (*c*) on the reciprocal-lattice system.
(*e*) An example for the reader.

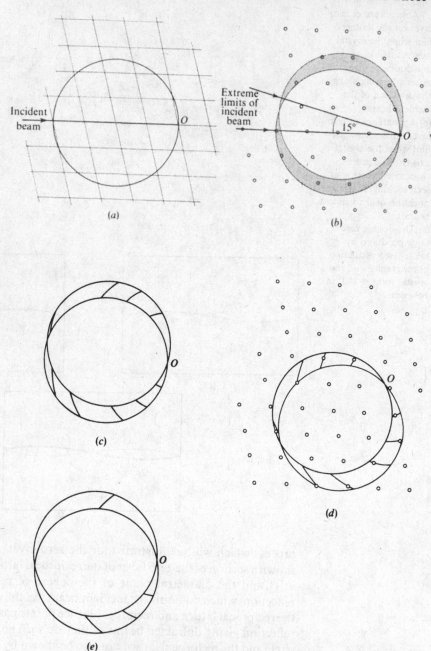

distances, at the scale of the reciprocal lattice, correspond to the $\sin\theta$ values of the observed reflections. If a transparent replica of fig. 5.14(c) is placed with O over the origin of a reciprocal-lattice net and is pivoted about O a position will be found when each of the sections of arc passes through a reciprocal-lattice point. Such a fit of arcs to points (not all perfect because of errors of measurement) is shown in fig. 5.14(d). The reader may try for himself the problem of fitting the arcs in fig. 5.14(e) on to the same net. Similar considerations apply to the indexing of non-zero layers although this introduces some additional complications.

If the principles of the oscillation camera have been dealt with in some detail it is because it is a simple camera, not difficult to understand, in which one can readily see the basic principles of X-ray diffraction from crystals. In addition it will have demonstrated how useful is the concept of the reciprocal lattice in interpreting the results which one obtains.

## 5.4  The Weissenberg camera

While it is quite feasible to collect complete sets of X-ray intensity data with an oscillation camera the inconvenience of indexing the photographs is such that they are hardly ever used for this purpose. We shall now examine a more convenient instrument, the Weissenberg camera, which makes the task of indexing photographs quite straightforward and which also enables a whole layer of information to be collected at one time without overlap difficulty.

In fig. 5.15(a) there is shown a crystal mounted on a goniometer head with an X-ray beam incident normal to the axis of rotation. There are also

Fig. 5.15.
(a) Layer-line screens set for the zero layer.
(b) Layer-line screens set for an upper layer.

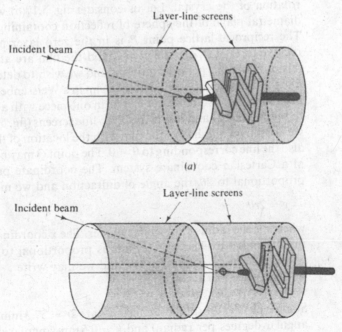

Incident beam

Layer-line screens

(a)

Incident beam

Layer-line screens

(b)

shown a pair of cylindrical metal screens which can be adjusted so that a circular slit occurs between them and this slit can be so arranged that one and only one of the layer lines passes through it. Diffracted beams in other layer lines will strike the screens and be absorbed. In fig. 5.15(a) the screens are adjusted to let through the zero layer and in fig. 5.15(b) a higher layer is allowed through.

A film, inserted into a cylinder and placed coaxial with the screens will record only a single layer but, if the crystal is completely rotated, there will be overlap of spots for diffracted beams which pass through the same points of the slit (same values of Bragg angle). It should be recalled that the diffracted beams occur wherever a reciprocal-lattice point passes through the sphere of reflection and although two reflections may have the same Bragg angle they will not be produced at the same time. In the Weissenberg camera the crystal is oscillated, the range of the oscillation being under the control of the user and may be greater than 180°. The film holder can move along the direction of its axis, which is also the axis of oscillation, and it is mechanically coupled to the rotation of the crystal in such a way that it slides backwards and forwards as the crystal oscillates to-and-fro. A succession of positions through one cycle of the motion is shown in fig. 5.16. Wherever the crystal passes through a position for a particular reflection to occur the film is always in the same relative position and the diffracted beam strikes the same point of the film. The general effect of the film motion is to spread the data in a single layer over the whole area of the film and no overlap of spots can possibly occur. If spots are produced at different times then they will be displaced from one another along the direction of the motion of the film.

The distance travelled by the film holder is linearly related to the angle of rotation of the crystal. Let us consider fig. 5.17(a) which represents the diametral plane of the sphere of reflection containing the incident beam. The reciprocal-lattice point $P$ is in the zero layer and has coordinates relative to $O$ of $(s, \phi)$. The crystal and the film are at one of the extreme positions of their periodic motions and we wish to determine where on the film the diffracted spot will fall. The film in a Weissenberg camera can either be mounted in two halves or can be in one piece with a beam stop attached to the collimator and inside the layer-line screens (fig. 5.17(b)). In fig. 5.17(c) we are shown one-half of the film with the location of the slit indicated and also the line corresponding to $\theta = 0$. The point $Q$ may be taken as the origin of a Cartesian coordinate system. The coordinate parallel to the slit is proportional to $2\theta$, the angle of diffraction and we may write

$$y = 2r\theta \tag{5.11}$$

where $r$ is the radius of the film cylinder. The $x$ coordinate, which measures the distance travelled by the film, is proportional to the angle through which the crystal rotates, say $\psi$, and we may write

$$x = k\psi. \tag{5.12}$$

Some typical values of these constants are $2r = 57.3$ mm (note there are 57.3 angular degrees per radian) and $k = 0.5$ mm/angular degree.

Turning back to fig. 5.17(a) we can see that the reflection will be

Fig. 5.16.
Diagram to show
successive positions of a
rotating crystal and the
linked motion of the
camera for a 180°
oscillation in a
Weissenberg camera.

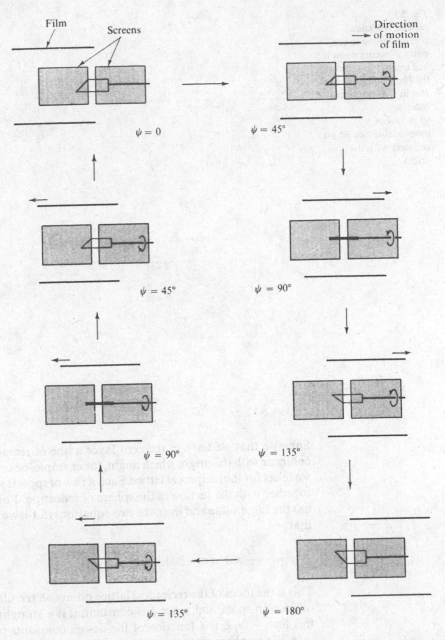

produced when the crystal has rotated into such a position that the reciprocal-lattice point originally at $P$ has moved to $P'$. The angle through which the crystal rotates is

$$\psi = \frac{\pi}{2} - \theta + \phi. \tag{5.13}$$

The spot will thus occur on the film with coordinates

$$y = 2r\theta$$

Fig. 5.17.
(a) The reflection
corresponding to the
point $P$ occurs when it
has rotated to point $P'$.
(b) The arrangement of
film in a Weissenberg
camera.
(c) A row of
reciprocal-lattice points
collinear with the
origin.

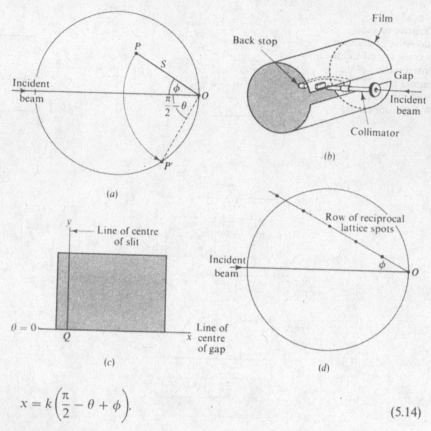

$$x = k\left(\frac{\pi}{2} - \theta + \phi\right). \tag{5.14}$$

Suppose that we have in the zero layer a line of reciprocal-lattice points collinear with the origin which might, for example, be one of the prime axes we select for the reciprocal lattice. Such a row of spots is shown in fig. 5.17(d) together with the section of the sphere of reflection. For all these points $\phi$ has the same value and from the two equations (5.14) we find, eliminating $\theta$, that

$$y = -\frac{2r}{k}x + r(\pi + 2\phi). \tag{5.15}$$

This is the locus of the reciprocal lattice points on the film, as shown by the diffraction spots, and, since $\phi$ is constant, it is a straight line. The slope of this line, $-2r/k$, is a function of the design constants of the camera.

A Weissenberg picture may be regarded as a distorted projection of the reciprocal lattice. In fig. 5.18(a) there is shown a representation of a polar coordinate grid in reciprocal-lattice space. The angular coordinate $\phi$ is measured from some arbitrary origin and the radial coordinate $s$ is the magnitude of the normal reciprocal-lattice vector. The corresponding grid in Weissenberg space is shown in fig. 5.18(b). The lines of constant $\phi$ at equal increments of $\phi$ appear as a set of equidistant, parallel straight lines of slope $-2r/k$. Since $s$ is proportional to $\sin\theta$ and the $y$ coordinate is proportional to $\theta$ (see equation (5.11)) the lines of constant $s$ at constant increments of $s$ in Weissenberg space are parallel but not equidistant. The two thicker lines in

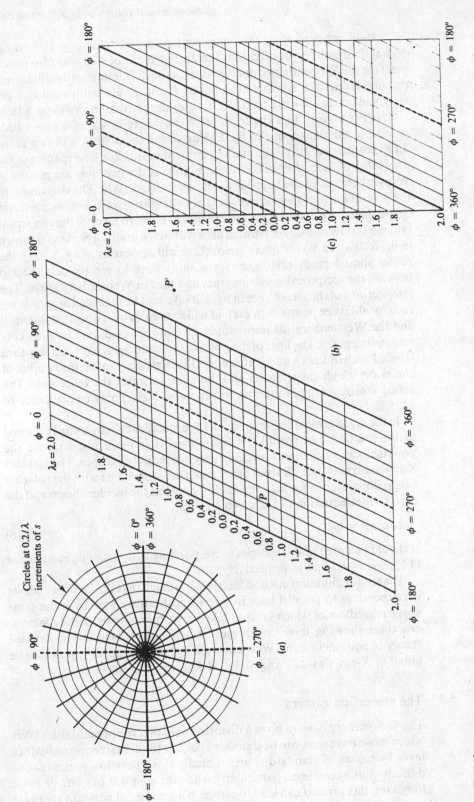

Fig. 5.18.
(a) The diametral plane of the sphere of reflection containing the incident beam marked with a polar-coordinate system.
(b) The transformation of the grid shown in (a) in Weissenberg space. (Half-scale for $2r = 57.3$ mm, $k = 0.5$ mm/angular degree.)
(c) The transformed grid within a region which could occur on a Weissenberg film. (Half-scale for $2r = 57.3$ mm, $k = 0.5$ mm/angular degree.)

fig. 5.18(b) correspond to the complete line $\phi = 0°$ plus $\phi = 180°$ in fig. 5.18(a). It should be noticed that the arrangement of slits and film movement for the Weissenberg camera results in the diffraction spots falling on a rectangular area of the film. The points $P$ and $P'$ are equivalent and the rectangular area shown in fig. 5.18(c) is equivalent to that shown in fig. 5.18(b).

It is possible to find the locus in Weissenberg space of a set of lines parallel to the thicker one in fig. 5.18(a); such a set of lines is shown in fig. 5.19(a) contained within the limiting circle of radius $2/\lambda$. The spacing of the lines is 0.1 of the radius of the limiting circle and when they are plotted in Weissenberg space they appear as shown in fig. 5.19(b). The derivation of the precise form of these curves is fairly straightforward but we shall not derive them here. Now let us suppose that in the zero layer of the reciprocal lattice we draw two sets of parallel lines within the limiting circle as is shown in fig. 5.20(a). In Weissenberg space these will appear as in fig. 5.20(b); the reader should study this figure for a short time to see the relationship between the reciprocal-space diagram and that in Weissenberg space. The intersections of the sets of lines in fig. 5.20(a) could be regarded as zero-layer reciprocal-lattice points with each of which there is an associated intensity and the Weissenberg diagram would give a diffracted spot at each corresponding point. The lines of the grid are shown in fig. 5.20(b) and these are labelled with indices $h$ and $k$. It is normally fairly easy to see the families of curves on which the spots lie and therefore to index the reflections. The actual Weissenberg photograph shown in fig. 5.20(c) may be compared to fig. 5.20(b).

In the arrangement shown in fig. 5.15 the incident X-ray beam is normal to the rotation axis for all the layer lines. It is more usual to use the 'equi-inclination method' for taking upper-layer photographs. The incident X-ray beam and the diffracted beam make the same angle with the rotation axis as illustrated in fig. 5.21; the angle $\mu$ between the incident beam and the axis is, from equation (3.12), given by

$$2a \cos \mu = h\lambda. \tag{5.16}$$

It should be noted that the angles corresponding to $\psi_{a,0}$ and $\psi_a$ in equation (3.12) are $\pi - \mu$ and $\mu$, respectively.

The equi-inclination method has the advantage that the family of curves corresponding to parallel lines in a reciprocal-lattice layer have the same shape regardless of which layer is taken and, by applying scaling factors, one chart showing these curves can be used for all layers. The detailed theory of equi-inclination Weissenberg diagrams for upper layers is to be found in *X-ray Crystallography* by M. J. Buerger (Wiley).

## 5.5   The precession camera

The Weissenberg camera gives a distorted map of a reciprocal-lattice layer which, as we have seen, can be translated back to the actual reciprocal-lattice layer by means of standard charts. Usually this operation presents little difficulty but, sometimes, particularly when the unit cell has one or more long axes, this process can lead to errors. If a number of adjacent spots are too weak to be seen in an otherwise highly populated row corresponding to

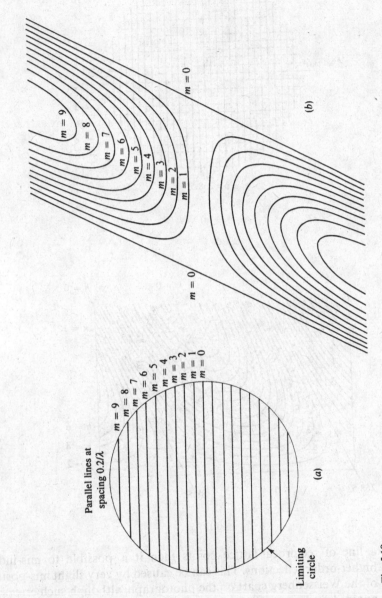

Fig. 5.19.
(a) A set of equidistant parallel lines in the diametral plane of the sphere of reflection.
(b) The transformation of (a) in Weissenberg space. (Half-scale for $2r = 57.3$ mm, $k = 0.5$ mm/angular degree.)

Fig. 5.20.
(a) Two sets of
equidistant parallel lines
in the diametral plane of
the sphere of reflection.
(b) Part of the
transformation of (a) in
Weissenberg space.
(Half-scale for
$2r = 57.3$ mm,
$k = 0.5$ mm/angular
degree.)

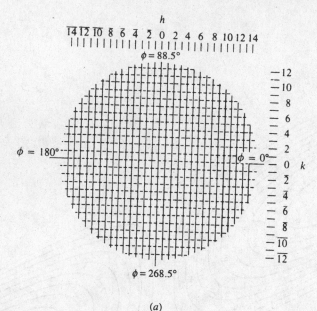

(a)

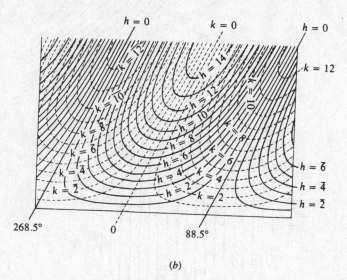

(b)

a line of reciprocal-lattice points then it is possible to mis-index the
higher-order reflections. This can be caused by very slight mis-positioning
of the Weissenberg chart on the photograph; although such errors can be
avoided by very careful measurement on the photograph this can be a
tedious process. In such a case it would be an advantage to have a camera
which would give an undistorted map of the reciprocal lattice. The location
of a spot in relation to its fellows may then be determined by a straight edge
alone and the possibility of error is much reduced. However even when the
unit-cell edges are all of moderate dimensions, it is convenient to obtain
directly an undistorted picture of the reciprocal lattice.

Fig. 5.20 (*cont.*)
(*c*) A Weissenberg film corresponding to the lattice shown in (*b*). Only one-half of the film is shown. Natural size.

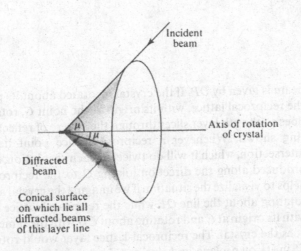

Fig. 5.21.
The equi-inclination arrangement for higher layer Weissenberg photographs. The incident and diffracted beams make angles $\pi - \mu$ and $\mu$, respectively, with the rotation axis.

Incident beam

Axis of rotation of crystal

Diffracted beam

Conical surface on which lie all diffracted beams of this layer line

It was shown by de Jong and Bouman (1938) that a camera could be designed which would represent without distortion a layer of the reciprocal lattice. We shall examine here a special case of the de Jong and Bouman arrangement – one which is simple to understand and illustrates the general principles of the method.

In fig. 5.22(*a*) there is represented a crystal at *C* with an incident X-ray beam normal to a prominent cell axis whose direction is *CD*. The sphere of reflection is shown and also the intersection of the sphere by an upper layer, say the *n*th, of the reciprocal lattice. If any reciprocal-lattice point lies on this ring of intersection, say at *P*, then the corresponding X-ray reflection is produced. If *O* is the centre of the sphere then the direction of the diffracted

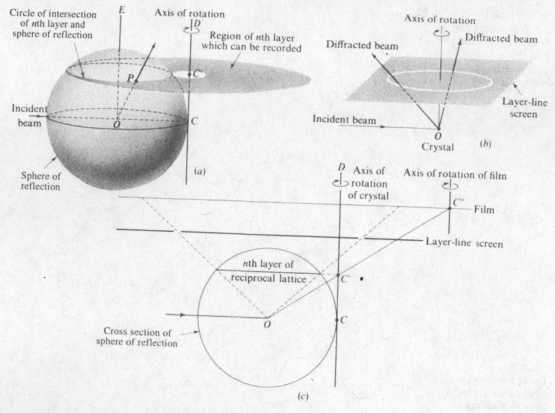

Fig. 5.22.
(a) The intersection of the nth layer of the reciprocal lattice with the sphere of reflection. The accessible area of the nth layer is the annular region shown shaded.
(b) A layer-line screen arranged to allow through only the nth layer.
(c) If the film is rotated about $C''$ with the same angular velocity as the crystal an undistorted projection of the reciprocal-lattice section is produced.

beam is given by $OP$. If the crystal is rotated about its prominent axis then the reciprocal lattice, with its origin at the point $C$, rotates about $CD$; as it does so the nth layer slices through the sphere of reflection cutting it in the ring shown. Whenever a reciprocal-lattice point lies on the circle of intersection, which it will do twice on each revolution, a diffracted beam is produced along the direction joining $O$ to the reciprocal-lattice point. It helps to visualize the situation if we imagine the crystal actually located at $O$ rotating about the line $OE$ while the reciprocal lattice is displaced from it with its origin at $C$ and rotating about $CD$ with the same angular velocity as does the crystal. The reciprocal-lattice layer would rotate about the point $C'$ in its own plane and one could record data corresponding to lattice points within the annular region shaded in the figure.

With the arrangement described all the diffracted beams from the nth layer would lie on the surface of a cone with apex $O$ and if a flat piece of film was placed perpendicular to $OE$ then the separate layers would appear as rings of spots on the film, each ring being the intersection of the appropriate conical surface with the plane of the film. It is possible to separate one layer line by interposing in the path of the X-rays a metal screen with an annular slit cut into it. The central ring of metal is kept in place by fine cellophane or plastic film which is almost completely transparent to X-rays. The relationship of this layer-line screen to the crystal and incident beam is shown in fig. 5.22(b).

The de Jong and Bouman method, like the Weissenberg method, spreads out the record of the diffraction data by moving the film. The way in which this is done may be followed by reference to fig. 5.22(c) which shows a cross-sectional view in the plane of the sphere of reflection containing the incident beam. The point $C''$ is the projection from $O$ of the point $C'$ on to the plane of the film. If the film is rotated about an axis through $C''$ parallel to $CD$ at the same angular velocity as the reciprocal lattice then the line joining $O$ to any reciprocal-lattice point in the $n$th layer *always* intersects the same point on the film. Thus whenever a particular reciprocal-lattice point passes through the circle of intersection the diffracted beam strikes the same place on the film. The exposed points on the film are thus a projection of the reciprocal-lattice layer from the point $O$ on to the plane of the film and the developed film will show a true undistorted picture of the reciprocal-lattice layer.

It will be noticed that the experimental arrangement which has been described is incapable of giving the zero-layer data and also presents considerable mechanical difficulties for layers close to the zero layer. Many other experimental arrangements are possible and their various advantages and disadvantages have been discussed by Buerger (*loc. cit.*). In fact the camera which is most frequently used to record diffraction data on an undistorted reciprocal-lattice net is the Buerger precession camera. This enables all layers to be recorded and for the layers to all be on the same scale. A typical film produced by a precession camera is shown in fig. 5.23. We shall not attempt to describe the complexities of the precession camera here; for our purpose a description of the de Jong and Bouman arrangement suffices to explain the basic theory of methods of this type.

For a given wavelength the precession camera records less of reciprocal space than does the oscillation or Weissenberg camera and therein lies its chief disadvantage. However, as will be clear from fig. 5.23, it enables angular measurements to be made more easily.

## 5.6  The photographic measurement of intensities

X-ray crystallographers have relied and, to some extent, still do rely on the photographic recording of diffraction data. Before very fast films were available it was customary to use an intensifying screen. This was a layer of fluorescent material placed behind the film in such a way that the X-rays passed through the film and then hit the screen. The bulk of the exposure of the film was caused by the photons of visible light released by the screen rather than by the direct X-ray beam itself. However, such screens are rarely used now; fast films, very sensitive to X-rays, are used and these are normally double-coated, that is covered with sensitive emulsion on both sides, to increase their sensitivity still further.

Usually one does not try to measure the intensities of the reflections by an absolute method. If the spots on the film are similar to one another in appearance then one can produce an intensity scale by exposing one reflection along a strip of film for various lengths of time with the X-ray tube running under constant conditions. The relative intensities of the diffraction spots can then be determined by comparison with the scale. The eye is very good at comparing intensities for spots in the intermediate range of density but is less efficient for spots which are either very faint or very dense.

Fig. 5.23.
A typical photograph
from a precession
camera.

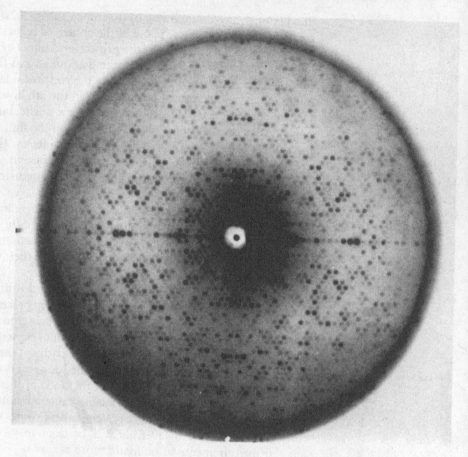

It is frequently found that the recorded diffraction spots are not all of similar appearance and then it is not the intensities of the spots we wish to measure but the total energies of the X-ray beams which have produced them. X-ray crystallographers are guilty of lax terminology here and use the word 'intensity' to denote total energy, since the total energy is proportional to the quantity $\mathscr{I}_{hkl}$ in equation (3.43). We shall use the term 'energy' to denote the total energy of the diffracted beam.

The reasons for the variation in spot appearances are manifold. For instance, the spots tend to become more diffuse at higher Bragg angles due to the finite bandwidth of the X-ray spectral lines. From Bragg's law, equation (3.37), by differentiation and rearrangement, we find

$$d\theta = \frac{d\lambda}{2d\cos\theta} \tag{5.17}$$

and since $\cos\theta$ decreases as $\theta$ increases so the diffraction image becomes more diffuse for high $\theta$. This effect may be seen in fig. 5.20(c) where at high Bragg angles the $K\alpha_1$ and $K\alpha_2$ wavelengths (see § 5.8) produce resolvable diffraction spots.

However a more important reason for the spread of diffraction spots is divergence in the X-ray beam. The beam divergence is usually defined by the X-ray source and the crystal (fig. 5.24); the collimator serves only to cut down the background radiation and so to reduce fogging. A rotating crystal, illuminated by radiation coming from a range of directions, will diffract X-rays from different parts of itself as it rotates. In fig. 5.25, for example, the crystal may diffract from point $A$ to give a diffracted beam $AA'$ and at some later time when it has rotated further it could diffract from point $B$ to give a diffracted beam $BB'$. A film placed in the position indicated would be blackened in the whole region from $A'$ to $B'$. When the formation of a diffraction spot in this way is considered together with motion of the film, as when one is using the Weissenberg camera, it is found to lead to the extension of some spots along the direction of film motion and a contraction of others. For example if $A'B' = l$ and the total time for formation of the diffraction spot is $\tau$ then the diffracted beam scans from $A'$ to $B'$ at an average velocity $l/\tau$. If the film moves in the direction from $A'$ to $B'$ at a velocity $v$ then the relative velocity of the scan relative to the film is $l/\tau - v$ (this may be negative). Hence the length of the blackened region is

$$l_1 = \tau \,|\, l/\tau - v \,| = |\,l - v\tau\,|. \tag{5.18a}$$

Similarly it can be found that if the film is moving in the opposite direction the length of the blackened region is

$$l_2 = l + v\tau. \tag{5.18b}$$

Thus spots of the same energy may appear in different forms – one may be contracted and more dense while the other may be extended and less dense.

We must now examine the problem of measuring the energy of reflections when the spots on the film can be so varied in appearance. If one were to use just the density of the spot then a contracted spot would be

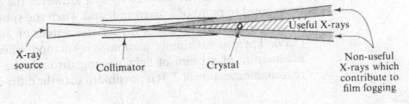

Fig. 5.24.
Divergence of the incident X-ray beam to the finite size of X-ray source and crystal.

X-ray source   Collimator   Crystal   Useful X-rays   Non-useful X-rays which contribute to film fogging

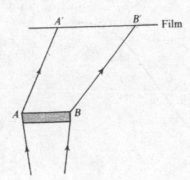

Fig. 5.25.
Different parts of the crystal reflect as the crystal rotates and the diffraction spot on the film grows from one end.

$A'$   $B'$ — Film

$A$   $B$

judged as more intense than an extended one. The quantity one is really trying to measure is the total energy $T$ of the radiation producing the spot or, what amounts to the same thing, the total number of photons falling on that region of the film. Film obeys the so-called *reciprocity law*; that is to say that the effect on the film is proportional to the total exposure $E$ as measured by the product of beam intensity and time. Since intensity is a measure of energy per unit area per unit time then $E$ is a measure of energy per unit area. The quantity we wish to determine is then given by

$$T = \int E \, dA \tag{5.19}$$

where the integration is over the whole area of the diffraction spot.

The blackening of a photographic emulsion is measured by its density defined by

$$D = \log_{10} \frac{1}{\text{fractional transmission of light}}. \tag{5.20}$$

Thus if a certain region of a film transmits 1% of incident light then, for that part of the film, $D = 2$. The behaviour of films on exposure to radiation depends on the wavelength of the radiation and also very critically on the developing process. However the general character, expressed as a relationship between $D$ and $E$, usually resembles that shown in fig. 5.26. The fog level may vary a great deal and depend on the duration and temperature of storage of the film, the cleanliness of the developing and fixing solutions and also exposure to ambient cosmic radiation before use. Fast films are particularly prone to a high fog level. Ideally when comparing diffraction spots with an intensity scale it is better to be in the range of $E$ corresponding to the steepest part of the curve for it is in this region that small differences of $E$ can most readily be detected by eye. However the quantity $E$ will vary from point to point of a given spot and when the spots are all of different areas it is not easy to assess the relative values of $T$ defined in equation (5.19). There is no simple photometric method whereby measuring the attenuation of a beam of light passing through the whole spot gives a reasonable measure of $T$. It is possible to scan the diffraction spot with a fine

Fig. 5.26.
A typical curve of density vs log (exposure) for an X-ray film.

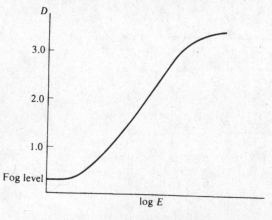

beam of light, to determine $D$ (and hence $E$ from the known $D$–$E$ relationship) at the points of a grid covering the spot and hence to find $T$ on a relative scale from the sum $\Sigma E$.

For accurate photographic work, one would normally use an integrating mechanism either with a Weissenberg or a precession camera. This is a device which, at the end of every cycle of operations of the camera, moves the film in such a way that consecutive images of the same diffraction spot fall on the film with displacements corresponding to the points of a small rectangular grid. This is illustrated in fig. 5.27 where 25 images of the single diffraction spot shown in (a) are represented with displacements on a 5 × 5 grid in (b). The multiple image will have a flat plateau at its centre the density of which will correspond to a total exposure proportional to $T$. The point $P$ in (b) for example has been subjected to exposures from each of the images and these exposures come from points similar to those shown in the grid in (a). Thus the total exposure at $P$ is $\Sigma E$ where the sum is over the points of the grid and, since this summation is, to a first approximation, proportional to the integral in equation (5.19), it will be proportional to $T$. The usual practice is to compare the densities of the plateaux with a scale the elements of which are small uniform spots made by exposure to the direct X-ray beam through a circular aperture.

There is shown in fig. 5.28 a Weissenberg picture, normal on one half and integrated on the other. Quite high precision can be reached by the mechanical–integration method. Because of the spreading out of the diffraction data longer exposures are required and weak spots may be missed altogether. However the very weak spots can be measured on non-integrated photographs and related to the integrated data by comparison with stronger spots with similar shapes.

Automatic densitometers are available which enable rapid and quite accurate assessments to be made of the quantity $T$ in equation (5.19). One such instrument, computer-controlled, scans the spots with a small source of light produced on the screen of a cathode-ray tube, the transmitted light being measured by a sensitive photocell. The background round the spot is also measured and the integral in equation (5.19) is then computed automatically. Once the film has been taken the actual measurements can be made at about one spot per second or even faster.

Fig. 5.27.
The superposition of several images of the diffraction spot on a film to produce a uniform plateau of intensity at the centre.

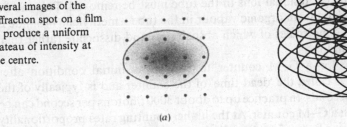

(a)

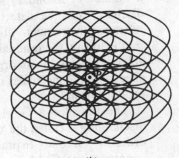

(b)

Fig. 5.28.
A Weissenberg film
showing unintegrated
spots on the right-hand
side and integrated spots
on the left-hand side.

## 5.7 Diffractometers

The energy of diffracted beams can also be measured by a counting device
which records individual X-ray photons. The Geiger–Muller counter has
been used for this purpose and fig. 5.29 shows a schematic G–M counter.
The cathode is a metallic cylinder and the anode a wire along the axis of the
cylindrical glass envelope. A potential is maintained between the electrodes,
of the order of 1000 volts, such that a discharge does just not take place but
so that the apparatus is in a critical condition. A single X-ray photon falling
on the gas in the tube can produce primary ions which in their turn produce
a cascade of secondary ions and a discharge through the counter. This
discharge can be detected electronically and amplified to operate a
recording instrument. After each discharge the counter must be quenched –
that is to say, the residual ions in the tube must be removed. This can be
achieved by having an organic vapour in the tube – methylene bromide is
effective – the molecules of which absorb ions and dissociate rather than
produce fresh ions.

The time for a G–M counter to restore its initial condition after
discharging is called the 'dead time' of the counter and is typically of the
order $10^{-4}$ seconds. In practice up to about 3000 photons per second can be
counted with a G–M counter. At the higher counting rates proportionality
between photon flux and counter discharges is lost because of the increasing

Fig. 5.29.
A schematic
Geiger–Muller counter.

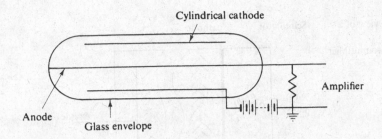

probability that photons will arrive in the dead time when they cannot be counted.

Another type of counter is the *proportional counter*. This is similar to a G–M counter but operates below the Geiger threshold. Consequently the voltage pulses produced are small ($\sim 1\,mV$) and require amplification.

An X-ray photon entering the counter produces a pulse of height proportional to the energy of the X-ray photon. This has the advantage that pulses due to unwanted photons (e.g. second-order reflections with wavelength $\lambda/2$) can be recognized and eliminated.

Proportional counters also have the advantage of a short dead time ($\sim 10^{-7}\,s$) and so give a linear response at high counting rates.

Discharge counters have largely been superseded by the scintillation counter where the radiation first falls on a fluorescent material the visible light from which is subsequently detected by a photomultiplier arrangement. In fig. 5.30 a typical set-up is illustrated. The photon of visible light first falls on a photoemissive surface which generates a photoelectron. The layers of 'venetian-blind' electrodes are each at about 100 volts higher potential than the one before. Each time an electron collides with an electrode it is energetic enough to produce several secondary electrons each of which is accelerated in the field to the next electrode and produces several more. This cascade effect can give a millionfold or more amplification factor and the final pulse of current can be amplified electronically and fed into a counting system. If a high-frequency electronic system is used the scintillation counter is capable of counting of the order of $10^7$ to $10^8$ events per second. This is so much greater than the rate at which X-ray photons would arrive in a diffracted beam that the scintillation counter gives a count proportional to the total energy of the beam – that is to say, that the probability of two X-ray photons arriving so close together that they cannot be counted separately becomes vanishingly small.

It is possible to incorporate a counter in a diffraction instrument based on any of the geometrical arrangements we have previously considered. The crystal can be placed in a diffracting position, the counter placed so as to receive the diffracted ray and the counting rate will then be a direct measure of the power of the beam. Devices incorporating counters, rather than taking a photographic record, are called diffractometers and, in § 5.2, we have already mentioned a powder diffractometer which records with a counter rather than film.

A typical fully-automatic four-circle diffractometer is shown schematically

Fig. 5.30.
A schematic view of a
scintillation
counter–photomultiplier
arrangement.

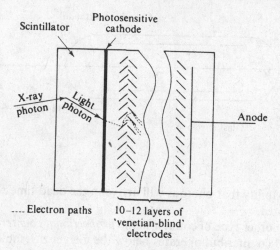

in fig. 5.31. The term 'four-circle' refers to the number of rotational motions available and in the instrument shown three of these are associated with the crystal and one with the counter. The counter rotates about a vertical axis so that the plane containing the incident and diffracted beam is always horizontal. For a particular reflection the counter is set at a position corresponding to the Bragg angle; the rotation axis corresponding to this motion is called the $2\theta$ axis. Once this has been done the diffracted beam can be produced by positioning the crystal so that the scattering vector **s** makes the correct angle with the incident and diffracted beams. Such positioning can be achieved by rotation of the crystal about the three axes shown in fig. 5.31 – the $\phi$, $\chi$ and $\Omega$ axes.

Modern single-crystal diffractometers are computer-controlled and are provided with sophisticated software to carry out many tasks. They can automatically determine the orientation of the crystal, move the crystal and detector to record each diffracted beam and also control the time each beam is measured to optimize the accuracy of the measurements. Using the stored intensity data the computer can then solve the crystal structure, usually by means of a so-called direct method as described in §8.8.

The accuracy which can be obtained with diffractometers depends on a number of factors. It is important that the X-ray tube output should be kept absolutely constant with time as the data is collected. This requires a stabilized power supply for the tube but in addition one should monitor a reference reflection from time to time to ensure that there is no variation in the intensity of the incident X-ray beam. Another possible source of error is the statistical fluctuations which are to be expected in the diffracted beam. If in a given time $t$, $N$ counts are made then the standard error in the number of counts is $\sqrt{N}$. This means that, from normal statistics, there is a 68% probability that the 'true' number of counts is in the range $N \pm \sqrt{N}$, a 95% probability that it is in the range $N \pm 2\sqrt{N}$ and a 99.7% chance of it being in the range $N \pm 3\sqrt{N}$. By the 'true' number we mean the average number of counts per unit time, as ascertained by counting for an infinite period, multiplied by $t$; we should note that this 'true' number is not available to us

Fig. 5.31.
A typical four-circle
diffractometer. The
counter rotates about
the $2\theta$ axis in one plane
and the crystal may be
orientated in any way
by the three axes of
rotation $\phi$, $\chi$ and $\Omega$.

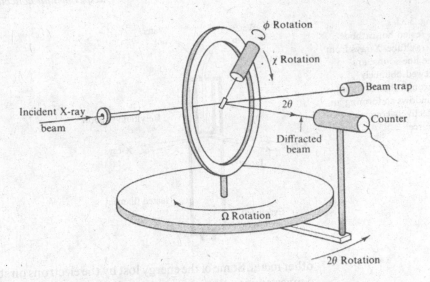

– only the number $N$. The diffracted beam power will thus be proportional to $N/t$ with a fractional standard error

$$\sigma_f = \frac{\sqrt{N}}{N} = N^{-1/2}. \tag{5.21}$$

Thus to have a standard error of 1% we must make 10000 counts, for 0.1% we need $10^6$ counts and so on. The number of counts which can be made is clearly limited by the time available to collect the data. It is also obvious that for weak reflections one must count over a considerable period of time to get reasonable accuracy. A typical count rate for a strong reflection would be about 30 000 counts per second, while a weak but observable reflection could give a count rate of 3–10 counts per second. A disturbing influence, especially for the weak reflections, is the inevitable background count; it acts as a noise in the presence of which the signal must be measured. However, whatever the difficulties with the use of diffractometers, they are potentially the most accurate and certainly a very convenient means of collecting diffraction data. For the study of the crystal structure of large biological molecules, where hundreds of thousands of items of data must be collected, diffractometers, or automatic densitometers, make possible what otherwise would border on the impossible.

A complete account of the use of diffractometers is given in *Single Crystal Diffractometry* by U. W. Arndt and B. M. T. Willis (Cambridge University Press).

## 5.8 X-ray sources

Over the years the most important source for the production of X-rays for the crystallographic community has been the sealed hot-cathode tube. In this device, illustrated in fig. 5.32, electrons from a heated filament are accelerated through a potential difference $V$ towards an anode on which there is a disk of target material – copper, molybdenum, cobalt or some

Fig. 5.32.
A sealed hot-cathode
X-ray tube. X-rays from
the line source are
viewed obliquely
through the beryllium
windows so forming an
effectively compact
source.

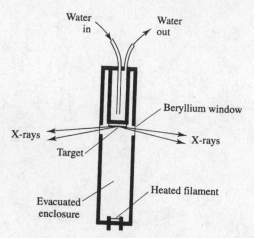

other metal. Some of the energy lost by the electrons on striking the target is
converted into electromagnetic radiation in the form of X-rays and, if the
accelerating potential $V$ is sufficiently low, then there will just be a
continuous intensity distribution of X-rays as shown in fig. 5.33(*a*). The
lower limit of X-ray wavelength $\lambda_{min}$, indicating the most energetic X-rays
produced, will have energy corresponding to all the energy of one electron,
$eV$, going into a single X-ray photon. If the accelerating potential is
increased then eventually a point will be reached where the impacting
electrons can eject inner core electrons from the target element. At this
point a different source of X-rays becomes available when an electron
moves from one level to another within a target atom, for example from an
L to a K shell. This gives rise to what is called *characteristic radiation*, the
wavelength of which will be given by

$$\lambda = \frac{hc}{E_1 - E_2} \tag{5.22}$$

where $h$ is Planck's constant, $c$ the speed of light and $E_1$ and $E_2$ are the
higher and lower energy levels of the shell electrons. The output from a
molybdenum X-ray tube operating at 35 000 V is shown in fig. 5.33(*b*). It
shows the characteristic Mo K$\alpha$ ($\lambda = 0.7097$ Å) and Mo K$\beta$ ($\lambda = 0.6323$ Å)
peaks derived from electrons moving from the L and M shells, respectively,
to the K shell. Within the L shell two slightly different energy levels occur so
that the Mo K$\alpha$ radiation consists of two close components Mo K$\alpha_1$
($\lambda = 0.70926$ Å) and Mo K$\alpha_2$ ($\lambda = 0.71354$ Å) with intensities in the ratio of
about 8:1. The target materials generally used in these sealed X-ray tubes,
with the wavelengths of their K$\alpha_1$ radiation shown in parentheses, are
Ag(0.5594 Å), Mo, Cu(1.5405 Å), Ni(1.6578 Å), Co(1.7889 Å), Fe(1.9360 Å)
and Cr(2.2896 Å). Elements of lower atomic number than chromium will
give longer-wavelength radiation that will be heavily absorbed by the exit
window of the X-ray tube while those with a larger atomic number than
silver will give a very high ratio of background continuous radiation to
characteristic radiation thus making them less convenient for crystallographic
use.

## 5.8 X-ray sources

Fig. 5.33.
(a) Continuous radiation
from a sealed X-ray tube
with a tungsten target at
an operating voltage of
30 000 V.
(b) Output from a
molybdenum tube at
35 000 V.

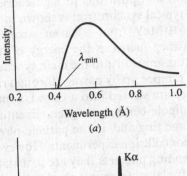

$\lambda_{min}$

Intensity

Wavelength (Å)

(a)

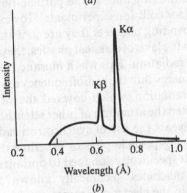

Kα

Kβ

Intensity

Wavelength (Å)

(b)

That part of the energy of the electrons striking the target which is not converted into X-rays, the great majority of it, is converted into heat, and to stop the target melting this heat is removed by water cooling, shown in fig. 5.32. The electrons fall on the target along a line, following the direction of the electron-emitting filament. This is viewed obliquely so that the effective source is compact and roughly square.

For most X-ray crystallographic purposes monochromatic, or near-monochromatic radiation is required and filters are used to eliminate as much as possible of the unwanted radiation, including the continuous background, so that only the desired characteristic radiation is transmitted. The principle by which the filters operate is explained in §6.2.

A normal Cu X-ray tube will run at 30 kV with an electron current of 15–20 mA, corresponding to about 600 W of power being dissipated at the target. This puts a very heavy demand on the water cooling system and should the cooling water be cut off for any reason then the target would quickly melt. In practice there are safety devices based on monitoring the water pressure and its flow which prevent this from happening. It is possible to produce even more powerful laboratory X-ray generators by using a rotating anode tube. In this device the target is in the form of a cylinder which spins about its axis so that the energy dissipated by the electron beam at the target is spread out over a much larger area. This makes the cooling problem soluble for currents of up to 200 mA. The target must be well shaped so that, as it spins, the appearance of the X-ray source remains constant.

Since about 1970 sources of X-radiation have become available which are many orders of magnitude more powerful than any laboratory source. These are synchrotron sources of which many now exist as national and international facilities. The first synchrotrons were built as machines for accelerating electrons and positrons to high energies for experiments in particle physics. A typical synchrotron is shown in fig. 5.34. Electrons are injected at about 10 MeV from a linear accelerator into a booster synchrotron which then increases the energy of the electrons to about 500 MeV, prior to transmission into the main synchrotron. The energies of these are gradually ramped up by passage through radiofrequency cavities to 2–4 GeV and they are steered into a closed, approximately polygonal, path by means of dipole bending magnets. Eventually the electrons are isolated in orbit in the ring and, in the particle-physics applications they could then be used for collision experiments. However, whenever electrons pass through the bending magnets they are in a state of acceleration and hence, according to the laws of classical physics, they lose energy in the form of electromagnetic radiation. This was a nuisance to the particle physicists who had to pump energy into the radiofrequency cavities to compensate for this loss but this radiation, which covered the range from infra-red to X-rays, soon attracted the attention of other scientists including crystallographers. For some years the use of synchrotron radiation was a parasitic activity accompanying particle-physics experiments but, later, new synchrotrons were built specifically designed to optimize the radiation output. These specialized machines, technically known as storage rings, are designed to maintain the electron currents over periods of typically 2–20 hours. The first storage ring, the SOR, was built in Tokyo and the first multi-user, high-energy storage ring for research purposes, the SRS, was built in the U.K. at Daresbury in 1980. This has been succeeded by many other machines, some of similar design to the Daresbury machine but with better working characteristics and others optimized either for softer (longer-wavelength) radiation or for harder radiation.

Synchrotron radiation has many interesting properties. It is strongly

Fig. 5.34.
A typical storage ring arrangement. The number of bending magnets is usually 16 or more and the diameter of the ring is of the order of 30 m for a major installation such as the SRS at Daresbury in the U.K.

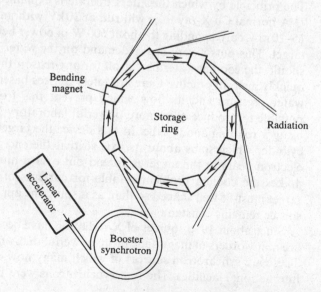

polarized, which is useful in many scientific applications and which has to be taken account of when interpreting X-ray intensities (see equation (6.18) et seq.). The electrons do not form a uniform stream but occur in bunches separated in time by $10^{-8}$–$10^{-9}$ s. The output of radiation is pulsed with a similar frequency and this gives the possibility of doing time-resolved experiments on nanosecond timescales. The bending magnet radiation is fanned out in a horizontal direction since it is emitted tangentially over the whole of its curved path through the magnet. On the other hand, the spread in the vertical direction is very small and is given in radians as

$$\delta\phi = mc^2/E \tag{5.23}$$

where $E$ is the electron-beam energy, $m$ the mass of an electron and $c$ the speed of light. For the SRS, $E = 2.0$ GeV and inserting this value in equation (5.23) we find that the vertical angular spread of the radiation is 0.25 milliradians or less than 1' of arc.

The power spectrum of radiation produced from a dipole bending magnet is always of the same shape, as shown in fig. 5.35, and is characterized by a critical photon wavelength $\lambda_c$ which divides the curve into two parts such that there is as much energy generated above $\lambda_c$ as below it. The peak of the curve is approximately at $1.4\lambda_c$, which corresponds to the greatest output of energy per unit wavelength. The critical wavelength, in ångstrom units, is given by

$$\lambda_c = 18.64/(BE^2) \tag{5.24}$$

where the electron-beam energy is in units of GeV and the dipole magnet field $B$ is in tesla. For the SRS, $E = 2.0$ GeV and $B = 1.2$ T so that $\lambda_c = 3.9$ Å. The absolute scale of the power spectrum will depend on other factors, in particular the electron current which for the SRS is 200 mA.

The output characteristics of a synchrotron source can be modified by what are called insertion devices, situated in the straight sections between the bending magnets. One such insertion device, called a *wiggler*, is illustrated in fig. 5.36(a). The beam passes through a series of dipole magnets of alternating polarity so that it undergoes small oscillations perpendicular

Fig. 5.35.
The universal power spectrum curve for synchrotron radiation.

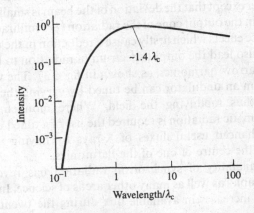

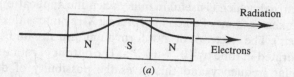

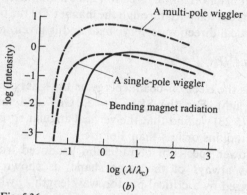

Fig. 5.36.
(*a*) Three dipole magnets giving a single-wiggle Wiggler. The magnetic fields are perpendicular to the figure with the nearer pole of each magnet indicated.
(*b*) The power spectrum in terms of $\lambda_c$, the critical wavelength of the bending magnets.

to its general direction of motion. The energy of the electron beam $E$ is unaffected but the magnets can be of the superconducting kind, capable of giving several times the field of the normal bending magnets and so decreasing the critical wavelength, as seen from equation (5.24). If there is only a single wiggle in the motion of the electrons then there is just a wavelength shift of the output curve. On the other hand, if there are several wiggles then each of them produces its own radiation and the total output is the sum of that from all of them. The output of a single-pole wiggler and a multi-pole wiggler, compared with that from the bending magnets, is shown in fig. 5.36(*b*).

An even more important insertion device is the *undulator* which is similar to a wiggler except that the deviation of the beam is smaller and constrained to be within the output cone of the radiation ($\sim 1$ milliradian). This leads to interference effects which firstly cause a reduction in the angle of the output cone but also lead the output spectrum of radiation to be in the form of a series of narrow harmonics, as shown in fig. 5.37. The wavelengths of the output from an undulator can be tuned by varying the gaps between the magnets, thus modifying the field. Where monochromatic, or near-monochromatic radiation is required the use of a tuned undulator can give greatly enhanced useful fluxes of X-rays by having the wavelength of interest at the centre of one of the harmonics.

The availability of synchrotron radiation has revolutionized X-ray crystallography as well as many other areas of science. In fig. 5.38 there are shown the increases in available flux during the twentieth century. The

Fig. 5.37.
The form of output from an undulator with the energy concentrated in harmonics.

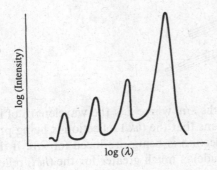

Fig. 5.38.
The increase in available brilliance since 1900. Brilliance is a measure of the intensity of well-collimated monochromatic radiation which can be delivered on to a specimen.

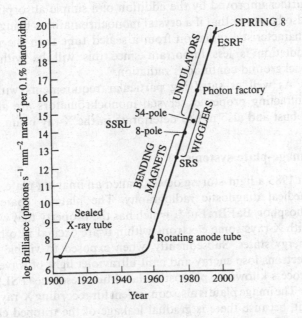

various machines indicated are the SRS at Daresbury, U.K., SSRL at Stanford in the U.S.A., Photon Factory at Tsukuba in Japan, ESRF at Grenoble in France and SPRING 8, an 8 GeV machine at Hirama Science Garden City in Japan.

For sealed X-ray tubes, filtering by metallic foils can produce a beam in which most of the radiation is within a characteristic line of the target. For synchrotron radiation where, if monochromatic radiation is required, the intensity varies smoothly around the wavelength of interest it is more common to use a crystal monochromator. This involves the use of a single crystal which has a very strongly reflecting set of Bragg planes so that for radiation of the required wavelength $\lambda$ the beam is directed at an angle $\frac{1}{2}\pi - \theta$ to the normal to the planes where $\theta$ is given by equation (3.37). Even the best monochromator will give a finite bandwidth of radiation in its output but if a pair of monochromating crystals is used as shown in fig. 5.39 then not only is the emergent beam parallel to the incident one but the quality of the monochromatic radiation is much improved. A difficulty with crystal monochromators is that they also

Fig. 5.39.
The action of a
two-crystal
monochromator.

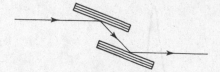

reflect wavelengths $\lambda/n$, where $\lambda$ is the wavelength of interest and $n$ is an integer. This means that the $(hkl)$ reflection is being produced for $\lambda$ while the $(nh\,nk\,nl)$ reflection is being produced for $\lambda/n$. If the intensity (or the structure amplitude) is much greater for the $(hkl)$ reflection than for any multiple of it then this may not be too serious and the situation can be further improved by the addition of a simple absorption filter. It should also be noted that if a crystal monochromator is being used to isolate the characteristic radiation from a sealed tube then the removal of the $\lambda/n$ radiation is less important since this will be within the lower-level background continuous radiation.

As well as satisfying particular requirements with respect to their diffracting properties, crystal monochromators are also required to be robust and also not to deteriorate in the X-ray beam.

## 5.9   Image-plate systems

In 1983 a light-storing device, called an imaging plate, was developed for medical diagnostic radiography. The plate is coated with a storage phosphor, $BaFBr:Eu^{2+}$, which has the property that when it is irradiated with X-rays some electrons within it are excited into higher quasi-stable energy states. Subsequently, when exposed to visible light, the trapped electrons lose energy and emit ultraviolet light of wavelength 390 nm – a process known as photostimulated luminescence (PSL).

The image plate is also convenient for recording X-ray diffraction images but, because there is gradual leakage of the trapped electrons, it is only suitable for experiments where all the data can be collected and the pattern read out within a few hours – a day at most – as otherwise the image will be greatly weakened. This gives the requirement for a powerful X-ray source, either of the rotating-anode variety or, better still, a synchrotron source; the latter will be able to record a pattern even from a weakly scattering biological crystal within the required timescale. It is in the solution of biological structures that the image-plate system has most to offer.

A common experimental arrangement is for the image plate to be the recording part of a precession camera. The BaFBr phosphor, in a layer about 150 μm thick, is a very efficient absorber of X-rays, as shown in fig. 5.40. The plate is organized in pixels, $100 \times 100$ μm in dimension, which is about the limiting resolution of the system. After the image is stored on the plate it is scanned by a laser beam, usually a He–Ne source, at a rate of 10 μs per pixel, the release time of PSL being about 0.8 μs from the time of light stimulation. The wavelength of the PSL radiation is sufficiently different from that of the stimulating light that the latter can easily be removed by a filter. The PSL from each illuminated pixel is guided by fibre optics to a photomultiplier detector, the signal from which goes into

Fig. 5.40.
The absorption of the
imaging phosphor
BaFBr:Eu²⁺ as a
function of wavelength.
The absorption edge at
0.330 Å is due to
barium.

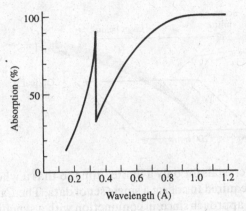

a digital storage and image-processing system. A representation of such a system is shown in fig. 5.41. The intensity of the PSL is linearly related to the X-ray intensity over a wide range of about 1:10⁵. There is a little background noise in the system, due to the photomultiplier tube for example, but it is no worse than for film. Again, there are usually some variations of sensitivity in different parts of the image plate but these are normally less than 1.6%.

The sensitivity, dynamic range and precision of image-plate systems give them a growing role in X-ray crystallography, particularly in the analysis of biological crystals.

## 5.10   The modern Laue method

The very first X-ray photograph, produced by Friedrich and Knipping in 1912, was taken with a stationary crystal and, effectively, a continuous source of X-rays. The picture was crude but was good enough to verify von Laue's idea that crystals could diffract X-rays. From the Bragg point of view in the Laue experimental arrangement the reflecting planes corresponding to the (*hkl*) reflection are fixed in direction, as is the normal to those planes. The angle between the direction of the incident radiation and the normal is also fixed but if that angle is $\frac{1}{2}\pi - \theta$ for some wavelength present in the incident beam then radiation of that wavelength will form a reflected (diffracted) beam.

In general, Laue photographs are very difficult to decipher and the only simple information they give is when the incident beam is aimed along a symmetry axis, when the diffraction pattern will clearly show a corresponding symmetry (see § 7.3). A characteristic of the Laue method is that all the diffracted beams are produced at the same time and this can have important advantages, particularly for protein crystallography. One of the great problems with biological crystals is that they are degraded by X-rays so that if the same crystal is used for a long time the quality of the data steadily gets worse. The degradation of the crystal is not just due to the total X-ray dose but also involves a time element. The X-rays produce free radicals within the crystal by breaking off extremities of the protein molecules and it is the diffusion of these through the crystal that seems to do the damage. Traditionally this problem of crystal decay has been met by using

Fig. 5.41.
The scanning mechanism for an image-plate system. At the end of each laser-beam scan the plate is moved in the direction shown, by the dimension of one pixel.

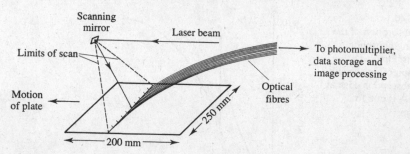

individual crystals for only a limited time so that a whole batch of crystals would be required to collect a single set of data. The Laue method offers an alternative approach since, in conjunction with a synchrotron source, it can give a data set in a fraction of a second. Another possible application is if an experiment is carried out on the crystal, say exposing it to a laser beam, causing it to undergo some short-lived structural change. This change may then be investigated by recording its Laue diffraction pattern while it is in the transient state.

A Laue diffraction photograph taken with synchrotron radiation for a crystal of the protein concanavalin A is shown in fig. 5.42. While it is quite a complicated pattern it can be seen that the reflections all lie on families of curves. Knowing the orientation of the crystal it is possible to calculate the position of each diffraction spot and hence to index every spot on the film. There are, nevertheless, some problems to be solved. The first, and the simplest, is that since each spot corresponds to a different wavelength of X-radiation then it is necessary to interpret the intensity of a spot relative to the power spectrum of the source, which is well known for synchrotron radiation and possibly also the quantum efficiency of the film for different X-ray wavelengths.

The second problem is a little more difficult and can be understood in terms of Bragg's law, equation (3.37). This can be written as

$$\sin \theta = \frac{2d_{hkl}}{\lambda} \tag{5.25}$$

where $d_{hkl}$ is the spacing of the reflecting planes for the Miller indices $(hkl)$. If we now consider the reflection with Miller indices $(nh\,nk\,nl)$ then

$$d_{nh,nk,nl} = d_{hkl}/n \tag{5.26}$$

and we may write

$$\sin \theta = \frac{2d_{nh,nk,nl}}{\lambda/n}. \tag{5.27}$$

Fig. 5.42.
Laue photograph of the
protein concanavalin A
taken with Daresbury
SRS wiggler radiation
with exposure time of
19 ms. Under the
conditions of the
experiment $\lambda_{min} = 0.5$ Å
and $\lambda_{max} = 2.0$ Å.
(Courtesy of J. R.
Helliwell.)

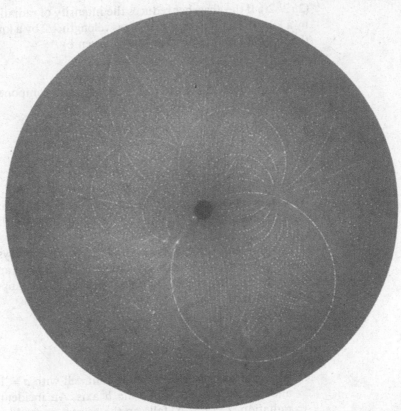

This means that in the Laue photograph all the spots for reflections
($nh\,nk\,nl$) will overlap for all $n$.

The problem is not quite as bad as it might seem at first sight. The X-rays
falling on the crystal have an upper and lower wavelength limit – 2.0 Å and
0.5 Å, respectively, for conconavalin A so that $n$ cannot be greater than 4 for
any reflection. Again if a reflection such as the (17 38 2) is recorded then no
reflection with smaller indices can be coincident with it, because 17 is a
prime number, and no reflection with higher indices will be present if the
maximum possible $k$ index is less than 76. In general, it has been shown by
Cruickshank and Helliwell that for most realistic experimental arrangements
more than 85% of reflections on Laue pictures should be single spots
without overlap and 95% should be either without overlap or singly
overlapped. Even when there is overlap the problem of separating the
intensities due to the components is possible in principle. This is achieved
by having a multiple film pack separated by layers of absorbing material.
The absorption of X-rays in materials is wavelength dependent (§ 6.2). Let
us assume that two diffraction spots overlap corresponding to reflection
($hkl$) with wavelength $\lambda$ and reflection ($2h\,2k\,2l$) with wavelength $\lambda/2$ and the
combined intensity on the top film is

$$I_T = I_1 + I_2 \tag{5.28}$$

where $I_1$ is the intensity due to reflection ($hkl$) and $I_2$ is that due to reflection

($2h\,2k\,2l$). If the absorber reduces the intensity of radiation of wavelength $\lambda$ by a known factor $a_1$ and that of wavelength $\lambda/2$ by a known factor $a_2$ then the intensity on the bottom film is given by

$$I_B = a_1 I_1 + a_2 I_2. \tag{5.29}$$

From equations (5.28) and (5.29) the separate components in the top film may be found as

$$I_1 = (a_2 I_T - I_B)/(a_2 - a_1)$$

and

$$I_2 = (a_1 I_T - I_B)/(a_1 - a_2). \tag{5.30}$$

With three films separated by two layers of absorbing material it should be possible similarly to unravel the intensities of three overlapped spots on the top film.

The problems of the Laue method are such that crystallographers will usually prefer to use methods involving moving crystals and monochromatic radiation but for special applications the Laue approach is a useful additional technique.

## Problems to Chapter 5

5.1 A crystal has an orthorhombic unit cell with $a = 11.4$ and $c = 8.9\,\text{Å}$ and can be rotated about the $b$ axis. An incident beam of Cu K$\alpha$ radiation ($\lambda = 1.54\,\text{Å}$) falls on the crystal normal to the rotation axis. The crystal is in the reflecting position for the (907) reflection. By what angle must it be rotated for the (702) reflection to be produced?

5.2 (a) An oscillation photograph is taken with Cu K$\alpha$ radiation the rotation being about an axis of length 7.2 Å. If the camera diameter is 57.3 mm find the distances of the layer lines $h = 1, 2, 3$ from the zero-layer line.

(b) An oscillation photograph is taken with Cu K$\alpha$ radiation and gives layer lines whose distances from the zero-layer line are as follows:

$$
\begin{aligned}
h &= 1 \qquad 4.5\,\text{mm},\\
h &= 2 \qquad 9.3\,\text{mm},\\
h &= 3 \qquad 14.9\,\text{mm}.
\end{aligned}
$$

The camera diameter is 57.3 mm. What is the length of the $a$ axis?

5.3 For a certain crystal $a^* = 0.15$, $b^* = 0.20\,\text{Å}^{-1}$ and $\gamma^* = 80°$. When a 15° oscillation photograph is taken with Cu K$\alpha$ radiation about the $c$ axis, reflections are produced in the zero-layer line making the following angles with the incident beam. To right looking in direction of beam – 13°, 41°, 61°, 75°, 85°, 102°, 125°, 152°. To left looking in direction of beam – 13°, 48°, 62°, 78°, 90°, 114°, 116°, 130°, 152°.

What are the indices of these reflections? Which reflections in the observable range are too weak to be observed?

5.4 A normal-beam zero-layer Weissenberg photograph is taken with Cu K$\alpha$ radiation giving the ($hk0$) reflections for an orthorhombic crystal for which $a = 8.0$ and $b = 6.0$Å. When the crystal is at the end of its traverse the X-ray beam points straight down the $a$ axis. Find the coordinates on the film of all possible reflections of the type $1k0$ and $\bar{1}k0$. Plot these on a replica of fig. 5.19($b$) and confirm that they follow the form of the curves shown in that figure.

Assume that the oscillation is through 180°, the diameter of the camera is 57.3 mm and the translation constant of the film is 0.5 mm per degree of rotation.

5.5 The energy of the electrons in a synchroton is 3.0 GeV and the bending magnet fields are 1.1 T. What is the critical wavelength and the vertical divergence of the emitted radiation?

It is required to reduce the critical wavelength to 0.5 Å with a wiggler. What field must be produced by the wiggler magnets?

5.6 A Laue photograph, which is known to have overlapping (231) and (462) reflections is recorded on two films separated by an absorbing layer. On the top film the sum of the intensities of the overlapped spots is 262 and on the lower film 106. The absorber allows through 30% of the shorter-wavelength radiation and 60% of the longer-wavelength radiation. What are the individual intensities of the two spots on the top film?

# 6 The factors affecting X-ray intensities

## 6.1 Diffraction from a rotating crystal

It has been seen that methods of recording X-ray intensities usually involve a crystal rotating in the incident X-ray beam. We shall now look at the problem of determining the total energy in a particular diffracted beam produced during one pass of the crystal through a diffracting position. In order to do this we must make some assumptions about the geometry of the diffraction process; the configuration we shall take is that the crystal is rotating about some axis with a constant angular velocity $\omega$ and that the incident and diffracted beams are both perpendicular to the axis of rotation.

Let us first look at the situation when we have a stationary crystal in a diffracting position. Associated with the crystal, and fixed relative to it, there is a reciprocal space within which is defined the Fourier transform, $F_x(\mathbf{s})$, of the electron density of the crystal. For a theoretically perfect crystal of infinite extent the value of $F_x(\mathbf{s})$ would be zero everywhere except at the nodes of a $\delta$-function reciprocal lattice, the weight associated with the point $(hkl)$ being $(1/V)F_{hkl}$. However, if the crystal is imperfect in some way there may be non-zero $F_x(\mathbf{s})$ well away from the reciprocal-lattice points and for a finite crystal there will be a small region of appreciable $F_x(\mathbf{s})$ around each of the reciprocal-lattice points. The imperfect-crystal case we shall not consider here but we shall be concerned with the size of the crystal, for this is a factor which must be present in every diffraction experiment.

Consider a crystal completely bathed in an incident beam of intensity $I_0$. The crystal will reflect over a small range of angles around the 'ideal' position for a reflection $(hkl)$; in a particular position corresponding to an angle $\theta$ let the power of the reflected beam be $(\mathrm{d}E/\mathrm{d}t)_\theta$. This power is clearly proportional to $I_0$ and one can define the *reflecting power* by

$$P(\theta) = \frac{(\mathrm{d}E_{hkl}/\mathrm{d}t)_\theta}{I_0} \tag{6.1}$$

where $P(\theta)$ is a quantity with the dimensions of area.

In fig. 6.1 there is shown a typical curve relating $P(\theta)$ to $\theta$. Because of various factors such as divergence of the incident beam and crystal imperfection the detailed shape of this curve is less important than the total area under the curve. This area $\rho_{hkl}$ is called the *integrated reflection*.

Thus when the crystal is rotated with an angular velocity $\omega$ ( $= \mathrm{d}\theta/\mathrm{d}t$)

$$\rho_{hkl} = \int P(\theta)\mathrm{d}\theta = \frac{1}{I_0}\int \left(\frac{\mathrm{d}E_{hkl}}{\mathrm{d}t}\right)_\theta \mathrm{d}\theta$$

Fig. 6.1.
A typical curve for
reflecting power as a
function of $\theta$.

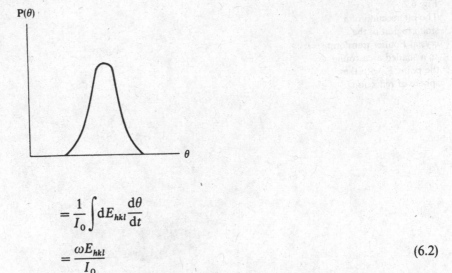

$$= \frac{1}{I_0} \int dE_{hkl} \frac{d\theta}{dt}$$

$$= \frac{\omega E_{hkl}}{I_0} \tag{6.2}$$

where $E_{hkl}$ is the total energy in the diffracted beam when the crystal sweeps through the diffracting position. We shall now examine the problem of finding $E_{hkl}$.

We saw in § 3.1 that a finite one-dimensional array of scatterers gives some scattering away from the directions of the main peaks and the same is true for a finite three-dimensional lattice. Let us assume that we have, say, an $N_a \times N_b \times N_c$ three-dimensional array of point electrons with the array defined by the vectors $\mathbf{a}$, $\mathbf{b}$ and $\mathbf{c}$. In a direction corresponding to a scattering vector $\mathbf{s}$ the intensity of scattering can be found by a simple extension of the results in § 3.1 and by applying equation (2.22) as

$$[\mathscr{I}_s]_{\text{electrons}} = I_0 \left( \frac{e^2}{4\pi\varepsilon_0 c^2 m} \right)^2 \frac{1 + \cos^2 2\theta}{2} \frac{\sin^2(\pi N_a \mathbf{a} \cdot \mathbf{s})}{\sin^2(\pi \mathbf{a} \cdot \mathbf{s})} \frac{\sin^2(\pi N_b \mathbf{b} \cdot \mathbf{s})}{\sin^2(\pi \mathbf{b} \cdot \mathbf{s})}$$

$$\times \frac{\sin^2(\pi N_c \mathbf{c} \cdot \mathbf{s})}{\sin^2(\pi \mathbf{c} \cdot \mathbf{s})}. \tag{6.3}$$

The component of equation (6.3)

$$\frac{\sin^2(\pi N_a \mathbf{a} \cdot \mathbf{s})}{\sin^2(\pi \mathbf{a} \cdot \mathbf{s})} \frac{\sin^2(\pi N_b \mathbf{b} \cdot \mathbf{s})}{\sin^2(\pi \mathbf{b} \cdot \mathbf{s})} \frac{\sin^2(\pi N_c \mathbf{c} \cdot \mathbf{s})}{\sin^2(\pi \mathbf{c} \cdot \mathbf{s})}$$

is a 'shape factor' associated with the distribution of intensity around each reciprocal-lattice point. As in the one-dimensional case, illustrated by fig. 3.2, this factor will give the same shape of distribution about each reciprocal-lattice point.

Clearly as $s \to 0$ so the height associated with this peak tends to $N_a^2 N_b^2 N_c^2$ while the width in each of the primary directions, i.e. the distance at which the factors first fall off to zero, is of order $(N_a a)^{-1}$, $(N_b b)^{-1}$ and $(N_c c)^{-1}$, respectively. Hence the total weight associated with the 'shape factor' is of order $(N_a N_b N_c)/(abc) \simeq N/V$ where $N$ is the total number of unit cells in the crystal and $V$ the volume of the unit cell. This rough assessment will be confirmed later.

Fig. 6.2.
The intersection of a
small region of the
crystal Fourier transform
as a shaded area round
the point P, with the
sphere of reflection.

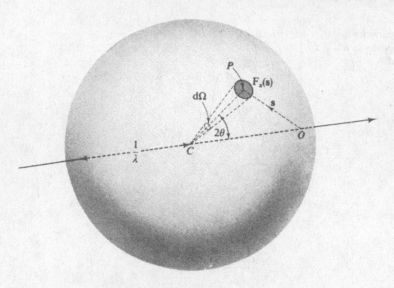

Let us now examine this question in more detail. In fig. 6.2 the crystal may be considered at $C$, the centre of the sphere of reflection, while the origin of $F_x(s)$ is situated at $O$. In general the value of $F_x(s)$ over the surface of the sphere will be small unless a reciprocal-lattice point lies on or very near the surface which is the condition we have previously noted for a diffracted beam to occur. We may consider that the incident X-ray beam is scattered in *all* directions; if at the point $P$ in fig. 6.2 the value of the transform is $F_x(s)$ then this means that in the direction $CP$ the intensity of scattering, $\mathscr{I}_s$, is $|F_x(s)|^2$ times that which would result from a point electron at $C$. Hence from equation (2.22) we may write

$$\mathscr{I}_s = I_0 \left( \frac{e^2}{4\pi\varepsilon_0 c^2 m} \right)^2 \frac{1 + \cos^2 2\theta}{2} |F_x(s)|^2 \tag{6.4}$$

and in a small solid angle $d\Omega$ in the direction $CP$ the total scattered energy per unit time is $\mathscr{I}_s d\Omega$.

The situation for the particular diffraction geometry we are considering when the crystal is passing through a diffracting position is shown in fig. 6.3($a$). The reciprocal-lattice point $P$, with indices ($hkl$), has associated with it a small volume, represented by the ellipsoidal surface, within which resides all the intensity associated with this particular reflection. Of course no such sharp boundary really exists; surfaces of constant intensity need not be ellipsoidal and what is shown is purely illustrative. The volume is shown just entering the sphere of reflection and its surface of intersection with the sphere is that shown shaded which has total area $\alpha$. Within this surface is shown a very small area $d\alpha$; this subtends an angle $\lambda^2 d\alpha$ at $C$ and the total energy scattered through it in time $dt$ is

$$dE_{hkl} = \lambda^2 \mathscr{I}_s d\alpha dt \tag{6.5}$$

and hence the total energy scattered during the total time $T$ of the passage of

Fig. 6.3.
(*a*) A schematic view of
the passage of the
region round a
reciprocal-lattice point
through the sphere of
reflection as the crystal
rotates.
(*b*) The small shaded
volume between $Q$ and
$Q'$ passes through the
area $\alpha$ on the sphere of
reflection as the crystal
rotates through an
angle $\omega\,dt$.

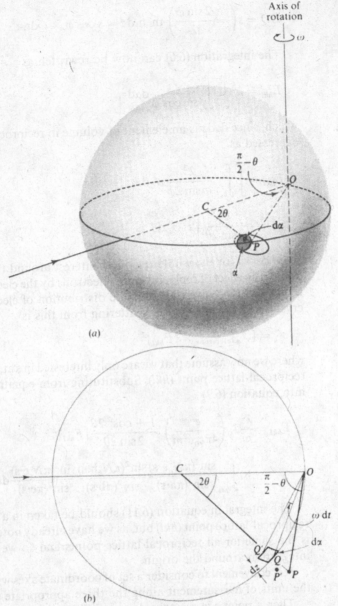

the small volume through the sphere of reflection is

$$E_{hkl} = \lambda^2 \int_T \int_\alpha \mathscr{I}_s \, d\alpha \, dt. \tag{6.6}$$

It should be noted that as time changes so different cross-sections of the volume pass through the sphere. In fig. 6.3(*b*) the plane $COP$ is shown and the area $d\alpha$ at $Q$ on the surface of the sphere. The crystal, and its associated reciprocal space, rotates by an angle $\omega\,dt$ in a time $dt$ and the transform surface $d\alpha$ moves to $Q'$ while $P$ moves to $P'$. It can be seen that a volume $d\alpha\,dz$ of the transform moves through the small element of surface of the sphere. From the diagram it is easily seen that if

$$OQ = s\left( = \frac{2\sin\phi}{\lambda}\right) \text{ then } dz = s\cos\phi \times \omega dt. \tag{6.7}$$

The integration (6.6) can now be rewritten as

$$E_{hkl} = \lambda^2 \int_z \int_\alpha \frac{\mathscr{I}_s}{s\omega\cos\phi} d\alpha dz, \tag{6.8}$$

which, since $d\alpha dz$ is an element of volume in reciprocal space, can also be expressed as

$$E_{hkl} = \int \int_{v_{hkl}} \frac{\lambda^3}{\omega\sin 2\phi}\mathscr{I}_s dv$$

$$\simeq \frac{\lambda^3}{\omega\sin 2\theta}\int_{v_{hkl}}\mathscr{I}_s dv \tag{6.9}$$

since $\phi \simeq \theta$ for the whole region of interest around the reciprocal-lattice point. The effect of replacing point electrons by the electron distribution in a unit cell is to give the complete distribution of electron density in the crystal and the intensity of scattering from this is

$$\mathscr{I}_s = [\mathscr{I}_s]_{electrons} \times |F_{hkl}|^2 \tag{6.10}$$

where we now assume that we are only interested in **s** at, or very close to, the reciprocal-lattice point ($hkl$). Substituting from equations (6.3) and (6.10) into equation (6.9)

$$E_{hkl} = \frac{\lambda^3}{\omega}I_0\left(\frac{e^2}{4\pi\varepsilon_0 c^2 m}\right)^2 \frac{1+\cos^2 2\theta}{2\sin 2\theta}|F_{hkl}|^2$$

$$\times \int_{v_{hkl}} \frac{\sin^2(\pi N_a \mathbf{a}\cdot\mathbf{s})}{\sin^2(\pi\mathbf{a}\cdot\mathbf{s})}\frac{\sin^2(\pi N_b \mathbf{b}\cdot\mathbf{s})}{\sin^2(\pi\mathbf{b}\cdot\mathbf{s})}\frac{\sin^2(\pi N_c \mathbf{c}\cdot\mathbf{s})}{\sin^2(\pi\mathbf{c}\cdot\mathbf{s})}dv. \tag{6.11}$$

The integral in equation (6.11) should be taken in a region around the reciprocal-lattice point ($hkl$) but, as we have already noted, the shape factor is the same for all reciprocal-lattice points and so we can carry out the integration around the origin.

It is convenient to consider a set of coordinate axes with **a***, **b*** and **c*** as the units of measurement along the three appropriate axial directions. Then a point **s** in reciprocal space will have coordinates ($\eta, \xi, \zeta$) so that

$$\mathbf{s} = \eta\mathbf{a}^* + \xi\mathbf{b}^* + \zeta\mathbf{c}^*. \tag{6.12a}$$

This leads to

$$\mathbf{a}\cdot\mathbf{s} = \eta, \mathbf{b}\cdot\mathbf{s} = \xi, \mathbf{c}\cdot\mathbf{s} = \zeta \tag{6.12b}$$

and a volume element is given by

$$dv = V^* d\eta d\xi d\zeta \tag{6.12c}$$

where $V^*$ is the volume of the reciprocal unit cell.
We may now rewrite equation (6.11) as

$$E_{hkl} = \frac{\lambda^3}{\omega} I_0 \left(\frac{e^2}{4\pi\varepsilon_0 c^2 m}\right)^2 \frac{1 + \cos^2 2\theta}{2\sin 2\theta} |F_{hkl}|^2 V^* \int_{\eta_2}^{\eta_1} \frac{\sin^2(\pi N_a \eta)}{\sin^2(\pi\eta)} d\eta$$

$$\times \int_{\xi_2}^{\xi_1} \frac{\sin^2(\pi N_b \xi)}{\sin^2(\pi\xi)} d\xi \times \int_{\zeta_2}^{\zeta_1} \frac{\sin^2(\pi N_c \zeta)}{\sin^2(\pi\zeta)} d\zeta \qquad (6.13)$$

where the limits of integration include the area of interest around the reciprocal-lattice point.

The individual integrals may be simplified in two stages. Firstly we know from § 3.1 that for large $N_a$, for example, we are only interested in the region in which $\pi\eta$ is small. Hence we may write

$$\int_{\eta_2}^{\eta_1} \frac{\sin^2(\pi N_a \eta)}{\sin^2(\pi\eta)} d\eta \simeq \int_{\eta_2}^{\eta_1} \frac{\sin^2(\pi N_a \eta)}{\pi^2\eta^2} d\eta. \qquad (6.14)$$

This new integral is non-periodic and the integrand falls off so rapidly with $\eta$ that the limits can be replaced by $\infty$ and $-\infty$ to give

$$\int_{\eta_2}^{\eta_1} \frac{\sin^2(\pi N_a \eta)}{\sin^2(\pi\eta)} d\eta \simeq \int_{-\infty}^{\infty} \frac{\sin^2(\pi N_a \eta)}{\pi^2\eta^2} d\eta. \qquad (6.15)$$

The definite integral on the right-hand side of equation (6.15) is well known in diffraction theory and it equals $N_a$. From equation (6.13) we now have

$$E_{hkl} = \frac{\lambda^3}{\omega} I_0 \left(\frac{e^2}{4\pi\varepsilon_0 c^2 m}\right)^2 \frac{1 + \cos^2 2\theta}{2\sin 2\theta} V^* N_a N_b N_c |F_{hkl}|^2. \qquad (6.16)$$

Since, from equation (3.26), $V^* = 1/V$, where $V$ is the volume of the unit cell, we see that the integrals have yielded a contribution $N/V$ where $N$ is the total number of unit cells in the crystal. This was the result earlier found by order-of-magnitude considerations.

The volume of the crystal is given by

$$V_x = N_a N_b N_c V \qquad (6.17)$$

so that the total energy in the diffracted beam is given by

$$E_{hkl} = \frac{\lambda^3}{\omega} I_0 \left(\frac{e^2}{4\pi\varepsilon_0 c^2 m}\right)^2 \frac{1 + \cos^2 2\theta}{2\sin 2\theta} \frac{V_x}{V^2} |F_{hkl}|^2. \qquad (6.18)$$

One can see in equation (6.18) the effect of various factors in the diffraction process. The term $\frac{1}{2}(1 + \cos^2 2\theta)$ is called the *polarization factor* and it should be noted that it depends on having an unpolarized incident X-ray beam. The other trigonometric factor $1/\sin 2\theta$ is called the *Lorentz factor* and depends on the diffraction geometry. The Lorentz and polarization factors, combined and referred to as the *Lp* factor, are tabulated for various diffraction geometries in Vol. B of the *International Tables for Crystallography*. For the geometry which led to equation (6.18), for example, if each of the intensities is multiplied by $2\sin 2\theta/(1 + \cos^2 2\theta)$ the effect of the polarization term and the diffraction geometry will be removed. Once the *Lp* factor has been applied to the observed intensities one is left with intensities proportional to $|F_{hkl}|^2$ (ignoring other corrections) and, as we shall see in chapter 8, these intensities are required if one is to solve the crystal structure.

It is instructive to look at the effect of the crystal size on the energy in the diffracted beam. This is seen in equation (6.18) to be proportional to the volume of the crystal; it must be pointed out that although equation (6.18) was derived for a regular parallelepiped-shaped crystal it is applicable to any shaped crystal. If the volume of the crystal is doubled then, from the results in §3.1, it appears that the intensity in the direction of the diffracted beam will be increased fourfold. However, with the larger crystal the distribution of intensity around the reciprocal-lattice point is more condensed and effectively occupies one-half the previous volume. This means that the overall increase in total energy is just doubled, i.e. is proportional to the volume of the crystal used.

Let us find the order of magnitude of the total energy in a diffracted beam for a typical situation. A typical copper X-ray tube of 1 kW rating probably delivers about 1 watt of characteristic Kα X-rays which is emitted over a solid angle of $2\pi$ steradians (a hemisphere). The crystal will be about 100 mm from the target and if we assume that the radiation is emitted uniformly and that the collimator allows the crystal to be completely exposed to X-rays, then the intensity of the incident X-ray beam is

$$I_0 = \frac{50}{\pi} \, \text{W m}^{-2}$$

$$\simeq 10 \, \text{W m}^{-2}.$$

We now take

$2\theta = \pi/2$
$\lambda = 1.54 \, \text{Å}$
$V_x = 10^{-12} \, \text{m}^{-3}$
$V = 500 \, \text{Å}^3$
$\omega = 0.05 \, \text{radians s}^{-1}$
$|F_{hkl}| = 30$

and we find

$$E_{hkl} \simeq 10^{-11} \, \text{J}.$$

This is equivalent to a number of X-ray photons

$$n = \frac{E_{hkl}\lambda}{hc}.$$

From the values of $h$, Planck's constant, $= 6.63 \times 10^{-34} \, \text{J s}$ and $c$, the velocity of light, $= 3 \times 10^8 \, \text{m s}^{-1}$ we find for this example $n \simeq 10^4$.

## 6.2  Absorption of X-rays

When X-rays pass through a material their intensity is attenuated by absorption, that is the conversion of the energy of the electromagnetic radiation to thermal energy, and also by scattering. In fig. 6.4 there is depicted an X-ray beam of intensity $I$ having its intensity changed to $I + dI$

Fig. 6.4.
The attenuation of an
X-ray beam after
passing through a small
distance in a material.

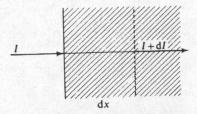

by passage through a thickness $dx$ of material. The law which will always be
found to hold is

$$\frac{dI}{I} = -\mu dx \tag{6.19}$$

or, integrating and putting $I = I_0$ when $x = 0$,

$$I = I_0 e^{-\mu x}. \tag{6.20}$$

The quantity $\mu$ is known as the *linear absorption coefficient* of the material
and it has the dimension of inverse distance. The quantity $\exp(-\mu x)$ is
known as the *attenuation factor*.

If we examine fig. 6.5 the effect of absorption on the intensity of the X-ray
reflection can be seen. An incident beam entering a crystal would, but for
the effect of absorption, have the same intensity at all points of the crystal –
for example at the points $P$ and $Q$. In addition, without absorption, small
equal volumes in the vicinity of $P$ and $Q$ would give equivalent contributions
to the emergent diffracted beam. When absorption is taken into account it
can be seen that the incident beam at $P$ will have a lesser intensity than at $Q$
because of the longer path to $P$ in the crystal and in fact the intensity at $P$
will be $I_0 e^{-\mu x_P}$ while that at $Q$ is $I_0 e^{-\mu x_Q}$. Similarly the diffracted beam from
$P$ will be more attenuated in going from $P$ to $P'$ than does that going from $Q$
to $Q'$, the attenuation factors being $e^{-\mu x'_P}$ and $e^{-\mu x'_Q}$, respectively. The
diffracted beam from $P'$ is reduced in intensity by absorption by a factor
$e^{-\mu(x_P + x'_P)}$ and the position is similar for all other points in the crystal.

If the crystal was an ideal, perfect crystal we should have to take account
of the coherence of the radiation scattered from different parts of the lattice.

Fig. 6.5.
The variation of the path
lengths of the incident
and diffracted beams for
diffraction from different
regions of the crystal.

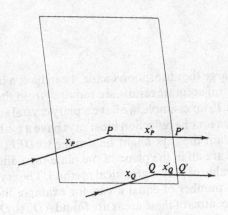

Fig. 6.6.
A schematic diagram of
a mosaic crystal.

This would mean a modification of the theory from which equation (6.18) was derived since, in this theory, absorption was ignored. However, as will be seen later, for such a perfect crystal the simple theory is, in any case, inadequate in a more fundamental way. In practice perfect crystals are rare and usually we have what approximate to ideal *imperfect* crystals of the type illustrated schematically in fig. 6.6. The crystal consists of a mosaic of small blocks, each block being perfect but separated by faults and cracks in the crystal from other blocks. The small blocks are usually of the order of a few microns in dimensions, the gaps between them are non-parallel and of random thickness so clearly there can be no systematic relationship between the phases of radiation scattered from different parts of the crystal which are in different blocks. Thus to get the total intensity of scattering from the complete crystal we add the *intensities* of the radiation scattered from different regions of the crystal. For the crystal shown in fig. 6.7, assumed ideally imperfect, the small volume d$v$ will contribute to the diffracted beam an intensity proportional to $e^{-\mu(x+x')}$d$v$. The fraction by which the total intensity is reduced by absorption is given by

$$A = \int_{V_x} e^{-\mu(x+x')} \mathrm{d}v \bigg/ \int_{V_x} \mathrm{d}v = \frac{1}{V_x} \int_{V_x} e^{-\mu(x+x')} \mathrm{d}v. \tag{6.21}$$

Fig. 6.7.
The contribution to the
diffracted beam of small
volume d$v$ is reduced by
absorption by a factor
$\exp\{-\mu(x+x')\}$.

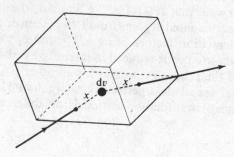

When $\mu$ is large the absorption factor $A$ can have a large effect and must be corrected for if accurate results are required from the eventual structure determination. If, for example, we have a platy crystal similar to that shown in fig. 6.8(*a*) then for one reflection most rays have a path like $ABC$ while for another reflection the rays might be similar to $DEF$. If the incident and diffracted rays are all in the plane of the plate then one can determine the approximate value of $A$ by a graphical method. The crystal cross-section is divided into a number of equal areas, for example the ten shown in fig. 6.8(*b*), and the centres of these areas are found – $O_1$ to $O_{10}$ in the figure. We

Fig. 6.8.
(*a*) For a crystal of very unequal dimensions absorption corrections can vary greatly from reflection to reflection.
(*b*) Approximate graphical method of calculating absorption factors by dividing crystal into small regions.

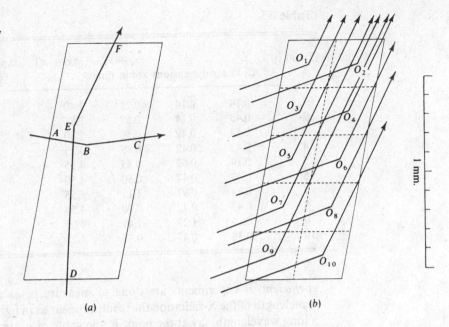

(*a*)                    (*b*)

now approximate the integral (6.21) by the summation

$$A = \frac{1}{M} \sum_{i=1}^{M} \exp\{-\mu(x_i + x_i')\} \tag{6.22}$$

where the total area is divided into $M$ smaller areas. For the example shown in fig. 6.8(*b*) the details of the calculation, assuming $\mu = 5 \times 10^3 \, \mathrm{m}^{-1}$, follow in table 6.1. From equation (6.22) this gives $A = 0.044$. The accuracy of this method of calculating absorption factors is increased if a greater number $M$ of points is taken but, of course, the labour involved in the calculation also increases. Computer programs have been written to calculate absorption factors; some of these are very sophisticated and will compute values of $A$ for crystals of somewhat irregular shape. However if it is possible to shape the crystal by grinding it into a sphere or a cylinder then absorption factors can be expressed in the form of tables and these are given in Vol. B of the *International Tables for Crystallography*. For a sphere of radius $R$, for example, the absorption factor is a function of $\mu R$ and $\sin \theta$ only and can be very concisely tabulated.

We shall now take a look at some general aspects of absorption with a view to discovering the general rules which govern it and also to finding how to determine $\mu$ for materials of known composition.

The simplest materials to consider are those which consist of a single element. For the attenuation of a beam going through such material one may write

$$\frac{\mathrm{d}I}{I} = -\mu_a \mathrm{d}n \tag{6.23}$$

where d$n$ is the number of atoms in the path of the beam per unit area and $\mu_a$, which depends on the element, is known as the *atomic absorption*

Table 6.1.

| Point ($i$) | $x_i$ | $x'_i$ | $x_i + x'_i$ | $\mu(x_i + x'_i)$ | $\exp\{-\mu(x_i + x'_i)\}$ |
|---|---|---|---|---|---|
| | All linear dimensions are in mm | | | | |
| 1 | 0.14 | 0.14 | 0.28 | 1.40 | 0.2471 |
| 2 | 0.43 | 0.14 | 0.57 | 2.85 | 0.0578 |
| 3 | 0.14 | 0.42 | 0.56 | 2.80 | 0.0608 |
| 4 | 0.43 | 0.42 | 0.85 | 4.25 | 0.0144 |
| 5 | 0.14 | 0.69 | 0.83 | 4.15 | 0.0159 |
| 6 | 0.43 | 0.47 | 0.90 | 4.50 | 0.0112 |
| 7 | 0.14 | 0.97 | 1.11 | 5.55 | 0.0039 |
| 8 | 0.43 | 0.47 | 0.90 | 4.50 | 0.0112 |
| 9 | 0.14 | 1.27 | 1.41 | 7.05 | 0.0009 |
| 10 | 0.38 | 0.47 | 0.85 | 4.25 | 0.0144 |

*coefficient*. If experiments are done to measure $\mu_a$ as a function of the wavelength of the X-radiation the results appear as in fig. 6.9. Starting with a long wavelength, say at the point $P$, the value of $\mu_a$ steadily reduces as $\lambda$ decreases. The X-ray photons, with energies $hc/\lambda$, will have more energy for smaller $\lambda$ and hence one should expect them to be more penetrating. However if the wavelength is reduced to the value of $Q$ the value of $\mu_a$ sharply increases and then, with a further reduction of $\lambda$, $\mu_a$ decreases again. These sudden discontinuities in atomic absorption coefficient – and there are several of them – are known as absorption edges. They can be understood in terms of the electronic structure of the atom. Within the atom, electrons exist in definite energy states and the absorption edges correspond to energies $hc/\lambda$ which are *just* sufficient to eject an atomic electron from the atom. At the point $P$ the photon energy is insufficient to eject any of the electrons of an unexcited atom and the photons either interact with the whole atom increasing its energy of motion, which is equivalent to increasing its temperature, or lose energy through the Compton effect. However at $Q$ the photons have just enough energy to eject an $L_{III}$ shell electron; a new mechanism for interaction of the photon and the atom becomes available and results in a sharp increase in absorption. Similar effects occur corresponding to the other atomic electrons and gives a whole series of absorption edges. Between the absorption edges the general fall-off of the absorption curves follows a form close to

$$\mu_a = CZ^4\lambda^{5/2} \tag{6.24}$$

where $Z$ is the atomic number of the material and $C$ is a constant which depends on the nature of the two flanking absorption edges.

Equation (6.23) can be transformed to be in terms of the distance traversed by the beam. In fig. 6.10 the slab of material has unit cross-sectional area, and thickness d$x$. If it contains d$n$ atoms then the total mass of material is $M \mathrm{d}n/N_A$ where $M$ is the atomic weight and $N_A$ is Avogadro's

Fig. 6.9.
Variation of absorption
coefficient with
wavelength for a typical
element. If the element is
nickel then the Cu Kα
and Cu Kβ wavelengths
are as shown.

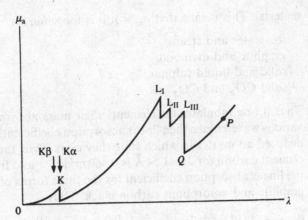

number. The density of the material is given by

$$\rho = \frac{M}{N_A}\frac{dn}{dx} \tag{6.25}$$

and we can now rewrite equation (6.23) as

$$\frac{dI}{I} = -\mu_a \frac{N_A\rho}{M}dx. \tag{6.26}$$

Comparing equations (6.19) and (6.26)

$$\mu_a \frac{N_A\rho}{M} = \mu$$

or

$$\mu_a = \frac{\mu}{\rho}\frac{M}{N_A} = \mu_m\frac{M}{N_A}. \tag{6.27}$$

We have written $\mu_m = \mu/\rho$ where $\mu_m$ is called the *mass absorption coefficient*. Although $\mu_m$ has been introduced by considering the absorption of elementary material it can also be used for mixed materials or compounds. The important property which makes the mass absorption coefficient a useful quantity is that it is completely independent of the state of the

Fig. 6.10.
An X-ray beam
traversing a slab of
material of unit
cross-sectional area, and
thickness dx.

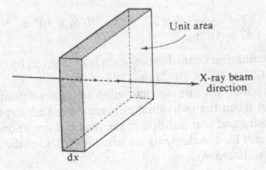

material. This means that $\mu_m = \mu/\rho$ is the same for

ice, water and steam;
graphite and diamond;
solid and liquid sulphur;
solid $CO_2$ and $CO_2$ gas.

Thus if one tabulates for elements their mass absorption coefficients for various wavelengths then linear absorption coefficients for elements can be deduced no matter in which state they exist. Thus the value of $\mu_m$ for the element carbon for $\lambda = 1.54\,\text{Å}$ is $0.460\,\text{m}^2\,\text{kg}^{-1}$ and from this we can find the linear absorption coefficient for the three forms of carbon – diamond, graphite and amorphous carbon black. Thus

$$\mu_{\text{diamond}} = (\mu_m)_C + \rho_{\text{diamond}} = 0.460 \times 3.52 \times 10^3\,\text{m}^{-1} = 1.62 \times 10^3\,\text{m}^{-1}$$
$$\mu_{\text{graphite}} = (\mu_m)_C + \rho_{\text{graphite}} = 0.460 \times 2.25 \times 10^3\,\text{m}^{-1} = 1.03 \times 10^3\,\text{m}^{-1}$$
$$\mu_{\text{amorphous}} = (\mu_m)_C + \rho_{\text{amorphous}} = 0.460 \times 1.88 \times 10^3\,\text{m}^{-1} = 0.86 \times 10^3\,\text{m}^{-1}$$

This same principle can be extended to determine the linear absorption coefficients of mixtures of elements or compounds by using the relationship

$$\mu_{\text{compound}} = \sum \mu_{\text{element}} = \sum \mu_m \rho_p \tag{6.28}$$

where the summations are taken over all the elements in the compound and $\rho_p$ represents the partial density of the element within the compound. An example will illustrate more clearly how one uses this equation. We shall determine the attenuation of a beam of Cu $K\alpha$ radiation ($\lambda = 1.54\,\text{Å}$) in passing through a slab of NaCl of thickness 0.1 mm.

The following information can be found in tables:

Density of NaCl $= 2.2 \times 10^3\,\text{kg m}^{-3}$
Na   atomic weight 23   $\mu_m(\lambda = 1.54\,\text{Å}) = 3.0\,\text{m}^2\,\text{kg}^{-1}$
Cl   atomic weight 35   $\mu_m(\lambda = 1.54\,\text{Å}) = 10.6\,\text{m}^2\,\text{kg}^{-1}$.

The proportion of the two elements by mass is Na:Cl $= 23:35$ so that the partial densities are

$$(\rho_p)_{\text{Na}} = 2.2 \times 10^3 \times \tfrac{23}{58}\,\text{kg m}^{-3}$$

and

$$(\rho_p)_{\text{Cl}} = 2.2 \times 10^3 \times \tfrac{35}{58}\,\text{kg m}^{-3}.$$

Thus

$$\mu = (2.2 \times \tfrac{23}{58} \times 3.0 + 2.2 \times \tfrac{35}{58} \times 10.6) \times 10^3\,\text{m}^{-1}$$
$$= 1.67 \times 10^4\,\text{m}^{-1}.$$

The attenuation factor for the beam is then given by equation (6.20) and is $\exp(-1.67 \times 10^4 \times 10^{-4})$ or 0.188.

By such means one can determine linear absorption coefficients for any material when the individual elemental mass absorption coefficients and the density and composition of the material are known. Mass absorption coefficients for the elements are listed in Vol. C of the *International Tables for Crystallography*.

In § 5.8 the use of absorption filters to remove unwanted radiation from an X-ray source was mentioned. For example, a sealed copper tube will give as its main characteristic output Cu Kα and Cu Kβ radiation of wavelengths 1.542 and 1.389 Å, respectively, and it is usually required to remove the shorter-wavelength component. The K absorption edge for copper, which is the energy required to eject an electron from the K shell, must be at a wavelength shorter than either the Kα or Kβ radiations. However, the element one less in atomic number than copper is nickel (atomic number 28) and its K absorption edge falls between the CuKα and CuKβ wavelengths. If fig. 6.9 represents the atomic absorption coefficient of nickel then the Cu Kα and Cu Kβ wavelengths are shown as Kα and Kβ. It is clear that the absorption is far less for the Kα radiation. For example, the mass absorption coefficients of Ni for Cu Kα and Cu Kβ radiations are 4.57 and 27.5 $m^2 kg^{-1}$, respectively. The density of Ni is $8.9 \times 10^3 kg m^{-3}$ and hence the linear absorption coefficients are $4.1 \times 10^4 m^{-1}$ and $24.5 \times 10^4 m^{-1}$. A foil of thickness 20 μm ($= 2 \times 10^{-5}$ m) transmits a fraction 0.44 of the Kα radiation but only 0.0075 of the Kβ and thus constitutes a very efficient filter.

## 6.3   Primary extinction

So far in our consideration of diffraction from a crystal we have assumed that the atoms are under the influence of an incident beam whose intensity has been modified only by absorption in the crystal. However if we look at fig. 6.11, in which is represented the passage of the incident X-ray beam through the material and the formation of the diffracted beam, it is clear that this simple picture is not really true. Every point in the crystal is under the influence of the incident beam and also some part of the diffracted beam which is moving in a different direction. If the crystal is in a strongly reflecting position the extra effect of the diffracted beam component will not be negligible.

We shall first investigate the propagation of a plane wave through a crystal in the direction of the incident radiation. We imagine that the X-radiation is 'established' in the crystal which is therefore occupied by a continuum of radiation. In fig. 6.12 there is represented a section in which

Fig. 6.11.
A schematic representation of the formation of a diffracted beam originating from points distributed throughout the crystal.

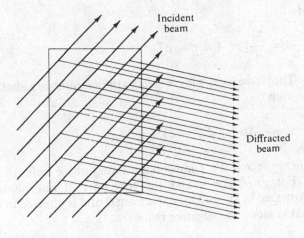

Incident beam

Diffracted beam

Fig. 6.12.
(*a*) The formation of a
plane wave front from
wavelets originating
from a plane of
scatterers.
(*b*) All radiation
originating in the
annular region arrive at
*P* with the same phase
relationship.

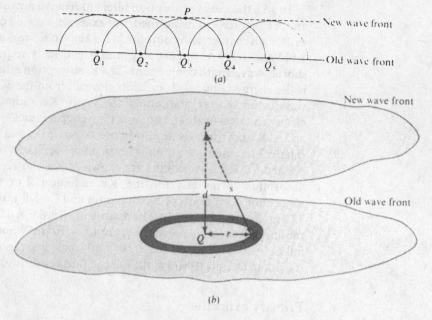

(*a*)

(*b*)

there are scatterers through a plane labelled $Q_1$ to $Q_5$. Most of the radiation
passing through the plane $Q_1Q_5$ will not be scattered and will eventually
pass through the plane containing the point $P$ in a time which depends on
the distance $Q_3P$. However some small part of the X-radiation will be
scattered by the scatterers $Q$ so that the point $P$, for example, is at any time
also receiving some radiation coherently scattered by each of the points $Q$.
We imagine that the points $Q$ are sufficiently densely packed so that the
plane containing them may be considered as a uniform partially scattering
plane.

It is clear from fig. 6.12(*b*) that radiation from all the points on the ring of
radius $r$ will arrive at $P$ with the same phase relationship. The phase lag
behind the radiation coming from $Q$ will be

$$\phi_r = \frac{2\pi}{\lambda}(s - d)$$

or

$$\phi_r = \frac{2\pi}{\lambda}[(r^2 + d^2)^{\frac{1}{2}} - d].$$

The phase lag $\alpha_r$ with respect to the *unscattered* radiation coming from $Q$
is given by

$$\alpha_r = \frac{2\pi}{\lambda}[(r^2 + d^2)^{\frac{1}{2}} - d] + \pi \tag{6.29a}$$

where the $\pi$ is the Thomson scattering phase shift.

For $r = 0, \alpha_r = \pi$ and the first part of our phase-vector diagram,
illustrated in fig. 6.13(*a*), starts from $O$ and is directed in the direction $AO$ –
that is along the negative real axis.

Fig. 6.13.
(a) Phase-vector
diagram for radiation
originating from a
plane of scatterers and
arriving at a plane wave
front. The resultant,
$OC$, is $\pi/2$ ahead in
phase of the radiation
arriving from the
nearest scatterer.
(b) Radiation going
along $OA$ without being
scattered is reduced in
intensity due to
destructive interference
with radiation doubly
scattered at $C$ and $O$.

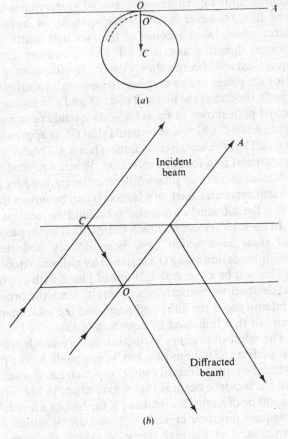

By differentiating equation (6.29a) we find

$$d\alpha_r = \frac{2\pi r\,dr}{\lambda(r^2 + d^2)^{\frac{1}{2}}}. \tag{6.29b}$$

On the phase-vector diagram the contribution of a small annular region
of radius $r$ and thickness $dr$ will be a line segment, $dq_r$, whose length will be
proportional to the area of the annulus and inversely proportional to the
distance $s$. This latter factor is equivalent to an inverse-square-law fall-off
for intensity.

Thus we have

$$dq_r = K\frac{2\pi r\,dr}{(r^2 + d^2)^{\frac{1}{2}}} \tag{6.30}$$

where $K$ is some constant. Taken together, equations (6.29b) and (6.30) give,
dropping the subscript $r$,

$$d\alpha = \frac{dq}{K\lambda}. \tag{6.31}$$

The geometrical shape corresponding to equation (6.31), where the
change of angle is proportional to arc length, is a circle. With only the

considerations given above this is what the phase-vector diagram would look like. However in the above analysis we have ignored an obliquity factor which takes account of the fact that scattering is strongest in the forward direction and falls off with increasing angle of scatter. This is equivalent to $K$ being a slowly decreasing function of $r$ and the result of this is for the phase-vector diagram to become a spiral rather than a circle. For $\alpha = 2\pi$ the end of the line will be at $O'$ and as $\alpha$ increases it moves along the dotted path shown in fig. 6.13($b$). As $r$ tends to infinity the end of the line approaches $C$ which is so situated that $OC$ is approximately at right angles to $OA$. Thus the scattered radiation has a $\pi/2$ phase lag with respect to that unscattered in a forward direction. When a crystal is producing a Bragg reflection there is no phase difference introduced by path differences from the scatterers arranged on a lattice. It can be shown that at any point in the crystal the diffracted wave is also $\pi/2$ behind the incident unscattered radiation.

In fig. 6.13($b$) there is shown the incident beam passing through a crystal and some rays which have been doubly 'reflected'. The unscattered radiation passing along $OA$ is joined by radiation doubly scattered at $C$ and $O$. This will be $\pi$ (i.e. $2 \times \pi/2$) out of phase with the unscattered radiation and destructive interference will result. Since the primary beam is reduced in intensity so is the diffracted beam and the total energy of the reflection is less than that indicated by equation (6.18).

The effect of primary extinction is very much reduced if the crystal is non-perfect and if the mosaic blocks are small. At any particular point in the crystal the only diffracted radiation which has a special phase relationship with the incident beam is that which arises in the same mosaic block and this will be of negligible intensity if the blocks are small. Crystallographers sometimes dip their crystals in liquid air to reduce the effect of primary extinction. The thermal shock to which the crystal is thus subjected produces the desired effect of breaking it up into a mosaic structure.

This theory, which is the one usually given in elementary texts, suffers from the drawback that energy seems not to be conserved. The primary beam is reduced in intensity and so is the diffracted beam and one might well ask where all the energy has gone.

A different theoretical approach by Zachariasen (1967) has the conservation of energy as one of its fundamental hypotheses. He considers that in some volume element of the crystal the intensity $I_0$ of the beam in the incident direction and the intensity $I$ of the beam in the diffracted direction are related by

$$\frac{\partial I_0}{\partial t_1} = -\sigma I_0 + \sigma I$$

and

$$\frac{\partial I}{\partial t_2} = -\sigma I + \sigma I_0 \tag{6.32}$$

where $t_1$ and $t_2$ are lengths in the directions of the incident and diffracted beams, respectively, and $\sigma$ is the diffracted power per unit distance and intensity. These equations express the fact that each beam is depleted by scattering into the other beam and enhanced by scattering from the other beam into itself. The sum of the two equations gives

$$\frac{\partial I_0}{\partial t_1} + \frac{\partial I}{\partial t_2} = 0 \tag{6.33}$$

which is the condition for the conservation of energy. These equations are augmented by the boundary conditions that, where the incident beam enters the crystal, $I_0$ equals the intensity of the incident X-ray beam and $I = 0$.

The coupling constant $\sigma$ is clearly a function of $\varepsilon$, the angle the beam makes with the 'ideal' direction as given by Bragg's equation and clearly $\sigma$ falls off very rapidly with $\varepsilon$ and is significantly only for small $\varepsilon$.

We cannot explore here the full depth of Zachariasen's treatment of these equations but a selection of the solutions he obtains will give the general pattern of his results.

The factor by which the diffracted intensity is reduced for a spherical specimen and for low scattering angles is given by

$$\phi(\sigma) = 1 - \sigma\bar{t} + \tfrac{16}{15}(\sigma\bar{t})^2 - \tfrac{80}{81}(\sigma\bar{t})^3 \tag{6.34}$$

where $\bar{t}$ is the mean path length in the crystal and equals $\tfrac{3}{2}r$ for radiation passing straight through a sphere.

For small $\sigma\bar{t}$ this is approximately

$$\phi(\sigma) = \frac{1}{1 + \sigma\bar{t}} \tag{6.35a}$$

which is an exact solution for an infinite plane parallel plate when the incident and reflected beams make the same angle to the normal to the plate.

When the crystal has a mosaic structure, with domain size small compared to that of the whole crystal, the result corresponding to equation (6.35a) is

$$\phi(\sigma) = \frac{1}{1 + \bar{\sigma}\bar{T}} \tag{6.35b}$$

where $\bar{T}$ is the mean path through the whole crystal and $\bar{\sigma}$ a mean $\sigma$ for the distribution of orientations of the domains. Since $\bar{\sigma}$ is less than the value of $\sigma$ for a perfect crystal a reduction in primary extinction results from mosaic structure.

## 6.4 Secondary extinction

It is found that even when a crystal is ideally imperfect, consisting of very small mosaic blocks, there is another type of extinction process which may occur. We have already found that the intensity of an X-ray beam is attenuated by its passage through a material by the absorption of some of the X-ray energy and its conversion to thermal energy. This process is quite independent of the phenomenon of diffraction and could be recorded by measuring the attenuation of a beam of X-rays passing through slabs of material. However when the crystal is in a diffracting position there is available another mechanism for removing energy from the incident beam – that is to say, that some of the energy goes into the diffracted beam. In fig. 6.14 the passage of the incident beam, of initial power $I_0\alpha$, through a small

Fig. 6.14.
Effective increase in
absorption coefficient
due to diffraction.

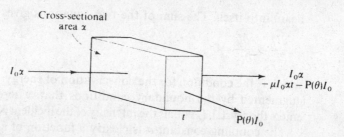

mosaic element of a crystal is illustrated. The emergent beam power is depleted firstly by an amount $\mu I_0 \alpha t$ due to ordinary absorption and secondly by an amount equal to $P(\theta)I_0$ (equation (6.1)) which is a measure of the power of the diffracted beam from this small element of the crystal. The overall result of this additional depletion is effectively to raise the linear absorption coefficient from $\mu$ to $\mu'(\theta)$ where

$$\mu'(\theta) = \mu + \frac{P(\theta)}{\alpha t}$$

$$= \mu + Q(\theta), \tag{6.36a}$$

where $Q(\theta)$ is the reflecting power per unit volume of the mosaic element. For a complete sweep of the crystal through a reflecting position there will be some average $\mu'$ given by

$$\mu' = \mu + \overline{Q(\theta)}. \tag{6.36b}$$

The magnitude of $\mu' - \mu$ can be very considerable; for example some experiments by Bragg, James and Bosanquet (1921) with rock salt crystals showed that, for the (200) reflection with Rh K$\alpha$ radiation ($\lambda = 0.613$ Å), the value of $\mu' = 1.630 \times 10^3$ m$^{-1}$ whereas the normal linear absorption coefficient is only $1.070 \times 10^3$ m$^{-1}$.

This phenomenon is known as *secondary extinction* and it is difficult to eliminate by any treatment of the crystal. It certainly can be reduced to some extent by increasing the imperfection of the crystal. Secondary extinction can be thought of as a shielding effect which exists at a point in a crystal due to diffraction by the layers of crystal above it – that is, closer to the incident X-ray beam. If the mosaic blocks were all precisely parallel then all of them would diffract simultaneously and the secondary extinction effect would be the same as that for a perfect crystal. However if the mosaic blocks have a range of orientations then they do not all diffract together as the crystal is rotated and the shielding effect is consequently reduced. One would not expect to find secondary extinction with a powder sample where only a tiny fraction of the crystallites are diffracting at any one time.

Since secondary extinction is equivalent to a change in the linear absorption coefficient of the material of the crystal its proportional effect on intensities is reduced if very small crystals are employed. This can be illustrated by the example shown in fig. 6.8(b) and table 6.1. If the value of $\mu'$ is $5.5 \times 10^3$ m$^{-1}$ ($\mu = 5.0 \times 10^3$ m$^{-1}$) then the numbers in the column $\mu(x_i + x_i')$ are all increased by 10% and the value of $A$ changes from 0.044 to 0.035 or by about 19%. If we now consider a crystal similar in shape but with its dimensions reduced by a factor of two we find $A_\mu = 0.165$ and $A_{\mu'} = 0.145$, a change of 12%. A further reduction by a factor of two in the dimensions gives $A_\mu = 0.381$ and $A_{\mu'} = 0.349$ and secondary extinction now gives a 6% reduction of intensity. However this reduction of crystal size and reduction of secondary extinction is bought at the expense of lowering the diffraction intensities, so in practice there is a limit to the smallness of the crystal which can be used. For the most part crystallographers accept that only a few very strong reflections will be affected by secondary extinction and these do not usually interfere with the processes of determining or refining crystal structures.

## 6.5   The temperature factor

The picture we have created of a crystal is that of a rigid stationary assemblage of atoms bound together in a periodic pattern. We must now consider how the picture is modified by the presence of thermal energy because thermal energy on the atomic scale is energy of motion.

Every atom in a crystal structure is bound to numbers of other atoms by bonding forces of various types and the position of the atom is that corresponding to the minimum potential energy. It is probably rather more accurate to say that the complete crystal structure corresponds to an arrangement of minimum potential energy and where, for example, a substance crystallizes in two different forms it means that two arrangements exist with potential energies virtually equal to each other and lower than that of any other arrangement. If an atom is disturbed from its equilibrium position it experiences a restoring force tending to move it back again and we can see that the atom can acquire thermal energy by oscillating about its equilibrium position. Thus we get a modified view of a crystalline solid in which all the atoms are vibrating about their mean positions with amplitudes which increase as does the temperature of the solid. These vibrations will affect the relative coordinates of the atoms and hence the diffraction pattern and we shall now investigate this phenomenon.

The first thing to note is that the frequencies of vibration of atoms are very low in relation to the sort of times in which we are interested – for example the time of transmission of X-rays through the crystal. For this reason we may imagine that, at any instant of time, the diffraction pattern being produced is that of a 'frozen' crystal in which all the atoms are stationary and displaced by some distance from their mean positions. The total intensity pattern that we get over any long period of time is a *time-average* of the patterns which are obtained at successive instants. We are going to assume that all the atomic vibrations are completely uncorrelated so that averaging processes can be carried out for individual atoms.

To begin with let us take one unit cell of a one-dimensional structure containing $N$ atoms per unit cell where the $j$th atom has mean *fractional* coordinate $x_j$ and, at some instant of time, an *absolute* displacement from that position $u_j$. If all the other unit cells were in a precisely similar state the structure factor of index $h$ would be given by

$$F_h = \sum_{j=1}^{N} f_j \exp\left\{2\pi i h\left(\frac{u_j}{a} + x_j\right)\right\}$$

$$= \sum_{j=1}^{N} f_j \exp\left(2\pi i h \frac{u_j}{a}\right) \exp(2\pi i h x), \qquad (6.37)$$

The actual structure amplitude in a direction corresponding to $h$ will be a time and space average of equation (6.37) since $u_j$ varies from one unit cell to the next and, within one unit cell, varies with time.

We write for the structure factor at some temperature $T$

$$[F_h]_T = \sum_{j=1}^{N} f_j \overline{\exp\left(2\pi i h \frac{u_j}{a}\right)} \exp(2\pi i h x_j) \qquad (6.38)$$

where $\exp\{2\pi i h(u_j/a)\}$ is the average value of the diplacement term. In all practical cases $u_j$ is small enough to be able to write approximately

$$\overline{\exp\left(2\pi i h \frac{u_j}{a}\right)} \simeq 1 + 2\pi i h \frac{\bar{u}_j}{a} - 2\pi^2 h^2 \frac{\overline{u_j^2}}{a^2}. \qquad (6.39)$$

For simple harmonic or other symmetrical vibrations $\bar{u}_j = 0$, so we have

$$\overline{\exp\left(2\pi i h \frac{u_j}{a}\right)} \simeq 1 - 2\pi^2 h^2 \frac{\overline{u_j^2}}{a^2} \qquad (6.40a)$$

or, in a more convenient form,

$$\overline{\exp\left(2\pi i h \frac{u_j}{a}\right)} \simeq \exp\left(-2\pi^2 h^2 \frac{\overline{u_j^2}}{a^2}\right). \qquad (6.40b)$$

The result (6.40b) substituted in equation (6.38) together with $h/a = 2\sin\theta/\lambda$ gives

$$[F_h]_T = \sum_{j=1}^{N} f_j \exp(-8\pi^2 \overline{u_j^2} \sin^2\theta/\lambda^2) \exp(2\pi i h x_j). \qquad (6.41)$$

It can be seen that the overall result of the thermal motion of the atoms is effectively to modify their scattering factors to

$$[f_j]_T = f_j \exp(-8\pi^2 \overline{u_j^2} \sin^2\theta/\lambda^2). \qquad (6.42)$$

If the three-dimensional case is now considered it will be found that the equation corresponding to (6.42), without the introduction of $\theta$, appears in the form

$$[f_j]_T = f_j \exp\{-2\pi^2 \overline{(\mathbf{s}\cdot\mathbf{u}_j)^2}\}. \qquad (6.43)$$

Now

$$\overline{(\mathbf{s}\cdot\mathbf{u}_j)^2} = s^2 u_j^2 \cos^2\phi \qquad (6.44)$$

where $\phi$ is the angle between the vectors $\mathbf{s}$ and $\mathbf{u}_j$ and since we are considering a particular $s$ and since $u_j$ and $\phi$ are independent of each other we may write

$$\overline{(\mathbf{s}\cdot\mathbf{u}_j)^2} = s^2 \overline{u_j^2}\, \overline{\cos^2\phi}. \qquad (6.45)$$

Fig. 6.15.
The annular ring on the
sphere of unit radius
gives the probability of
a random angle in the
range 0–π being
between $\phi$ and d$\phi$.

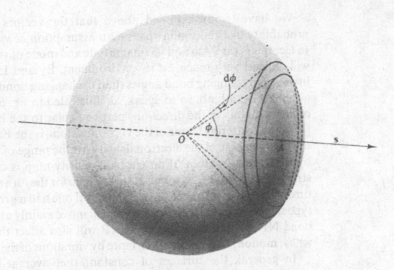

The value of $\overline{\cos^2 \phi}$ is the average value of $\cos^2 \phi$ where $\phi$ is an angle
between some random direction and a particular direction. In fig. 6.15 the
direction of **s** is shown along $OA$ and a sphere of unit radius centred on $O$.
There is a uniform probability distribution for the ends of a randomly
oriented unit vector to fall on any point of the surface of the sphere so that
the probability that it falls within any specified area, d$\alpha$, is d$\alpha/4\pi$. The
probability of the vector terminal being within the shaded annular region is
$(2\pi \sin \phi \, d\phi)/4\pi$ so that the average value of $\cos^2 \phi$ is

$$\overline{\cos^2 \phi} = \int_0^\pi \tfrac{1}{2}\cos^2 \phi \sin \phi \, d\phi = \tfrac{1}{3}.$$

Equation (6.43) may now be written

$$[f_j]_T = f_j \exp\left( -\frac{8}{3}\pi^2 \overline{u_j^2} \frac{\sin^2 \theta}{\lambda^2} \right) \tag{6.46}$$

where $\overline{u_j^2}$ is the mean square displacement of the atom from its mean position

The three-dimensional and one-dimensional formulae can be expressed
in similar terms if $\mathbf{s} \cdot \mathbf{u}_j$ is replaced by $su_{\perp j}$ where $u_{\perp j}$ is the magnitude of $\mathbf{u}_j$
projected on $s$.

One may then write

$$[f_j]_T = f_j \exp(-8\pi^2 \overline{u_{\perp j}^2} \sin^2 \theta/\lambda^2) \tag{6.47}$$

where $\overline{u_{\perp j}^2}$ is the mean square displacement of the atom perpendicular to the
reflecting planes.

If

$$B_j = 8\pi^2 u_{\perp j}^2 \tag{6.48}$$

then

$$[f_j]_T = f_j \exp(-B_j \sin^2 \theta/\lambda^2) \tag{6.49}$$

and the quantity $B_j$ is known as the temperature factor of the $j$th atom.

We have tacitly assumed above that the vectors $\mathbf{u}_j$ have an equal probability distribution in space – an assumption of vibrational isotropy. In fact this is rarely so and the magnitude and mode of vibration of an atom will depend very much on its environment. By and large, less energy is involved in changing bond angles than in changing bond lengths, so that an atom out on a limb, so to speak, as illustrated in fig. 6.16(*a*), will tend to oscillate mainly in the directions perpendicular to the bond. The quantity $[f_j]_T$, the scattering factor of the vibrating atom, is the Fourier transform of the time-average of the electron density for the range of positions taken up by the vibrating atom. If an electron-density map is computed with the structure factors $[F_{hkl}]_T$ for all $h, k$ and $l$ then for the $j$th atom one will see the time-averaged electron density. Similarly if one had a group of atoms of the type shown in fig. 6.16(*b*) the atom P will move mainly at right angles to the bond NP. Vibration of the bond NP will also affect the atoms Q and R whose motion is augmented even more by vibrations of the bonds PQ and PR.

In general, the surfaces of constant time-averaged electron density approximate to ellipsoids and in the most precise crystal-structure analyses the parameters which describe these ellipsoids are determined (see §9.3). However for all but the most accurate work it is usually sufficient to assume that the thermal vibrations are isotropic and that the values of $B$ are the same for all atoms. In that case we have

$$[F_{hkl}]_T = F_{hkl}\exp(-B\sin^2\theta/\lambda^2) \tag{6.50}$$

or for intensities

$$[\mathscr{I}_{hkl}]_T = \mathscr{I}_{hkl}\exp(-2B\sin^2\theta/\lambda^2). \tag{6.51}$$

The factor by which the observed intensities are reduced by thermal vibrations, $\exp(-2B\sin^2\theta/\lambda^2)$, is known as the *Debye–Waller factor*.

Fig. 6.16.
(*a*) Atom B vibrates most strongly in the plane perpendicular to AB.
(*b*) Atoms Q and R should have a larger amplitude of vibration than does atom P.

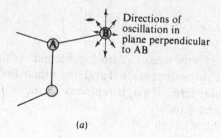

(*a*)

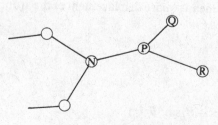

(*b*)

Table 6.2.

| | Debye–Waller factor, $\exp(-2B\sin^2\theta/\lambda)$, for three values of $B$ (in $\text{Å}^2$) | | |
|---|---|---|---|
| $\sin\theta/\lambda$ | 2.0 | 2.5 | 3.0 |
| 0.2 | 0.8521 | 0.8187 | 0.7866 |
| 0.4 | 0.5273 | 0.4493 | 0.3829 |
| 0.6 | 0.2369 | 0.1653 | 0.1153 |
| 0.8 | 0.0773 | 0.0408 | 0.0215 |
| 1.0 | 0.0183 | 0.0067 | 0.0025 |

Typical values of temperature factors fall in the range 2–3 $\text{Å}^2$ and, to illustrate their order or magnitude, Debye–Waller factors are given in table 6.2 for various values of $B$ and $\sin\theta/\lambda$.

For $B = 3\,\text{Å}^2$, for example, and for the limit of Cu K$\alpha$ radiation ($\sin\theta/\lambda \simeq 0.65$), it will be seen that the intensity is reduced to about 10% of its unmodified value.

The dependence of $B$ on the absolute temperature $T$ has been theoretically investigated by Debye who produced a formula whose application is limited to materials which are elements. If $\Theta$ is the characteristic temperature of the material and $x = \Theta/T$ then the formula is

$$B = \frac{6h^2}{mk_B\Theta}\left\{\frac{\phi(x)}{x} + \frac{1}{4}\right\} \tag{6.52}$$

where

$m$ is the atomic mass,
$h$ is Planck's constant,
$k_B$ is Boltzmann's constant

and

$$\phi(x) = \frac{1}{x}\int_0^x \frac{\xi}{\exp\xi - 1}\,d\xi.$$

In practice the value of $B$ can be found from the experimentally determined values of $\mathscr{I}_{hkl}$; the method of doing this is described in §7.5.

## 6.6 Anomalous scattering

In determining scattering factor equations as in §2.6 it has been assumed that all the atomic electrons are unrestrained and undamped so that the scattering phase shift is $\pi$ for each of them. If we consider one atomic electron, moving independently of all the others, then its vibrational motion in the alternating electric field of an incident electromagnetic wave might be governed by an equation of the form

$$m\ddot{x} + g\dot{x} + kx = E_0 e \exp(i\omega t). \tag{6.53}$$

In equation (6.53) $m$ is the electron mass, $k$ the restoring force per unit displacement, $E_0$ and $\omega/2\pi$ are the amplitude and frequency of the incident wave and it is assumed that there is a damping force $g\dot{x}$ proportional to the velocity.

The steady-state solution of this equation is

$$x = A\exp(i\omega t) \tag{6.54a}$$

where

$$A = \frac{E_0 e}{(k - \omega^2 m) + ig\omega} = \frac{E_0 e[(k - \omega^2 m) - ig\omega]}{(k - \omega^2 m)^2 + g^2\omega^2}. \tag{6.54b}$$

Another form of solution, without recourse to complex notation, is

$$x = R\cos(\omega t + \varepsilon) \tag{6.55a}$$

where

$$R = \frac{E_0 e}{[(k - \omega^2 m)^2 + g^2\omega^2]^{\frac{1}{2}}} \tag{6.55b}$$

and

$$\tan \varepsilon = -\frac{g\omega}{k - \omega^2 m}. \tag{6.55c}$$

For an unrestrained and undamped electron one would have $k = 0$ and $g = 0$ giving

$$x = -\frac{E_0 e}{\omega^2 m}\exp(i\omega t). \tag{6.56}$$

This represents an oscillation of the electron with the same frequency as the oncoming wave but $\pi$ out of phase with it – as is shown by the negative amplitude. By writing equation (6.56) as

$$x = \frac{E_0 e}{\omega^2 m}\exp\{i(\omega t + \pi)\} \tag{6.57}$$

the phase shift of $\pi$ with respect to the oncoming wave $E = E_0\exp(i\omega t)$ is made more evident.

When $k \neq 0$ there is a resonance frequency corresponding to $\omega = \omega_0$ where

$$k - \omega_0^2 m = 0$$

so that we may write

$$k = \omega_0^2 m. \tag{6.58}$$

Substituting this in equation (6.54b) we obtain

$$A = \frac{(\omega_0^2 - \omega^2)mE_0 e}{(\omega_0^2 - \omega^2)^2 m^2 + g^2\omega^2} - \frac{E_0 eg\omega}{(\omega_0^2 - \omega^2)^2 m^2 + g^2\omega^2}i. \tag{6.59}$$

When $\omega = \omega_0$ and the damping constant $g$ is small it can be seen that the amplitude of oscillation of the electron would be very large. In our present classical model the amplitudes of vibration of atomic electrons are continuously and infinitely variable. However, from a quantum-theory point of view, the angular frequency $\omega_0$ is that for an X-ray photon with just enough energy to eject the electron from the atom, i.e. that which has a wavelength corresponding to the absorption edge.

The critical wavelength $\lambda_0$ is given by

$$\lambda_0 = \frac{2\pi c}{\omega_0} \tag{6.60}$$

and $\lambda_0$ would correspond to one of the absorption-edge wavelengths shown in fig. 6.9.

The electrons which are most tightly bound to the nucleus and therefore have the highest value of $k$ are the K electrons but, even for these, we can see from equation (6.59) that when $\omega \gg \omega_0$ their behaviour is not very different from that of a free electron. Thomson-scattering theory is only truly applicable for $\omega \to \infty$ and in the vicinity of an absorption edge the vibration of the electron, and hence its scattering, will be very different from that predicted by the Thomson formula. The amplitude of the scattered wave will be proportional to the amplitude of the electron vibration; we can express the real and imaginary components of the amplitude of the scattered wave as fractions of that from a free electron by dividing the components in equation (6.59) by the amplitude given in equation (6.56). These fractions are, for the real part,

$$\phi_A = \frac{-\omega^2(\omega_0^2 - \omega^2)m^2}{(\omega_0^2 - \omega^2)^2 m^2 + g^2\omega^2} \tag{6.61a}$$

and, for the imaginary part,

$$\phi_B = \frac{g\omega^3 m}{(\omega_0^2 - \omega^2)^2 m^2 + g^2\omega^2}. \tag{6.61b}$$

For a Thomson-scattering electron one would have $\phi_A = 1$ and $\phi_B = 0$ so that, if the absolute Thomson contribution to the scattering factor of this electron is $p_0$, then the actual contribution is

$$\begin{aligned} p_0' &= p_0(\phi_A + \phi_B i) \\ &= p_0 + p_1 + p_2 i \end{aligned} \tag{6.62}$$

where

$$p_1 = p_0(\phi_A - 1) = \frac{\omega_0^2(\omega^2 - \omega_0^2)m^2 - g^2\omega^2}{(\omega_0^2 - \omega^2)^2 m^2 + g^2\omega^2} p_0 \tag{6.63a}$$

and

$$p_2 = \phi_B p_0 = \frac{g\omega^3 m}{(\omega_0^2 - \omega^2)^2 m^2 + g^2\omega^2} p_0. \tag{6.63b}$$

Let us consider a diffraction situation where we have an iron atom in the crystal and Cu K$\alpha$ radiation ($\lambda = 1.54$ Å) is being used. The K absorption edge for iron is at $1.74$ Å and is at a greater wavelength for all the other absorption edges so that we might expect the two K electrons to scatter *anomalously* (i.e. other than as free electrons) while the others would be closer in behaviour to free electrons. For the two wavelengths concerned

$$\omega = 1.224 \times 10^{19} \, \text{s}^{-1} \text{ and } \omega_0 = 1.083 \times 10^{19} \, \text{s}^{-1}.$$

If we write

$$\frac{g}{m} = \psi \frac{\omega_0^2 - \omega^2}{\omega} \tag{6.64}$$

then we find

$$\phi_A = \frac{4.61}{1 + \psi^2} \tag{6.65a}$$

and

$$\phi_B = \frac{4.61\psi}{1 + \psi^2}. \tag{6.65b}$$

The actual behaviour of the electrons will depend very much on the value of $\psi$ or, what amounts to the same thing, on the value of $g$.

One possible mechanism which would give us an estimate of the value of $g$ is the radiation of energy by the electron – that is to say, that $g$ is a radiation-damping constant. In fact this does not tie in with our picture of a harmonically vibrating electron because the greatest radiation of energy, and therefore the greatest damping resistance, occurs when the acceleration is greatest which, for a simple-harmonic motion, is when the velocity is least. By writing the damping term as $g\dot{x}$ in equation (6.53) we have implied, on the contrary, that the damping force is proportional to the velocity. However we can calculate an order-of-magnitude value for a mean damping constant $\bar{g}$. It can be shown by a simple extension of equation (2.18) that an electron vibrating with an amplitude $R$ emits energy at an average rate

$$Q = \frac{e^2 R^2 \omega^4}{12\pi\varepsilon_0 c^3}. \tag{6.66}$$

The total distance it travels in unit time is

$$D = 4R \times \text{frequency} = \frac{2R\omega}{\pi} \tag{6.67}$$

and therefore the mean force it experiences due to the radiation of energy would be

$$\bar{F} = \frac{Q}{D} = \frac{e^2 R \omega^3}{24\varepsilon_0 c^3}. \tag{6.68}$$

The mean velocity of the electron also equals $D$ so that the average force per unit velocity becomes

$$\bar{g} = \frac{\bar{F}}{D} = \frac{\pi e^2 \omega^2}{48\varepsilon_0 c^3}. \tag{6.69}$$

We can test this mean value of damping constant to see whether or not it gives sensible values for the scattering fractions $\phi_A$ and $\phi_B$ when the incident radiation has the wavelength of the absorption edge, or is very close to that wavelength. For $\omega = \omega_0$ we have

$$\phi_A = 0$$

and

$$\phi_B = -\frac{\omega_0 m}{\bar{g}} = \frac{48\varepsilon_0 c^3 m}{\pi e^2 \omega} \tag{6.70}$$

or, when numerical values are substituted, $\phi_B = 1.2 \times 10^4$. This very large value, which implies that the electron is vibrating with $1.2 \times 10^4$ times the amplitude of a Thomson electron, shows that radiation damping is far too feeble to offer an explanation of the anomalous scattering effects which are actually observed.

Another approach which can give us a rough estimate of $g$ is by consideration of the mass absorption coefficient of the element in question. If the K absorption-edge wavelength is being scattered then the two K electrons should be oscillating more strongly than any others. Let us assume that the damping force, whatever its origin, is the mechanism by which X-radiation is absorbed by the atom and transformed into heat. The damping force acting on the electron is $g\dot{x}$ and when it moves through a distance $dx$ does work

$$dW = g\dot{x}dx = g\dot{x}^2 dt$$

or, from equation (6.55a),

$$dW = gR^2\omega^2 \sin^2(\omega t + \varepsilon)dt. \tag{6.71}$$

Since the average value of $\sin^2(\omega t + \varepsilon)$ is $\frac{1}{2}$ we have as the rate of loss of energy through damping forces for each K electron

$$\frac{dW}{dt} = \frac{1}{2}gR^2\omega^2. \tag{6.72}$$

When $\omega = \omega_0$, $R = E_0 e/g\omega_0$ so that

$$\frac{dW}{dt} = \frac{E_0^2 e^2}{2g}. \tag{6.73}$$

Now let us consider the situation when the radiation passes through a thin slab of material of thickness $dx$ and of unit cross-sectional area. If there are $N$ absorbing electrons per unit volume then the total energy absorbed per unit time by passing of the beam is

$$N dx \frac{dW}{dt} = \frac{NE_0^2 e^2}{2g} dx. \tag{6.74}$$

If the intensity of the incident beam is $I$ then this is also the energy passing into the block per unit time and equation (6.74) represents the change of this intensity, say $dI$. For an electromagnetic wave the electric-field strength and intensity are related by

$$I = E^2 c \varepsilon_0 \tag{6.75}$$

so that

$$dI = -\frac{Ne^2}{2gc\varepsilon_0} I \, dx. \tag{6.76}$$

The negative sign is put into equation (6.76) to indicate a reduction of the intensity of the beam; we can now compare equations (6.76) and (6.19) to obtain

$$\frac{Ne^2}{2gc\varepsilon_0} = \mu$$

or

$$g = \frac{Ne^2}{2c\mu\varepsilon_0} \tag{6.77}$$

where $\mu$ is the linear absorption coefficient of the material.

If the density of the material is $\rho$ and the atomic weight $M$ then the number of atoms per unit volume is

$$n = \frac{\rho}{M} N_A \tag{6.78}$$

where $N_A$ is Avogadro's number and if we assume two fully effective K electrons for each atom then $N = 2n$ and

$$g = \frac{\rho N_A e^2}{Mc\mu\varepsilon_0}$$

$$= \frac{N_A e^2}{Mc\mu_m\varepsilon_0}. \tag{6.79}$$

For an iron atom at its K absorption edge $\mu_m = 46 \, \text{m}^2 \, \text{kg}^{-1}$ and $M = 56$ so that

$$g = 2.26 \times 10^{-12} \, \text{kg s}^{-1}.$$

If this value of $g$ is used for all frequencies then $\phi_A$ and $\phi_B$ can be computed as a function of frequency or wavelength. The results of this calculation are shown in table 6.3.

Now the K electrons are the ones most tightly bound to the nucleus of the atom and the ones which most closely resemble point electrons. It will be noticed in fig. 2.12(b) that the scattering contribution of the K electrons of carbon do not fall off very rapidly with $\sin\theta/\lambda$. For an iron atom where the K electrons are even more tightly bound the fall-off will be even less and, for a reasonable range of $\sin\theta/\lambda$, may be neglected altogether.

From equation (2.38), if we refer to the normal atomic scattering factor as $f_0$, then we have

$$f_0 = \sum_{j=1}^{z} (p_s)_j.$$

If the non-anomalous contribution of the two K electrons is taken to be unity for each then we can write the actual scattering factor as

Table 6.3.

| $\lambda/\lambda_K$ | $\phi_A$ | $\phi_B$ | $f_1 = 2(\phi_A - 1)$ | $f_2 = 2\phi_B$ |
|---|---|---|---|---|
| 0.00 | 1.00 | 0.00 | 0.00 | 0.00 |
| 0.25 | 1.06 | 0.06 | 0.12 | 0.12 |
| 0.50 | 1.31 | 0.20 | 0.62 | 0.40 |
| 0.75 | 1.99 | 0.77 | 1.98 | 1.54 |
| 1.00 | 0.00 | 4.42 | −2.00 | 8.84 |
| 1.25 | −1.42 | 0.71 | −4.84 | 1.42 |
| 1.50 | −0.75 | 0.20 | −3.50 | 0.40 |
| 1.75 | −0.47 | 0.09 | −2.94 | 0.18 |
| 2.00 | −0.33 | 0.08 | −2.66 | 0.16 |

$$f = f_0 + f_1 + if_2 \tag{6.80}$$

where

$$f_1 = 2(\phi_A - 1) \tag{6.81a}$$

and

$$f_2 = 2\phi_B. \tag{6.81b}$$

The terms $f_1$ and $f_2$ are, respectively, the real and imaginary changes in the scattering factor due to anomalous scattering and these are shown in table 6.3 and in fig. 6.17.

It should not be expected that this rather crude theory based on classical ideas should give good agreement with experiment but it does give a reasonable qualitative picture. A quantum-mechanical treatment of anomalous scattering has been given by Hönl (1933) and the results obtained for iron from this theory are shown in fig. 6.18. The general form of the classical solution is quite good except that it predicts a finite $f_2$ for $\lambda > \lambda_K$ whereas the quantum-mechanical theory gives $f^2 = 0$ under this condition.

An important consequence of anomalous scattering is that it causes a breakdown of Friedel's law. Let us consider a non-centrosymmetrical structure containing five atoms in the unit cell. The separate contributions of the five atoms to the $(hkl)$ and $(\bar{h}\bar{k}\bar{l})$ structure factors can be shown on a phase-vector diagram as in fig. 6.19(a) and the equality of $|F_{hkl}|$ and $|F_{\bar{h}\bar{k}\bar{l}}|$ can be seen. However if the scattering factor of the fifth atom has anomalous components then $|F_{hkl}|$ is no longer equal to $|F_{\bar{h}\bar{k}\bar{l}}|$. Let the contribution to $F_{hkl}$ of the other four atoms be $R\exp(i\phi)$ and the scattering factor of the fifth atom be $f + if'$. Then

$$\begin{aligned} F_{hkl} &= R\exp(i\phi) + (f + if')\exp(2\pi i\mathbf{r}_5\cdot\mathbf{s}) \\ &= [R\cos\phi + f\cos(2\pi\mathbf{r}_5\cdot\mathbf{s}) - f'\sin(2\pi\mathbf{r}_5\cdot\mathbf{s})] \\ &\quad + i[R\sin\phi + f'\cos(2\pi\mathbf{r}_5\cdot\mathbf{s}) + f\sin(2\pi\mathbf{r}_5\cdot\mathbf{s})] \end{aligned} \tag{6.82}$$

and similarly

$$\begin{aligned} F_{\bar{h}\bar{k}\bar{l}} &= [R\cos\phi + f\cos(2\pi\mathbf{r}_5\cdot\mathbf{s}) + f'\sin(2\pi\mathbf{r}_5\cdot\mathbf{s})] \\ &\quad + i[R\sin\phi + f'\cos(2\pi\mathbf{r}_5\cdot\mathbf{s}) - f\sin(2\pi\mathbf{r}_5\cdot\mathbf{s})]. \end{aligned} \tag{6.83}$$

Fig. 6.17.
The anomalous
scattering components
$f_1$ and $f_2$ for iron from
the classical theory.

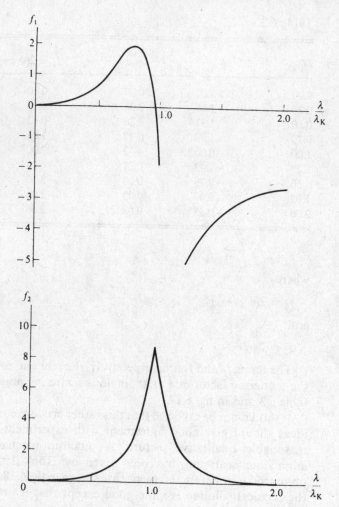

From equations (6.82) and (6.83) it will be seen that $|F_{hkl}| \neq |F_{\bar{h}\bar{k}\bar{l}}|$. This breakdown of Friedel's law is also shown in fig. 6.19(b) where it will be seen that the effect of the anomalous-scattering component, apart from a small change in the magnitude of the $f$, is to rotate the contributions to $F_{hkl}$ and $F_{\bar{h}\bar{k}\bar{l}}$ both in the same sense. The reader may readily confirm by drawing similar diagrams that, for centrosymmetric structures, Friedel's law is retained.

Anomalous scattering can be an important aid to structure determination and the method by which this is done is described in §8.5.

Fig. 6.18.
The anomalous
scattering components
$f_1$ and $f_2$ for iron from
the quantum-
mechanical theory.

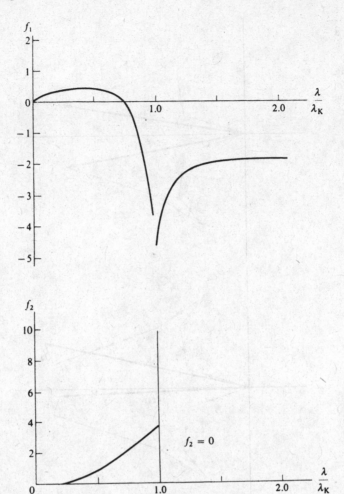

Fig. 6.19.
(*a*) Friedel's law for a
non-centrosymmetric
structure without
anomalous scattering.
(*b*) The breakdown of
Friedel's law with one
anomalous scatterer.

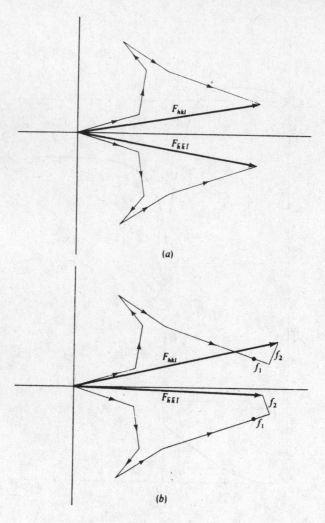

(*a*)

(*b*)

## Problems to Chapter 6

6.1 A crystal of volume $10^{-9}\,m^{-3}$ is rotated about its $c$ axis at an angular velocity of 0.02 radians sec$^{-1}$. A beam of Cu K$\alpha$ radiation ($\lambda = 1.54\,\text{Å}$) of intensity 10 W m$^{-2}$ is incident on the crystal at right angles to the axis of rotation. The structure is orthorhombic with $a = 6.0$, $b = 12.0$ and $c = 10.0\,\text{Å}$. How many photons will there be in the (240) diffracted beam for a single passage through the reflecting position if $|F_{240}| = 120$?

6.2 A material of density $1.28 \times 10^3\,kg\,m^{-3}$ has the chemical composition $C_{14}H_{22}S_2P_2$. What is the linear absorption coefficient of the material for Cu K$\alpha$ radiation?

| Element | $\mu_m(\lambda = 1.54\,\text{Å})$ | Atomic weight |
|---------|-----------------------------------|---------------|
| H | $0.04\,\text{m}^2\,\text{kg}^{-1}$ | 1 |
| C | 0.46 | 12 |
| P | 7.41 | 31 |
| S | 9.81 | 32 |

6.3 The crystal in Problem 6.1 is a block with sides parallel to $a, b, c$ of lengths 2.00, 1.00 and 0.50 mm, respectively. If the linear absorption coefficient of the material is $5.0 \times 10^3\,\text{m}^{-1}$ what is the absorption factor for the (240) reflection? Use the graphical method described in § 6.2.

6.4 Recalculate the structure factors in Problem 3.4 if the overall temperature factor for the structure is $3.0\,\text{Å}^2$. The program STRUCFAC (appendix I) can be used for this calculation.

6.5 A crystal of space group $P1$ with $a = 10.00$, $b = 7.00$ and $\gamma = 110°$ has a projection with the following structure:

| Atom | $x$ | $y$ |
|------|-------|-------|
| Br | 0.625 | 0.250 |
| N | 0.300 | 0.643 |
| C | 0.400 | 0.286 |
| C | 0.300 | 0.429 |
| C | 0.425 | 0.714 |
| C | 0.525 | 0.571 |
| C | 0.525 | 0.393 |

Calculate the structure factors of type ($hk0$) with STRUCFAC for Cu Kα radiation ($\lambda = 1.542\,\text{Å}$), given that the overall temperature factor is $3.0\,\text{Å}^2$ and that bromine has anomalous scattering components $f_1 = -0.96$ and $f_2 = 1.46$. Note the breakdown of Friedel's law.

# 7 The determination of space groups

## 7.1 Tests for the lack of a centre of symmetry

As we have seen in §1.4 measurements of crystals with an optical goniometer can, in favourable circumstances, reveal the crystal class. Such measurements should be carried out on many crystals for there is a tendency for crystals to exist in a number of slightly different forms in each of which some facets may not be present. By-and-large, when incorrect conclusions are drawn from optical goniometry it is in the direction of assigning too high a symmetry to the crystal. However it is sometimes possible to be reasonably sure by means of such measurements that a crystal structure either has or does not have a centre of symmetry.

There are a number of physical properties of crystals, the measurements of which can be used unequivocally to detect the *lack* of a centre of symmetry in a crystal structure. We shall briefly consider these, the physical principles involved and the apparatus which may be used for the tests.

### Piezoelectric effect

The first physical phenomenon we shall consider is that of piezoelectricity. This is the process whereby when a material is placed in an electric field it undergoes a mechanical strain and, conversely, when the material is mechanically strained it becomes electrically polarized and produces a field in its environment. Let us see what mechanism can produce this effect. In fig. 7.1(*a*) there is a schematic representation of a pair of atoms with their surrounding electron density. The two atoms are bonded together and the electron density is distorted from the configuration it would have for the superposition of that from two isolated atoms. The distortion is in the sense of building up extra electron density in the bond between the atoms which tends to attract the nuclei inwards and hence keep them together. If the pair of atoms is placed in an electric field this will lead to a displacement of the negatively charged electron cloud with respect to the positively charged nucleus. It may happen that the outer electrons of atom *B* are more shielded from the nucleus than are those of atom *A* and hence they may more easily be displaced. Consequently the net migration of electrons, as shown in fig. 7.1(*b*), may be such as to enhance the build-up of electron density in the bond. This will have the effect of attracting the two nuclei inwards and so of shortening the bond. This diminution in bond length would be small and dependent on the first power of the field; if the direction of the field was reversed then some electron density would move out from between the nuclei and the bond length would increase.

For a crystalline material the field will produce an induced polarization and all the bonds would be affected in one or other of the two possible ways, and in varying degrees, so that the aggregate effect would be to change the physical dimensions of the crystal. The sign of the change of dimensions

Fig. 7.1.
(a) Representation of
electron density
associated with a pair
of bonded atoms.
(b) Change of electron
density in an electric
field.

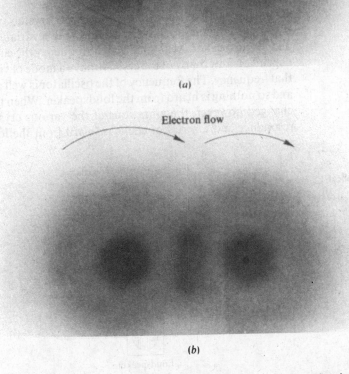

(a)

Electron flow

(b)

would depend on whether the shortening or lengthening of bonds predominates. However it should be noted that an overall effect can only occur for a non-centrosymmetric crystal. In fig. 7.2(a) a one-dimensional crystal structure is illustrated and it will be seen that there are four different types of bond which, depending on their nature, can give an overall increase or decrease in the dimension of the crystal. On the other hand, for the centrosymmetric structure shown in fig. 7.2(b) there will always be pairs of bonds, such as $AB$ and $A'B'$, whose contributions to the change of dimensions are equal and opposite so that there is no externally observable effect.

The piezoelectric effect is reversible so that, if the material is physically strained, the electronic charge will flow in and out of the regions between nuclei. The net flow of charge will lead to a polarization of the material and hence a field will be produced around it. This effect is used in the quartz-controlled oscillator, the natural frequency of one of the modes of mechanical vibration of the crystal controlling the electrical oscillation of a circuit coupled to it.

A very simple type of piezoelectric-effect tester is shown in fig. 7.3. The

Fig. 7.2.
(*a*) A one-dimensional non-centrosymmetric structure; no bonds in one unit cell are related.
(*b*) A one-dimensional centrosymmetric structure; bonds exist in pairs in opposite directions.

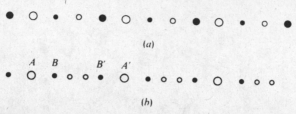

sample, in the form of a large number of crystals, is placed between the plates of a condenser which is in series with a variable-frequency oscillator and a loudspeaker. At any particular frequency the current in the circuit will depend on how many of the crystals have a mode of vibration at, or close to, that frequency. The frequency of the oscillator is well above the audio range and so nothing is heard from the loudspeaker. When the frequency is slowly changed however, the vibrations of the various crystals go in and out of resonance and a series of clicks is heard from the loudspeaker.

Fig. 7.3.
A piezoelectric-effect detector.

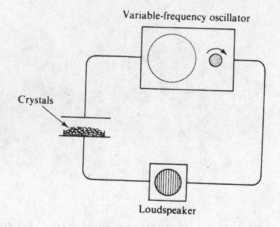

It must be emphasized that while the detection of the piezoelectric effect certainly denotes the lack of a centre of symmetry no conclusion is possible from a negative result as the piezoelectric effect may be present but just be too feeble to be detected. In addition Wooster (1938) showed that the symmetry of one point group, 4 3 2, does not permit a piezoelectric effect.

## Pyroelectric effect

Another physical property which can be detected in non-centrosymmetric crystals is that of pyroelectricity. This is the electrical polarization of a crystal shown by the development of opposite charges on two opposite faces of a crystal when it is heated. In fact since a crystal must always be at some temperature it suggests that pyroelectric crystals should be polarized even at room temperature and, if this is not observed, it is because the charges on the surface are usually neutralized by stray charges on particles of dust, etc. Thus it is inferred that in a pyroelectric crystal there exists permanent polarization which suggests the presence of electric dipoles within the crystal. If this is so then the additional polarization due to heating can be ascribed to asymmetric oscillation of the dipoles. In fig. 7.4 the situation is depicted where on heating the crystal the mean distance apart of the charged atomic groups, and therefore the mean dipole moment is increased by raising the temperature.

A very simple form of pyroelectric test is to insert into a specimen consisting of small crystals a heated glass rod. If the crystals are pyroelectric they will develop charges on their surfaces, induce opposite charges on the rod, and so adhere to it. The precaution should be taken of repeating the experiment with a cold rod to ensure that some other effect is not causing the adhesion.

The above test is not a very sensitive one and other types of test are possible. However tests for pyroelectricity are rarely performed by crystallographers so we shall pursue this topic no further.

Fig. 7.4.
Asymmetric oscillations of charged groups give rise to increased mean dipole moment as temperature increases.

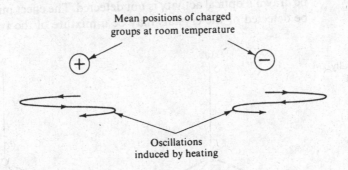

Mean positions of charged
groups at room temperature

Oscillations
induced by heating

## Optical activity

A material is said to be optically active if, when plane-polarized light passes through it the plane of polarization is rotated. Some materials, such as quartz and certain inorganic salts, are optically active in the crystalline state but not when fused or dissolved in solution. This means that it is the specific arrangement of atoms in the crystal that determines the optical activity. There are other substances, notably organic molecular substances, which are optically active when crystalline, molten, vaporized or in solution, and for these the optical activity is clearly a property of the molecular structure.

When materials are optically active they often exist in two forms which rotate the plane of polarization in opposite senses. If looking towards the source of light gives a clockwise rotation the crystal is said to be dextro-rotatory; otherwise it is laevo-rotatory. The two contrary-rotating forms of the material are enantiomorphs, that is to say that structurally they are mirror images of each other. Sometimes this shows itself in the crystals themselves, the sets of normals to the faces being mirror related in the two types of crystal. Clearly centrosymmetrical structures cannot be optically active since centrosymmetrically related objects are also enantiomorphically related and a centrosymmetric structure is its own enantiomorph. The same is true of structures with planes of symmetry or inversion axes.

For organic materials the optical activity is associated with the tetrahedral arrangement of bonds around a carbon atom bonded to four other atoms. This is illustrated in fig. 7.5(*a*) for two carbon atoms with four different atoms attached to each; the two arrangements are enantiomorphically related. In fig. 7.5(*b*) two enantiomorphically related molecules of tartaric acid are shown.

It is not too simple to observe optical rotation from crystals because, as we shall see in the following section, light passing through any other than a cubic crystal is usually split into two components polarized at right angles. If the crystals are soluble then the solution can be tested by means of the apparatus shown in fig. 7.6. The light is plane-polarized by a polarizer *P* (a Nicol prism or polaroid sheet for example) and after passing through the tube of solution traverses the analyser *A*. The polarizer and analyser are initially adjusted so as to give no passage of light when the tube is full of solvent. If light passes through when the solvent is replaced by solution then the dissolved crystals must consist of molecules of one enantiomorph and so the space group cannot be centrosymmetric. However no conclusion can be drawn if optical activity is not detected. The effect might be too small to be detected or the crystals may be a mixture of the two enantiomorphic

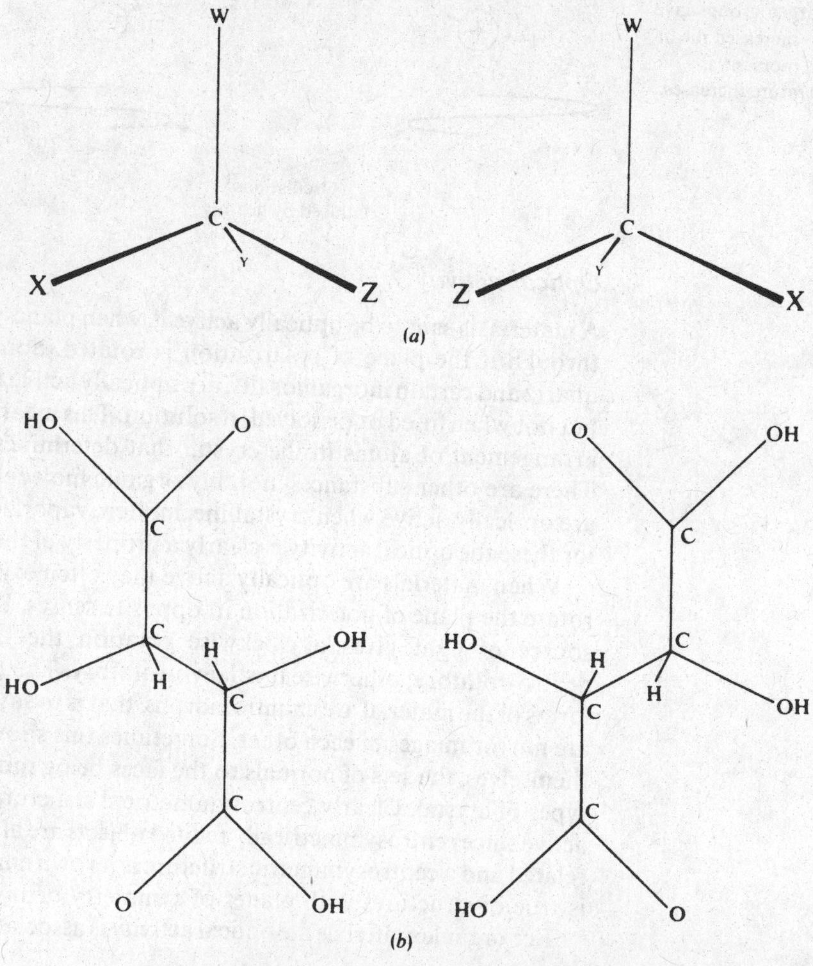

Fig. 7.5.
(*a*) Enantiomorphically related four-bonded carbon atoms.
(*b*) Enantiomorphic forms of tartaric acid.

(*a*)

(*b*)

Fig. 7.6.
A simple polarimeter.

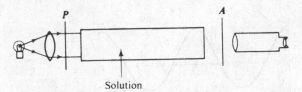

Solution

forms (racemic mixture) and hence neutralize the effect of each other in solution.

### Frequency doubling

Light is propagated in materials by inducing vibrations of the charged components, mainly electrons, within it. If for a region round a small element of charge there is a lack of symmetry then the vibration of the charge element in an electric field due to a light wave can be anharmonic. This is only likely to occur appreciably at high fields (high intensity of light) and the general form of the vibration is shown in fig. 7.7(a). It is due to the fact that the restoring force per unit displacement of charge is a variable and is less for displacements in one direction than for the other. Let us assume that the vibration has the form

$$x = A \cos \omega t (1 + \alpha \cos \omega t) \tag{7.1}$$

which is that shown in fig. 7.7(a) with $\alpha = 0.2$. Then

$$x = \tfrac{1}{2} A \alpha + A \cos \omega t + \tfrac{1}{2} A \alpha \cos 2\omega t \tag{7.2}$$

and it can be seen that the displacement has a constant component which is equivalent to a polarization of the material and also one of twice the incident frequency.

The effect of this is that when a beam of light, of frequency $v$, passes into the crystal then a second harmonic of frequency $2v$ is generated. The form of displacement, equation (7.1), is rather special and for more general displacement–time relationships third and higher-order harmonic generation is possible. However since the non-linear component will only be appreciable for high intensities it requires the light from a pulsed laser to easily observe these effects.

It should be noted that harmonic generation may occur only for a non-centrosymmetric crystal. When there is a centre of symmetry then to any displacement form such as fig. 7.7(a) there is another, coherent in-phase displacement, like fig. 7.7(b) and, in combining these two, the second and higher harmonics do not appear.

There is a snag connected with harmonic generation; due to dispersion the different frequencies of light travel at different speeds in the crystal and hence there is a loss of phase correlation for the second harmonic generated at different points along the path of the incident beam. In some crystals there are certain directions in which the phase correlation is preserved and then there can be a very high efficiency for second harmonic generation – for example 20% for potassium dihydrogen phosphate.

This provides a very sensitive means of detecting the lack of a centre of

Fig. 7.7.
(a) Asymmetric
oscillations of electrons
giving rise to harmonic
generation.
(b) Oscillation of
electrons in
centrosymmetric
relationship to those in
(a).

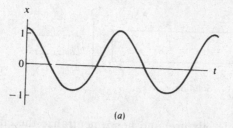

(a)

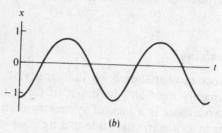

(b)

symmetry. A typical pulse from a moderate-sized ruby laser contains 1 joule of energy or about $10^{18}$–$10^{19}$ photons. A fairly straightforward photocell can detect 1000 photons so that a second harmonic conversion efficiency of about 1 part in $10^{15}$ should be detectable. In fig. 7.8 the specimen is a quantity of small crystals randomly oriented so that some may be well placed for most efficient harmonic generation. The light passing through the crystals is filtered to remove most of the primary frequency and then passed into a grating spectrometer set at the second harmonic frequency. A second filter will remove any primary radiation which reaches the slit $S$ by random scattering and the second harmonic can be detected by the photocell $P$.

A commercial instrument incorporating these ideas, the SHA or Second Harmonic Analyser, has been produced by the North American Philips Corporation.

## 7.2 The optical properties of crystals

We have, in the previous section, already considered two optical properties associated with non-centrosymmetrical crystals – optical activity and frequency doubling. Another revealing property of a crystal is its refractive index or, what amounts to the same thing, the velocity of light in the crystal. It is found that the refractive index of a crystal sometimes varies with direction and the way that this happens is related to the crystal system.

There are many excellent text books, some of which are referred to in the bibliography, dealing fully with the subject of crystal optics. In this section there is given a very brief account of the subject which, inevitably, must fall short in clarity from that of a longer treatment. For those readers with some preliminary knowledge about the behaviour of light in crystals it is hoped that the material given here will be informative. For other readers there is appended to this section a short summary of the information which can be elicited by the examination of crystals with a microscope.

One way of envisaging the variation with direction of the refractive index

Fig. 7.8.
Harmonic-generation
detector.

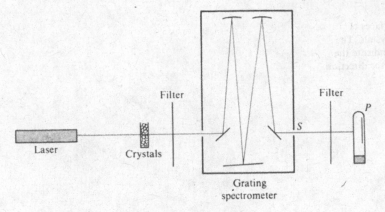

of a crystal is to consider a point source of light within the material and to note the limits of the disturbance after some time $\tau$. For an isotropic material such as glass the disturbance will, as one would expect, extend out to a spherical surface. All points on this surface will act as a new source of light and so the wave front spreads out preserving its spherical symmetry.

It is found that a crystal belonging to the cubic system behaves in a similar way. The cubic symmetry implies that any physical properties of the crystal are identical in the three mutually orthogonal directions along the axes. In addition the properties in sets of directions related by the triad or tetrad axes would also have to be identical; if the surface of the disturbance was not a sphere it would have to be something else having cubic symmetry.

The next type of crystal we shall consider is the uniaxial crystal which is a crystal having a unique axis and belonging to the trigonal, tetragonal or hexagonal systems. The light from a point source within such a crystal will be found to split into two components which are polarized at right angles to each other and which in general have different refractive indices. The *ray-velocity* (or *wave*) *surfaces* may be of two types which are shown in one octant of a set of cartesian axes in fig. 7.9($a$) and the planes through the surfaces containing the unique axis are shown in fig. 7.9($b$). The following points will be noted:

(i)  One of the wave surfaces is spherical while the other is an oblate spheroid (negative uniaxial crystal) or a prolate spheroid (positive uniaxial crystal). A spheroid is an ellipsoid with two equal axes and in this case the equal axes are perpendicular to the unique axis so that the spheroid has circular sections perpendicular to the unique axis. The axis of the spheroid along the unique axis equals the radius of the spherical surface.

(ii) The surface has been called a ray-velocity (or wave) surface and this is not to be confused with a *wave front* which is a surface of constant phase moving through the crystal. If we consider a plane wave front, $OO'O''$, moving through the medium, as in fig. 7.10, then it is possible to determine, by Huygens' construction, the new position of the wave front, $PP'P''$, after a time $\Delta\tau$. Thus it would be possible to interpret the

Fig. 7.9.
(a) Ray surfaces of
uniaxial crystals. The
striations indicate the
electric-vector direction.

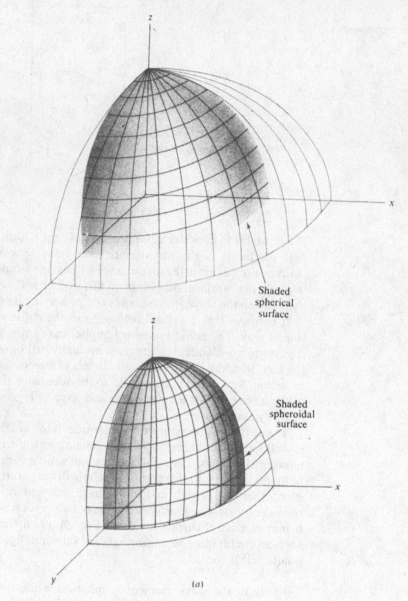

Shaded
spherical
surface

Shaded
spheroidal
surface

(a)

spheroidal wave surface as the envelope of the wave fronts moving
outwards in various directions in the crystal as is shown in fig. 7.11.

(iii) The striations which define the surfaces in fig. 7.9(*a*) also indicate the
directions of the electric vector for the plane wave moving in the
appropriate direction in the crystal. Thus in fig. 7.12 the initial unpolarized wave front, *OO'*, whose normal is in the direction *OP* is split
into two components, *PP'* and *QQ'*, moving at different velocities and
with their electric vectors in the directions shown. It will be noted that
the unique axis is also a direction in which a plane wave of any state of
polarization (or unpolarized) will always move with the same velocity
and for this reason it is termed the *optic axis* of the crystal. It must be

Fig. 7.9 (*b*) Sections of
the ray surfaces
containing the optic axis.

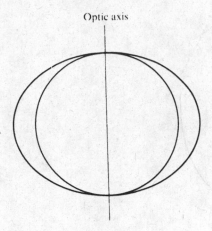

Optic axis

Optic axis

(*b*)

stressed that the optic axis represents a *direction* in the crystal *not* a
particular line.

Fig. 7.10.
Wave-front generation
by extraordinary ray
component.

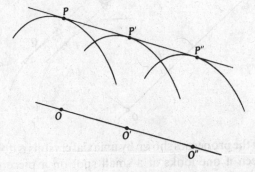

There is a certain self-consistency in the form of these surfaces. The
cross-sections of the spheroidal surface perpendicular to the optic axis are
circular; in particular this implies that all directions are optically equivalent

Fig. 7.11.
Ray front as envelope
of wave fronts.

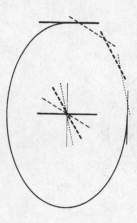

perpendicular to the optic axis. For trigonal, tetragonal or hexagonal
symmetry one would expect sets of directions related by these symmetries
to be equivalent and hence it can be seen that if the surface is to be any sort
of ellipsoid at all it has to be a spheroid. Again, if one looks at the spherical
surface one sees that the electric-vector directions are all perpendicular to
the optic axis, and hence optically equivalent, no matter in which direction
in space the light travels.

Fig. 7.12.
Generation of wave
fronts and direction of
electric vectors for the
two components in a
uniaxial crystal.

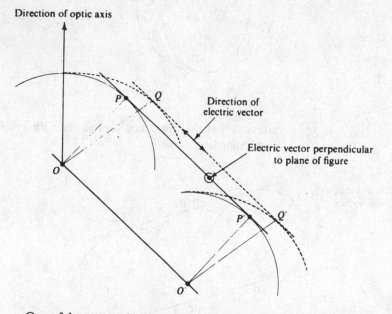

One of the properties shown by uniaxial crystals is double refraction. This
can be seen if one looks at a small spot on a piece of paper through a
parallel-sided slab of calcite. Two images of the spot are seen – one where
expected along the normal to the slab but the other displaced from the
normal. A piece of Polaroid held above the slab and rotated extinguishes the
spot images in turn at positions separated by $\pi/2$, showing that the light
forming each of the images is plane-polarized and that the planes of
polarization are perpendicular to each other. If the calcite slab is rotated as

illustrated in fig. 7.13(a) then the expected image, O, does not move whereas the unexpected one, E, rotates about O. The reason for this behaviour may be followed in fig. 7.13(b) which shows a principal section of the calcite slab, that is to say, a section containing the optic axis, which is also perpendicular to the surface of the slab. A plane wave front OO' falls normally on the surface and it can be seen that it is split into two plane waves, one moving normal to the surface and the other at some angle to the normal. This latter wave front is *not* normal to its direction of motion. The wave front is a surface of constant phase while the direction of motion of the light is the direction in which the energy moves and these need not be perpendicular to each other. However the electric vector is always in the plane of the wave front or is tangential to it if the wave front is not planar. The rays and waves which behave like those in an isotropic medium and give rise to the image O are called the *ordinary* rays and waves; the others are called *extraordinary* rays and waves.

An experiment which clearly identifies a uniaxial crystal is to examine it in convergent light between crossed Polaroids (or other devices for producing plane-polarized light). The experimental arrangement is shown schematically in fig. 7.14(a); light from an extended source S passes through

Fig. 7.13.
(a) Path of extraordinary image when crystal rotates.
(b) Production of ordinary and extraordinary images in a uniaxial crystal.

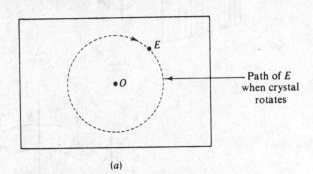

(a)

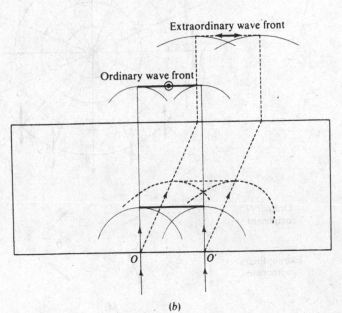

(b)

Fig. 7.14.
(a) Experimental
arrangement for
observing a crystal in
divergent light.
(b) Cone of light of
semi-angle $\theta$ entering the
point $O$ of a crystal.
(c) The division of the
cone of light into two
components.
(d) The relative
amplitudes and
directions of the
components for various
rays in the cone.

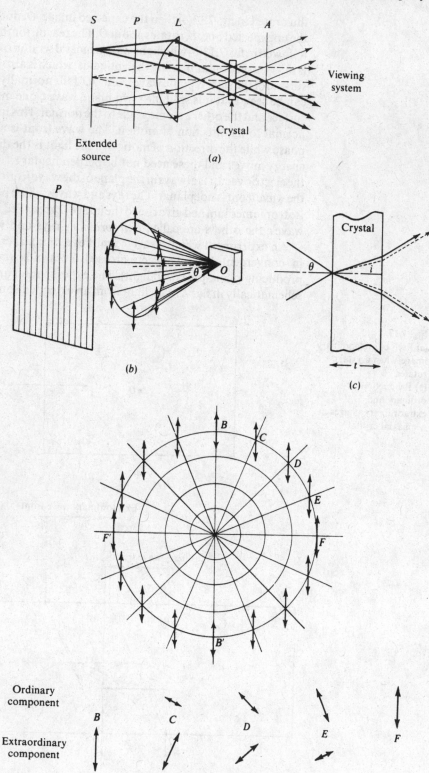

a polarizer $P$ and a lens $L$ so that, at all points of the crystal, light passes through for a whole range of angles contained within a fairly wide-angle cone. After passing through the crystal and a crossed Polaroid analyser $A$ the light is viewed by what amounts to a telescope focussed at infinity so that what is seen at any point of the field of view is representative of the light passing in some direction through the crystal over the whole of its area.

In fig. 7.14(b) there is represented a cone of light passing in at the point $O$ of the crystal and making an angle $\theta$ with the optic axis which is here assumed to be normal to the crystal face. The direction of the electric vector for light passing through $P$ is shown by the striations and this is also shown by the double-headed arrows at the base of the cone. Within the crystal the original cone will split up into two separate cones corresponding to the two wave surfaces in a uniaxial crystal but once they emerge from the crystal they will again each form a cone of semi-angle $\theta$. This is shown in fig. 7.14(c); the relative displacement of the two emergent cones will, in practice, be very small.

We now consider the separation of the original cone into the separate cones. In fig. 7.14(d) we look at the situation in (b) straight down the optic axis and we also show the striations corresponding to the wave surfaces in fig. 7.9(a). The circles correspond to striations on the sphere, the ordinary wave surface, and the radial lines to striations on the spheroid, the extraordinary wave surface. It is clear from (d) that a ray passing through the point $B$ of the cone produces only an extraordinary ray in the crystal since there is no component of the electric vector in the direction of the circular striations. When this ray reaches the crossed Polaroid analyser $A$ it will be completely absorbed and so no light gets through the system in a direction corresponding to $B$. Similarly a ray passing through the point $F$ of the cone produces only an ordinary ray and again is blocked by the analyser. It should be noticed that this conclusion for the points $B$ and $F$ (also $B'$ and $F'$) is independent of $\theta$ and so this produces in the field of view a dark cross as is shown in fig. 7.15.

If, on the other hand, the direction of the light corresponds to the point $D$ where $OD$ makes an angle $\pi/4$ with $OB$ then two components pass through the crystal which are of equal magnitude. It can be seen in fig. 7.14(c) that there will be a phase difference between the components introduced by passage through the crystal given by

$$\phi = \frac{2\pi}{\lambda}(\mu - \mu')t \sec i \tag{7.3}$$

Fig. 7.15.
Appearance of uniaxial
crystal in convergent
light between crossed
Polaroids with optic axis
perpendicular to crystal
face.

where $\mu$ and $\mu'$ are the refractive indices for the two polarization directions and $i$ is assumed to be the same for both rays. It is clear that as $i$ increases from zero so $\phi$ monotonically increases from zero ($\mu = \mu'$ when $i = 0$). We can see in fig. 7.16 how the two components recombine for various values of $\phi$ between 0 and $2\pi$. When $\phi$ is a multiple of $2\pi$ the recombined beams give a resultant equivalent to that entering the crystal – and this cannot go through the crossed Polaroid $A$. This gives a dark point in the field of view. For other ray directions, with the same value of $\theta$ but given by $C$ and $E$ on either side of $D$, the two components in the crystal are unequal in intensity and one does not get complete blackness when $\phi$ is a multiple of $2\pi$. The complete resultant field of view is shown in fig. 7.15 and when this is seen it denotes a uniaxial crystal. We have assumed that the light is monochromatic and that one obtains only patterns of light and dark regions. If white light is used then the condition for blackness in the field of view varies with the wavelength and the result is a series of coloured rings although the dark cross is retained.

Fig. 7.16.
Resultant of adding two perpendicularly plane-polarized beams of light with various phase differences.

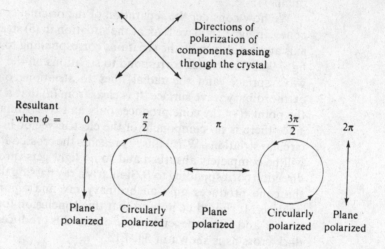

Directions of
polarization of
components passing
through the crystal

Resultant
when $\phi$ =

If the crystal surface is parallel to the optic axis then the appearance of the field under the same experimental conditions as in fig. 7.14($a$) is shown in fig. 7.17.

Fig. 7.17.
Appearance of uniaxial crystal in convergent light between crossed Polaroids with optic axis contained in the crystal face.

Optic axis

Crystals belonging to the triclinic, monoclinic and orthorhombic systems are known as biaxial crystals and their ray surfaces are much more complicated than those for uniaxial crystals; a typical ray surface is shown in fig. 7.18(*a*). The surface is represented in one octant of a set of rectangular Cartesian axes and is completed by mirror reflection in each of the principal planes. The striations represent the direction of the electric vector for a plane wave tangential to the surface at the point in question.

The surface characteristics can also be appreciated by their intersections with the principal planes which are shown in fig. 7.18(*b*). In the *XZ* plane, for example, the curves of intersection are a circle and an ellipse. It can be seen that rays travelling out to the circular boundary all travel with the same velocity and this is consistent with what is seen in fig. 7.18(*a*) – that the electric-vector direction is equivalent for each of them and along the *Y* direction. The same consistency will be found for the circular boundaries in the *XY* and *YZ* planes.

In the *XY* plane the two curves cross over and so a ray moving in the direction *OP* has the same velocity for any plane of polarization. This is evident from fig. 7.18(*a*) which shows that all directions of the electric vector occur at *P*. Biaxial crystals have two optic axes but they are *not* in the directions *OP* and *OP'*. These latter are known as *ray axes*; on the other hand, the optic axes are directions in which there is a single *phase velocity* and they are the normals to the common tangent planes to the two parts of the surface. The optic axes are shown as *OQ* and *OQ'* in fig. 7.18(*b*). This point is illustrated in fig. 7.19 where two points, $O_1$ and $O_2$, on a plane wave front are used as origin points for Huygens-type wavelets to construct a new position of the wave front. It can clearly be seen that if the wave front is parallel to the common tangent plane to the two parts of the surface then there is only a single phase velocity. This definition of an optic axis also applies to the uniaxial crystal for which the ray axis and optic axis are identical.

If a biaxial crystal is used in the experimental arrangement of fig. 7.14 then the resultant field of view shows the presence of the two optic axes. If the crystal face is perpendicular to the bisector of the acute angle between the optic axes then the appearance will take a symmetrical form dependent on the orientation of the polarizer to the line joining the optic axes. Two examples of these forms are shown in fig. 7.20.

*Summary*

1. In terms of crystal optics crystals can be divided into three types:
   (*a*) Cubic – these are isotropic in their optical properties and behave optically like a non-crystalline material such as glass.
   (*b*) Uniaxial – these are crystals belonging to the trigonal, tetragonal and hexagonal systems. Along the unique axis of the crystal, called the optic axis, the refractive index of light is independent of its state of polarization. For other directions light breaks up into two components polarized at right angles to each other and moving with different velocities.

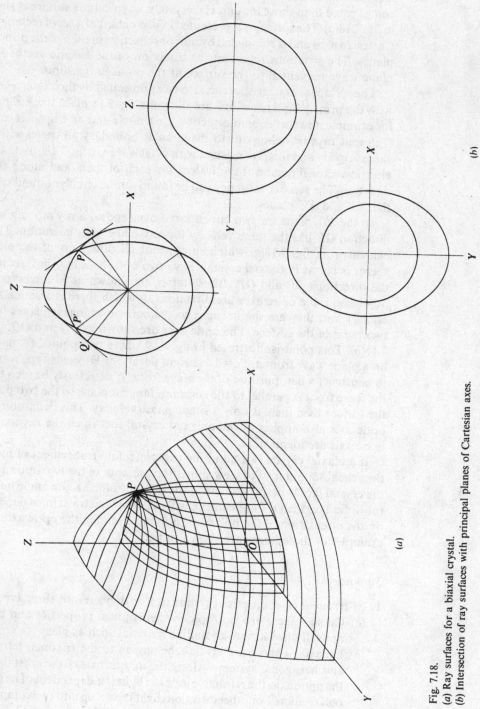

Fig. 7.18.
(a) Ray surfaces for a biaxial crystal.
(b) Intersection of ray surfaces with principal planes of Cartesian axes.

Fig. 7.19.
The optic axis is along
the direction in which
both rays have the same
phase velocity. It is
perpendicular to the
common tangent surface
to two parts of the ray
surface.

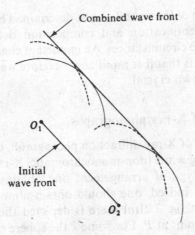

Fig. 7.20.
Appearance of a biaxial
crystal in convergent
light between crossed
Polaroids for two
different orientations of
the crystal.

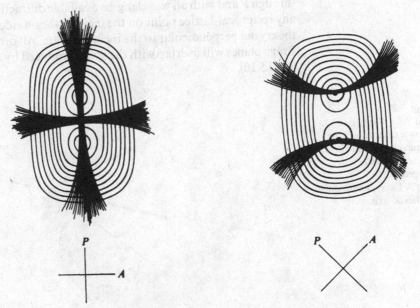

(c) Biaxial – these are crystals belonging to the triclinic, monoclinic or orthorhombic systems. There are two directions in which light of any polarization moves with the same phase velocity. These are the two optic axes of the crystal.

2. If a crystal is viewed in convergent light between crossed Polaroids as in fig. 7.14(a) then one can usually distinguish which type of crystal one is dealing with. Some characteristic patterns are shown:

(a) Fig. 7.15 – a uniaxial crystal with the optic axis normal to the crystal face.

(b) Fig. 7.17 – a uniaxial crystal with the optic axis contained in the crystal face.

(c) Fig. 7.20 – a biaxial crystal with the normal to the crystal face bisecting the acute angle between the optic axes.

Other forms of these patterns exist and can be recognized.

3. The orientation of the crystal can be determined by examination with a microscope; identification and composition determination are also possible in some circumstances. An important characteristic of this type of examination is that it is rapid and therefore well worth doing with a new and unknown crystal.

## 7.3 The symmetry of X-ray photographs

The very first type of X-ray diffraction photograph, the Laue picture, was made by diffracting white (non-monochromatic) X-rays from a stationary crystal. For this type of arrangement one obtains a large number of diffraction spots – indeed, one should obtain almost a complete set of diffraction data. In fig. 7.21(a) there is depicted the crystal at $O$ and a reciprocal lattice point at $P$. For some $\lambda$ the sphere of reflection will pass through $P$ and with all wavelengths available diffraction may take place for any reciprocal-lattice point on the incident-beam side of the plane through the crystal perpendicular to the incident beam. All orders of the same set of Bragg planes will overlap with the spots produced by different wavelengths (see § 5.10).

Fig. 7.21.
(a) With X-ray beam down triad axis three equivalent points fall on sphere of reflection together.
(b) A Laue picture down a hexad axis.

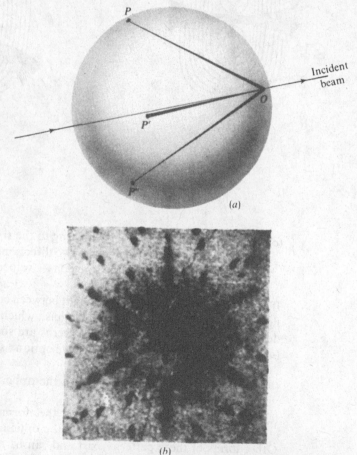

(a)

(b)

The Laue photograph is particularly suited to detecting symmetry axes if the X-ray incident beam is directed along, or very close to, the symmetry axis. In fig. 7.21(*a*) the incident beam is directed along a trigonal axis so that there are two other reciprocal-lattice points $P'$ and $P''$ equivalent to $P$ and related to it by the trigonal axis. It is evident that the value of $\lambda$ which gives the reflection $P$ will also give those corresponding to the symmetry-related ones, $P'$ and $P''$. These will form an equilateral triangle of spots on the film, all of the same intensity. Even if the axis is slightly offset from the incident beam the trigonal symmetry will still be evident. The triangles of spots will not be equilateral and they may vary slightly in intensity since they are produced by different wavelengths but the overall evidence from the whole photograph is usually unmistakable. A typical Laue photograph taken down a hexad axis is shown in fig. 7.21(*b*).

One can also detect on other kinds of X-ray photographs sets of equivalent reflections (other than $F_{hkl}$ and $F_{\bar{h}\bar{k}\bar{l}}$) which give information about symmetry elements in the crystal. For example if a unit-cell axis, say $b$, is a diad axis, a twofold screw axis or is perpendicular to a mirror plane then

$$|F_{hkl}| = |F_{h\bar{k}l}| = |F_{\bar{h}kl}| = |F_{\bar{h}k\bar{l}}|. \tag{7.4}$$

We can deduce this relationship from the basic structure-factor relationship, equation (3.41), by noting that for these symmetry elements there are pairs of atoms with related positions as follows:

diad axis      $(x, y, z); (\bar{x}, y, \bar{z})$
screw diad axis $(x, y, z); (\bar{x}, \frac{1}{2} + y, \bar{z})$
mirror plane    $(x, y, z); (x, \bar{y}, z).$

Thus with a diad axis

$$F_{hkl} = \sum_{j=1}^{N} f_j \exp\{2\pi i(hx_j + ky_j + lz_j)\}$$

$$= \sum_{j=1}^{\frac{1}{2}N} f_j [\exp\{2\pi i(hx_j + ky_j + lz_j)\} + \exp\{2\pi i(-hx_j + ky_j - lz_j)\}]$$

$$= \sum_{j=1}^{\frac{1}{2}N} 2f_j \exp(2\pi iky_j)\cos\{2\pi(hx_j + lz_j)\} \tag{7.5}$$

from which relationship (7.4) follows. The reader may confirm that relationship (7.4) also follows for the other two symmetry elements and also for any tetrad or hexad axis.

If an oscillation photograph is taken about a 2 or $2_1$ axis or about an axis perpendicular to a mirror plane then the photograph will have the zero layer as a line of mirror symmetry. This will be seen in fig. 7.22 which shows that the reciprocal-lattice points $(hkl)$ and $(h\bar{k}l)$ cut through the sphere of reflection simultaneously. When an oscillation photograph shows a mirror line then it is certain that the space group is at least monoclinic and if mirror lines are located about two different axes the symmetry must be at least orthorhombic (a third axis giving a mirror line must also then be present).

The symmetry of the weighted reciprocal lattice, which can be obtained from X-ray photographs, will reveal the presence of all the symmetry

Fig. 7.22.
With X-ray beam
perpendicular to 2, $\bar{2}$ or
$2_1$ axis the points $(hkl)$
and $(h\bar{k}l)$ touch the
sphere of reflection
simultaneously.

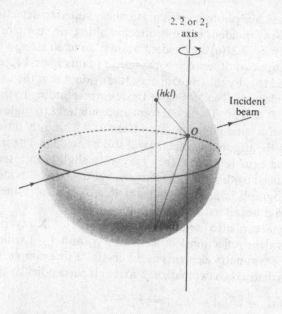

elements associated with the various crystal classes except for a centre of symmetry. Because of Friedel's law all weighted reciprocal lattices have a centre of symmetry whether or not the crystal structure has one. It should be noted that Friedel's law does not imply that every X-ray photograph will have a centre of symmetry since, in general, only a limited amount of data is recorded at any one time and this is usually from some asymmetric region of the reciprocal lattice. There are only eleven possible symmetry arrangements for diffraction data; these are known as the eleven Laue groups. They are associated with the crystal classes in accordance with table 7.1.

## 7.4 Information from systematic absences

The types of information we have considered so far have enabled deductions to be made about the crystal system or even the crystal class. If the X-ray diffraction data are examined in greater detail there can often be found sufficient information to enable the space group to be determined unambiguously or at least to be narrowed down to a choice of two or three.

The type of observation we shall consider in this section is that of systematic absences – reflections which are systematically zero because of space-group considerations rather than accidentally zero (or unobservably small) due to the particular arrangement of atoms in an asymmetric unit. Such systematic absences are always found whenever there is present a symmetry element which involves a translation – for example screw axes, glide planes and non-primitive lattices. We shall consider the systematic absences due to some of the more commonly occurring types of symmetry element.

### C-face centring

Atoms occur in pairs with coordinates $(x, y, z)$ and $(x + \frac{1}{2}, y + \frac{1}{2}, z)$. The structure-factor equation can be written as

Table 7.1.

| Laue group | Crystal classes | | | |
|---|---|---|---|---|
| $\bar{1}$ | 1, | $\bar{1}$ | | |
| $2/m$ | 2, | $m$, | $2/m$ | |
| $mmm$ | 222, | $mm2$, | $mmm$ | |
| $4/m$ | 4, | $\bar{4}$, | $4/m$ | |
| $4/mmm$ | 422, | $4mm$, | $\bar{4}2m$, | $4/mmm$ |
| $\bar{3}$ | 3, | $\bar{3}$ | | |
| $\bar{3}m$ | 32, | $3m$, | $\bar{3}m$ | |
| $6/m$ | 6, | $\bar{6}$, | $6/m$ | |
| $6/mmm$ | 622, | $6mm$, | $\bar{6}2m$, | $6/mm$ |
| $m3$ | 23, | $m3$ | | |
| $m3m$ | 432, | $\bar{4}3m$, | $m3m$ | |

$$F_{hkl} = \sum_{j=1}^{\frac{1}{2}N} f_j [\exp\{2\pi i(hx_j + ky_j + lz_j)\}$$

$$+ \exp\{2\pi i(hx_j + ky_j + lz_j) + \pi i(h+k)\}]. \tag{7.6}$$

Now

$$\exp(i\phi) = \exp i(\phi + 2n\pi) \tag{7.7a}$$

and

$$\exp(i\phi) = -\exp i\{\phi + (2n+1)\pi\} \tag{7.7b}$$

where $n$ is an integer. From this we see that for $C$-face centring there are systematic absences given by

$$F_{hkl} = 0 \text{ for } h + k \text{ odd for any } l.$$

When this type of systematic absence is found the presence of $C$-face centring is established.

### $2_1$ axis parallel to $b$

Atoms occur in pairs with coordinates $(x, y, z)$ and $(\bar{x}, \frac{1}{2} + y, \bar{z})$. The structure-factor equation can be written as

$$F_{hkl} = \sum_{j=1}^{\frac{1}{2}N} f_j [\exp\{2\pi i(hx_j + ky_j + lz_j)\}$$

$$+ \exp\{2\pi i(-hx_j + ky_j - lz_j) + \pi ik\}]. \tag{7.8}$$

From equation (7.7) we find, for $k$ even,

$$F_{hkl} = \sum_{j=1}^{\frac{1}{2}N} 2f_j \exp(2\pi iky_j)\cos\{2\pi(hx_j + lz_j)\} \tag{7.9a}$$

and, for $k$ odd,

$$F_{hkl} = \sum_{j=1}^{\frac{1}{2}N} 2f_j\exp(2\pi iky_j)\sin\{2\pi(hx_j + lz_j)\} \tag{7.9b}$$

An examination of these expressions will show that the type of systematic absence which can occur is that

$$F_{0k0} = 0 \text{ for } k \text{ odd.}$$

Whenever this systematic absence is found without others affecting reflections for which both $h$ and $l$ are not zero then the existence of a $2_1$ axis parallel to $b$ is established.

### *c-glide plane (perpendicular to b)*

Atoms occur in pairs with coordinates $(x, y, z)$ and $(x, \bar{y}, \frac{1}{2} + z)$. This gives, for $l$ even,

$$F_{hkl} = \sum_{j=1}^{\frac{1}{2}N} 2f_j\exp\{2\pi i(hx_j + lz_j)\}\cos(2\pi ky_j) \tag{7.10a}$$

and, for $l$ odd,

$$F_{hkl} = \sum_{j=1}^{\frac{1}{2}N} 2f_j\exp\{2\pi i(hx_j + lz_j)\}\sin(2\pi ky_j). \tag{7.10b}$$

The systematic absences which arise from these equations are that

$$F_{h0l} = 0 \text{ for } l \text{ odd.}$$

This means that for the layer $k = 0$ only those reflections are observed for $l$ even; if one had only these data available then one would judge that the $c^*$ spacing was twice as large as it really was or that the cell parameter was one-half of its true value. The reason for this is that the $(h0l)$ data corresponds to that from a two-dimensional projection of the structure down the $y$ axis. The two-dimensional coordinates for this projection are $(x, z)$ and $(x, \frac{1}{2} + z)$; if one breaks the projected cell into two by drawing the line $z = \frac{1}{2}$ then the two parts are equivalent. This is shown in fig. 7.23; in the projection there is an effective unit cell with $c' = \frac{1}{2}c$ and this gives rise to the systematic absences which are observed.

It should be noticed that a $c$-glide plane can only be diagnosed if a lattice absence does not interfere. For example, for an $A$-face centred lattice

$$F_{hkl} = 0 \text{ for } k + l \text{ odd for any } h \text{ and, in particular, this gives}$$
$$F_{h0l} = 0 \text{ for } l \text{ odd for any } h.$$

Thus the systematic absences due to the $c$-glide plane would be masked.

There follows in table 7.2 a selection of some of the more common lattices and symmetry elements and the systematic absences which follow from them.

We shall now look at a few cases where systematic absences give space-group information.

*Case* 1. Monoclinic. All reflections observed except $0k0$ for $k$ odd. The cell must be primitive with a $2_1$ axis along $b$. The space group is $P2_1$.

Table 7.2.

| Lattice | Absences | Symmetry element | Absences |
|---|---|---|---|
| $P$ | None | $\bar{1}, \bar{3}, \bar{4}, \bar{6}, m$ \} | |
| $A$ | $k + l = 2n + 1$ | $2, 3, 4, 6$ \} | None |
| $B$ | $h + l = 2n + 1$ | $a \perp$ to $b$ axis | $h0l$ for $h = 2n + 1$ |
| $C$ | $h + k = 2n + 1$ | $a \perp$ to $c$ axis | $hk0$ for $h = 2n + 1$ |
| $F$ | $h, k, l$ not all odd | $b \perp$ to $a$ axis | $0kl$ for $k = 2n + 1$ |
| | or all even | $b \perp$ to $c$ axis | $hk0$ for $k = 2n + 1$ |
| $I$ | $h + k + l = 2n + 1$ | $c \perp$ to $a$ axis | $0kl$ for $l = 2n + 1$ |
| | | $c \perp$ to $b$ axis | $h0l$ for $l = 2n + 1$ |
| Symmetry element | | $n \perp$ to $a$ axis | $0kl$ for $k + l = 2n + 1$ |
| | | $n \perp$ to $b$ axis | $h0l$ for $h + l = 2n + 1$ |
| $2_1 \parallel$ to $a$ axis | $h00$ for $h = 2n + 1$ | $n \perp$ to $c$ axis | $hk0$ for $h + k = 2n + 1$ |
| $2_1 \parallel$ to $b$ axis | $0k0$ for $k = 2n + 1$ | | |
| $2_1 \parallel$ to $c$ axis | $00l$ for $l = 2n + 1$ | | |

Fig. 7.23.
Crystal projection down
$b$ axis perpendicular to
which there is a $c$-glide
plane.

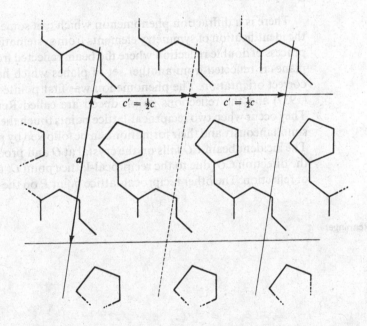

*Case* 2. Monoclinic. The observed absences are

$hkl$ for $h + k = 2n + 1$.

The cell must be $C$-face centred but otherwise no other symmetry element is indicated by the systematic absences. However since the system is monoclinic, there must be along one axis, say $b$, a 2 or $\bar{2}\,(\equiv m)$ symmetry axis or both.

The possible space groups are $C2$, $Cm$ or $C2/m$. A monoclinic cell with no systematic absences would be $P2$, $Pm$ or $P2/m$; the way of distinguishing these is discussed in §7.6 and similar methods can be used for the face-centred cells.

*Case* 3. Orthorhombic. The observed absences are for

$h00$ with $h$ odd
$0k0$ with $k$ odd
$00l$ with $l$ odd.

The cell is primitive with a $2_1$ axis along each cell axis. The space group is therefore $P2_12_12_1$.

*Case* 4. Orthorhombic. The systematic absences are for

$hkl$ for $h + k + l = 2n + 1$
$0kl$ for $\quad\quad k = 2n + 1$
$h0l$ for $\quad\quad l = 2n + 1$
$hk0$ for $\quad\quad h = 2n + 1$.

The systematic absences tell us that the lattice is $I$-centred, that there is a $b$-glide plane perpendicular to the $a$ axis, a $c$-glide plane perpendicular to the $b$ axis and an $a$-glide plane perpendicular to the $c$ axis. Hence the space group is $Ibca$.

There is a diffraction phenomenon which can sometimes interfere with the identification of symmetry elements from systematic absences. This is a process of double reflection where the beam reflected from one set of lattice planes is reflected from another set of planes which happens to be in the correct orientation. The phenomenon was first pointed out by Renninger (1937) and the reflections so produced are called Renninger reflections. They occur when two reciprocal-lattice points touch the sphere of reflection simultaneously and their formation can be followed by reference to fig. 7.24. The incident beam $CO$ falls on the crystal at $O$ and produces a reflection in the direction $CO'$ due to the reciprocal-lattice point $O'$ touching the sphere of reflection. The other reciprocal-lattice point $P$ on the sphere of reflection

Fig. 7.24.
Production of Renninger reflections.

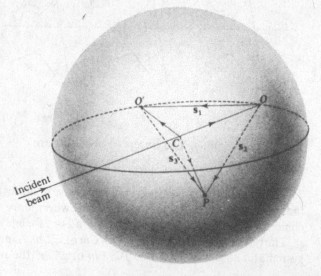

corresponds to a systematic absence and so will not give rise to a reflection due to the incident beam $CO$.

Now let us assume that an incident beam falls on the crystal in the direction $CO'$. We now centre the reciprocal lattice at $O'$ and it is clear from the figure that $O'P$ corresponds to a reciprocal-lattice vector $s_3 (= s_2 - s_1)$. Thus a reflection would be produced in the direction $CP$.

The reflected beam $CO'$ due to the reciprocal-lattice vector $s_1$ can thus be reflected again, the beam leaving the crystal along $CP$, the direction of a forbidden reflection for the incident beam $CO$. If the intensities corresponding to the reciprocal-lattice vectors $s_1$ and $s_3$ are both strong the Renninger reflection can be very prominent. Renninger reflections can usually be recognized by their very sharp, non-diffuse appearance; the double-reflection process will happen only for a restricted range of wavelengths of the characteristic radiation being used. In practice if one has overwhelming evidence of a given type of systematic absence with one or possibly two faint reflections disturbing the pattern then one should suspect that the systematic absences are genuine and that Renninger reflections are occurring. Of course if $P$ was not a point corresponding to a systematic absence a Renninger reflection would still occur and this would enhance the intensity of the primary reflection and would appear as an experimental error in the data.

## 7.5   Intensity statistics

None of the techniques we have previously considered, with the possible exception of optical goniometry, has enabled a centre of symmetry to be positively detected. In fact, as was shown by Wilson (1949), the presence or absence of a centre of symmetry does impress itself on the diffraction data but in a way which does not reveal itself without a certain amount of analysis – through the distribution of the structure amplitudes or intensities.

In order to derive these distributions it will be necessary to use an important theorem of statistical theory – the central-limit theorem. This states that:

'The sum of a large number of independent random variables will have a normal probability distribution with mean equal to the sum of the means of the independent variables and variance equal to the sum of their variances.'

Let us see what this means. We consider a number $N$ of independent random variables $x_j$ ($j = 1$ to $N$) which have means $\bar{x}_j$ and variances $\alpha_j^2$. There are no restrictions on the distribution functions of the variables $x$ and the distributions can all be different. The sum of the variables

$$X = \sum_{j=1}^{N} x_j \tag{7.11}$$

has a mean

$$\bar{X} = \sum_{j=1}^{N} \overline{x_j} \tag{7.12}$$

and a variance

$$\sigma^2 = \sum_{j=1}^{N} \alpha_j^2. \tag{7.13}$$

The probability distribution of $X$ is given by the normal distribution

$$P(X) = (2\pi\sigma^2)^{-\frac{1}{2}}\exp\{-(X - \bar{X})^2/2\sigma^2\} \tag{7.14}$$

which means that the probability that the sum lies between $X$ and $X + \mathrm{d}X$ is $P(X)\mathrm{d}X$.

This theorem can now be applied to the distribution of the structure factors for a centrosymmetric structure. The structure factor equation can be written

$$F_\mathbf{h} = \sum_{j=1}^{N} f_j\cos(2\pi\mathbf{h}\cdot\mathbf{r}_j)$$

where $\mathbf{h}$ is the reciprocal-lattice vector. However all the terms in the summation are not independent because for an atom at $r_j$ there is also one at $-r_j$. The equation can be rewritten as the sum of a number of independent terms as

$$F_\mathbf{h} = \sum_{j=1}^{\frac{1}{2}N} 2f_j\cos(2\pi\mathbf{h}\cdot\mathbf{r}_j) \tag{7.15}$$

where the summation is over one asymmetric unit. Comparing equation (7.15) with equation (7.11) it will be seen that we should equate $2f_j\cos(2\pi\mathbf{h}\cdot\mathbf{r}_j)$ to $x_j$ so that

$$\overline{x_j} = \overline{2f_j\cos(2\pi\mathbf{h}\cdot\mathbf{r}_j)} = 2f_j\overline{\cos(2\pi\mathbf{h}\cdot\mathbf{r}_j)}. \tag{7.16}$$

If we assume a random distribution of atomic positions then all positions in the unit cell are equally probable and

$$\overline{\cos(2\pi\mathbf{h}\cdot\mathbf{r}_j)} = 0 \tag{7.17a}$$

and hence

$$\overline{x_j} = 0. \tag{7.17b}$$

The variance of $x_j$ is given by the well-known expression

$$\alpha_j^2 = \overline{x_j^2} - \bar{x}_j^2. \tag{7.18}$$

Since $\overline{x_j} = 0$ then

$$\alpha_j^2 = 4f_j^2\overline{\cos^2(2\pi\mathbf{h}\cdot\mathbf{r}_j)}$$
$$= 2f_j^2 \tag{7.19}$$

since the mean value of $\cos^2\theta$ is 0.5.

Thus the mean value of the structure factor is given by

$$\bar{F} = \sum_{j=1}^{\frac{1}{2}N} \overline{x_j} = 0 \tag{7.20}$$

and its variance is given by

$$\sigma^2 = \sum_{j=1}^{\frac{1}{2}N} 2f_j^2$$

$$= \sum_{j=1}^{N} f_j^2 = \Sigma, \text{ say.} \tag{7.21}$$

The distribution function of $F$ is now given by

$$P_{\bar{1}}(F) = (2\pi\Sigma)^{-\frac{1}{2}} \exp(-F^2/2\Sigma). \tag{7.22}$$

The form of this function, the normal Gaussian curve, is shown in fig. 7.25. It can only be an approximation to the true distribution since we know that $F$ has a maximum possible value of $\sum_{j=1}^{N} f_j$, whereas the Gaussian curve extends from $-\infty$ to $+\infty$. A more exact theory has been given by Klug (1958).

Before looking at the properties of equation (7.22) any further to see how one could use it to distinguish a centrosymmetric structure we shall first find the corresponding distribution for a non-centrosymmetric structure. For such a structure we have

$$F_{\mathbf{h}} = A_{\mathbf{h}} + iB_{\mathbf{h}}$$

where

$$A_{\mathbf{h}} = \sum_{j=1}^{N} f_j \cos(2\pi\mathbf{h}\cdot\mathbf{r}_j)$$

$$B_{\mathbf{h}} = \sum_{j=1}^{N} f_j \sin(2\pi\mathbf{h}\cdot\mathbf{r}_j).$$

The distributions of $A$ and $B$ can be found separately – thus

$$\bar{A} = \sum_{j=1}^{N} f_j \overline{\cos(2\pi\mathbf{h}\cdot\mathbf{r}_j)} = 0$$

Fig. 7.25.
The centric distribution
function.

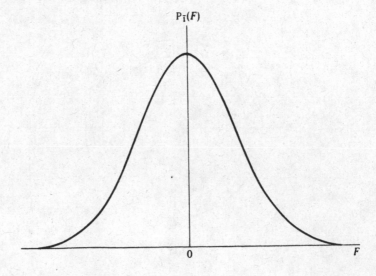

$$\sigma_A^2 = \sum_{j=1}^{N} f_j^2 \,\overline{\cos^2(2\pi\mathbf{h}\cdot\mathbf{r}_j)} = \tfrac{1}{2}\Sigma$$

so that

$$P(A) = (\pi\Sigma)^{-\frac{1}{2}}\exp(-A^2/\Sigma) \qquad\qquad (7.23a)$$

and similarly

$$P(B) = (\pi\Sigma)^{-\frac{1}{2}}\exp(-B^2/\Sigma). \qquad\qquad (7.23b)$$

Since $A$ and $B$ are the real and imaginary parts of the structure factor they can be plotted on an Argand diagram. The probability that $A$ lies between $A$ and $A + dA$, and also that $B$ lies between $B$ and $B + dB$, is given by

$$P(A)P(B)dA\,dB = (\pi\Sigma)^{-1}\exp\{-(A^2 + B^2)/\Sigma\}dA\,dB$$
$$= (\pi\Sigma)^{-1}\exp(-|F|^2/\Sigma)dA\,dB. \qquad\qquad (7.24)$$

The region defined by the above probability is shown in fig. 7.26(a) and since $dAdB$ is an element of area in the diagram we can say that the

Fig. 7.26.
(a) Region defined between $A$ and $A + dA$ and between $B$ and $B + dB$ for a complex structure factor.
(b) Region of complex space for which structure amplitude is between $|F|$ and $|F| + d|F|$.

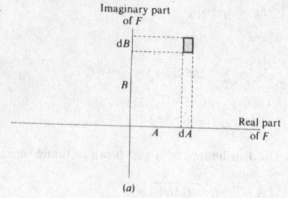

(a)

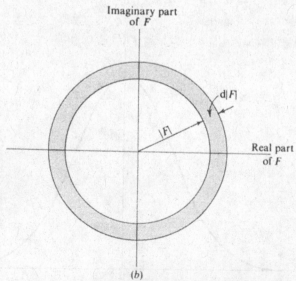

(b)

probability that a structure factor lies in a small region of area $dS$ round a point distant $|F|$ from the origin is

$$(\pi\Sigma)^{-1}\exp(-|F|^2/\Sigma)dS. \tag{7.25}$$

Hence the probability that the structure amplitude lies in the range $|F|$ to $|F| + d|F|$, that is within the shaded area shown in fig. 7.26(b), is

$$P_1(|F|)d|F| = (\pi\Sigma)^{-1}\exp(-|F|^2/\Sigma)2\pi|F|d|F|$$

which gives

$$P_1(|F|) = \frac{2}{\Sigma}|F|\exp(-|F|^2/\Sigma). \tag{7.26}$$

The form of this distribution is shown in fig. 7.27 and is quite different from that which obtains for a centrosymmetric structure. The two distributions are referred to as *centric* and *acentric* respectively; it should be noted that these terms are used only for the distribution functions and *not* for the symmetry of the structures which produce them which are centrosymmetric and non-centrosymmetric.

How can the distributions represented by equations (7.22) and (7.26) be used to distinguish the two types of structure? There is one snag which prevents one from plotting the distribution curves directly, which is that the value of $\Sigma$ is not constant over the whole of reciprocal space. In fact what these distribution functions mean is either:

(i) for a particular structure factor at a point of reciprocal space where $\Sigma$ has a known value the probability that its value (or magnitude) lines between $F$ and $F + dF$ (or $|F|$ and $|F| + d|F|$) is given by $P_1(F)dF$ (or $P_1(|F|)d|F|$), or

(ii) for a region of reciprocal space where $\Sigma$ is constant (or approximately so) the proportion of structure factors (or amplitudes) which lie between $F$ and $F + dF$ (or $|F|$ and $|F| + d|F|$) is $P_1(F)dF$ (or $P_1|F|d|F|$).

Fig. 7.27.
The acentric distribution function.

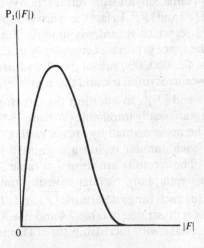

There is a quantity which enables one to distinguish the two types of distribution and is independent of $\Sigma$ and that is

$$M = \frac{\overline{|F|}^2}{\overline{|F|^2}}. \tag{7.27}$$

For the centric distribution

$$\overline{|F|} = \int_0^\infty 2F P_{\bar{1}}(F) dF = \left(\frac{2\Sigma}{\pi}\right)^{\frac{1}{2}} \tag{7.28}$$

and

$$\overline{|F|^2} = \int_{-\infty}^\infty 2F^2 P_{\bar{1}}(F) dF = \Sigma \tag{7.29}$$

so that

$$M_{\bar{1}} = \frac{2}{\pi} = 0.637. \tag{7.30}$$

For the acentric distribution

$$\overline{|F|} = \int_0^\infty |F| P_1(|F|) d|F| = \tfrac{1}{2}(\pi\Sigma)^{\frac{1}{2}} \tag{7.31}$$

and

$$\overline{|F|^2} = \int_0^\infty |F|^2 P_1(|F|) d|F| = \Sigma \tag{7.32}$$

so that

$$M_1 = \frac{\pi}{4} = 0.785. \tag{7.33}$$

The two values given in equations (7.30) and (7.33) are sufficiently different to give a discriminating test. As an example in fig. 7.28 there is shown one quadrant of a rectangular reciprocal-lattice net for the $h = 0$ layer for dicyclopentadienyldi-iron tetracarbonyl (Mills, 1958) together with the values of $|F|$ and $|F|^2$. In fact the space group can be unambiguously determined as $P2_1/c$, which is centrosymmetric, but we shall treat the $0kl$ data as though the space group was completely unknown. Very often axial structure factors, i.e. $0k0$, $00l$, are atypical and are real even when the projection is non-centrosymmetric and these will be ignored in determining the values of $\overline{|F|}$ and $\overline{|F|^2}$. In addition the structure factors at low $\sin\theta$ values tend to be statistically unreliable (Wilson, 1949) and will not be used for $\sin\theta < 0.2$. The net is divided by arcs of radii which increase by 0.1 in $\sin\theta$ and within each annular region are counted the number of points, $\Sigma|F|$ and $\Sigma|F|^2$. These results are shown in table 7.3.

We now take overlapping regions covering the $\sin\theta$ ranges 0.2–0.4, 0.3–0.5, etc., and for each range determine $\overline{|F|}$ and $\overline{|F|^2}$ and then the value of $M$. This process is illustrated in table 7.4 and it will be seen that the value of $M$ seems to decrease with increasing $\sin\theta$. This is probably due to the

Table 7.3.

| sin θ range | No. of points | Σ\|F\| | Σ\|F\|² |
|---|---|---|---|
| 0.2–0.3 | 3 | 60 | 1814 |
| 0.3–0.4 | 6 | 133 | 3843 |
| 0.4–0.5 | 8 | 181 | 5027 |
| 0.5–0.6 | 9 | 104 | 2096 |
| 0.6–0.7 | 12 | 130 | 2094 |
| 0.7–0.8 | 16 | 166 | 3434 |
| 0.8–0.9 | 11 | 84 | 1238 |
| 0.9–1.0 | 18 | 49 | 267 |

Fig. 7.28.
$|F|$ and $|F|^2$ for $(0kl)$ data of dicyclopentadienyldi-iron tetracarbonyl.

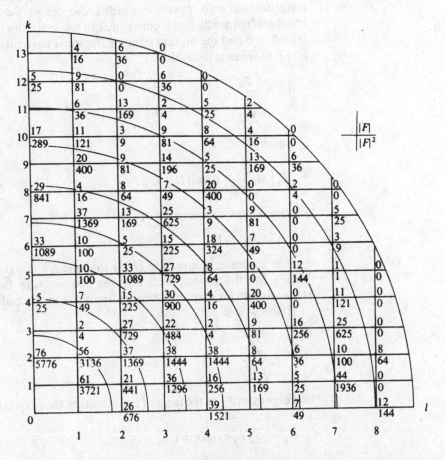

large number of unobserved reflections at high angles whose contribution, if measurable, would change substantially the values of $\overline{|F|}$ and $\overline{|F|^2}$. However the average value of $M$ is 0.612, giving a fairly clear indication that the projection is centrosymmetric

Table 7.4.

| $\sin\theta$ range | No. of points | $\Sigma\lvert F\rvert$ | $\Sigma\lvert F\rvert^2$ | $\overline{\lvert F\rvert}$ | $\overline{\lvert F\rvert^2}$ | $M$ |
|---|---|---|---|---|---|---|
| 0.2–0.4 | 9 | 193 | 5657 | 21.4 | 629 | 0.728 |
| 0.3–0.5 | 14 | 314 | 8870 | 22.4 | 634 | 0.791 |
| 0.4–0.6 | 17 | 285 | 7123 | 16.8 | 419 | 0.674 |
| 0.5–0.7 | 21 | 234 | 4190 | 11.1 | 200 | 0.616 |
| 0.6–0.8 | 28 | 296 | 5528 | 10.6 | 197 | 0.570 |
| 0.7–0.9 | 27 | 250 | 4672 | 9.3 | 173 | 0.500 |
| 0.8–1.0 | 29 | 133 | 1505 | 4.6 | 52 | 0.407 |

In order to get meaningful values of $M$ it was necessary to divide the reciprocal net into regions in which $\Sigma$ did not vary overmuch. Crystallographers frequently find it convenient to use structure factors corrected for fall-off in $\theta$ and the unitary structure factor is particularly useful (see §§ 8.7 and 8.8). This is defined by

$$U_{\mathbf{h}} = \frac{F_{\mathbf{h}}}{\sum\limits_{j=1}^{N} f_j}. \tag{7.34}$$

Since

$$\lvert F_{\mathbf{h}}\rvert \leqslant \sum_{j=1}^{N} f_j \text{ then } \lvert U_{\mathbf{h}}\rvert \leqslant 1$$

and for a real structure factor

$$-1 \leqslant U_{\mathbf{h}} \leqslant 1. \tag{7.35}$$

It is assumed in equation (7.34) that the $f_j$'s are the actual scattering factors for the atoms, including the temperature factor, and not the stationary-atom scattering factors one finds in the published tables.

If one writes

$$n_j = \frac{f_j}{\sum\limits_{j=1}^{N} f_j} \tag{7.36}$$

and expresses $F_{\mathbf{h}}$ in the form of a summation then equation (7.34) becomes

$$U_{\mathbf{h}} = \sum_{j=1}^{N} n_j \exp(2\pi i \mathbf{h}\cdot\mathbf{r}_j). \tag{7.37}$$

When the structure contains atoms of more than one type the quantities $n_j$, known as the unitary scattering factors, will vary with $\sin\theta$ since the ratio of scattering factors for different atoms does not remain constant; in particular the higher the atomic number the more tightly bound are the electrons and the less rapid is the relative fall-off of $f$. However one usually makes the assumption that the $n_j$'s are constant over the whole of reciprocal space.

By analogy with what we have had previously, the distribution functions of the unitary structure factor for centrosymmetric and non-centrosymmetric structures, respectively, are

$$P_{\bar{1}}(U) = (2\pi\varepsilon)^{-\frac{1}{2}}\exp(-U^2/2\varepsilon) \tag{7.38}$$

and

$$P_1(|U|) = \frac{2}{\varepsilon}|U|\exp(-|U|^2/\varepsilon) \tag{7.39}$$

where

$$\varepsilon = \sum_{j=1}^{N} n_j^2. \tag{7.40}$$

For a structure with $N$ equal atoms in the unit cell

$$n_j = \frac{f_j}{Nf_j} = \frac{1}{N} \tag{7.41}$$

and

$$\varepsilon = \sum_{j=1}^{N} \frac{1}{N^2} = \frac{1}{N}. \tag{7.42}$$

In order to determine $U$'s from observed data we note from equations (7.29) and (7.32) that for both the centric and acentric distributions we shall have

$$\overline{|U|^2} = \varepsilon. \tag{7.43}$$

The $|U|^2$'s can be obtained from the $|F|^2$'s by applying some factor $\phi$ which depends upon $\sin\theta$. We can determine this factor from the condition that, for an annular region in reciprocal space after the factor has been applied, $\overline{|U|^2} = \varepsilon$.

Dicyclopentadienyldi-iron tetracarbonyl has two molecules of $Fe_2(CO)_4(C_5H_5)_2$ in the unit cell. We shall ignore hydrogen since it contributes negligibly to the scattering and we take $f_{Fe}:f_O:f_C = 21:4:3$ over the whole range of $\sin\theta$ which gives $n_{Fe} = 0.105, n_O = 0.020, n_C = 0.015$ and $\varepsilon = 0.0536$.

In table 7.4 in the $\sin\theta$ range 0.2–0.4, corresponding approximately to $\overline{\sin\theta} = 0.3$, the value of $\overline{|F|^2}$ is 629. Hence the best estimate of the factor $\phi$ which one should apply to $|F|^2$'s at $\sin\theta = 0.3$ to convert them into $|U|^2$'s is 0.0536/629. The values of $\phi$ for the other values of $\overline{\sin\theta}$ may be found similarly. The factor $\sqrt{\phi}$ applied to the $|F|$'s will give $|U|$'s and $\sqrt{\phi}$ is shown plotted in fig. 7.29. A smooth curve is drawn as close to the points as possible and the curve is extrapolated at both ends. From this curve a linear scale, also shown in fig. 7.29, can be prepared. If the point $O$ of the scale is pivoted at the origin of the reciprocal lattice then the reading of the scale at the reciprocal-lattice points gives the factor which changes $|F|$ into $|U|$.

Fig. 7.29.
Curve for $\sqrt{\phi}$ and
chart for deriving $|U|$
from $|F|$.

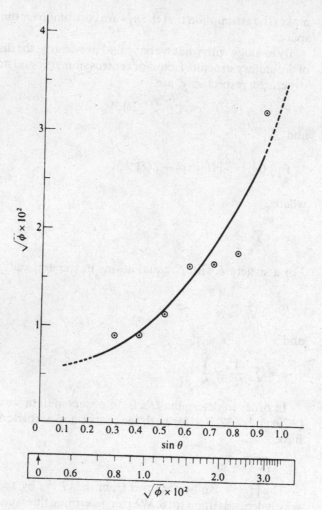

The complete set of $U$'s for dicyclopentadienyldi-iron tetracarbonyl is given in fig. 7.30.

Another type of structure factor corrected for $\sin \theta$ fall-off which is sometimes used is the normalized structure factor defined by

$$|E_\mathbf{h}|^2 = \frac{|F_\mathbf{h}|^2}{\Sigma}.$$

(7.44)

The normalized structure factors have distribution functions

$$P_{\bar{1}}(E) = (2\pi)^{-\frac{1}{2}}\exp(-E^2/2)$$

(7.45)

and

$$P_1(|E|) = 2|E|\exp(-|E|^2).$$

(7.46)

For either distribution $\overline{|E|^2} = 1$ and it can also be seen that the distribution functions are completely independent of the structural complexity. The values of $E$ can be obtained from the $U$'s by

Fig. 7.30.
100 | U | for (0kl) data of
dicyclopentadienyldi-iron
tetracarbonyl.

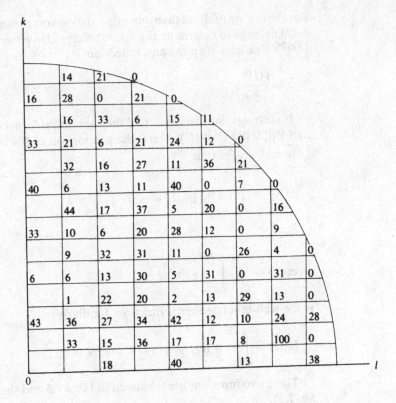

$$E = \frac{U \sum\limits_{j=1}^{N} f_j}{\Sigma^{\frac{1}{2}}}. \tag{7.47}$$

In appendix V there is given a FORTRAN program, FTOUE, which converts | F |'s to either | U |'s or | E |'s by calculating the factor $\sqrt{\phi}$. The values of $\overline{|F|^2}$ are calculated in overlapping regions of $\sin\theta$ and, for calculating | U |'s, the values of $\varepsilon \times (\overline{|F|^2})^{-\frac{1}{2}}$ (i.e. $\phi$) which converts $|F|^2$ to $|U|^2$ are found and associated with the average value of $\sin\theta$ in each region. Coefficients are found for a best-fit parabola of the form

$$\phi^{\frac{1}{2}} = a + b\sin\theta + c\sin^2\theta \tag{7.48}$$

which is then used for finding the conversion factor from | F | to | U | for any reflection from its associated $\sin\theta$. If | E |'s are required then a different scaling factor is used.

In practice it may be found that the values of $\sqrt{\phi}$ deviate considerably from the best parabola, both above and below the curve. This is particularly true for protein structures where strong structural features, for example helices or planes, impose a variation of average intensities with $(\sin\theta)/\lambda$.

A test which is frequently used to detect a centre of symmetry is the N(z) test suggested by Howells, Phillips and Rogers (1950). This is a cumulative distribution curve for intensities and N(z) is the fraction of reflections with intensities less than or equal to z times the mean intensity. If one is using data with a $\sin \theta$ fall-off then one must always compare an intensity with the local average to determine the appropriate $z$. However if one has $|U|$'s or $|U|^2$'s available then for any reflection

$$z = \frac{|U|^2}{\varepsilon}.$$

(7.49)

It is straightforward to derive the theoretical forms of $N(z)$ from $P_{\bar{1}}(U)$ and $P_1(|U|)$. For $P_{\bar{1}}(U)$, $N(z)$ is the proportion of $U$'s with values between $\sqrt{z\varepsilon}$ and $-\sqrt{z\varepsilon}$ thus

$$N_{\bar{1}}(z) = \int_{-\sqrt{z\varepsilon}}^{\sqrt{z\varepsilon}} P_{\bar{1}}(U)\mathrm{d}U = \mathrm{erf}\left(\sqrt{\frac{z}{2}}\right)$$

(7.50)

where

$$\mathrm{erf}(x) = \sqrt{\frac{2}{\pi}} \int_0^{\sqrt{2x}} \mathrm{e}^{-\frac{1}{2}t^2}\mathrm{d}t$$

is the well-tabulated error function. Similarly

$$N_1(z) = \int_0^{\sqrt{z\varepsilon}} P_1(|U|)\mathrm{d}|U| = 1 - \exp(-z).$$

(7.51)

These two functions are tabulated in table 7.5 and shown graphically in fig. 7.31.

For dicyclopentadienyldi-iron tetracarbonyl the values of $|U|$ corresponding to various values of $z$ and the numbers of reflections with $|U|^2$ less than $z\varepsilon$ are also shown in table 7.5 and $N(z)$ is also plotted in fig. 7.31. It is clear from this that the projection is centrosymmetric.

The general quality of $N_1(z)$ and $N_{\bar{1}}(z)$ can be seen in the distributions $P_1(|F|)$ and $P_{\bar{1}}(F)$ plotted in figs. 7.27 and 7.25 where, in particular, it will be noticed that the centric distribution has a higher proportion of small $F$'s. It was shown by Lipson and Woolfson (1952) and by Rogers and Wilson (1953) that other types of distribution could occur when there were molecules with non-crystallographic centres of symmetry in the unit cell. In fig. 7.31 is shown the $N(z)$ curve for a hypercentric distribution which occurs when there are in a unit cell two centrosymmetric molecules related by a crystallographic centre of symmetry (fig. 7.32). It will be noticed that the distribution has an even higher proportion of small intensities than the centric one. The form of the distribution has been given by Lipson and Woolfson; one important property which it has, as have all other distributions, is that $\overline{|F|^2} = \Sigma$, so that the procedure described for determining $U$'s is still valid.

The various distribution functions described in this section depend on randomness of the atomic positions. When there is a very heavy atom in the unit cell, equivalent to a non-statistical aggregation of scattering power, or when atoms are in special positions (§ 8.1), the distributions can depart markedly from those which have been described. In such cases tests involving the moments of the distributions can be used (Foster & Hargreaves, 1963a, 1963b; Pradhan, Ghosh & Nigam, 1985).

Table 7.5.

| $z$ | $N_1(z)$ | $N_{\bar{1}}(z)$ | $\lvert U \rvert$ | For $Fe_2(CO)_4(C_5H_5)_2$ No. $\leqslant \lvert F \rvert$ | $N(z)$ |
|-----|----------|----------|---------|---------------|--------|
| 0.0 | 0.00 | 0.00 | 0.00 | 0 | 0.00 |
| 0.1 | 0.09 | 0.25 | 0.07 | 25 | 0.29 |
| 0.2 | 0.18 | 0.35 | $0.10_3$ | 30 | 0.35 |
| 0.3 | 0.26 | 0.42 | $0.12_7$ | 37 | 0.44 |
| 0.4 | 0.33 | 0.47 | $0.14_6$ | 42 | 0.49 |
| 0.5 | 0.39 | 0.52 | $0.16_4$ | 47 | 0.55 |
| 0.6 | 0.45 | 0.56 | $0.17_9$ | 50 | 0.59 |
| 0.7 | 0.50 | 0.60 | $0.19_4$ | 50 | 0.59 |
| 0.8 | 0.55 | 0.63 | $0.20_7$ | 53 | 0.62 |
| 0.9 | 0.59 | 0.66 | $0.22_0$ | 59 | 0.69 |
| 1.0 | 0.63 | 0.68 | $0.23_1$ | 59 | 0.69 |
|     |      |      | 1.00 | 85 |      |

Fig. 7.31.
The cumulative
distribution curves for
the following
distributions: acentric
(equation (7.51)); centric
(equation (7.50)); and
hypercentric (from
symmetry shown in fig.
7.32).

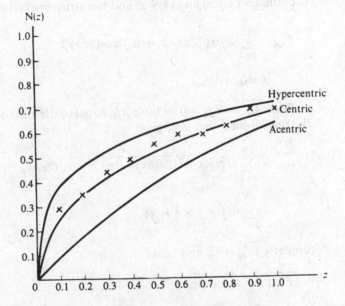

## 7.6   Detection of mirror planes and diad axes

The only common symmetry elements for which detection techniques using diffraction data have not been described are the diad axis and the mirror plane. In fact one frequently needs to distinguish these two. For example, the space group of a monoclinic structure with no systematic absences is one of $P2$, $Pm$ or $P2/m$. The last of these is centrosymmetric and can therefore be isolated by an $N(z)$ test but, if the test indicated a non-centrosymmetric structure, one would still be left with the choice of $P2$ or $Pm$. It can be shown that when there is a mirror plane, perpendicular to the $b$

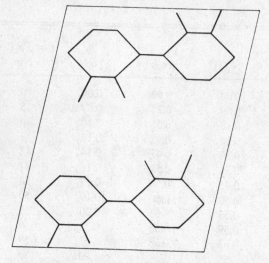

axis let us say, then the value of $\overline{|F|_{h0l}|^2}$ is $2\Sigma$ or twice the average value for general reflection intensities. For a mirror plane atoms occur in pairs with coordinates $(x, y, z)$ and $(x, \bar{y}, z)$ and the structure factor equation become

$$F_{hkl} = \sum_{j=1}^{\frac{1}{2}N} 2f_j \exp\{2\pi i(hx_j + lz_j)\}\cos(2\pi ky_j)$$

$$= A_{hkl} + iB_{hkl}. \tag{7.52}$$

Since $\overline{A_{hkl}} = \overline{B_{hkl}} = 0$ we have $\overline{A_{hkl}^2}$ equal to the variance of $A_{hkl}$ which, by the central-limit theorem, gives

$$\overline{A_{hkl}^2} = \sum_{j=1}^{\frac{1}{2}N} 4f_j^2 \times \overline{\cos^2\{2\pi(hx_j + lz_j\}} \times \overline{\cos^2(2\pi ky_j)}$$

$$= \sum_{j=1}^{\frac{1}{2}N} 4f_j^2 \times \tfrac{1}{2} \times \tfrac{1}{2} = \tfrac{1}{2}\Sigma.$$

Similarly $\overline{B_{hkl}^2} = \tfrac{1}{2}\Sigma$ and hence

$$\overline{|F_{hkl}|^2} = \overline{A_{hkl}^2} + \overline{B_{hkl}^2} = \Sigma. \tag{7.53}$$

However for $k = 0$

$$\overline{A_{h0l}^2} = \sum_{j=1}^{\frac{1}{2}N} 4f_j^2 \times \overline{\cos^2\{2\pi(hx_j + lz_j)\}} = \Sigma = \overline{B_{h0l}^2}$$

and

$$\overline{|F_{h0l}|^2} = \overline{A_{h0l}^2} + \overline{B_{h0l}^2} = 2\Sigma. \tag{7.54}$$

Another, and easier, way of deducing this relationship is to see that the $b$-axis projection will show complete overlap of pairs of atoms due to the mirror plane. Thus, in projection, there are $\frac{1}{2}N$ double-weight atoms and the effective value of $\Sigma$, $\Sigma_m$, becomes

$$\Sigma_m = \sum_{j=1}^{\frac{1}{2}N} (2f_j)^2 = 2\Sigma. \tag{7.55}$$

In a similar way the presence of a diad axis along $b$ shows itself in that $|F_{0k0}|^2$ is $2\Sigma$. Pairs of atoms are related with coordinates $(x, y, z)$ and $(\bar{x}, y, \bar{z})$ so that projected on to the $b$ axis they overlap at $(0, y, 0)$. Since, in general, there are fewer reflections involved this test for a diad axis is normally less reliable than that for a mirror plane.

In order to apply these tests all the observed data must be on the same relative scale and the method of achieving this is described in §9.2. It is also necessary to compare the values of $|F|^2$ for similar narrow ranges of $\sin\theta$ and not for all the data together, since $\Sigma$ will not have the same average over the whole three-dimensional reciprocal lattice as it has for a one- or two-dimensional projection.

## Problems to Chapter 7

7.1 What systematic absences are given by the following symmetry elements?
  (a) $I$-centred lattice,
  (b) $n$-glide plane perpendicular to $b$,
  (c) $3_1$ axis.

7.2 Identify the space groups given by the following sets of systematic absences:
  (a) Primitive monoclinic lattice:
    $0k0$ only present for $k = 2n$,
    $h0l$ only present for $l = 2n$.
  (b) Primitive orthorhombic lattice:
    $0kl$ only present for $k + l = 2n$,
    $h0l$ only present for $h = 2n$,
    $h00$ only present for $h = 2n$,
    $0k0$ only present for $k = 2n$,
    $00l$ only present for $l = 2n$.

7.3 In table 7.6 are given the $(h0l)$ and $(0kl)$ data for $m$-toluidine dihydrochloride (Fowweather & Hargreaves, 1950). By comparing intensities with the local averages, determine $N(z)$ for each projection and determine whether they are centrosymmetrical or non-centrosymmetrical. The data are given in order of $\sin\theta$ and hence may be taken in overlapping groups of consecutive items.

7.4 Using the output file generated in the solution of Problem 3.4, use the program FTOUE to find the normalized structure factors ($|E|$'s) for the $(h0l)$ reflections of the phenylethylammonium salt of fosfomycin. Plot the values of $\sqrt{\phi}$ found by averaging in $\sin\theta$ regions and compare with the curve given by equation (7.48). You may request output of the values of $a$, $b$ and $c$ (equation (7.48)) when running FTOUE.

Table 7.6. Data for m-toluidine dihydrochloride.

| h0l | \|F\| | h0l | \|F\| | h0l | \|F\| | h0l | \|F\| | 0kl | \|F\| | 0kl | \|F\| | 0kl | \|F\| |
|---|---|---|---|---|---|---|---|---|---|---|---|---|---|
| 002 | 8 | 305 | 4 | 1̄,0,21 | 0 | 2,0,24 | 3 | 002 | 7 | 046 | 12 | 0,5,15 | 16 |
| 004 | 16 | 2̄,0,12 | 24 | 4,0,10 | 42 | 0,0,26 | 12 | 011 | 36 | 048 | 0 | 0,0,24 | 4 |
| 1̄01 | 34 | 2,0,12 | 9 | 1,0,10 | 7 | 4,0,18 | 4 | 004 | 13 | 0,3,13 | 11 | 068 | 5 |
| 101 | 0 | 3̄07 | 5 | 3̄,0,17 | 11 | 5,0,11 | 0 | 013 | 8 | 0,1,17 | 10 | 0,6,10 | 10 |
| 1̄03 | 69 | 1̄,0,15 | 17 | 2,0,20 | 7 | 5̄,0,13 | 8 | 006 | 27 | 0,2,16 | 29 | 0,4,20 | 13 |
| 103 | 25 | 1,0,15 | 11 | 4,0,12 | 0 | 3,0,23 | 9 | 015 | 28 | 0,0,18 | 6 | 0,2,24 | 14 |
| 006 | 33 | 307 | 5 | 0,0,22 | 3 | 4̄,0,20 | 5 | 020 | 50 | 0,4,10 | 19 | 0,1,25 | 16 |
| 1̄05 | 69 | 0,0,16 | 19 | 2,0,20 | 28 | 5,0,13 | 3 | 022 | 47 | 0,3,15 | 4 | 0,5,17 | 11 |
| 105 | 27 | 309 | 6 | 3,0,17 | 26 | 3̄,0,23 | 3 | 008 | 25 | 051 | 10 | 0,3,23 | 23 |
| 008 | 31 | 2̄,0,14 | 17 | 4,0,12 | 14 | 1̄,0,27 | 4 | 017 | 39 | 053 | 4 | 0,6,12 | 22 |
| 1̄07 | 46 | 309 | 3 | 4̄,0,14 | 4 | 2̄,0,26 | 5 | 024 | 38 | 0,1,19 | 26 | 0,0,26 | 13 |
| 107 | 51 | 2,0,14 | 9 | 3,0,19 | 10 | 1,0,27 | 5 | 026 | 20 | 0,4,12 | 20 | 071 | 12 |
| 200 | 17 | 1̄,0,17 | 25 | 1̄,0,23 | 0 | 4,0,20 | 7 | 019 | 29 | 055 | 19 | 073 | 0 |
| 2̄02 | 29 | 3,0,11 | 9 | 501 | 5 | 5,0,15 | 0 | 0,0,10 | 13 | 0,2,18 | 15 | 0,4,22 | 5 |
| 202 | 29 | 1,0,17 | 9 | 503 | 13 | 2,0,26 | 4 | 028 | 18 | 0,0,20 | 0 | 0,5,19 | 20 |
| 0,0,10 | 15 | 0,0,18 | 8 | 4,0,14 | 5 | 0,0,28 | 10 | 031 | 12 | 057 | 21 | 0,6,14 | 18 |
| 1̄09 | 28 | 3,0,11 | 11 | 1,0,23 | 3 | 600 | 13 | 0,1,11 | 16 | 0,3,17 | 4 | 075 | 13 |
| 204 | 5 | 2̄,0,16 | 11 | 2,0,22 | 15 | 5,0,15 | 11 | 033 | 16 | 0,4,14 | 29 | 0,7,26 | 10 |
| 2̄04 | 34 | 2,0,16 | 16 | 503 | 35 | 6̄02 | 5 | 0,0,12 | 53 | 059 | 18 | 0,1,27 | 3 |
| 109 | 74 | 400 | 0 | 505 | 8 | 3,0,25 | 7 | 035 | 12 | 0,1,21 | 28 | 077 | 16 |
| 206 | 16 | 402 | 33 | 3,0,19 | 6 | 602 | 10 | 0,2,10 | 22 | 0,2,20 | 18 | 0,3,25 | 7 |
| 2̄06 | 26 | 3,0,13 | 20 | 0,0,24 | 3 | 6̄04 | 9 | 037 | 35 | 0,5,11 | 12 | 0,6,16 | 7 |
| 1̄,0,11 | 17 | 402 | 8 | 2,0,22 | 4 | 4̄,0,22 | 6 | 0,1,13 | 29 | 0,0,22 | 2 | 079 | 8 |
| 0,0,12 | 65 | 404 | 10 | 505 | 14 | 604 | 3 | 0,0,14 | 32 | 0,4,16 | 24 | 0,0,28 | 11 |
| 1,0,11 | 14 | 404 | 9 | 507 | 18 | 6̄06 | 5 | 0,2,12 | 38 | 0,3,19 | 18 | 0,5,21 | 10 |
| 208 | 12 | 3,0,13 | 21 | 4,0,16 | 15 | 5̄,0,17 | 3 | 039 | 32 | 060 | 39 | 0,4,24 | 6 |
| 2̄08 | 2 | 1̄,0,19 | 46 | 507 | 7 | 3,0,25 | 7 | 040 | 29 | 062 | 20 | 0,7,11 | 8 |
| 2̄,0,10 | 58 | 4̄06 | 10 | 4,0,16 | 29 | 6̄08 | 0 | 042 | 37 | 064 | 5 | 0,2,28 | 11 |
| 1̄,0,13 | 9 | 406 | 8 | 509 | 3 | 606 | 9 | 0,1,15 | 21 | 0,5,13 | 9 | 0,6,18 | 4 |
| 1,0,13 | 11 | 1,0,19 | 14 | 3,0,21 | 17 | 4,0,22 | 13 | 044 | 23 | 0,2,22 | 7 | 0,1,29 | 7 |
| 2,0,10 | 11 | 406 | 0 | 1,0,25 | 6 | 1̄,0,29 | 2 | 0,3,11 | 15 | 0,1,23 | 19 | 0,3,27 | 6 |
| 0,0,14 | 37 | 3,0,15 | 5 | 509 | 6 | 5,0,17 | 8 | 0,2,14 | 28 | 066 | 7 | 0,7,13 | 8 |
| 301 | 5 | 4̄,0,10 | 0 | 2̄,0,24 | 12 | 1,0,29 | 5 | 0,0,16 | 16 | 0,4,18 | 4 | 0,5,23 | 11 |
| 3̄01 | 29 |  |  | 3,0,21 | 14 | 608 | 5 |  |  | 0,3,21 | 22 | 0,4,26 | 5 |
| 3̄03 | 19 |  |  | 1,0,25 | 5 | 2,0,28 | 4 |  |  |  |  |  |  |
| 303 | 25 |  |  | 4,0,18 | 15 | 6̄,0,10 | 2 |  |  |  |  |  |  |
| 305 | 19 |  |  | 5,0,11 | 8 |  |  |  |  |  |  |  |  |

# 8 The determination of crystal structures

## 8.1 Trial-and-error methods

The object of a crystal-structure determination is to locate the atomic positions within the unit cell and thus completely to define the whole structure. Sometimes there are special features in the diffraction pattern, the space group or the suspected chemical configuration of the material under investigation which enable a guess to be made of the crystal structure or at least restrict it to a small number of possibilities. In the early days of the subject, when methods of structure determination were poorly developed, only the simpler types of structure could be tackled and trial-and-error methods based on such special features were commonly used. That is not to say that such techniques are now outmoded – no crystallographer would ignore the information from special features if it was available, but he does not rely on such information as much as hitherto.

One type of situation which is of great importance and is always sought by the crystallographer is when space-group considerations lead to the fixing or restricting of the positions of atoms or whole groups of atoms. If a centrosymmetric unit cell has only one atom of a particular species (or an odd number) then that atom (or one of them) must be at a centre of symmetry. In a case with an odd number of atoms in a cell with a diad axis one of the atoms would have to lie on the diad axis. Similarly, in some situations, an $SO_4$ group may have to be symmetrically arranged on a triad axis as shown in fig. 8.1. It is also possible to gain stereochemical information from space-group requirements so that a molecule may be found necessarily to have a centre of symmetry or a diad axis of symmetry within itself.

An interesting example of a structure determination using this sort of information is provided by the hydrocarbon fluorene, a molecule of which is shown in fig. 8.2. This was solved by Brown and Bortner (1954) whose X-ray data showed the crystal system to be orthorhombic, of space group $Pna2_1$ or $Pnam$ with four molecules per unit cell. The unit-cell dimensions were $a = 8.50$, $b = 5.71$ and $c = 19.00$ Å. Subsidiary evidence showed that the space group was probably $Pnam$ for which in general there are eight symmetry-related positions. The only way in which four fluorene molecules can be accommodated in the cell is if the molecule itself has a mirror plane which coincides with the crystallographic mirror plane. This could mean either that the molecule is precisely planar and lies completely in the mirror plane or that the line $AB$ in fig. 8.2 is a mirror plane for the molecule. The latter alternative was considered the more likely and, if the molecular shape was assumed to be known – approximately at any rate – then the crystal structure could be defined in terms of three parameters. These could be the

Fig. 8.1.
An SO$_4$ group arranged
on a triad axis.

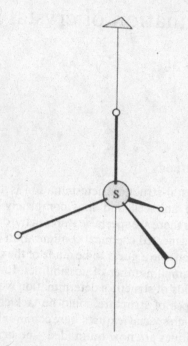

$x$ and $y$ coordinates of the point $A$ and the angle made by the line $AB$ with the $y$ axis. For the $a$ or $b$ axis projections there are only two parameters. Structure factors were calculated for various positions of the model molecule and reasonable agreement of calculated and observed ($h0l$) and ($0kl$) structure factors was found for the point $A$ at position (0.353, 0.431, 0.250) with $AB$ making an angle 30.5° with the $y$ axis.

Fig. 8.2.
An outline of the carbon
atoms of fluorene.

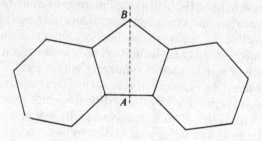

As a measure of agreement between observed and calculated structure factors crystallographers use a quantity called the *reliability index* or *residual* defined by

$$R = \frac{\Sigma ||F_o| - |F_c||}{\Sigma |F_o|}.$$

(8.1)

The $F_o$'s and $F_c$'s are the observed and calculated structure factors respectively and, of course, for them to be comparable the $F_c$'s must include the temperature factor. The way that one can determine the temperature factor from the observed data is described in §9.2.

Once a trial structure has been found which is close enough to the true structure to give a low value of $R$ then one can refine the structure by the routine procedures which are described in chapter 9. The whole crystal-structure problem consists of getting the first trial structure; once the crystallographer has his foot in the door, so to speak, the crystal structure is all but solved.

If there are a few very strong reflections with high unitary structure factors then these alone can reveal the structure, particularly if one has some initial idea of the stereochemistry of the substance under investigation. For a centrosymmetric structure, if a particular $|U_{\mathbf{h}}|$ is very large, then the atoms will tend to have coordinates $\mathbf{r}$ such that $\cos(2\pi\mathbf{h}\cdot\mathbf{r})$ is either near $+1$ for each of them or near $-1$. One can try various combinations of signs for the few structure factors and, in two dimensions, lines can be drawn corresponding to $\cos(2\pi\mathbf{h}\cdot\mathbf{r}) = \pm 1$ depending on which sign is being tried. In fig. 8.3($a$) there is shown a diagram given by Dunitz (1949) demonstrating how six strong reflections revealed the position of the molecule in the $b$-axis projection of 1,2,3,4-tetraphenylcyclobutane. The sets of lines corresponding to $\cos\{2\pi(hx + lz)\} = \pm 1$ are shown, the sign taken in each case being indicated in the diagram. To show how well this indicated the correct position, the final electron-density map is shown in fig. 8.3($b$). Another example in which information was gained from only one very strong reflection is shown in fig. 8.3($c$) where it can be seen that in the $b$-axis projection of *para*-nitroaniline (Abrahams & Robertson, 1948), all the atoms lie along the direction of the (202) planes, where the (202) reflection is the one which is very strong.

When the structure is a molecular one then the shape of the unit cell and the restrictions on location imposed by symmetry elements may enable a trial structure to be found from packing considerations alone. The separate molecules may be linked by hydrogen bonds or may have to be separated by van der Waals distances. It can occasionally be very profitable to play about with wire models of the unit cell and of the molecules. A computer can also be used systematically to explore various methods of packing to achieve sensible intermolecular distances and structure-factor agreement (Milledge, 1962).

A word of warning should be given about pseudo-symmetry arising as a result of statistical disorder. This occurs when the structure consists of 'unit cells' in which the atoms are arranged in two or more possible ways and where the arrangement from cell to cell varies randomly. Such a statistical-disorder situation is illustrated in two dimensions in fig. 8.4. The intensities of the Bragg peaks will correspond very closely to those from a 50:50 mixed structure and there will be an apparent centre of symmetry. However one would know from packing considerations that the two arrangements of atoms could not occur simultaneously in one cell and this would also be revealed by the electron-density peaks which would be half the normal height.

## 8.2 The Patterson function

We have seen that the lack of knowledge of the phases of structure factors prevents us from directly computing an electron-density map and so

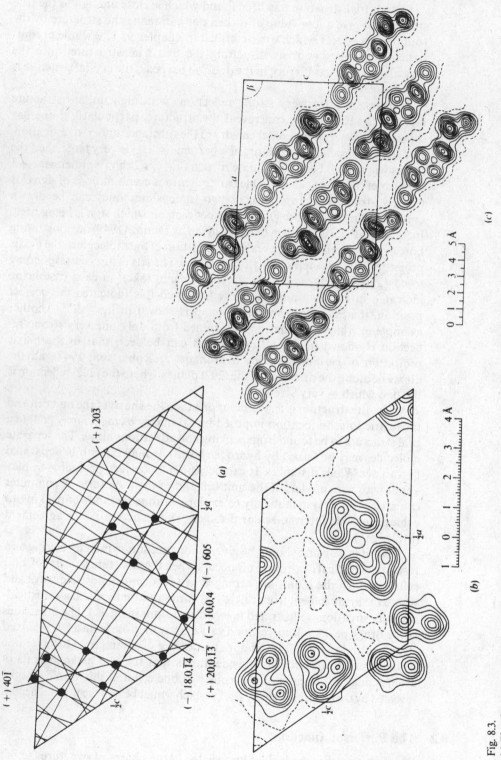

Fig. 8.3.
(a) The six strong reflections which showed the *b*-axis projection of 1,2,3,4-tetraphenylcyclobutane.
(b) The *b*-axis projected electron-density map for 1,2,3,4-tetraphenylcyclobutane (both (a) and (b) from Dunitz, 1949).
(c) The *b*-axis projection of *p*-nitroaniline showing the molecule lying along the (202) direction (from Abrahams & Robertson, 1948).

Fig. 8.4.
A two-dimensional
statistically disordered
structure.

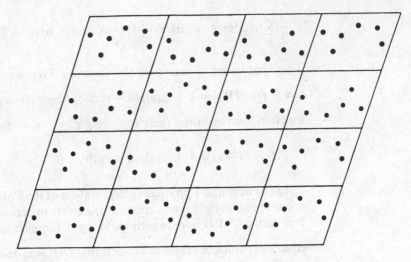

showing the positions of the atoms in the unit cell. Patterson (1934) suggested as an aid to structure determination the use of the function

$$P(\mathbf{r}) = \frac{1}{V}\sum_{\mathbf{h}}|F_{\mathbf{h}}|^2\exp(-2\pi i\mathbf{h}\cdot\mathbf{r}) \qquad (8.2)$$

where $\mathbf{r} = x\mathbf{a} + y\mathbf{b} + z\mathbf{c}$ is a position in unit-cell space and

$$\mathbf{h} = h\mathbf{a}^* + k\mathbf{b}^* + l\mathbf{c}^*$$

is a reciprocal-lattice point coordinate.

With the operation of Friedel's law we have $|F_{\mathbf{h}}|^2 = |F_{\bar{\mathbf{h}}}|^2$ and so the Patterson function is real and may be written as

$$P(\mathbf{r}) = \frac{1}{V}\sum_{\mathbf{h}}|F_{\mathbf{h}}|^2\cos(2\pi\mathbf{h}\cdot\mathbf{r}). \qquad (8.3)$$

Ideally, the values of $F_{\mathbf{h}}$ and $|F_{\mathbf{h}}|^2$ are finite only at the points of the reciprocal lattice. The transform of $(1/V)F_{\mathbf{h}}$ is $\rho(\mathbf{r})$, and a few moments study of the electron-density equation (4.72) will show that the transform of $(1/V)F_{\bar{\mathbf{h}}}$ is $\rho(-\mathbf{r}) = \psi(\mathbf{r})$, say. But we found in §4.6 that the transform of a product of two functions is the convolution of their separate transforms and this gives

$$\int_V \rho(\mathbf{u})\psi(\mathbf{r} - \mathbf{u})\mathrm{d}v = \frac{1}{V^2}\sum_{\mathbf{h}}F_{\mathbf{h}}F_{\bar{\mathbf{h}}}\exp(-2\pi i\mathbf{h}\cdot\mathbf{r}). \qquad (8.4)$$

The right-hand side can be simplified because, from equation (4.81), it can be seen that $F_{\bar{\mathbf{h}}}$ is the complex conjugate of $F_{\mathbf{h}}$ so that $F_{\mathbf{h}}F_{\bar{\mathbf{h}}} = |F_{\mathbf{h}}|^2$. From equations (8.4) and (8.2) it now appears that

$$\frac{1}{V}\sum_{\mathbf{h}}|F_{\mathbf{h}}|^2\exp(-2\pi i\mathbf{h}\cdot\mathbf{r}) = V\int_{V}\rho(\mathbf{u})\psi(\mathbf{r}-\mathbf{u})\,\mathrm{d}v = \mathrm{P}(\mathbf{r}). \qquad (8.5)$$

This enables us to interpret the function P(**r**); we see that $\psi(\mathbf{r}-\mathbf{u}) = \rho(\mathbf{u}-\mathbf{r})$ and $\mathrm{P}(\mathbf{r}) = V\int_{V}\rho(\mathbf{u})\rho(\mathbf{u}-\mathbf{r})\,\mathrm{d}v$ or, since P(**r**) is a centrosymmetrical function from equation (8.3), replacing **r** by −**r** we find

$$\mathrm{P}(\mathbf{r}) = \mathrm{P}(-\mathbf{r}) = V\int_{V}\rho(\mathbf{u})\rho(\mathbf{u}+\mathbf{r})\,\mathrm{d}v. \qquad (8.6)$$

Let us now find a physical interpretation of the Patterson function. We suppose that in fig. 8.5 we represent the electron density with the origin of one unit cell at $O$ and superimpose on it the electron density with the equivalent unit-cell origin at $O'$ such that $\overline{OO'} = \mathbf{r}$. Then $\int_{V}\rho(\mathbf{u})\rho(\mathbf{u}+\mathbf{r})\mathrm{d}v$ represents the product of the staggered electron-density functions integrated over the complete unit cell outlined boldly in the figure.

A brief word about the units of P(**r**) should be given here. The electron density $\rho(\mathbf{r})$ will be in units of electrons per unit volume (or area or length depending on the number of dimensions of the representation) while the structure factor $F_{\mathbf{h}}$ will be in electron units. This gives a consistent dimensional system if fractional coordinates are always used and volume integrals are taken over the unit cell in dimensionless units of volume which are expressed in terms of fractions of the unit-cell volume. The reader should examine equations (4.72) and (4.75) to see that they are consistent with this set of dimensions. If the integration in equation (8.6) is over a dimensionless volume then both equation (8.2) and (8.6) indicate that the dimensions of P(**r**) are (electrons)$^2$ per unit volume.

Fig. 8.5.
A representation of two similar electron-density maps displaced by a vector **r**.

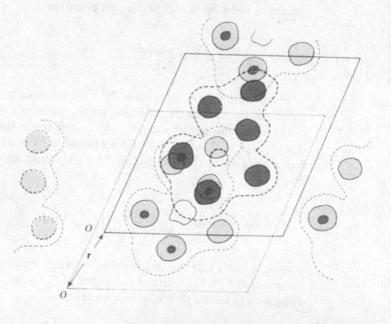

What should the Patterson function P(**r**) show? Clearly it will be large if strong regions of electron density overlap in the displaced electron-density maps illustrated in fig. 8.5. In fact this is equivalent to saying that the value of P(**r**) will be large if strong regions of electron density are separated by a vector **r** in the unit cell. If there are several strong regions of electron density separated by the vector **r** the P(**r**) will show the total effect of this and will consequently have a large value.

A well-resolved electron-density map shows regions of isolated electron density symmetrically disposed about the atomic centres. The contribution to the Patterson function of two atoms separated by a vector **r** will be a diffuse peak centred on **r**. In fig. 8.6(*a*) there are seen two atomic peaks of a one-dimensional electron-density map. When two such maps are overlapped with a displacement **r** the value of P(**r**), as given by equation (8.6), depends on the product of the $\rho$'s over the range for which they are non-zero. This will be the integrated product for the shaded region in fig. 8.6(*b*). An examination of this figure should make it clear that the maximum integrated product will occur when the atomic centres exactly overlap and that the width of a Patterson peak will be wider than that of an electron-density peak.

The Patterson function will be a superposition of peaks derived from all the pairs of atoms in the unit cell; if there was no overlap of Patterson peaks, or the overlap was slight, then the function P(**r**) would show the positions of all the interatomic vectors in the cell. In fact the resolution of Patterson maps is usually rather poor and this is what should be expected. The $N$ peaks in an electron-density map, which *will* be well resolved, give rise to $N(N-1)$ vectors (ignoring the null vectors of an atomic position to itself) and the corresponding Patterson peaks are more diffuse than the electron-density peaks.

Thus if there is a set of atoms in positions $\mathbf{r}_j (j = 1$ to $N)$ they will give rise to a set of Patterson peaks at positions $\mathbf{r}_i - \mathbf{r}_j$ for $i = 1$ to $N$ and $j = 1$ to $N$. The $N$ null vectors give a large peak at the origin, which can be seen from equation (8.3), since

$$P(0) = \frac{1}{V}\sum_{\mathbf{h}} |F_{\mathbf{h}}|^2. \tag{8.7}$$

Fig. 8.6.
(*a*) Two peaks of
electron density.
(*b*) Overlap of peaks
when two
electron-density maps
are displaced by a vector
**r**.

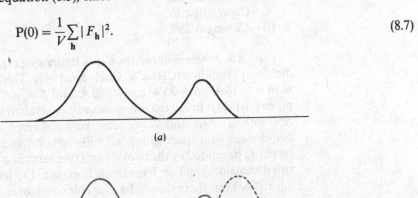

The other vectors occur in centrosymmetric pairs $r_i - r_j$ and $r_j - r_i$, again in conformity with the centrosymmetric nature of $P(r)$ as shown by equation (8.3).

In the event that the structure is centrosymmetrical there is a systematic overlap of some peaks to give double-weight peaks. For atoms with coordinates $r_i$, $-r_i$, $r_j$ and $-r_j$ the interatomic vectors, apart from the null vectors, are

$$
\begin{array}{ccc}
r_i - (-r_i) & r_i - r_j & r_i - (-r_j) \\
-r_i - r_i & -r_i - r_j & -r_i - (-r_j) \\
r_j - r_i & r_j - (-r_i) & r_j - (-r_j) \\
-r_j - r_i & -r_j - (-r_i) & -r_j - r_j
\end{array}
$$

or, to summarize, single peaks at $\pm 2r_i$, $\pm 2r_j$ and double peaks at $\pm(r_i - r_j)$ and $\pm(r_i + r_j)$.

Despite the overcrowded nature of Patterson maps useful information can frequently be derived from them and, indeed, quite complex crystal structures can be solved from an interpretation of the Patterson function. Some common situations which occur and give rise to recognizable Patterson-map features will now be considered.

## (i) Peaks due to heavy atoms

If there are a few atoms (perhaps only two) which have a higher atomic number than the remainder then the peaks due to these may well stand out in the ruck of minor peaks and give clear interatomic vectors. The total Patterson density associated with a vector between two atoms of atomic numbers $Z_i$ and $Z_j$ is equal to $Z_i Z_j$; thus if a structure contains carbon atoms $(Z = 6)$ and chlorine atoms $(Z = 17)$ the types of peaks present and their weights are

C—C   weight   36
C—Cl   weight 102
Cl—Cl weight 289.

In fig. 8.7(a) there is shown the $b$-axis Patterson projection for 2-amino-4,6-dichloropyrimidine (Clews & Cochran, 1948). The space group is $P2_1/a$ with $a = 16.45$, $b = 3.84$, $c = 10.28\,\text{Å}$ and $\beta = 108°$. Due to the $a$-glide plane the $b$-axis projected cell shows two equivalent parts so that a simpler cell with $a' = \frac{1}{2}a$ and $c' = c$ can be considered. This simple cell, of two-dimensional space group $p2$, will contain two molecules which must therefore be related by the twofold axis (the same as a centre of symmetry in two dimensions) and each molecule is expected to look like that shown in fig. 8.7(b). Thus there should be four chlorine atoms in the simple cell with coordinates $\pm(x_1, z_1)$ and $\pm(x_2, z_2)$ and hence one should expect single-weight peaks at $\pm(2x_1, 2z_1)$ and $\pm(2x_2, 2z_2)$ and double-weight peaks at $\pm(x_1 - x_2, z_1 - z_2)$ and $\pm(x_1 + x_2, z_1 + z_2)$. The two heaviest peaks of the Patterson map are those marked $C$ and $D$ which have heights of 326 and 323 units, respectively. In terms of the cell with sides $a'$ and $c$ the coordinates

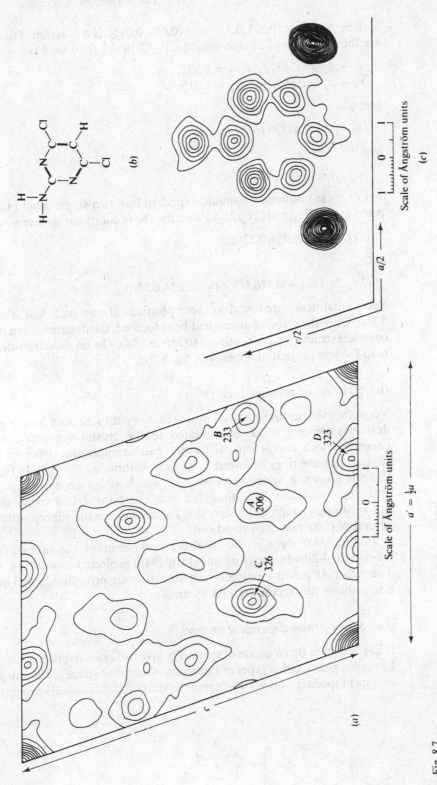

Fig. 8.7.

(a) The b-axis Patterson projection for 2-amino-4,6-dichloropyrimidine.

(b) A molecule of 2-amino-4,6-dichloropyrimidine. (Parts (a) and (b) from Clews & Cochran, 1948.)

(c) The electron-density map for the b-axis projection of 2-amino-4,6-dichloropyrimidine.

of these peaks are $(0.228, 0.321)$ and $(0.665, 0.015)$. If it is assumed that they are the two types of double-weight Cl—Cl peaks then we have

$$x_1 + x_2 = 0.288, \; z_1 + z_2 = 0.321,$$
$$x_1 - x_2 = 0.665, \; z_1 - z_2 = 0.015,$$

giving

$$(x_1, z_1) = (0.477, 0.168)$$

and

$$(x_2, z_2) = (-0.188, 0.153).$$

For these positions we should expect to find two single-weight Cl—Cl peaks at $(2x_1, 2z_1)$ and $(2x_2, 2z_2)$. From the above coordinates these would be

$$(2x_1, 2z_1) = (0.954, 0.336)$$

and

$$(2x_2, 2z_2) = (-0.376, 0.306) = (0.624, 0.306).$$

Substantial peaks are found in these positions shown as $B$ and $A$ in fig. 8.7($a$). Once the chlorine atoms had been located, the determination of the complete structure was a routine matter (see § 8.4). The final electron-density map for this projection appears in fig. 8.7($c$).

### (ii) Overlap of parallel vectors

Very often the arrangements of atoms in a crystal structure have a great deal of symmetry which is unrelated to the crystal symmetry but is concerned with stereochemical factors. For example the structure may contain benzene rings or linked rings as in anthracene (fig. 8.8). In fig. 8.8 there is shown a selection of vectors which occur repeatedly in the anthracene molecule. Even though a single carbon–carbon vector might not be strong enough to be seen in a Patterson map the superposition of several of them may well stand out.

In fig. 8.9($a$) there is shown the arrangement of the molecules of 4,6-dimethyl-2-hydroxypyrimidine (Pitt, 1948) projected down the $a$ axis. The vector $AB$ occurs repeatedly and the Patterson projection, given in fig. 8.9($b$), shows this vector marked by an $X$.

### (iii) Space-group-dependent vectors

It has been seen that a centre of symmetry gives an exact overlap of pairs of Patterson peaks. Other types of symmetry element or space group can give vectors in special positions. For example, the two-dimensional space group

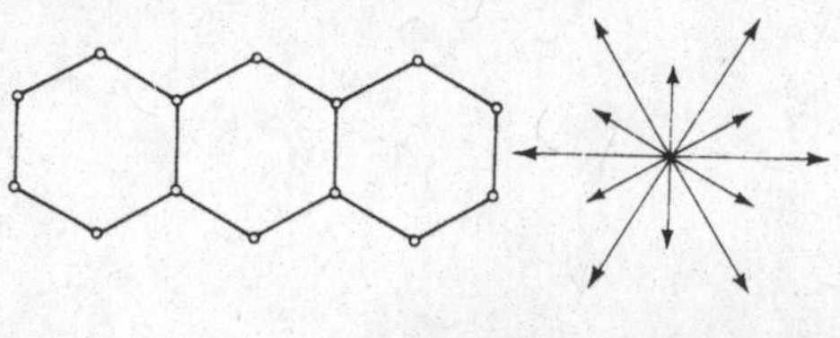

Fig. 8.8.
The carbon atoms of the anthracene molecule and some of the interatomic vectors which occur repeatedly.

Fig. 8.9.
(a) The arrangement of
molecules in the a-axis
projection of
4,6-dimethyl-2-
hydroxypyrimidine.
(b) The Patterson peak
marked with an X
corresponds to the
repeated vector AB in
(a) (from Pitt, 1948).

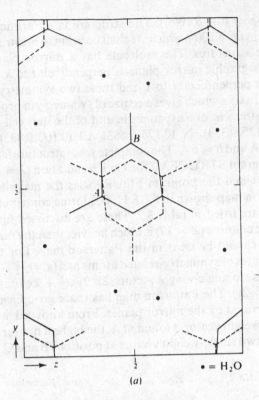

(a)

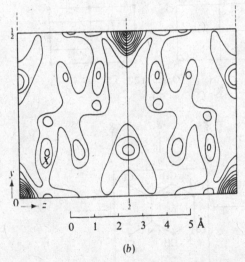

(b)

*pgg* has atoms in general symmetry-related positions $\pm(x, y)$, $\pm(\frac{1}{2} - x, \frac{1}{2} + y)$. These four related positions give vectors (except null vectors) –

single weight $\pm(2x, 2y)$; $\pm(-2x, 2y)$
double weight $\pm(\frac{1}{2} + 2x, \frac{1}{2})$; $\pm(\frac{1}{2}, \frac{1}{2} + 2y)$.

We now look at an example of the finding of heavy atoms from space-group-dependent vectors for a structure in projection with space group *pmg* (fig. 8.10(*a*)). The structure is an artificial one, $(C_2H_2NCl)_2$, shown in fig. 8.10(*b*), which we shall be using to illustrate various methods of solving structures. The molecule has a mirror plane which sits on the crystallographic mirror plane *m* perpendicular to *x*. There is also a glide plane *g* perpendicular to *y* and these two symmetry elements generate a twofold axis – which gives a centre of symmetry in projection. The positions of the atoms in one asymmetric unit of the unit cell are: $C_1$ (0.175, 0.083), $C_2$ (0.125, 0.333), N (0.175, 0.583), Cl (0.100, 0.833) and we also have $a = 10\,\text{Å}$ and $b = 6\,\text{Å}$. The complete set of structure factors, calculated with the program STRUCFAC for Cu K$\alpha$ radiation ($\lambda = 1.542\,\text{Å}$), are given in table 8.1 and the program FOUR2 uses the magnitudes to generate the Patterson map shown in fig. 8.11. [Chlorine contributions to the structure factors are listed in table 8.2. These are discussed further in §8.3.]

Since chlorine ($Z = 17$) is much heavier than the other atoms the Cl—Cl vectors should be clear in the Patterson map. For this space group the coordinates of symmetry-related atoms are: $(x, y)$; $(\frac{1}{2} - x, y)$; $(\bar{x}, \bar{y})$; $(\frac{1}{2} + x, \bar{y})$ giving rise to single-weight vectors $(2x, 2y)$, $(\frac{1}{2} + 2x, 0)$ and the double-weight vector $(\frac{1}{2}, 2y)$. The Patterson map has space group *pmm* and other vectors are generated by the mirror planes. From knowledge of the structure, the double-weight vector is found at A, the highest non-origin peak in fig. 8.11, and the two single-weight vectors at positions B and C. The vector is not on

Fig. 8.10.
(*a*) The two-dimensional space group *pmg*(*p2mg*).
(*b*) The artificial structure $C_4H_4N_2Cl_2$. The mirror and glide planes are indicated *m* and *g*, respectively.

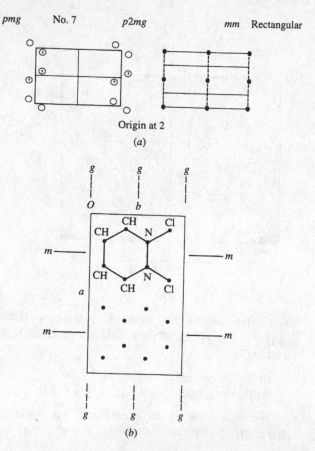

*pmg*    No. 7    *p2mg*    *mm*    Rectangular

Origin at 2

(*a*)

(*b*)

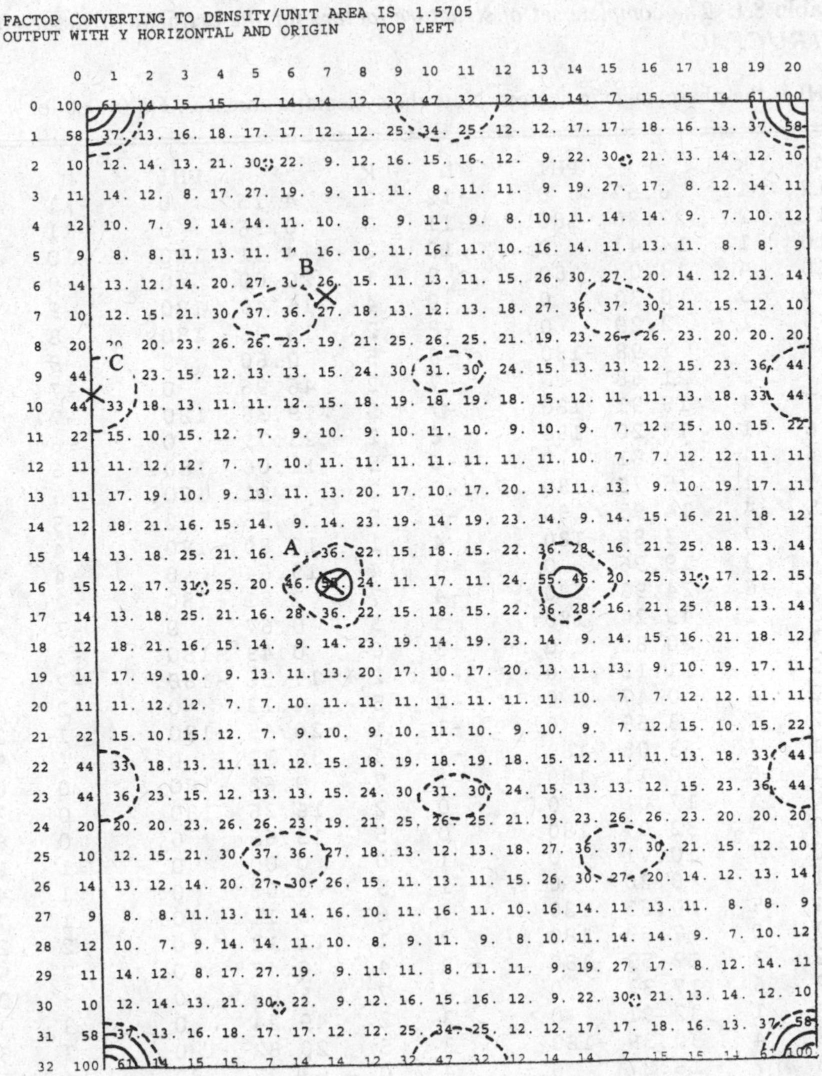

Fig. 8.11.
The Patterson function
for $C_4H_4N_2Cl_2$.

```
FACTOR CONVERTING TO DENSITY/UNIT AREA IS  1.5705
OUTPUT WITH Y HORIZONTAL AND ORIGIN    TOP LEFT

      0   1   2   3   4   5   6   7   8   9  10  11  12  13  14  15  16  17  18  19  20

 0  100  61  14  15  15   7  14  14  12  32  47  32  12  14  14   7  15  15  14  61 100
 1   58  37  13  16  18  17  17  12  12  25  34  25  12  12  17  17  18  16  13  37  58
 2   10  12  14  13  21  30  22   9  12  16  15  16  12   9  22  30  21  13  14  12  10
 3   11  14  12   8  17  27  19   9  11  11   8  11  11   9  19  27  17   8  12  14  11
 4   12  10   7   9  14  14  11  10   8   9  11   9   8  10  11  14  14   9   7  10  12
 5    9   8   8  11  13  11  14  16  10  11  16  11  10  16  14  11  13  11   8   8   9
 6   14  13  12  14  20  27  30  26  15  11  13  11  15  26  30  27  20  14  12  13  14
 7   10  12  15  21  30  37  36  27  18  13  12  13  18  27  36  37  30  21  15  12  10
 8   20  20  20  23  26  26  23  19  21  25  26  25  21  19  23  26  26  23  20  20  20
 9   44  33  23  15  12  13  13  15  24  30  31  30  24  15  13  13  12  15  23  36  44
10   44  33  18  13  11  11  12  15  18  19  18  19  18  15  12  11  11  13  18  33  44
11   22  15  10  15  12   7   9  10   9  10  11  10   9  10   9   7  12  15  10  15  22
12   11  11  12  12   7   7  10  11  11  11  11  11  11  10   7   7  12  12  11  11
13   11  17  19  10   9  13  11  13  20  17  10  17  20  13  11  13   9  10  19  17  11
14   12  18  21  16  15  14   9  14  23  19  14  19  23  14   9  14  15  16  21  18  12
15   14  13  18  25  21  16  36  22  15  18  15  22  36  28  16  21  25  18  13  14
16   15  12  17  31  25  20  36  36  24  11  17  11  24  55  45  20  25  31  17  12  15
17   14  13  18  25  21  16  28  36  22  15  18  15  22  36  28  16  21  25  18  13  14
18   12  18  21  16  15  14   9  14  23  19  14  19  23  14   9  14  15  16  21  18  12
19   11  17  19  10   9  13  11  13  20  17  10  17  20  13  11  13   9  10  19  17  11
20   11  11  12  12   7   7  10  11  11  11  11  11  11  10   7   7  12  12  11  11
21   22  15  10  15  12   7   9  10   9  10  11  10   9  10   9   7  12  15  10  15  22
22   44  33  18  13  11  11  12  15  18  19  18  19  18  15  12  11  11  13  18  33  44
23   44  36  23  15  12  13  13  15  24  30  31  30  24  15  13  13  12  15  23  36  44
24   20  20  20  23  26  26  23  19  21  25  26  25  21  19  23  26  26  23  20  20  20
25   10  12  15  21  30  37  36  27  18  13  12  13  18  27  36  37  30  21  15  12  10
26   14  13  12  14  20  27  30  26  15  11  13  11  15  26  30  27  20  14  12  13  14
27    9   8   8  11  13  11  14  16  10  11  16  11  10  16  14  11  13  11   8   8   9
28   12  10   7   9  14  14  11  10   8   9  11   9   8  10  11  14  14   9   7  10  12
29   11  14  12   8  17  27  19   9  11  11   8  11  11   9  19  27  17   8  12  14  11
30   10  12  14  13  21  30  22   9  12  16  15  16  12   9  22  30  21  13  14  12  10
31   58  37  13  16  18  17  17  12  12  25  34  25  12  12  17  17  18  16  13  37  58
32  100  61  14  15  15   7  14  14  12  32  47  32  12  14  14   7  15  15  14  61 100
```

the centre of peak B because of other vectors nearby. However, the special peaks along $y = 0$ and $x = \frac{1}{2}$ at A and C are sufficient to determine the position of the chlorine atoms and peak B can then just be used as confirmation.

The vectors at the special positions A and C in fig. 8.11 are space-group-dependent vectors and these generally occur when a symmetry element involving a translation is present. For example, if the symmetry element $2_1$ is present along the $b$ direction then atoms will be related in pairs with coordinates $(x, y, z)$ and $(\bar{x}, \frac{1}{2} + y, \bar{z})$. The vector between such a pair of atoms is $(2x, \frac{1}{2}, 2z)$ and hence the section of the three-dimensional Patterson map $y = \frac{1}{2}$ should show all the vectors of this type. There may, of course, be other peaks in this section due to the accidental occurrence of vectors with a component $y = \frac{1}{2}$, but a high proportion of the peaks should be associated with the systematic peaks which show the structure projected down the $b$

Table 8.1. *The complete set of structure factors for the artificial structure* $(C_2NCl)_2$ *as output by* STRUCFAC

PHI is the phase angle in degrees. Note the systematic absences, $F(h,0) = 0$ for $h$ odd.

| h | k | F | PHI | h | k | F | PHI | h | k | F | PHI |
|---|---|---|---|---|---|---|---|---|---|---|---|
| -12 | 1 | 6.67 | 0. | -12 | 2 | 4.15 | 0. | -11 | 1 | 10.02 | 180. |
| -11 | 2 | 22.80 | 180. | -11 | 3 | 0.36 | 0. | -11 | 4 | 11.74 | 0. |
| -10 | 1 | 14.43 | 0. | -10 | 2 | 14.31 | 180. | -10 | 3 | 27.56 | 180. |
| -10 | 4 | 13.07 | 180. | -9 | 1 | 20.00 | 0. | -9 | 2 | 5.00 | 0. |
| -9 | 3 | 0.30 | 0. | -9 | 4 | 15.49 | 180. | -9 | 5 | 16.81 | 180. |
| -8 | 1 | 2.29 | 0. | -8 | 2 | 15.08 | 180. | -8 | 3 | 2.28 | 180. |
| -8 | 4 | 1.98 | -180. | -8 | 5 | 0.60 | 0. | -8 | 6 | 23.89 | 0. |
| -7 | 1 | 21.68 | 0. | -7 | 2 | 46.96 | 0. | -7 | 3 | 1.65 | 180. |
| -7 | 4 | 15.91 | 180. | -7 | 5 | 19.38 | 180. | -7 | 6 | 0.53 | 0. |
| -6 | 1 | 17.20 | 180. | -6 | 2 | 23.79 | 0. | -6 | 3 | 26.36 | 0. |
| -6 | 4 | 4.81 | 0. | -6 | 5 | 11.06 | 180. | -6 | 6 | 34.28 | 180. |
| -5 | 1 | 5.75 | -180. | -5 | 2 | 7.91 | 180. | -5 | 3 | 1.69 | 0. |
| -5 | 4 | 14.93 | 180. | -5 | 5 | 4.55 | 0. | -5 | 6 | 0.06 | -180. |
| -5 | 7 | 3.88 | -180. | -4 | 1 | 10.80 | -180. | -4 | 2 | 18.56 | 0. |
| -4 | 3 | 19.96 | 0. | -4 | 4 | 19.64 | 0. | -4 | 5 | 9.15 | 180. |
| -4 | 6 | 24.93 | 180. | -4 | 7 | 7.51 | -180. | -3 | 1 | 32.91 | 180. |
| -3 | 2 | 49.24 | 180. | -3 | 3 | 0.67 | 0. | -3 | 4 | 30.39 | 0. |
| -3 | 5 | 20.82 | 0. | -3 | 6 | 0.43 | -180. | -3 | 7 | 18.27 | 180. |
| -2 | 1 | 11.12 | 0. | -2 | 2 | 17.56 | -180. | -2 | 3 | 12.53 | 180. |
| -2 | 4 | 0.47 | 0. | -2 | 5 | 4.13 | 0. | -2 | 6 | 17.33 | 0. |
| -2 | 7 | 3.69 | 0. | -1 | 1 | 20.65 | 180. | -1 | 2 | 8.20 | -180. |
| -1 | 3 | 3.08 | -180. | -1 | 4 | 39.32 | 0. | -1 | 5 | 10.63 | 0. |
| -1 | 6 | 0.11 | -180. | -1 | 7 | 9.68 | -180. | 0 | 0 | 144.00 | 0. |
| 0 | 1 | 17.01 | 0. | 0 | 2 | 16.26 | 180. | 0 | 3 | 30.58 | 180. |
| 0 | 4 | 32.07 | 180. | 0 | 5 | 13.65 | 0. | 0 | 6 | 21.38 | 0. |
| 0 | 7 | 10.71 | 0. | 1 | 0 | 0.00 | 0. | 1 | 1 | 20.65 | 0. |
| 1 | 2 | 8.20 | 0. | 1 | 3 | 3.08 | 0. | 1 | 4 | 39.32 | -180. |
| 1 | 5 | 10.63 | 180. | 1 | 6 | 0.11 | 0. | 1 | 7 | 9.68 | 0. |
| 2 | 0 | 7.69 | 180. | 2 | 1 | 11.12 | 0. | 2 | 2 | 17.56 | -180. |
| 2 | 3 | 12.53 | -180. | 2 | 4 | 0.47 | 0. | 2 | 5 | 4.13 | 0. |
| 2 | 6 | 17.33 | 0. | 2 | 7 | 3.69 | 0. | 3 | 0 | 0.00 | 0. |
| 3 | 1 | 32.91 | 0. | 3 | 2 | 49.24 | 0. | 3 | 3 | 0.67 | -180. |
| 3 | 4 | 30.39 | -180. | 3 | 5 | 20.82 | 180. | 3 | 6 | 0.43 | 0. |
| 3 | 7 | 18.27 | 0. | 4 | 0 | 63.27 | -180. | 4 | 1 | 10.80 | -180. |
| 4 | 2 | 18.56 | 0. | 4 | 3 | 19.96 | 0. | 4 | 4 | 19.64 | 0. |
| 4 | 5 | 9.15 | 180. | 4 | 6 | 24.93 | -180. | 4 | 7 | 7.51 | -180. |
| 5 | 0 | 0.00 | 0. | 5 | 1 | 5.75 | 0. | 5 | 2 | 7.91 | 0. |
| 5 | 3 | 1.69 | -180. | 5 | 4 | 14.93 | 0. | 5 | 5 | 4.55 | 180. |
| 5 | 6 | 0.06 | 0. | 5 | 7 | 3.88 | 0. | 6 | 0 | 8.70 | -180. |
| 6 | 1 | 17.20 | 180. | 6 | 2 | 23.79 | 0. | 6 | 3 | 26.36 | 0. |
| 6 | 4 | 4.81 | 0. | 6 | 5 | 11.06 | -180. | 6 | 6 | 34.28 | -180. |
| 7 | 0 | 0.00 | 0. | 7 | 1 | 21.68 | -180. | 7 | 2 | 46.96 | -180. |
| 7 | 3 | 1.65 | 0. | 7 | 4 | 15.91 | 0. | 7 | 5 | 19.38 | 0. |
| 7 | 6 | 0.53 | 180. | 8 | 0 | 3.71 | 0. | 8 | 1 | 2.29 | 0. |
| 8 | 2 | 15.08 | 180. | 8 | 3 | 2.28 | -180. | 8 | 4 | 1.98 | -180. |
| 8 | 5 | 0.60 | 0. | 8 | 6 | 23.89 | 0. | 9 | 0 | 0.00 | 0. |
| 9 | 1 | 20.00 | -180. | 9 | 2 | 5.00 | 180. | 9 | 3 | 0.30 | 180. |
| 9 | 4 | 15.49 | 0. | 9 | 5 | 16.81 | 0. | 10 | 0 | 29.16 | 0. |
| 10 | 1 | 14.43 | 0. | 10 | 2 | 14.31 | -180. | 10 | 3 | 27.56 | -180. |
| 10 | 4 | 13.07 | -180. | 11 | 0 | 0.00 | 0. | 11 | 1 | 10.02 | 0. |
| 11 | 2 | 22.80 | 0. | 11 | 3 | 0.36 | 180. | 11 | 4 | 11.74 | -180. |
| 12 | 0 | 12.53 | 0. | 12 | 1 | 6.67 | 0. | 12 | 2 | 4.15 | 0. |

Table 8.2. *Chlorine contributions to the structure factors of* $(C_2H_2NCl)_2$

PHI is the phase angle in degrees. Comparison with table 8.1 shows that the indications of sign given by chlorine are mostly correct.

| h | k | F | PHI | h | k | F | PHI | h | k | F | PHI |
|---|---|---|---|---|---|---|---|---|---|---|---|
| -12 | 1 | 4.07 | 0. | -12 | 2 | 4.04 | -180. | -11 | 1 | 14.08 | 180. |
| -11 | 2 | 13.80 | 180. | -11 | 3 | 0.10 | 0. | -11 | 4 | 12.99 | 0. |
| -10 | 1 | 14.43 | 0. | -10 | 2 | 14.31 | 180. | -10 | 3 | 27.56 | 180. |
| -10 | 4 | 13.07 | 180. | -9 | 1 | 15.39 | 0. | -9 | 2 | 15.03 | 0. |
| -9 | 3 | 0.11 | -180. | -9 | 4 | 14.08 | 180. | -9 | 5 | 13.26 | 180. |
| -8 | 1 | 4.91 | 0. | -8 | 2 | 4.82 | 180. | -8 | 3 | 9.21 | 180. |
| -8 | 4 | 4.35 | -180. | -8 | 5 | 4.26 | 0. | -8 | 6 | 7.87 | 0. |
| -7 | 1 | 27.96 | 0. | -7 | 2 | 26.87 | 0. | -7 | 3 | 0.19 | 180. |
| -7 | 4 | 24.43 | 180. | -7 | 5 | 22.88 | 180. | -7 | 6 | 0.31 | 0. |
| -6 | 1 | 14.87 | 180. | -6 | 2 | 14.18 | 0. | -6 | 3 | 26.37 | 0. |
| -6 | 4 | 12.15 | 0. | -6 | 5 | 11.85 | -180. | -6 | 6 | 21.81 | 180. |
| -5 | 1 | 0.00 | 0. | -5 | 2 | 0.00 | 0. | -5 | 3 | 0.00 | 0. |
| -5 | 4 | 0.00 | 0. | -5 | 5 | 0.00 | 0. | -5 | 6 | 0.00 | 0. |
| -5 | 7 | 0.00 | 0. | -4 | 1 | 18.57 | -180. | -4 | 2 | 16.89 | 0. |
| -4 | 3 | 29.34 | 0. | -4 | 4 | 13.02 | 0. | -4 | 5 | 12.38 | 180. |
| -4 | 6 | 22.73 | 180. | -4 | 7 | 10.29 | -180. | -3 | 1 | 43.26 | 180. |
| -3 | 2 | 37.17 | 180. | -3 | 3 | 0.23 | 0. | -3 | 4 | 27.83 | 0. |
| -3 | 5 | 25.09 | 0. | -3 | 6 | 0.34 | -180. | -3 | 7 | 22.01 | 180. |
| -2 | 1 | 8.93 | 0. | -2 | 2 | 7.61 | -180. | -2 | 3 | 12.54 | 180. |
| -2 | 4 | 5.26 | 180. | -2 | 5 | 4.90 | 0. | -2 | 6 | 8.92 | 0. |
| -2 | 7 | 4.03 | 0. | -1 | 1 | 31.39 | 180. | -1 | 2 | 26.46 | 180. |
| -1 | 3 | 0.16 | 0. | -1 | 4 | 17.99 | 0. | -1 | 5 | 15.93 | 0. |
| -1 | 6 | 0.21 | -180. | -1 | 7 | 13.81 | -180. | 0 | 0 | 68.00 | 0. |
| 0 | 1 | 31.33 | 0. | 0 | 2 | 26.78 | 180. | 0 | 3 | 42.49 | 180. |
| 0 | 4 | 17.44 | 180. | 0 | 5 | 16.08 | 0. | 0 | 6 | 29.16 | 0. |
| 0 | 7 | 13.13 | 0. | 1 | 0 | 0.00 | 0. | 1 | 1 | 31.39 | 0. |
| 1 | 2 | 26.46 | 0. | 1 | 3 | 0.16 | -180. | 1 | 4 | 17.99 | -180. |
| 1 | 5 | 15.93 | 180. | 1 | 6 | 0.21 | 0. | 1 | 7 | 13.81 | 0. |
| 2 | 0 | 18.95 | 0. | 2 | 1 | 8.93 | 0. | 2 | 2 | 7.61 | -180. |
| 2 | 3 | 12.54 | -180. | 2 | 4 | 5.26 | 180. | 2 | 5 | 4.90 | 0. |
| 2 | 6 | 8.92 | 0. | 2 | 7 | 4.03 | 0. | 3 | 0 | 0.00 | 0. |
| 3 | 1 | 43.26 | 0. | 3 | 2 | 37.17 | 0. | 3 | 3 | 0.23 | -180. |
| 3 | 4 | 27.83 | -180. | 3 | 5 | 25.09 | -180. | 3 | 6 | 0.34 | 0. |
| 3 | 7 | 22.01 | 0. | 4 | 0 | 38.83 | -180. | 4 | 1 | 18.57 | -180. |
| 4 | 2 | 16.89 | 0. | 4 | 3 | 29.34 | 0. | 4 | 4 | 13.02 | 0. |
| 4 | 5 | 12.38 | 180. | 4 | 6 | 22.73 | -180. | 4 | 7 | 10.29 | -180. |
| 5 | 0 | 0.00 | 0. | 5 | 1 | 0.00 | 0. | 5 | 2 | 0.00 | 0. |
| 5 | 3 | 0.00 | 0. | 5 | 4 | 0.00 | 0. | 5 | 5 | 0.00 | 0. |
| 5 | 6 | 0.00 | 0. | 5 | 7 | 0.00 | 0. | 6 | 0 | 30.55 | -180. |
| 6 | 1 | 14.87 | 180. | 6 | 2 | 14.18 | 0. | 6 | 3 | 26.37 | 0. |
| 6 | 4 | 12.15 | 0. | 6 | 5 | 11.85 | -180. | 6 | 6 | 21.81 | -180. |
| 7 | 0 | 0.00 | 0. | 7 | 1 | 27.96 | -180. | 7 | 2 | 26.87 | -180. |
| 7 | 3 | 0.19 | 0. | 7 | 4 | 24.43 | 0. | 7 | 5 | 22.88 | 0. |
| 7 | 6 | 0.31 | 180. | 8 | 0 | 9.98 | 0. | 8 | 1 | 4.91 | 0. |
| 8 | 2 | 4.82 | 180. | 8 | 3 | 9.21 | -180. | 8 | 4 | 4.35 | -180. |
| 8 | 5 | 4.26 | 0. | 8 | 6 | 7.87 | 0. | 9 | 0 | 0.00 | 0. |
| 9 | 1 | 15.39 | -180. | 9 | 2 | 15.03 | -180. | 9 | 3 | 0.11 | 0. |
| 9 | 4 | 14.08 | 0. | 9 | 5 | 13.26 | 0. | 10 | 0 | 29.16 | 0. |
| 10 | 1 | 14.43 | 0. | 10 | 2 | 14.31 | -180. | 10 | 3 | 27.56 | -180. |
| 10 | 4 | 13.07 | -180. | 11 | 0 | 0.00 | 0. | 11 | 1 | 14.08 | 0. |
| 11 | 2 | 13.80 | 0. | 11 | 3 | 0.10 | 180. | 11 | 4 | 12.99 | -180. |
| 12 | 0 | 8.21 | 0. | 12 | 1 | 4.07 | 0. | 12 | 2 | 4.04 | -180. |

axis at double scale. A section of the Patterson function which shows space-group-dependent vectors is called a Patterson–Harker section (Harker, 1936). An example of a Patterson–Harker section is given in fig. 8.12; this was used to solve the structure of *para*-chlor-idoxy benzene (Archer, 1948). The X-ray measurements gave the space group as $P2_1/a$ with $a = 14.4$, $b = 6.50$, $c = 8.11$ Å and $\beta = 98.5°$ and four molecules of $ClC_6H_4IO_2$ per unit cell. The symmetry gives related coordinates $\pm(x, y, z)$ and $\pm(\frac{1}{2} + x, \frac{1}{2} - y, z)$ and the vectors between related atoms, apart from null vectors, are:

single weight $\pm(2x, 2y, 2z)$; $\pm(2x, -2y, 2z)$
double weight $\pm(\frac{1}{2}, \frac{1}{2} + 2y, 0)$; $\pm(\frac{1}{2} + 2x, \frac{1}{2}, 2z)$.

The two heavy peaks in fig. 8.12 are the iodine–iodine vectors corresponding to $\pm(\frac{1}{2} + 2x, \frac{1}{2}, 2z)$ and hence the iodine position may be derived.

It should be noted that the symmetry of the Patterson function need not be the same as that of the structural group. The effects of space-group symmetry which lead to relationships between the phases of equivalent reflections, are lost in the Patterson function for which $|F|^2$ is used as a coefficient. For all orthorhombic structures one has

$$|F_{hkl}|^2 = |F_{\bar{h}kl}|^2 = |F_{h\bar{k}l}|^2 = |F_{hk\bar{l}}|^2 = |F_{\bar{h}\bar{k}l}|^2 = |F_{\bar{h}k\bar{l}}|^2 = |F_{h\bar{k}\bar{l}}|^2 = |F_{\bar{h}\bar{k}\bar{l}}|^2$$

and the Patterson function of all primitive orthorhombic space groups is *Pmmm*. It can be shown (Buerger, 1950a) that for the 230 space groups there are only 24 possible Patterson symmetries.

A very powerful technique for making use of Patterson maps is by the use of superposition methods. It was shown by Wrinch (1939) that if the complete set of vectors between atoms was available it would be possible to recover the original atomic positions. To see how this can be done there is shown in fig. 8.13(*a*) a set of five points and in (*b*) the vector set derived from

Fig. 8.12.
The Patterson–Harker section $y = \frac{1}{2}$ for *p*-chlor-idoxy benzene (from Archer, 1948).

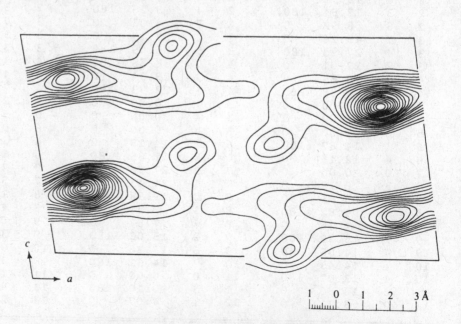

$c$

$a$

1   0   1   2   3 Å

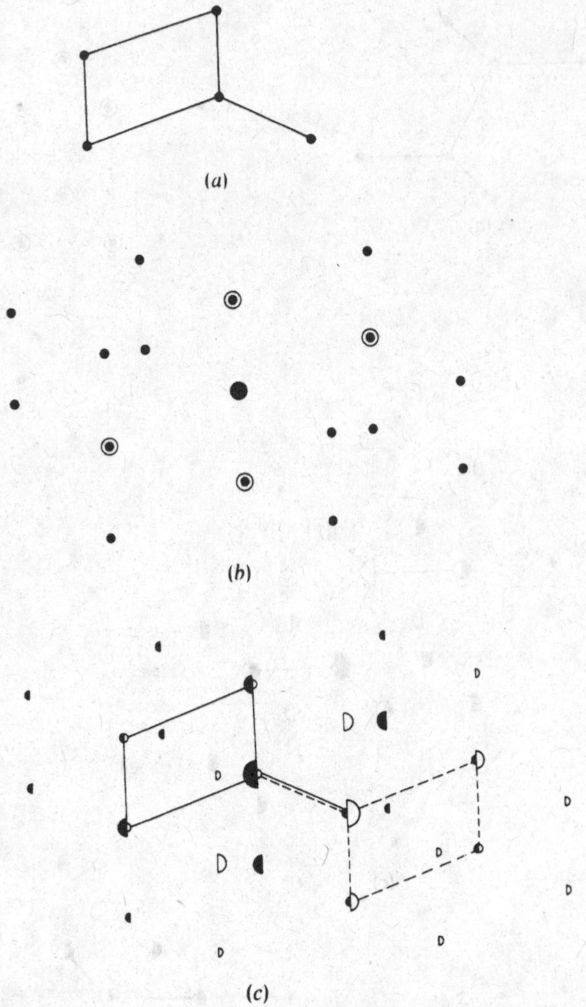

Fig. 8.13.
(*a*) A
non-centrosymmetric set
of points.
(*b*) The vector set.
(*c*) Superposition on a
single-weight vector.
For one vector set:
◖   origin peak,
◂   single peak,
◖   double peak.
For second set:
◗   origin peak,
◦   single peak,
◗   double peak.

(*a*)

(*b*)

(*c*)

these points. If two of the vector sets are now superimposed with a relative displacement equal to that of one of the single vectors then a number of points exactly overlap. These are shown in (*c*) and they show the form of the original set of points *plus* the centrosymmetrically related set connected with dashed lines. On the other hand, if the original unit is centrosymmetric as in fig. 8.14(*a*) giving the vector set (*b*) and if the overlap is made on a single-weight peak then a single image of the set of points is produced.

It will be appreciated that this method depends on having available the vector set and this is *not* what one usually has from a Patterson function. The Patterson peaks are spread out and overlap, and errors in the observed data can lead to random fluctuations which are as big as some of the smaller peaks. Incidentally it should be noticed that if the displacement of two vector maps corresponds to a double-weight vector, as in fig. 8.14(*d*), then *two* images of the original set of points result. Similarly superposition with an *n*-fold vector displacement gives *n* distinct images.

If the relative positions of more than two points of the original set are

Fig. 8.14.
(a) A centrosymmetric
set of points.
(b) The vector set.
(c) Superposition on a
single-weight vector.
(d) Superposition on a
double-weight vector
(symbols as in fig. 8.13).

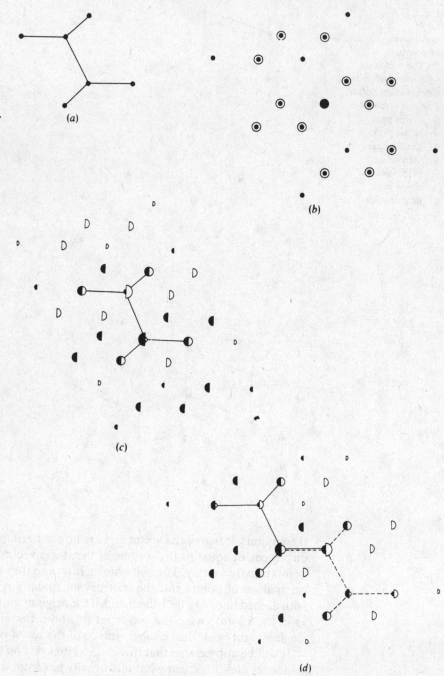

(a)

(b)

(c)

(d)

known then the image of the set of points can be recovered by multiple superposition. This type of procedure is very necessary when one is using a real Patterson function; firstly, it is usually difficult to find a single-weight peak and, secondly, there will always be a number of chance peaks which will confuse the image obtained from the overlap function. There are various functions of the overlapped Patterson maps which can be taken. If the displacement of the two map origins is $\mathbf{R}$ then one can find the product function $P(\mathbf{r})P(\mathbf{r} + \mathbf{R})$ or the sum function $P(\mathbf{r}) + P(\mathbf{r} + \mathbf{R})$. A function proposed by Buerger (1950b) is the minimum function, $M\{P(\mathbf{r}), P(\mathbf{r} + \mathbf{R})\}$, which means the least of the two quantities $P(\mathbf{r})$ and $P(\mathbf{r} + \mathbf{R})$. If $\mathbf{r}$ and $\mathbf{r} + \mathbf{R}$ correspond to the positions of peaks of interest then each of the peaks should at least have a single weight but might be heavier because of the accidental overlapping of peaks. By taking the minimum function one reduces the amount of spurious information in the result. When the relative positions of a number of atoms are known the use of the minimum function can lead to the solution of quite complicated crystal structures.

The quality of Patterson functions can be improved by various sharpening procedures. Ideally one would like the Patterson map corresponding to point atoms but if one tries to achieve this by using $|U|^2$ instead of $|F|^2$ then the termination of the Fourier series leads to considerable diffraction ripple around the peaks. The negative regions thus introduced may reduce or eliminate other peaks. However to take the observed $|F|^2$ will probably lead to very diffuse peaks giving considerable peak overlap. In fact if one modifies the $|F|^2$'s by any smooth function of $s$ ($= 2\sin\theta/\lambda$) then the individual peaks retain their spherical symmetry. A 'rule-of-thumb' which gives reasonable results is so to modify the $|F|^2$ that the average value of the modified $|F|^2$ for the region near the edge of the sphere of reflection is about 0.1 times the average of $|F|^2$ near the centre of the sphere. The modification can be made with an analytical expression, such as $\exp(Ks^2)$, or by a hand-drawn function on graph paper – it is not a very critical operation. With judicious sharpening one can pick out useful vectors which would otherwise be lost. Oversharpening, on the other hand, leads to the *appearance* of greater resolution but the extra peaks can be of diffraction-ripple origin and some of them can be quite spurious.

## 8.3   The heavy-atom method

When a structure contains a large number of light atoms and a few heavy atoms the structure can frequently be solved in a straightforward way. Let us consider a centrosymmetric structure containing in each unit cell $n$ heavy atoms whose positions can be located through the use of the Patterson function. If the contribution of the heavy atoms to the structure factor of index $\mathbf{h}$ is $C_{\mathbf{h}}$ then we have

$$F_{\mathbf{h}} = C_{\mathbf{h}} + \sum_{j=1}^{N-n} f_j \cos(2\pi\mathbf{h}\cdot\mathbf{r}_j) = C_{\mathbf{h}} + K_{\mathbf{h}}, \text{ say,} \tag{8.8}$$

where $N$ is the total number of atoms in the unit cell. Sometimes the contribution of the heavy atoms can dominate to the extent that most of the $F$'s will have the same signs as the $C$'s. Either a Fourier synthesis can be calculated with the signs of the $C$'s completely accepted or discretion can be exercised and the sign only accepted if the magnitude of $C$ is sufficiently high.

A standard technique of solving a stereochemical problem is to prepare a heavy-atom derivative of the substance in question. In selecting the heavy atom a careful balance has to be struck. If the heavy atom is too light it will be a poor determiner of signs and the structure will not be soluble. On the other hand, the contribution of too heavy an atom will so dominate the intensities that the errors in intensity measurement may become comparable to the contribution of the lighter atoms. A rule-of-thumb which works well if there are a number $n_L$ of almost-equal light atoms and a number $n_H$ of heavy atoms is to have

$$\sum_{\text{heavy atoms}} f^2 = \sum_{\text{light atoms}} f^2 \tag{8.9}$$

which, from equations (7.29) and (7.32), indicates that the average contributions of the light atoms and heavy atoms to an intensity should be equal.

The principle of the heavy-atom method can be illustrated with the synthetic structure $(C_2H_2NCl)_2$ shown earlier in fig. 8.10(b). The position of the chlorine atom could be determined from the Patterson function (fig. 8.11) and the chlorine atom forms a substantial part of the total scattering power. The left- and right-hand sides of equation (8.9), ignoring the hydrogen atoms, are 289 and 121, respectively. The program STRUCFAC was used to calculate structure factors from the chlorine positions alone and these are shown in table 8.2 (p. 245). Of the 156 structure factors there are 14 for which the chlorine contribution is zero (excluding 6 structure factors which are systematically zero) so that no sign indication is given. Of the remaining 136 structure factors all but 8 have their signs indicated correctly by the chlorine contribution and 4 of these 8 have very small structure amplitudes. The program HEAVY listed in appendix VI, enables the output file from STRUCFAC to be modified and this has been used to make the structure amplitudes zero for the 20 reflections for which there are no sign indications and to change the signs (phases) of the 8 incorrectly indicated by the chlorine contribution. The modified data file was then input to FOUR2 to give fig. 8.15, a Fourier map which shows clearly the positions of the light atoms.

In this example all the signs of the structure factors were taken as those of the chlorine contributions but, in fact, it is possible to determine the probability that the sign of the structure factor is correctly given by the heavy-atom contribution. From equation (8.8) it can be deduced that if $F_h$ and $C_h$ have the same sign then

$$|K_h| = ||F_h| - |C_h|| \tag{8.10a}$$

whereas if $F_h$ and $C_h$ have opposite signs then

$$|K_h| = |F_h| + |C_h|. \tag{8.10b}$$

From equation (7.22) it can be seen that, from structure-factor statistics alone and not taking account of any other information, the probability distribution of $K_h$ is given by

$$P(K) = (2\pi\Sigma')^{-\frac{1}{2}}\exp(-K^2/2\Sigma') \tag{8.11}$$

Fig. 8.15.
Fourier map for
$C_4H_4N_2Cl_2$ with signs
given by the chlorine
atoms. The light atoms
are in the positions
outlined by a single
contour.

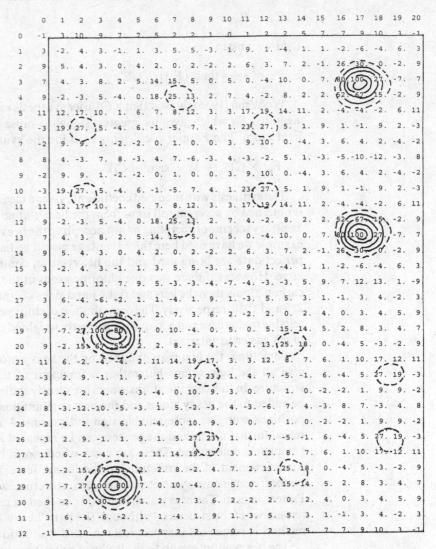

FACTOR CONVERTING TO DENSITY/UNIT AREA IS 0.1955
OUTPUT WITH Y HORIZONTAL AND ORIGIN TOP LEFT

where $\Sigma' = \sum_{j=1}^{N-n} f_j^2$. The probability that $F_h$ and $C_h$ have the same sign, $P_+$, will be related to the probability that they will have opposite signs, $P_-$, by

$$\frac{P_+}{P_-} = \frac{\exp\{-(|F_h| - |C_h|)^2/2\Sigma'\}}{\exp\{-(|F_h| + |C_h|)^2/2\Sigma'\}}. \tag{8.12}$$

The expression (8.12) uses the distribution (8.11) to determine the relative probability of having the two values of $|K_h|$ given by equation (8.10). Expression (8.12) can be simplified to

$$\frac{P_+}{P_-} = \exp(2|F_h||C_h|/\Sigma') \tag{8.13}$$

and if we use the relationship

$$P_+ + P_- = 1 \tag{8.14}$$

then we obtain

$$P_+ = \tfrac{1}{2} + \tfrac{1}{2}\tanh(|F_{\mathbf{h}}||C_{\mathbf{h}}|/\Sigma'). \tag{8.15}$$

Woolfson (1956) showed that a Fourier synthesis can be calculated with weighted $F$'s which gives the greatest signal-to-noise ratio in showing the light atom. The Fourier coefficients used for the synthesis are

$$|F_{\mathbf{h}}| \times (2P_+ - 1) \times (\text{sign of } C_{\mathbf{h}}).$$

The synthesis with modified coefficients gives a clearer density map than is obtained by accepting all signs from the heavy-atom contribution or any arbitrary scheme of accepting or not accepting the heavy-atom sign. For the example we have used here the advantage would be hard to detect because perfect data are being used and the heavy-atom contribution is relatively large. However if there was one chlorine atom in the presence of, say, 30 carbon atoms then the left- and right-hand sides of equation (8.9) would be 289 and 1080, respectively. The contribution of chlorine would be much less reliable in this case and a map with properly weighted coefficients would be much easier to interpret.

Sometimes the first map obtained by the heavy-atom method does not give all the light atoms but rather a fragment of the remainder of the structure. In this case the atoms which are clearly indicated in the map can be added to the heavy atoms and the new combination of atoms in known positions will give more reliable sign indications. A new map may then reveal further atoms and the process can be repeated until the structure is completely determined. There are cases where, for a light-atom structure some fragment of the structure can be found, for example by matching vectors from some symmetric part of a structure with the peaks of a Patterson map. The heavy-atom method can then be used to try to find the remainder of the structure – although the term 'heavy atom' applied to the method is somewhat inappropriate in such a situation.

The heavy-átom method can also be of value for solving non-centrosymmetric structures. However if there is only one heavy atom then this may be arbitrarily placed at the origin and all the phases are indicated as zero. This will give rise to a Fourier synthesis which has a false centre of symmetry and it may be difficult to pick out the peaks corresponding to the structure. If there are two equal heavy atoms then the situation is similar since the centre of the pair may be taken as origin and all the phases will be 0 or $\pi$.

When there is a non-centrosymmetric heavy-atom group which can be located by Patterson methods then the phase from this group can be used as a first phase for a Fourier synthesis. Sim (1957, 1959, 1960) has made a statistical analysis of the errors in the phase angles obtained from the heavy atoms alone and has also found the weighting scheme for structure amplitudes which gives the best Fourier synthesis – in the sense of having the highest signal-to-noise ratio. The map calculated is

$$\rho'(\mathbf{r}) = \sum_{\mathbf{h}} W(\mathbf{h})F(\mathbf{h})\cos\{2\pi\mathbf{h}\cdot\mathbf{r} - \phi(\mathbf{h})\}. \tag{8.16}$$

The weight imposed on the structure amplitude is

$$W(\mathbf{h}) = I_1(X)/I_0(X) \tag{8.17}$$

where $I_1$ and $I_0$ are modified Bessel functions and $X = 2|F(\mathbf{h})||C(\mathbf{h})|/\Sigma'$. The form of $I_1(X)/I_0(X)$ as a function of $X$ is shown in fig. 8.16.

Fig. 8.16.
The Sim weighting curve.

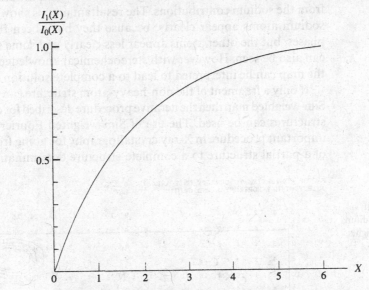

An illustration of the use of Sim weighting is given by application to the synthetic structure $C_3H_5N_2ONa_2$, space group $pg$ with $a = 8\,\text{Å}$, $b = 6\,\text{Å}$, illustrated in fig. 8.17. The coordinates of atoms in one asymmetric unit are:
Na (0.937, 0.250),   Na (0.437, 0.583),   O (0.063, 0.667),   N (0.156, 0.417),
N (0.219, 0.958),   C (0.125, 0.167),   C (0.375, 0.125),   C (0.344, 0.375).

Fig. 8.17.
The synthetic structure $C_3H_5N_2ONa_2$.

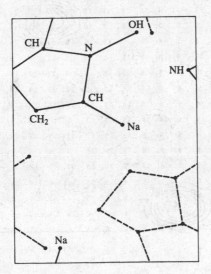

The sodium atoms are regarded as 'heavy' and the left- and right-hand sides of equation (8.9) are, excluding hydrogen, 242 and 270, respectively. In this case the heavy atoms are not so dominant as in the previous synthetic structure $(C_2H_2NCl)_2$, and tables 8.3 and 8.4 show the structure factors for Cu K$\alpha$ radiation ($\lambda = 1.542$ Å) for the complete structure and the contributions of the sodium atoms, respectively. It will be seen that the sodium atoms tend to give reasonable phase estimates, mostly within 60° or so, but with some notable exceptions. The program HEAVY was used to produce an input file for FOUR2 which had Sim-weighted Fourier amplitudes and the phases from the sodium contributions. The resultant map is shown in fig. 8.18; the sodium atoms appear clearly, because they have been the source of the phases, but the other atoms appear less clearly and some spurious density can also be seen. However with stereochemical knowledge of the molecule the map can be interpreted to lead to a complete solution of the structure.

If only a fragment of the non-heavy-atom structure can be found from a Sim-weighted map then the iterative procedure described for centrosymmetric structures can be used. The use of Sim-weighted Fourier syntheses is an important procedure in X-ray crystallography for going from a knowledge of a partial structure to a complete structure determination.

Fig. 8.18.
Fourier map for
$C_3H_5N_2ONa_2$ with
phases from the sodium
atoms and Sim weights.

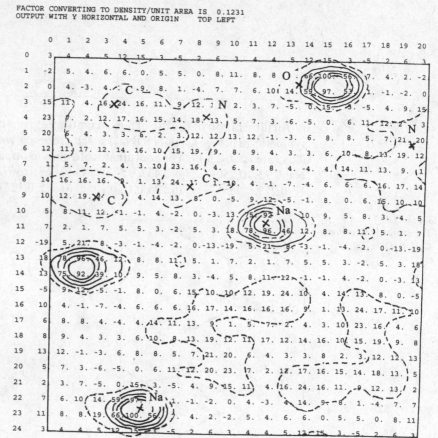

## 8.4 Isomorphous replacement

Very often it is possible to have chemically similar materials in which one or more atoms or groups in one compound are replaced by one or more similar atoms or groups in the other. Sometimes the substitution can be fairly straightforward – one halogen atom for another or one alkali metal for another – but can be less simple. For example, in fig. 8.19 there are shown the molecular forms of 2-amino-4-methyl-6-chloropyrimidine and 2-amino-4,6-dichloropyrimidine (Clews & Cochran, 1948) which differ in that the methyl group of the first is replaced by a chlorine atom in the second.

The crystals of the two materials look rather similar (long needles) and both have four molecules in a monoclinic unit cell of space group $P2_1/a$. The cell dimensions of the two structures are:

|  | $a$ | $b$ | $c$ | $\beta$ |
|---|---|---|---|---|
| methyl | 16.43 Å | 4.00 Å | 10.31 Å | 109° |
| chloro | 16.45 Å | 3.85 Å | 10.28 Å | 108° |

An examination of the X-ray data reveals that the diffraction patterns are strikingly similar in appearance and it is clear that, in terms of crystal structure, the atoms must be similarly arranged except that in one case a methyl group is replaced by chlorine.

Two crystal structures bearing this sort of relationship to each other are termed *isomorphous* and from the differences in their diffraction patterns it is often possible to obtain a complete structure determination.

Let us take the case where we have two isomorphous structures which differ in having $n$ atoms of type $A$ in one replaced by atoms of type $B$ in the other. Then the structure factors of vector index $\mathbf{h}$ for the two compounds can be written as

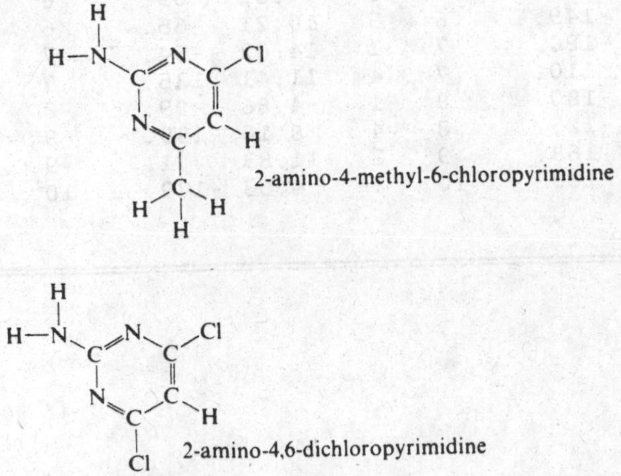

Fig. 8.19.
The molecular forms of the isomorphous compounds 2-amino-4-methyl-6-chloropyrimidine and 2-amino-4,6-dichloropyrimidine.

2-amino-4-methyl-6-chloropyrimidine

2-amino-4,6-dichloropyrimidine

Table 8.3. *Structure factors for the synthetic non-centrosymmetric structure* $C_3H_5N_2ONa_2$

PHI is the phase angle in degrees.

| h | k | F | PHI | h | k | F | PHI | h | k | F | PHI |
|---|---|---|---|---|---|---|---|---|---|---|---|
| -10 | 1 | 4.53 | 1. | -10 | 2 | 3.03 | -20. | -9 | 1 | 12.08 | -12. |
| -9 | 2 | 11.83 | 1. | -9 | 3 | 6.69 | 41. | -8 | 1 | 4.86 | 81. |
| -8 | 2 | 6.65 | -46. | -8 | 3 | 5.96 | 55. | -8 | 4 | 8.10 | -139. |
| -7 | 1 | 14.02 | 126. | -7 | 2 | 13.38 | 22. | -7 | 3 | 6.84 | -70. |
| -7 | 4 | 11.41 | 135. | -7 | 5 | 4.62 | -73. | -6 | 1 | 14.15 | -132. |
| -6 | 2 | 17.82 | -69. | -6 | 3 | 15.13 | -34. | -6 | 4 | 8.23 | -149. |
| -6 | 5 | 10.21 | 115. | -6 | 6 | 5.30 | 48. | -5 | 1 | 36.83 | 170. |
| -5 | 2 | 14.61 | 0. | -5 | 3 | 14.63 | -140. | -5 | 4 | 5.30 | 138. |
| -5 | 5 | 5.76 | 179. | -5 | 6 | 7.16 | -128. | -4 | 1 | 14.76 | -161. |
| -4 | 2 | 6.12 | -45. | -4 | 3 | 26.37 | -14. | -4 | 4 | 11.95 | 16. |
| -4 | 5 | 11.51 | 93. | -4 | 6 | 1.71 | -108. | -4 | 7 | 10.33 | 80. |
| -3 | 1 | 30.08 | 132. | -3 | 2 | 15.87 | 174. | -3 | 3 | 8.19 | 11. |
| -3 | 4 | 3.49 | 75. | -3 | 5 | 11.05 | -153. | -3 | 6 | 4.15 | 45. |
| -3 | 7 | 10.95 | -41. | -2 | 1 | 16.01 | -159. | -2 | 2 | 32.24 | 125. |
| -2 | 3 | 18.01 | -10. | -2 | 4 | 9.63 | 77. | -2 | 5 | 3.03 | 46. |
| -2 | 6 | 7.36 | 161. | -2 | 7 | 4.57 | 24. | -1 | 1 | 4.93 | 50. |
| -1 | 2 | 28.72 | -167. | -1 | 3 | 8.97 | -35. | -1 | 4 | 28.27 | -53. |
| -1 | 5 | 15.37 | -147. | -1 | 6 | 4.50 | -4. | -1 | 7 | 5.24 | -131. |
| 0 | 0 | 124.00 | 0. | 0 | 1 | 0.00 | 0. | 0 | 2 | 19.66 | 105. |
| 0 | 3 | 0.00 | 0. | 0 | 4 | 12.51 | -132. | 0 | 5 | 0.00 | 0. |
| 0 | 6 | 13.57 | -164. | 0 | 7 | 0.00 | 0. | 1 | 0 | 17.82 | 0. |
| 1 | 1 | 4.93 | -130. | 1 | 2 | 28.72 | -167. | 1 | 3 | 8.97 | 145. |
| 1 | 4 | 28.27 | -53. | 1 | 5 | 15.37 | 33. | 1 | 6 | 4.50 | -4. |
| 1 | 7 | 5.24 | 49. | 2 | 0 | 17.23 | 0. | 2 | 1 | 16.01 | 21. |
| 2 | 2 | 32.24 | 125. | 2 | 3 | 18.01 | 170. | 2 | 4 | 9.63 | 77. |
| 2 | 5 | 3.03 | -134. | 2 | 6 | 7.36 | 161. | 2 | 7 | 4.57 | -156. |
| 3 | 0 | 2.97 | -180. | 3 | 1 | 30.08 | -48. | 3 | 2 | 15.87 | 174. |
| 3 | 3 | 8.19 | -169. | 3 | 4 | 3.49 | 75. | 3 | 5 | 11.05 | 27. |
| 3 | 6 | 4.15 | 45. | 3 | 7 | 10.95 | 139. | 4 | 0 | 16.68 | -180. |
| 4 | 1 | 14.76 | 19. | 4 | 2 | 6.12 | -45. | 4 | 3 | 26.37 | 166. |
| 4 | 4 | 11.95 | 16. | 4 | 5 | 11.51 | -87. | 4 | 6 | 1.71 | -108. |
| 4 | 7 | 10.33 | -100. | 5 | 0 | 2.28 | 0. | 5 | 1 | 36.83 | -10. |
| 5 | 2 | 14.61 | 0. | 5 | 3 | 14.63 | 40. | 5 | 4 | 5.30 | 138. |
| 5 | 5 | 5.76 | -1. | 5 | 6 | 7.16 | -128. | 6 | 0 | 14.98 | 180. |
| 6 | 1 | 14.15 | 48. | 6 | 2 | 17.82 | -69. | 6 | 3 | 15.13 | 146. |
| 6 | 4 | 8.23 | -149. | 6 | 5 | 10.21 | -65. | 6 | 6 | 5.30 | 48. |
| 7 | 0 | 8.71 | -180. | 7 | 1 | 14.02 | -54. | 7 | 2 | 13.38 | 22. |
| 7 | 3 | 6.84 | 110. | 7 | 4 | 11.41 | 135. | 7 | 5 | 4.62 | 107. |
| 8 | 0 | 14.95 | -180. | 8 | 1 | 4.86 | -99. | 8 | 2 | 6.65 | -46. |
| 8 | 3 | 5.96 | -125. | 8 | 4 | 8.10 | -139. | 9 | 0 | 0.59 | -180. |
| 9 | 1 | 12.08 | 168. | 9 | 2 | 11.83 | 1. | 9 | 3 | 6.69 | -139. |
| 10 | 0 | 16.02 | 180. | 10 | 1 | 4.53 | -179. | 10 | 2 | 3.03 | -20. |

## Table 8.4. *Contributions to the structure factors of the Na atoms of the synthetic non-centrosymmetric structure* $C_3H_5N_2ONa_2$

PHI is the phase angle in degrees.

| h | k | F | PHI | h | k | F | PHI | h | k | F | PHI |
|---|---|---|---|---|---|---|---|---|---|---|---|
| -10 | 1 | 4.64 | 60. | -10 | 2 | 4.19 | -60. | -9 | 1 | 5.16 | -30. |
| -9 | 2 | 11.09 | 30. | -9 | 3 | 0.02 | -90. | -8 | 1 | 0.21 | 60. |
| -8 | 2 | 8.01 | -60. | -8 | 3 | 0.37 | 180. | -8 | 4 | 6.83 | -120. |
| -7 | 1 | 6.15 | 150. | -7 | 2 | 15.00 | 30. | -7 | 3 | 0.02 | 90. |
| -7 | 4 | 12.27 | 150. | -7 | 5 | 4.24 | -150. | -6 | 1 | 7.91 | -120. |
| -6 | 2 | 7.66 | -60. | -6 | 3 | 13.43 | 0. | -6 | 4 | 6.20 | -120. |
| -6 | 5 | 5.11 | 120. | -6 | 6 | 9.27 | 0. | -5 | 1 | 20.83 | 150. |
| -5 | 2 | 8.43 | 30. | -5 | 3 | 0.06 | 90. | -5 | 4 | 6.55 | 150. |
| -5 | 5 | 13.01 | -150. | -5 | 6 | 0.03 | 90. | -4 | 1 | 14.87 | -120. |
| -4 | 2 | 0.17 | -60. | -4 | 3 | 24.50 | 0. | -4 | 4 | 0.13 | -120. |
| -4 | 5 | 8.80 | 120. | -4 | 6 | 0.19 | 0. | -4 | 7 | 6.38 | 60. |
| -3 | 1 | 26.54 | 150. | -3 | 2 | 9.87 | -150. | -3 | 3 | 0.08 | 90. |
| -3 | 4 | 7.39 | -30. | -3 | 5 | 15.37 | -150. | -3 | 6 | 0.04 | -90. |
| -3 | 7 | 10.57 | -30. | -2 | 1 | 12.93 | -120. | -2 | 2 | 11.51 | 120. |
| -2 | 3 | 20.46 | 0. | -2 | 4 | 8.62 | 60. | -2 | 5 | 7.09 | 120. |
| -2 | 6 | 11.62 | 180. | -2 | 7 | 4.94 | 60. | -1 | 1 | 12.95 | 150. |
| -1 | 2 | 27.74 | -150. | -1 | 3 | 0.04 | 90. | -1 | 4 | 20.18 | -30. |
| -1 | 5 | 6.96 | -150. | -1 | 6 | 0.10 | -90. | -1 | 7 | 4.66 | -30. |
| 0 | 0 | 44.00 | 0. | 0 | 1 | 0.00 | 0. | 0 | 2 | 17.60 | 120. |
| 0 | 3 | 0.00 | 0. | 0 | 4 | 12.92 | 60. | 0 | 5 | 0.00 | 0. |
| 0 | 6 | 17.16 | 180. | 0 | 7 | 0.00 | 0. | 1 | 0 | 0.00 | 0. |
| 1 | 1 | 12.95 | -30. | 1 | 2 | 27.74 | -150. | 1 | 3 | 0.04 | -90. |
| 1 | 4 | 20.18 | -30. | 1 | 5 | 6.96 | 30. | 1 | 6 | 0.10 | -90. |
| 1 | 7 | 4.66 | 150. | 2 | 0 | 26.48 | 0. | 2 | 1 | 12.93 | 60. |
| 2 | 2 | 11.51 | 120. | 2 | 3 | 20.46 | 180. | 2 | 4 | 8.62 | 60. |
| 2 | 5 | 7.09 | -60. | 2 | 6 | 11.62 | 180. | 2 | 7 | 4.94 | -120. |
| 3 | 0 | 0.00 | 0. | 3 | 1 | 26.54 | -30. | 3 | 2 | 9.87 | -150. |
| 3 | 3 | 0.08 | -90. | 3 | 4 | 7.39 | -30. | 3 | 5 | 15.37 | 30. |
| 3 | 6 | 0.04 | -90. | 3 | 7 | 10.57 | 150. | 4 | 0 | 0.38 | -180. |
| 4 | 1 | 14.87 | 60. | 4 | 2 | 0.17 | -60. | 4 | 3 | 24.50 | 180. |
| 4 | 4 | 0.13 | -120. | 4 | 5 | 8.80 | -60. | 4 | 6 | 0.19 | 0. |
| 4 | 7 | 6.38 | -120. | 5 | 0 | 0.00 | 0. | 5 | 1 | 20.83 | -30. |
| 5 | 2 | 8.43 | 30. | 5 | 3 | 0.06 | -90. | 5 | 4 | 6.55 | 150. |
| 5 | 5 | 13.01 | 30. | 5 | 6 | 0.03 | 90. | 6 | 0 | 16.78 | 180. |
| 6 | 1 | 7.91 | 60. | 6 | 2 | 7.66 | -60. | 6 | 3 | 13.43 | 180. |
| 6 | 4 | 6.20 | -120. | 6 | 5 | 5.11 | -60. | 6 | 6 | 9.27 | 0. |
| 7 | 0 | 0.00 | 0. | 7 | 1 | 6.15 | -30. | 7 | 2 | 15.00 | 30. |
| 7 | 3 | 0.02 | -90. | 7 | 4 | 12.27 | 150. | 7 | 5 | 4.24 | 30. |
| 8 | 0 | 17.15 | -180. | 8 | 1 | 0.21 | -120. | 8 | 2 | 8.01 | -60. |
| 8 | 3 | 0.37 | 0. | 8 | 4 | 6.83 | -120. | 9 | 0 | 0.00 | 0. |
| 9 | 1 | 5.16 | 150. | 9 | 2 | 11.09 | 30. | 9 | 3 | 0.02 | 90. |
| 10 | 0 | 8.81 | 180. | 10 | 1 | 4.64 | -120. | 10 | 2 | 4.19 | -60. |

$$(F_h)_A = K_h + f_A \sum_{j=1}^{n} \exp(2\pi i h \cdot r_j)$$

and

$$(F_h)_B = K_h + f_B \sum_{j=1}^{n} \exp(2\pi i h \cdot r_j), \tag{8.18}$$

where $K_h$ is the contribution of the atoms common to both structures and the summation is over the positions of isomorphously-exchangeable atoms. If this summation is written as $C_h$ we have

$$(F_h)_A = K_h + f_A C_h$$

and

$$(F_h)_B = K_h + f_B C_h. \tag{8.19}$$

For centrosymmetric structures these equations can lead directly to phase information. The values of $f_A$, $f_B$, $|(F_h)_A|$ and $|(F_h)_B|$ will be known; if the value of $C_h$ is known as, for example, when the isomorphously-replaceable atoms are in special positions, then the signs of the $F$'s may be found.

Let us consider

$$|(F_h)_A| = 92, |(F_h)_B| = 78, f_A = 12, f_B = 20, C_h = -2.$$

The only reasonable solution of equations (8.19), allowing for some errors of measurement, is

$$(F_h)_A = +92, (F_h)_B = +78, K_h = +117.$$

The signs of $(F_h)_A$ and $(F_h)_B$ will both be established as positive in this example. Another set of values might be

$$|(F_h)_A| = 8, |(F_h)_B| = 10, f_A = 12, f_B = 20, C_h = -2.$$

Since, from equations (8.15), we have

$$(F_h)_A - (F_h)_B = (f_A - f_B)C_h \tag{8.20}$$

and the value of the right-hand side is $+16$, the only reasonable values for the structure factors are $(F_h)_A = +8$ and $(F_h)_B = -10$.

The situation where the $F$'s have opposite signs is quite rare if the isomorphously-replaced (abbreviated to *i-r*) atoms are not too powerful scatterers. It is possible with careful measurements to record structure factors with sufficient accuracy to measure the changes of scattering due to replacements of Na by K $(Z_B - Z_A = 8)$, Cl by Br $(Z_B - Z_A = 18)$ or O by S $(Z_B - Z_A = 8)$. Clearly it is important that the two sets of data should be reasonably well scaled together.

Let us now consider the situation for a centrosymmetric structure when we do not know the sign or magnitude of $C_h$. From equation (8.20) it can be seen that $(F_h)_A - (F_h)_B$ $(= \psi_h)$ is the structure factor for scatterers of scattering factors $f_A - f_B$ at the positions of the *i-r* atoms. If we accept that only rarely will a sign change occur then we know the magnitude of $\psi_h$ but not its sign. However a Fourier synthesis with $\psi_h^2$ as coefficients will give a Patterson map of the difference of the two structures which should show peaks only at positions corresponding to vectors between the *i-r* atoms. If this simple Patterson map can be interpreted then the positions of the *i-r* atoms will be found and the signs of the structure factors can be determined.

This process can be followed by using the data from the artificial two-dimensional structure $(C_2H_2NCl)_2$ shown earlier in fig. 8.10($b$) together with data for the isomorphous structure $(C_2H_2NF)_2$, the structure factors for which are given in table 8.5. A comparison of tables 8.1 and 8.5 shows that only 18 of the 156 structure-factor signs are different. In appendix VII there is given the computer program ISOFILE which takes output files from STRUCFAC for two isomorphous compounds and prepares an input file for the Fourier synthesis program FOUR2 with Fourier coefficients of magnitude $(|F_A| - |F_B|)^2$ with zero phase. This has been used for the present pair of isomorphous compounds and the difference Patterson function is shown in fig. 8.20. There are three independent vectors indicated

Fig. 8.20.
Difference Patterson function for the isomorphous structures $(C_2H_2NCl_2)_2$ and $(C_2H_2NF_2)_2$.

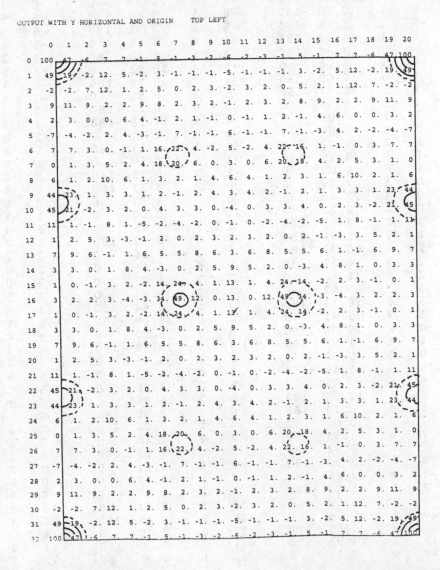

Table 8.5. *The complete set of structure factors for the artificial structure* $(C_2H_2NF)_2$ *as output by* STRUCFAC

PHI is the phase angle in degrees. This structure and $(C_2H_2NCl)_2$ are isomorphous. Compare the structure factors with those given in table 8.1 and note that 18 of the 156 structure factors have different signs.

| h | k | F | PHI | h | k | F | PHI | h | k | F | PHI |
|---|---|---|---|---|---|---|---|---|---|---|---|
| -12 | 1 | 4.01 | 0. | -12 | 2 | 6.81 | 0. | -11 | 1 | 1.07 | 180. |
| -11 | 2 | 13.93 | 180. | -11 | 3 | 0.30 | 0. | -11 | 4 | 3.11 | 0. |
| -10 | 1 | 5.64 | 0. | -10 | 2 | 5.43 | -180. | -10 | 3 | 10.02 | 180. |
| -10 | 4 | 4.55 | -180. | -9 | 1 | 11.15 | 0. | -9 | 2 | 3.89 | -180. |
| -9 | 3 | 0.36 | 0. | -9 | 4 | 6.52 | 180. | -9 | 5 | 8.12 | 180. |
| -8 | 1 | 0.36 | 180. | -8 | 2 | 12.39 | 180. | -8 | 3 | 3.11 | 0. |
| -8 | 4 | 0.71 | 0. | -8 | 5 | 2.15 | -180. | -8 | 6 | 18.67 | 0. |
| -7 | 1 | 7.68 | 0. | -7 | 2 | 32.91 | 0. | -7 | 3 | 1.55 | 180. |
| -7 | 4 | 1.40 | 180. | -7 | 5 | 4.98 | 180. | -7 | 6 | 0.32 | 0. |
| -6 | 1 | 10.20 | 180. | -6 | 2 | 16.85 | 0. | -6 | 3 | 12.50 | 0. |
| -6 | 4 | 2.10 | -180. | -6 | 5 | 3.80 | 180. | -6 | 6 | 20.19 | 180. |
| -5 | 1 | 5.75 | -180. | -5 | 2 | 7.91 | 180. | -5 | 3 | 1.69 | 0. |
| -5 | 4 | 14.93 | 180. | -5 | 5 | 4.55 | 0. | -5 | 6 | 0.06 | -180. |
| -5 | 7 | 3.88 | -180. | -4 | 1 | 2.58 | -180. | -4 | 2 | 10.96 | 0. |
| -4 | 3 | 6.00 | 0. | -4 | 4 | 12.82 | 0. | -4 | 5 | 1.98 | 180. |
| -4 | 6 | 10.69 | 180. | -4 | 7 | 0.74 | -180. | -3 | 1 | 13.52 | 180. |
| -3 | 2 | 32.77 | 180. | -3 | 3 | 0.56 | 0. | -3 | 4 | 16.33 | 0. |
| -3 | 5 | 6.64 | 0. | -3 | 6 | 0.22 | -180. | -3 | 7 | 3.91 | 180. |
| -2 | 1 | 7.06 | 0. | -2 | 2 | 14.18 | -180. | -2 | 3 | 6.85 | 180. |
| -2 | 4 | 3.06 | 0. | -2 | 5 | 1.41 | 0. | -2 | 6 | 11.88 | 0. |
| -2 | 7 | 1.07 | 0. | -1 | 1 | 6.23 | 180. | -1 | 2 | 3.65 | 0. |
| -1 | 3 | 3.15 | -180. | -1 | 4 | 30.59 | 0. | -1 | 5 | 1.90 | 0. |
| -1 | 6 | 0.02 | 0. | -1 | 7 | 0.75 | -180. | 0 | 0 | 112.00 | 0. |
| 0 | 1 | 2.56 | 0. | 0 | 2 | 4.23 | 180. | 0 | 3 | 11.58 | 180. |
| 0 | 4 | 23.65 | 180. | 0 | 5 | 4.87 | 0. | 0 | 6 | 3.74 | 0. |
| 0 | 7 | 2.22 | 0. | 1 | 0 | 0.00 | 0. | 1 | 1 | 6.23 | 0. |
| 1 | 2 | 3.65 | -180. | 1 | 3 | 3.15 | 0. | 1 | 4 | 30.59 | -180. |
| 1 | 5 | 1.90 | 180. | 1 | 6 | 0.02 | 180. | 1 | 7 | 0.75 | 0. |
| 2 | 0 | 16.39 | -180. | 2 | 1 | 7.06 | 0. | 2 | 2 | 14.18 | -180. |
| 2 | 3 | 6.85 | -180. | 2 | 4 | 3.06 | 0. | 2 | 5 | 1.41 | 0. |
| 2 | 6 | 11.88 | 0. | 2 | 7 | 1.07 | 0. | 3 | 0 | 0.00 | 0. |
| 3 | 1 | 13.52 | 0. | 3 | 2 | 32.77 | 0. | 3 | 3 | 0.56 | -180. |
| 3 | 4 | 16.33 | -180. | 3 | 5 | 6.64 | 180. | 3 | 6 | 0.22 | 0. |
| 3 | 7 | 3.91 | 0. | 4 | 0 | 46.09 | -180. | 4 | 1 | 2.58 | -180. |
| 4 | 2 | 10.96 | 0. | 4 | 3 | 6.00 | 0. | 4 | 4 | 12.82 | 0. |
| 4 | 5 | 1.98 | 180. | 4 | 6 | 10.69 | -180. | 4 | 7 | 0.74 | -180. |
| 5 | 0 | 0.00 | 0. | 5 | 1 | 5.75 | 0. | 5 | 2 | 7.91 | 0. |
| 5 | 3 | 1.69 | -180. | 5 | 4 | 14.93 | 0. | 5 | 5 | 4.55 | 180. |
| 5 | 6 | 0.06 | 0. | 5 | 7 | 3.88 | 0. | 6 | 0 | 5.52 | 0. |
| 6 | 1 | 10.20 | 180. | 6 | 2 | 16.85 | 0. | 6 | 3 | 12.50 | 0. |
| 6 | 4 | 2.10 | -180. | 6 | 5 | 3.80 | 180. | 6 | 6 | 20.19 | -180. |
| 7 | 0 | 0.00 | 0. | 7 | 1 | 7.68 | -180. | 7 | 2 | 32.91 | -180. |
| 7 | 3 | 1.55 | 0. | 7 | 4 | 1.40 | 0. | 7 | 5 | 4.98 | 0. |
| 7 | 6 | 0.32 | 180. | 8 | 0 | 1.62 | 180. | 8 | 1 | 0.36 | 180. |
| 8 | 2 | 12.39 | 180. | 8 | 3 | 3.11 | 0. | 8 | 4 | 0.71 | 0. |
| 8 | 5 | 2.15 | -180. | 8 | 6 | 18.67 | 0. | 9 | 0 | 0.00 | 0. |
| 9 | 1 | 11.15 | -180. | 9 | 2 | 3.89 | 0. | 9 | 3 | 0.36 | 180. |
| 9 | 4 | 6.52 | 0. | 9 | 5 | 8.12 | 0. | 10 | 0 | 11.51 | 0. |
| 10 | 1 | 5.64 | 0. | 10 | 2 | 5.43 | -180. | 10 | 3 | 10.02 | -180. |
| 10 | 4 | 4.55 | -180. | 11 | 0 | 0.00 | 0. | 11 | 1 | 1.07 | 0. |
| 11 | 2 | 13.93 | 0. | 11 | 3 | 0.30 | 180. | 11 | 4 | 3.11 | 180. |
| 12 | 0 | 7.18 | 0. | 12 | 1 | 4.01 | 0. | 12 | 2 | 6.81 | 0. |

in the map which for the space group *pmg* are at $(2x, 2y)$, $(\frac{1}{2} + 2x, 0)$ and double-weighted $(\frac{1}{2}, 2y)$. It will be left as an exercise for the reader to confirm that the three vectors have the required relationship and that the peaks indicate correctly the position of the Cl (or F) atoms in the structure. Once the positions of the *i-r* atoms have been located then it is possible to find probable signs for the structure factors in the way indicated previously. If the *i-r* atoms in one or other of the isomorphs form a sufficiently high proportion of the total scattering power of the structure then the *heavy-atom method* described in § 8.3 can be used to complete the structure determination.

For non-centrosymmetric structures it is not possible precisely to define the phases of the structure factors. If there is a single *i-r* atom then one can arbitrarily choose this as an origin for the unit cell and, from equation (8.19), the corresponding structure factors for the two isomorphs appear as

$$(F_{\mathbf{h}})_A = f_A + K_{\mathbf{h}}$$

and

$$(F_{\mathbf{h}})_B = f_B + K_{\mathbf{h}}. \tag{8.21}$$

These equations can be expressed in the form of the diagrams given in fig. 8.21(*a*). The quantities which are known in these diagrams are $f_A$ and $f_B$ and also $|(F_{\mathbf{h}})_A|$ and $|(F_{\mathbf{h}})_B|$. If the two diagrams are drawn together, as in (*b*), with $PQ$ coincident with $P'Q'$, the appearance of the joint diagram suggests a method of determining the phases. From a point $P$ one draws two lines $PO$ and $PO'$ equal in length to $f_A$ and $f_B$. With centres $O$ and $O'$ one draws the arcs of circles of radius $|(F_{\mathbf{h}})_A|$ and $|(F_{\mathbf{h}})_B|$, respectively. These intersect at $Q$ and define the complete diagram. However the solution is not unique for the arcs also cross at $R$ and so there is an ambiguity of phase – in this case the magnitude of the phase is determined but not its sign.

What are determined unambiguously are the real parts of the structure factors $(F_{\mathbf{h}})_A$ and $(F_{\mathbf{h}})_B$. If one computes a Fourier synthesis as for a centrosymmetric structure with these real parts as coefficients then one obtains an image of the structure with the addition of a centre of symmetry. The Fourier map has twice as many peaks as atoms and for each pair of centrosymmetrically related peaks one must be selected and the other rejected. From a knowledge of the stereochemistry of the material a consistent set of peaks can often be selected and the structure thus solved. This process is made more difficult by the overlap of peaks in the map and also by false peaks due to experimental error. If the difference between $|(F_{\mathbf{h}})_A|$ and $|(F_{\mathbf{h}})_B|$ is small compared with $K_{\mathbf{h}}$ then the arcs of circles intersect at a very acute angle and the error in the phase due to errors in the experimental determination of structure factors may be very high.

If instead of one *i-r* atom there are several which are all replaced when going from one structure to the other then the situation is somewhat different. If the difference of weight of the *i-r* atoms is considerable then a Fourier summation with coefficients $|(F_{\mathbf{h}})_A|^2 - |(F_{\mathbf{h}})_B|^2$ will contain peaks corresponding to vectors between the *i-r* atoms of weight $Z_A^2 - Z_B^2$ and other peaks due to vectors between the *i-r* atoms and other atoms of weight $(Z_A - Z_B)Z_j$. The peaks due to vectors between the non *i-r* atoms will not appear in this difference Patterson function. Thus if the *i-r* atoms are

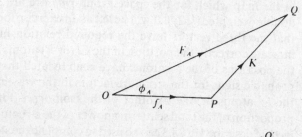

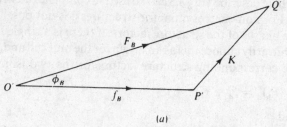

(a)

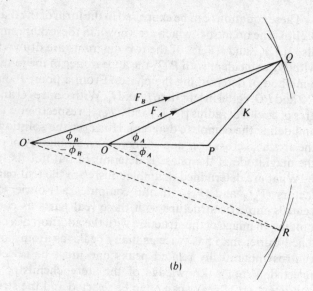

(b)

oxygen and sulphur in the two compounds and the remaining atoms are carbon then the weights of these peaks are

$$Z_A^2 - Z_B^2 = 256 - 64 = 192$$

and

$$(Z_A - Z_B)Z_j = 8 \times 6 = 48.$$

If the Patterson function of the sulphur compound is computed there will be S—S peaks of weight 256, S—C peaks of weight 96, and there will also be a number of C—C peaks of weight 36. The superiority of the difference Patterson for picking out vectors between the *i-r* atoms should be evident from these figures. Once the configuration of the *i-r* atoms has been determined, and an origin specified with respect to this group of atoms, then the contribution of these atoms, $\chi_A$ and $\chi_B$, can be determined. The phases of these contributions with respect to the origin, $\phi_\chi$, are identical and the resolution of the structure factors into the contributions of the $i$-$r$ and non $i$-$r$ atoms is shown in fig. 8.22($a$). The method of producing a diagram for determining phases is shown in fig. 8.22($b$) and follows closely the principle used in fig. 8.21; in this case the ambiguity of phase is not simply just one of sign and the real part of the structure factor is not determined.

Since the positions of the $i$-$r$ atoms are assumed known then completing the structure determination consists of finding the remaining atoms, with contribution $K_\mathbf{h}$ as shown in fig. 8.22. From fig. 8.22($b$) the phase of $K_\mathbf{h}$ is seen to be either $\phi_{\chi,\mathbf{h}} + \Delta\phi_\mathbf{h}$ or $\phi_{\chi,\mathbf{h}} - \Delta\phi_\mathbf{h}$. The angle $\Delta\phi_\mathbf{h}$ is found as follows:

(i) From the triangle $O'OQ$

$$\cos\psi = \frac{(|\chi_{B,\mathbf{h}}| - |\chi_{A,\mathbf{h}}|)^2 + |F_{B,\mathbf{h}}|^2 - |F_{A,\mathbf{h}}|^2}{2|F_B|(|\chi_{B,\mathbf{h}}| - |\chi_{A,\mathbf{h}}|)}; \qquad (8.22)$$

(ii) from the triangle $O'PQ$

$$|K_\mathbf{h}|^2 = |F_{B,\mathbf{h}}|^2 + |\chi_{B,\mathbf{h}}|^2 - 2|F_{B,\mathbf{h}}||\chi_{B,\mathbf{h}}|\cos\psi; \qquad (8.23)$$

(iii) from the triangle $O'PQ$ again

$$\cos\Delta\phi_\mathbf{h} = \frac{|F_{B,\mathbf{h}}|^2 - |K_\mathbf{h}|^2 - |\chi_{B,\mathbf{h}}|^2}{2|K_\mathbf{h}||\chi_{B,\mathbf{h}}|}. \qquad (8.24)$$

Fig. 8.22.
($a$) The components of $F_A$ and $F_B$ due to a group of isomorphously-replaceable atoms and the remainder of the structure.
($b$) Construction to determine the phase angles showing the phase ambiguity.

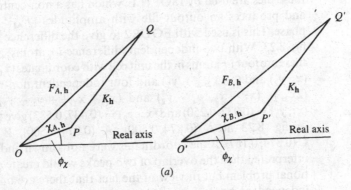

($a$)

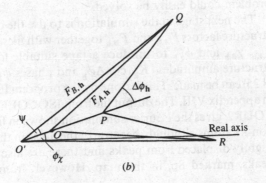

($b$)

Bokhoven, Schoone & Bijvoet (1951) and others have recommended dealing with the single-isomorphous-replacement (SIR) ambiguity by the calculation of a synthesis in which both possible phases are used. In our particular case it would involve calculating

$$\delta(\mathbf{r}) = \frac{1}{V}\sum_{\mathbf{h}}|K_{\mathbf{h}}|\{\cos(2\pi\mathbf{h}\cdot\mathbf{r} - \phi_{\chi,\mathbf{h}} - \Delta\phi_{\mathbf{h}}) + \cos(2\pi\mathbf{h}\cdot\mathbf{r} - \phi_{\chi,\mathbf{h}} + \Delta\phi_{\mathbf{h}})\}$$

(8.25a)

$$= \frac{2}{V}\sum_{\mathbf{h}}|K_{\mathbf{h}}|\cos\Delta\phi_{\mathbf{h}}\cos(2\pi\mathbf{h}\cdot\mathbf{r} - \phi_{\chi,\mathbf{h}}).$$

(8.25b)

One of the two terms for each $\mathbf{h}$ in equation (8.25a) will be correct. The correct terms will form an image of the structure without the *i-r* atoms and the incorrect terms will give a random background noise. It is often possible to recognize the correct structure from such a map. In the form of equation (8.25b) it is seen that the terms of the Fourier summation have phases corresponding to the heavy-atom contributions and structure amplitudes weighted according to the values of $\Delta\phi_{\mathbf{h}}$. If $\Delta\phi_{\mathbf{h}}$ is small, so that the two possible phases are not very different, then the structure amplitude is modified very little. On the other hand, if $\Delta\phi_{\mathbf{h}} = \pi/2$, so that the two possible phases differ by $\pi$ then the corresponding term is eliminated.

We shall look at how this works in practice by means of a simulation. The structures we shall use are the one shown earlier in fig. 8.17, $C_3H_5N_2ONa_2$, and the supposedly isomorphous structure, $C_3H_5N_2OAl_2$. Structure factors are calculated for both of these using STRUCFAC and these files are used by ISOFILE, which has a non-centrosymmetric option and produces an output file with amplitudes $|F_{Al}|^2 - |F_{Na}|^2$ and zero phase. This is used with FOUR2 to give the difference Patterson shown in fig. 8.23. With two independent 'difference-*i-r* atoms' and space group *pg* there are four *i-r* atoms in the unit cell with coordinates $(x_1, y_1), (-x_1, \frac{1}{2} + y_1)$, $(x_2, y_2)$ and $(-x_2, \frac{1}{2} + y_2)$ and four independent non-null vectors $(2x_1, \frac{1}{2})$, $(2x_2, \frac{1}{2})$, $(x_1 - x_2, y_1 - y_2)$ and $(x_1 + x_2, \frac{1}{2} + y_1 - y_2)$. The coordinates $(x_1, y_1) = (0.937, 0.250)$ and $(x_2, y_2) = (0.437, 0.583)$ give the Patterson peaks in fig. 8.23 at A $(0.874, 0.500)$, A $(0.874, 0.500)$, B $(0.500, 0.667)$ and C $(0.374, 0.167)$. If the coordinates were not known and the map had to be interpreted then the overlap of two peaks would create an extra interpretational problem but in view of the fact that there are only three significant independent non-origin peaks and no other peak on the line $y = \frac{1}{2}$ the problem could easily be solved.

The next stage in the simulation is to use the data files giving the sets of structure factors $F_{A,\mathbf{h}}$ and $F_{B,\mathbf{h}}$ together with files from STRUCFAC giving $\chi_{A,\mathbf{h}}, \chi_{B,\mathbf{h}}$ and $\phi_{\chi,\mathbf{h}}$ to produce a tape suitable for input to FOUR2 with structure amplitudes $|K_{\mathbf{h}}|\cos\Delta\phi_{\mathbf{h}}$ and phases $\phi_{\chi,\mathbf{h}}$ so that the summation (8.25) can be made. This is done by the program ISOCOEFF which is given in appendix VIII. The output file from ISOCOEFF, used as an input file for FOUR2, gives the Fourier summation shown in fig. 8.24. This shows all the non-*i-r* atoms although, because of the noise in the map, their positions are slightly displaced from peaks and there are also two significant spurious peaks, marked Sp, in the map. However, from this map the complete

Fig. 8.23.
The difference Patterson
for the isomorphous
compounds
$C_3H_5N_2ONa_2$ and
$C_3H_5N_2OAl_2$. There are
four Patterson peaks
with A being an overlap
of two peaks.

```
FACTOR CONVERTING TO DENSITY/UNIT AREA IS  1.5705
OUTPUT WITH Y HORIZONTAL AND ORIGIN    TOP LEFT
```

|    | 0 | 1 | 2 | 3 | 4 | 5 | 6 | 7 | 8 | 9 | 10 | 11 | 12 | 13 | 14 | 15 | 16 | 17 | 18 | 19 | 20 |
|----|---|---|---|---|---|---|---|---|---|---|----|----|----|----|----|----|----|----|----|----|----|
| 0  | 100 | 59 | 13 | 14 | 12 | 8 | 10 | 7 | 15 | 18 | 14 | 18 | 15 | 7 | 10 | 8 | 12 | 14 | 13 | 59 | 100 |
| 1  | 43 | 22 | 4 | 12 | 11 | 8 | 12 | 11 | 12 | 10 | 6 | 10 | 12 | 11 | 12 | 8 | 11 | 12 | 4 | 22 | 43 |
| 2  | -1 | 2 | 9 | 14 | 13 | 10 | 13 | 15 | 13 | 21 | 29 | 21 | 13 | 15 | 13 | 10 | 13 | 14 | 9 | 2 | -1 |
| 3  | 13 | 16 | 14 | 10 | 14 | 11 | 9 | 11 | 11 | 33 | 54 | 33 | 11 | 11 | 9 | 11 | 14 | 10 | 14 | 16 | 13 |
| 4  | 12 | 14 | 9 | 7 | 15 | 11 | 9 | 13 | 6 | 15 | 29 | 15 | 6 | 13 | 9 | 11 | 15 | 7 | 9 | 14 | 12 |
| 5  | 12 | 15 | 16 | 14 | 14 | 9 | 10 | 16 | 12 | 6 | 6 | 6 | 12 | 16 | 10 | 9 | 14 | 14 | 16 | 15 | 12 |
| 6  | 9 | 8 | 10 | 11 | 10 | 12 | 14 | 13 | 15 | 11 | 7 | 11 | 15 | 13 | 14 | 12 | 10 | 11 | 10 | 8 | 9 |
| 7  | 12 | 7 | 7 | 13 | 10 | 12 | 16 | 14 | 16 | 14 | 10 | 14 | 16 | 14 | 16 | 12 | 10 | 13 | 7 | 7 | 12 |
| 8  | 13 | 9 | 14 | 23 | 13 | 4 | 9 | 14 | 15 | 10 | 6 | 10 | 15 | 14 | 9 | 4 | 13 | 23 | 14 | 9 | 13 |
| 9  | 8 | 11 | 22 | 33 | 26 | 13 | 12 | 14 | 12 | 10 | 10 | 10 | 12 | 14 | 12 | 13 | 26 | 33 | 22 | 11 | 8 |
| 10 | 10 | 13 | 21 | 24 | 17 | 8 | 8 | 9 | 5 | 8 | 13 | 8 | 5 | 9 | 8 | 8 | 17 | 24 | 21 | 13 | 10 |
| 11 | 11 | 11 | 13 | 12 | 5 | 4 | 20 | 27 | 13 | 8 | 12 | 8 | 13 | 27 | 20 | 4 | 5 | 12 | 13 | 11 | 11 |
| 12 | 12 | 11 | 13 | 15 | 11 | 16 | 42 | 60 | 28 | 13 | 15 | 13 | 28 | 50 | 42 | 16 | 11 | 15 | 13 | 11 | 12 |
| 13 | 11 | 11 | 13 | 12 | 5 | 4 | 20 | 27 | 13 | 8 | 12 | 8 | 13 | 27 | 20 | 4 | 5 | 12 | 13 | 11 | 11 |
| 14 | 10 | 13 | 21 | 24 | 17 | 8 | 8 | 9 | 5 | 8 | 13 | 8 | 5 | 9 | 8 | 8 | 17 | 24 | 21 | 13 | 10 |
| 15 | 8 | 11 | 22 | 33 | 26 | 13 | 12 | 14 | 12 | 10 | 10 | 10 | 12 | 14 | 12 | 13 | 26 | 33 | 22 | 11 | 8 |
| 16 | 13 | 9 | 14 | 23 | 13 | 4 | 9 | 14 | 15 | 10 | 6 | 10 | 15 | 14 | 9 | 4 | 13 | 23 | 14 | 9 | 13 |
| 17 | 12 | 7 | 7 | 13 | 10 | 12 | 16 | 14 | 16 | 14 | 10 | 14 | 16 | 14 | 16 | 12 | 10 | 13 | 7 | 7 | 12 |
| 18 | 9 | 8 | 10 | 11 | 10 | 12 | 14 | 13 | 15 | 11 | 7 | 11 | 15 | 13 | 14 | 12 | 10 | 11 | 10 | 8 | 9 |
| 19 | 12 | 15 | 16 | 14 | 14 | 9 | 10 | 16 | 12 | 6 | 6 | 6 | 12 | 16 | 10 | 9 | 14 | 14 | 16 | 15 | 12 |
| 20 | 12 | 14 | 9 | 7 | 15 | 11 | 9 | 13 | 6 | 15 | 29 | 15 | 6 | 13 | 9 | 11 | 15 | 7 | 9 | 14 | 12 |
| 21 | 13 | 16 | 14 | 10 | 14 | 11 | 9 | 11 | 11 | 33 | 54 | 33 | 11 | 11 | 9 | 11 | 14 | 10 | 14 | 16 | 13 |
| 22 | -1 | 2 | 9 | 14 | 13 | 10 | 13 | 15 | 13 | 21 | 29 | 21 | 13 | 15 | 13 | 10 | 13 | 14 | 9 | 2 | -1 |
| 23 | 43 | 22 | 4 | 12 | 11 | 8 | 12 | 11 | 12 | 10 | 6 | 10 | 12 | 11 | 12 | 8 | 11 | 12 | 4 | 22 | 43 |
| 24 | 100 | 59 | 13 | 14 | 12 | 8 | 10 | 7 | 15 | 18 | 14 | 18 | 15 | 7 | 10 | 8 | 12 | 14 | 13 | 59 | 100 |

structure determination would be straightforward.

If there are several different isomorphs then the phase ambiguity can be resolved. If, for example, there were isomers $A$, $B$ and $C$ and the pair $AB$ gave phase indications $\phi_{AB}(1)$ and $\phi_{AB}(2)$ and for other pairs of isomorphs there were pairs of phase indications $\phi_{BC}(1)$, $\phi_{BC}(2)$ and $\phi_{CA}(1)$, $\phi_{CA}(2)$ then the correct phase should occur as one member of each pair of phase indications and so be recognized. In practice, because of errors in the data, the situation may not be so straightforward but a 'best' phase can nevertheless be found.

The multiple-isomorphous-replacement (MIR) method has been very important in the development of protein crystallography. If a native protein crystal is soaked in solutions containing heavy-atom compounds then these are sometimes absorbed by the crystal at specific sites in the protein molecules. Because proteins are such large structures the addition of a heavy-atom-containing group will only disturb its immediate environment and the bulk of the structure will be undisturbed. In this case the heavy-atom derivative will be isomorphous with the native protein although, because some atoms are displaced in the derivative, the isomorphism may only be valid to a limited resolution – say 2–3 Å. Different derivatives may have the heavy atoms attached at different sites in the

Fig. 8.24.
A Fourier summation, derived from the heavy-atom positions in $C_3H_5N_2ONa_2$ and $C_3H_5N_2OAl_2$, which shows the positions of the light atoms. Spurious peaks are marked Sp.

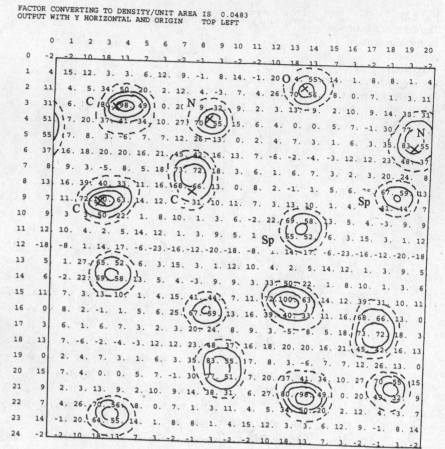

protein; each derivative gives two possible values for the phase of the protein but with several different derivatives a unique 'best' phase for the protein can usually be found.

A general approach to combining information from MIR experiments, considered in the form of a number of SIR experiments, from each of which there is a phase ambiguity, was suggested by Blow and Crick (1959). By considering the probable errors of measurement each experiment gives a probability curve for the phase looking something like that in fig. 8.25. The peaks of the map correspond to the ambiguity and the spread around each of the peaks is related to the probable errors of measurement. For very precise measurements the two peaks become sharp and narrow; if the measurements are very poor then the probability may vary little over the whole $2\pi$ range. Again, if the $\Delta\phi$ associated with the ambiguity is small then the two peaks may coalesce to give a unimodal distribution. Curves from four different experiments and the product of the four curves are shown in fig. 8.26. It will be seen that, although each individual experiment gives a rather diffuse curve the overall probability curve can define the phase quite well.

A complete review of methods dealing with multiple isomorphous replacement (MIR) has been given by Woolfson and Fan (1995).

Fig. 8.25.
A Blow and Crick
probability curve for a
phase from
single-isomorphous-
replacement (SIR) data.

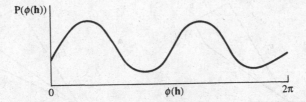

Fig. 8.26.
(*a–d*) Crick and Blow
probability curves from
different pairs of
isomorphs multiplied
together to give (*e*) a
unimodal probability
curve.

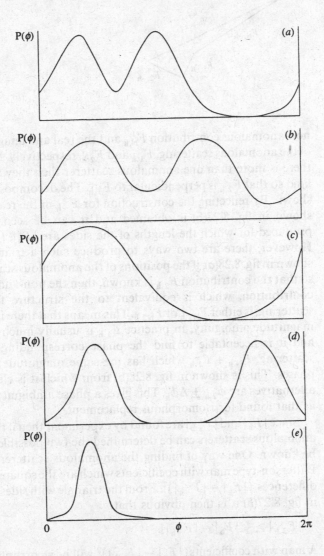

## 8.5   The application of anomalous scattering

In §6.6 anomalous scattering was described as a phenomenon which influences the intensities of X-ray reflections and causes a breakdown of Friedel's law so that $|F_{\mathbf{h}}| \neq |F_{-\mathbf{h}}|$. We shall now see how to exploit this difference to solve crystal structures.

In fig. 8.27(*a*) there is shown the way that $F_{\mathbf{h}}$ may be decomposed into the

Fig. 8.27.
(a) The decomposition of
$F_h$ and $F_{-h}$ in terms of
the non-anomalous
scattering, subscript N,
the real part of the
anomalous scattering,
subscript 1, and the
imaginary part of the
anomalous scattering,
subscript 2.
(b) Reflecting $F_{-h}$ in the
real axis produces a
triangle with sides
$F_{-h}^*$, $2F_{2,h}$ and $F_h$.

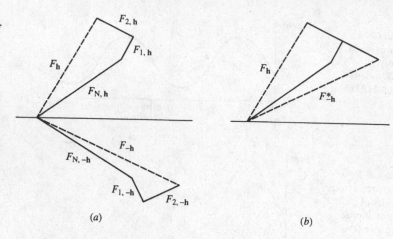

(a)                                    (b)

non-anomalous contribution $F_{N,h}$ and the real and imaginary components
of the anomalous scattering, $F_{1,h}$ and $F_{2,h}$, respectively. It is assumed that if
there is more than one anomalous scatterer then they are all of the same
kind so that $F_{2,h}$ is perpendicular to $F_{1,h}$. The decomposition of $F_{-h}$ is also
shown. By reflecting the construction for $F_{-h}$ in the real axis the diagram
shown in fig. 8.27(b) is obtained and it can be seen that a triangle is
produced for which the lengths of the sides are $|F_h|$, $|F_{-h}|$ and $2|F_{2,h}|$.
However, there are two ways to produce such a triangle and these are
shown in fig. 8.28(a); if the positions of the anomalous scatterers are known,
so that the contribution $F_{2,h}$ is known, then the non-anomalous scattering
contribution, which is equivalent to the structure factor for normal
scattering, is either $F_{N1,h}$ or $F_{N2,h}$. This means that there is both a phase and
magnitude ambiguity. In practice $F_{1,h}$ is usually much smaller than $F_{N,h}$
and it is acceptable to find the phase corresponding to the total real
scattering, $F_{N,h} + F_{1,h}$ which has the same magnitude for both possible
phases. This is shown in fig. 8.28(b) from which it is clear that the phase
alternatives are $\phi_{2,h} \pm \Delta\phi_h$. This gives a phase ambiguity of the same form
as that found for isomorphous replacement.

Since $|F_h|$ and $|F_{-h}|$ are found by experiment then if the positions of the
anomalous scatterers can be determined the two possible phase values will
be known. One way of finding the anomalous scatterers is to compute a
Patterson-type map with coefficients which are the square of the anomalous
differences, $(|F_h| - |F_{-h}|)^2$. From the triangle with sides $2F_{2,h}$, $F_h$ and $F_{-h}$
in fig. 8.27(b) it is then obvious that

$$|F_{2,h}| \geq \tfrac{1}{2}||F_h| - |F_{-h}||. \tag{8.26}$$

A map with coefficients $(|F_h| - |F_{-h}|)^2$ will be a corrupted Patterson map
for the anomalous scatterers where each Patterson coefficient has been
multiplied by an unknown factor which reduces its value. If the anomalous
difference is large then the true Patterson map must have a large coefficient
but some large coefficients for the true Patterson map will be small in the
corrupted map. In practice it is found preferable to use only some of the
largest anomalous differences for these are most likely to have magnitudes
close to the correct ones.

Fig. 8.28.
(a) If the positions of
the anomalous
scatterers are known, so
that $F_{2,h}$ is known in
both magnitude and
phase then finding the
normal structure factor
$F_{N,h}$ gives an ambiguity
both in magnitude and
phase.
(b) Finding the total
real scattering gives
only a phase ambiguity.

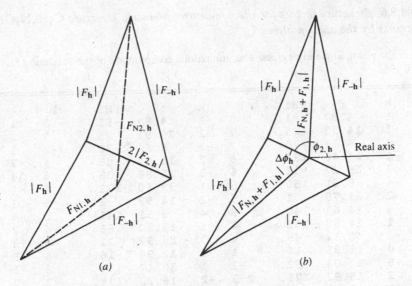

We now illustrate this process using the synthetic structure $C_3H_5N_2ONa_2$, shown earlier in fig. 8.17 where the Na atoms are the anomalous scatterers. Table 8.6 shows the structure factors calculated for Fe $K\alpha_1$ radiation ($\lambda = 1.936$ Å) for which, for Na, $f_1 = 0.186$ and $f_2 = 0.196$. This table was produced by STRUCFAC which contains a facility for including anomalous scattering. The anomalous components of the scattering factor are, in fact, very small and normally, in a real case, the anomalous scatterers would be heavier elements with the wavelength tuned to give large anomalous effects. For example, if Co $K\alpha_1$ radiation ($\lambda = 1.789$ Å) was used with Fe as the anomalous scatterer then $f_1 = -3.331$ and $f_2 = 0.490$. However, since we are dealing with idealised calculated structure factors we can use the very small differences in $|F_h|$ and $|F_{-h}|$ seen in table 8.6 for our illustration.

The output file from STRUCFAC was used as the input file for ANOFILE, given in appendix IX. This gives an output data file which can be input to FOUR2 to give the required anomalous-difference Patterson map, which is shown in fig. 8.29. Only one-quarter of the largest anomalous differences are used, the remainder being made equal to zero by ANOFILE. Ideally the map should show only the vectors linking the anomalous scatterers but, since the coefficients of the map are not those of a true Patterson function the map obtained is somewhat corrupted. Comparing fig. 8.29 with fig. 8.23 we can see that peaks A, B and C appear but B and C are slightly displaced and that there are also smaller false peaks present. It is possible to interpret fig. 8.29 although the coordinates found will be slightly different from the true coordinates. It should be noted that the X-ray wavelength used here for anomalous scattering, chosen to give reasonable values of $f_1$ and $f_2$, also gives lower resolution for fig. 8.29 than for fig. 8.23, for which Cu $K\alpha$ radiation ($\lambda = 1.542$ Å) was used.

Having determined the positions of the anomalous scatterers, the quantities $F_{2,h}$ for different **h** can be calculated so defining the phase alternatives for each reflection and then a double synthesis can be calculated as given in equation (8.25). Another interesting way of resolving

Table 8.6. *Structure factors for the non-centrosymmetric structure* $C_3H_5N_2ONa_2$ *with anomalous scattering by the sodium atoms*

PHI is the phase angle in degrees. The anomalous components of the scattering factor are $f_1 = 0.186$ and $f_2 = 0.196$.

| h | k | F | PHI | h | k | F | PHI | h | k | F | PHI |
|---|---|---|---|---|---|---|---|---|---|---|---|
| -8 | 1 | 4.87 | 81. | 8 | -1 | 4.87 | -81. | -7 | 1 | 14.13 | 127. |
| 7 | -1 | 14.33 | -125. | -7 | 2 | 13.91 | 25. | 7 | -2 | 14.07 | -20. |
| -7 | 3 | 6.84 | -70. | 7 | -3 | 6.84 | 70. | -6 | 1 | 14.35 | -131. |
| 6 | -1 | 14.46 | 133. | -6 | 2 | 18.04 | -68. | 6 | -2 | 18.12 | 69. |
| -6 | 3 | 15.27 | -31. | 6 | -3 | 15.86 | 35. | -6 | 4 | 8.34 | -147. |
| 6 | -4 | 8.61 | 150. | -5 | 1 | 37.60 | 171. | 5 | -1 | 37.18 | -169. |
| -5 | 2 | 14.70 | 1. | 5 | -2 | 14.97 | 1. | -5 | 3 | 14.63 | -140. |
| 5 | -3 | 14.63 | 140. | -5 | 4 | 5.51 | 141. | 5 | -4 | 5.61 | -136. |
| -4 | 1 | 14.80 | -158. | 4 | -1 | 15.30 | 161. | -4 | 2 | 6.12 | -45. |
| 4 | -2 | 6.12 | 45. | -4 | 3 | 26.92 | -12. | 4 | -3 | 27.29 | 15. |
| -4 | 4 | 11.95 | 16. | 4 | -4 | 11.94 | -16. | -4 | 5 | 11.68 | 96. |
| 4 | -5 | 12.02 | -92. | -3 | 1 | 30.47 | 134. | 3 | -1 | 30.85 | -131. |
| -3 | 2 | 15.92 | 175. | 3 | -2 | 16.22 | -174. | -3 | 3 | 8.18 | 11. |
| 3 | -3 | 8.19 | -11. | -3 | 4 | 3.68 | 70. | 3 | -4 | 3.19 | -72. |
| -3 | 5 | 11.64 | -150. | 3 | -5 | 11.70 | 156. | -2 | 1 | 16.04 | -158. |
| 2 | -1 | 16.39 | 159. | -2 | 2 | 32.53 | 125. | 2 | -2 | 32.48 | -124. |
| -2 | 3 | 18.44 | -8. | 2 | -3 | 18.63 | 11. | -2 | 4 | 9.97 | 78. |
| 2 | -4 | 9.81 | -75. | -2 | 5 | 2.86 | 53. | 2 | -5 | 3.37 | -49. |
| -2 | 6 | 7.71 | 166. | 2 | -6 | 8.04 | -159. | -1 | 1 | 4.64 | 52. |
| 1 | -1 | 5.15 | -53. | -1 | 2 | 29.12 | -166. | 1 | -2 | 29.48 | 168. |
| -1 | 3 | 8.97 | -35. | 1 | -3 | 8.97 | 35. | -1 | 4 | 28.59 | -51. |
| 1 | -4 | 29.05 | 53. | -1 | 5 | 15.64 | -146. | 1 | -5 | 15.61 | 148. |
| -1 | 6 | 4.50 | -4. | 1 | -6 | 4.49 | 4. | 0 | 0 | 124.75 | 0. |
| 0 | 0 | 124.75 | 0. | 0 | 1 | 0.00 | 0. | 0 | -1 | 0.00 | 0. |
| 0 | 2 | 19.92 | 106. | 0 | -2 | 20.12 | -104. | 0 | 3 | 0.00 | 0. |
| 0 | -3 | 0.00 | 0. | 0 | 4 | 12.23 | -134. | 0 | -4 | 12.07 | 131. |
| 0 | 5 | 0.00 | 0. | 0 | -5 | 0.00 | 0. | 0 | 6 | 14.52 | -162. |
| 0 | -6 | 14.10 | 168. | 1 | 0 | 17.82 | 0. | -1 | 0 | 17.82 | 0. |
| 1 | 1 | 4.64 | -128. | -1 | -1 | 5.15 | 127. | 1 | 2 | 29.12 | -166. |
| -1 | -2 | 29.48 | 168. | 1 | 3 | 8.97 | 145. | -1 | -3 | 8.97 | -145. |
| 1 | 4 | 28.59 | -51. | -1 | -4 | 29.05 | 53. | 1 | 5 | 15.64 | 34. |
| -1 | -5 | 15.61 | -32. | 1 | 6 | 4.50 | -4. | -1 | -6 | 4.49 | 4. |
| 2 | 0 | 17.76 | 2. | -2 | 0 | 17.76 | 2. | 2 | 1 | 16.04 | 22. |
| -2 | -1 | 16.39 | -21. | 2 | 2 | 32.53 | 125. | -2 | -2 | 32.48 | -124. |
| 2 | 3 | 18.44 | 172. | -2 | -3 | 18.63 | -169. | 2 | 4 | 9.97 | 78. |
| -2 | -4 | 9.81 | -75. | 2 | 5 | 2.86 | -127. | -2 | -5 | 3.37 | 131. |
| 2 | 6 | 7.71 | 166. | -2 | -6 | 8.04 | -159. | 3 | 0 | 2.97 | -180. |
| -3 | 0 | 2.97 | -180. | 3 | 1 | 30.47 | -46. | -3 | -1 | 30.85 | 49. |
| 3 | 2 | 15.92 | 175. | -3 | -2 | 16.22 | -174. | 3 | 3 | 8.18 | -169. |
| -3 | -3 | 8.19 | 169. | 3 | 4 | 3.68 | 70. | -3 | -4 | 3.19 | -72. |
| 3 | 5 | 11.64 | 30. | -3 | -5 | 11.70 | -24. | 4 | 0 | 16.69 | -180. |
| -4 | 0 | 16.69 | -180. | 4 | 1 | 14.80 | 22. | -4 | -1 | 15.30 | -19. |
| 4 | 2 | 6.12 | -45. | -4 | -2 | 6.12 | 45. | 4 | 3 | 26.92 | 168. |
| -4 | -3 | 27.29 | -165. | 4 | 4 | 11.95 | 16. | -4 | -4 | 11.94 | -16. |
| 4 | 5 | 11.68 | -84. | -4 | -5 | 12.02 | 88. | 5 | 0 | 2.28 | 0. |
| -5 | 0 | 2.28 | 0. | 5 | 1 | 37.60 | -9. | -5 | -1 | 37.18 | 11. |
| 5 | 2 | 14.70 | 1. | -5 | -2 | 14.97 | 1. | 5 | 3 | 14.63 | 40. |
| -5 | -3 | 14.63 | -40. | 5 | 4 | 5.50 | 141. | -5 | -4 | 5.61 | -136. |
| 6 | 0 | 15.53 | -178. | -6 | 0 | 15.53 | -178. | 6 | 1 | 14.35 | 49. |
| -6 | -1 | 14.46 | -47. | 6 | 2 | 18.04 | -68. | -6 | -2 | 18.12 | 69. |
| 6 | 3 | 15.27 | 149. | -6 | -3 | 15.86 | -145. | 6 | 4 | 8.34 | -147. |
| -6 | -4 | 8.61 | 150. | 7 | 0 | 8.71 | -180. | -7 | 0 | 8.71 | -180. |
| 7 | 1 | 14.13 | -53. | -7 | -1 | 14.33 | 55. | 7 | 2 | 13.91 | 25. |
| -7 | -2 | 14.07 | -20. | 7 | 3 | 6.84 | 110. | -7 | -3 | 6.84 | -110. |
| 8 | 0 | 15.71 | -177. | -8 | 0 | 15.71 | -177. | 8 | 1 | 4.87 | -99. |
| -8 | 1 | 4.87 | 99. | | | | | | | | |

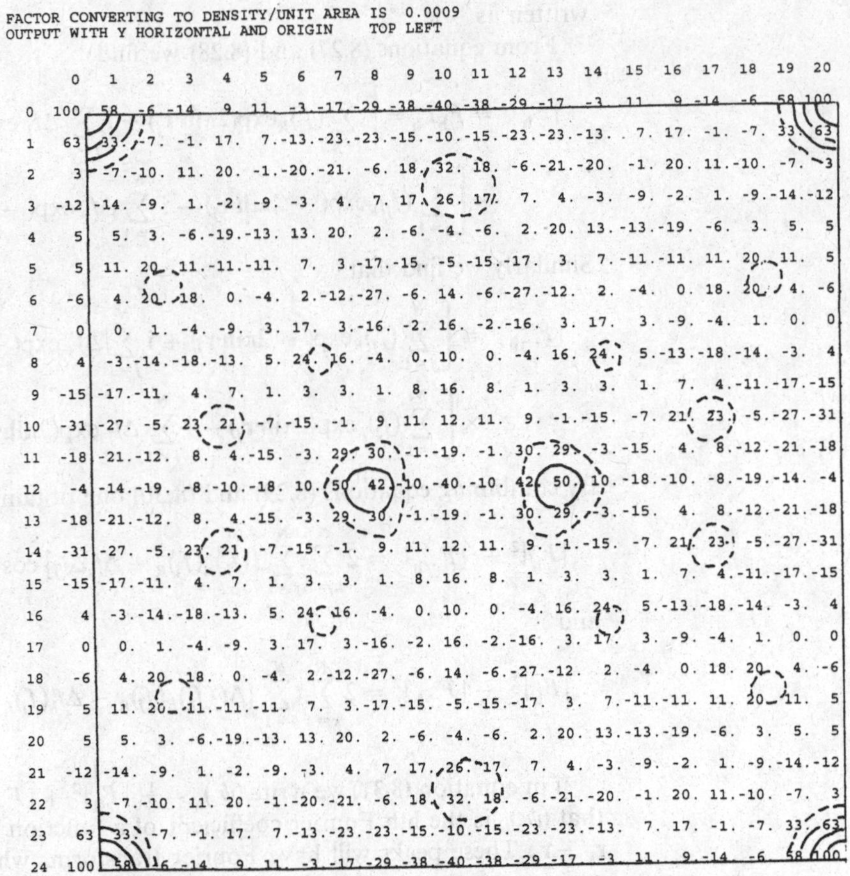

Fig. 8.29.
The
anomalous-difference
Patterson map for
$C_3H_5N_2ONa_2$.

```
FACTOR CONVERTING TO DENSITY/UNIT AREA IS  0.0009
OUTPUT WITH Y HORIZONTAL AND ORIGIN     TOP LEFT
```

|    | 0 | 1 | 2 | 3 | 4 | 5 | 6 | 7 | 8 | 9 | 10 | 11 | 12 | 13 | 14 | 15 | 16 | 17 | 18 | 19 | 20 |
|----|---|---|---|---|---|---|---|---|---|---|----|----|----|----|----|----|----|----|----|----|----|
| 0  | 100 | 58 | -6 | -14 | 9 | 11 | -3 | -17 | -29 | -38 | -40 | -38 | -29 | -17 | -3 | 11 | 9 | -14 | -6 | 58 | 100 |
| 1  | 63 | 33 | -7 | -1 | 17 | 7 | -13 | -23 | -23 | -15 | -10 | -15 | -23 | -23 | -13 | 7 | 17 | -1 | -7 | 33 | 63 |
| 2  | 3 | -7 | -10 | 11 | 20 | -1 | -20 | -21 | -6 | 18 | 32 | 18 | -6 | -21 | -20 | -1 | 20 | 11 | -10 | -7 | 3 |
| 3  | -12 | -14 | -9 | 1 | -2 | -9 | -3 | 4 | 7 | 17 | 26 | 17 | 7 | 4 | -3 | -9 | -2 | 1 | -9 | -14 | -12 |
| 4  | 5 | 5 | 3 | -6 | -19 | -13 | 13 | 20 | 2 | -6 | -4 | -6 | 2 | 20 | 13 | -13 | -19 | -6 | 3 | 5 | 5 |
| 5  | 5 | 11 | 20 | -11 | -11 | -11 | 7 | 3 | -17 | -15 | -5 | -15 | -17 | 3 | 7 | -11 | -11 | -11 | 20 | 11 | 5 |
| 6  | -6 | 4 | 20 | -18 | 0 | -4 | 2 | -12 | -27 | -6 | 14 | -6 | -27 | -12 | 2 | -4 | 0 | -18 | 20 | 4 | -6 |
| 7  | 0 | 0 | 1 | -4 | -9 | 3 | 17 | 3 | -16 | -2 | 16 | -2 | -16 | 3 | 17 | 3 | -9 | -4 | 1 | 0 | 0 |
| 8  | 4 | -3 | -14 | -18 | -13 | 5 | 24 | 16 | -4 | 0 | 10 | 0 | -4 | 16 | 24 | 5 | -13 | -18 | -14 | -3 | 4 |
| 9  | -15 | -17 | -11 | 4 | 7 | 1 | 3 | 3 | 1 | 8 | 16 | 8 | 1 | 3 | 3 | 1 | 7 | 4 | -11 | -17 | -15 |
| 10 | -31 | -27 | -5 | 23 | 21 | -7 | -15 | -1 | 9 | 11 | 12 | 11 | 9 | -1 | -15 | -7 | 21 | 23 | -5 | -27 | -31 |
| 11 | -18 | -21 | -12 | 8 | 4 | -15 | -3 | 29 | 30 | -1 | -19 | -1 | 30 | 29 | -3 | -15 | 4 | 8 | -12 | -21 | -18 |
| 12 | -4 | -14 | -19 | -8 | -10 | -18 | 10 | 50 | 42 | -10 | -40 | -10 | 42 | 50 | 10 | -18 | -10 | -8 | -19 | -14 | -4 |
| 13 | -18 | -21 | -12 | 8 | 4 | -15 | -3 | 29 | 30 | -1 | -19 | -1 | 30 | 29 | -3 | -15 | 4 | 8 | -12 | -21 | -18 |
| 14 | -31 | -27 | -5 | 23 | 21 | -7 | -15 | -1 | 9 | 11 | 12 | 11 | 9 | -1 | -15 | -7 | 21 | 23 | -5 | -27 | -31 |
| 15 | -15 | -17 | -11 | 4 | 7 | 1 | 3 | 3 | 1 | 8 | 16 | 8 | 1 | 3 | 3 | 1 | 7 | 4 | -11 | -17 | -15 |
| 16 | 4 | -3 | -14 | -18 | -13 | 5 | 24 | 16 | -4 | 0 | 10 | 0 | -4 | 16 | 24 | 5 | -13 | -18 | -14 | -3 | 4 |
| 17 | 0 | 0 | 1 | -4 | -9 | 3 | 17 | 3 | -16 | -2 | 16 | -2 | -16 | 3 | 17 | 3 | -9 | -4 | 1 | 0 | 0 |
| 18 | -6 | 4 | 20 | -18 | 0 | -4 | 2 | -12 | -27 | -6 | 14 | -6 | -27 | -12 | 2 | -4 | 0 | -18 | 20 | 4 | -6 |
| 19 | 5 | 11 | 20 | -11 | -11 | -11 | 7 | 3 | -17 | -15 | -5 | -15 | -17 | 3 | 7 | -11 | -11 | -11 | 20 | 11 | 5 |
| 20 | 5 | 5 | 3 | -6 | -19 | -13 | 13 | 20 | 2 | -6 | -4 | -6 | 2 | 20 | 13 | -13 | -19 | -6 | 3 | 5 | 5 |
| 21 | -12 | -14 | -9 | 1 | -2 | -9 | -3 | 4 | 7 | 17 | 26 | 17 | 7 | 4 | -3 | -9 | -2 | 1 | -9 | -14 | -12 |
| 22 | 3 | -7 | -10 | 11 | 20 | -1 | -20 | -21 | -6 | 18 | 32 | 18 | -6 | -21 | -20 | -1 | 20 | 11 | -10 | -7 | 3 |
| 23 | 63 | 33 | -7 | -1 | 17 | 7 | -13 | -23 | -23 | -15 | -10 | -15 | -23 | -23 | -13 | 7 | 17 | -1 | -7 | 33 | 63 |
| 24 | 100 | 58 | -6 | -14 | 9 | 11 | -3 | -17 | -29 | -38 | -40 | -38 | -29 | -17 | -3 | 11 | 9 | -14 | -6 | 58 | 100 |

the phase ambiguity for one-wavelength anomalous scattering (OAS) by using anomalous differences will now be described.

Let us write the scattering factor of an atom as

$$f_j = (f_j)_R + i\Delta f_j$$

where the imaginary component is non-zero only for an anomalous scatterer and is the $f_2$ of equation (6.80) and the real component for an anomalous scatterer, $(f_j)_R$, differs slightly from the non-anomalous value and is the $f_0 + f_1$ of equation (6.80).

The structure factor of index $\mathbf{h}$ can now be written as

$$F_{\mathbf{h}} = \sum_{j=1}^{N} (f_j)_R \exp(2\pi i \mathbf{h} \cdot \mathbf{r}_j) + i \sum_{j=1}^{N} \Delta f_j \exp(2\pi i \mathbf{h} \cdot \mathbf{r}_j). \qquad (8.27)$$

The complex conjugate of $F_{\mathbf{h}}$ is given by

$$F_{\mathbf{h}}^* = \sum_{j=1}^{N} (f_j)_R \exp(-2\pi i \mathbf{h} \cdot \mathbf{r}_j) - i \sum_{j=1}^{N} \Delta f_j \exp(-2\pi i \mathbf{h} \cdot \mathbf{r}_j). \qquad (8.28)$$

The form of the second summation follows from the result that the complex conjugate of $ie^{i\phi}$ ($= i\cos\phi - \sin\phi$) is $-i\cos\phi - \sin\phi$ which can be written as $-ie^{-i\varphi}$.

From equations (8.27) and (8.28) we find

$$|F_{\mathbf{h}}|^2 = F_{\mathbf{h}}F_{\mathbf{h}}^* = \left[\sum_{j=1}^{N}(f_j)_R\exp(2\pi i\mathbf{h}\cdot\mathbf{r}_j) + i\sum_{j=1}^{N}\Delta f_j\exp(2\pi i\mathbf{h}\cdot\mathbf{r}_j)\right]$$
$$\times\left[\sum_{j=1}^{N}(f_j)_R\exp(-2\pi i\mathbf{h}\cdot\mathbf{r}_j) - i\sum_{j=1}^{N}\Delta f_j\exp(-2\pi i\mathbf{h}\cdot\mathbf{r}_j)\right]. \quad (8.29)$$

Similarly we find that

$$|F_{-\mathbf{h}}|^2 = \left[\sum_{j=1}^{N}(f_j)_R\exp(-2\pi i\mathbf{h}\cdot\mathbf{r}_j) + i\sum_{j=1}^{N}\Delta f_j\exp(-2\pi i\mathbf{h}\cdot\mathbf{r}_j)\right]$$
$$\times\left[\sum_{j=1}^{N}(f_j)_R\exp(2\pi i\mathbf{h}\cdot\mathbf{r}_j) - i\sum_{j=1}^{N}\Delta f_j\exp(2\pi i\mathbf{h}\cdot\mathbf{r}_j)\right]. \quad (8.30)$$

By combining equations (8.29) and (8.30) one obtains

$$|F_{\mathbf{h}}|^2 + |F_{-\mathbf{h}}|^2 = 2\sum_{k=1}^{N}\sum_{j=1}^{N}\{(f_k)_R(f_j)_R + \Delta f_k\Delta f_j\}\cos\{2\pi\mathbf{h}\cdot(\mathbf{r}_k - \mathbf{r}_j)\} \quad (8.31)$$

and

$$|F_{\mathbf{h}}|^2 - |F_{-\mathbf{h}}|^2 = 2\sum_{k=1}^{N}\sum_{j=1}^{N}\{\Delta f_j(f_k)_R(f_j)_R - \Delta f_k(f_j)_R\}\sin\{2\pi\mathbf{h}\cdot(\mathbf{r}_k - \mathbf{r}_j)\}. \quad (8.32a)$$

If in equation (8.31) we write $(\psi_c)_{\mathbf{h}} = \frac{1}{2}\{|F_{\mathbf{h}}|^2 + |F_{-\mathbf{h}}|^2\}$ then it is clear that $(\psi_c)_{\mathbf{h}}$ is the hth Fourier coefficient of a function which has peaks at $\mathbf{r}_k - \mathbf{r}_j$. These peaks will have Fourier transforms which, at the point in reciprocal space corresponding to the index $\mathbf{h}$, have values $(f_k)_R(f_j)_R + \Delta f_k\Delta f_j$. This peak will be like a normal Patterson peak except that when one or both of the atoms is an anomalous scatterer the peak is very slightly modified in weight and form. The Fourier synthesis with $V^{-1}(\psi_c)_{\mathbf{h}}$ as Fourier coefficients gives a function $P_c(\mathbf{u})$ which will thus be similar to a normal Patterson synthesis obtained with non-anomalous scattering data except at points corresponding to the vectors involving anomalous scatterers.

Equation (8.32a) is the basis of a much more interesting function first discussed by Okaya, Saito and Pepinsky (1955). It simplifies our consideration of this function if it is assumed that there is only one anomalous scatterer in the unit cell, i.e. the one for which $j = 1$. Equation (8.32a) then becomes

$$|F_{\mathbf{h}}|^2 - |F_{-\mathbf{h}}|^2 = 2\sum_{k=2}^{N}\Delta f_1(f_k)_R\sin\{2\pi\mathbf{h}\cdot(\mathbf{r}_k - \mathbf{r}_1)\}. \quad (8.32b)$$

If we write $(\psi_s)_{\mathbf{h}} = |F_{\mathbf{h}}|^2 - |F_{-\mathbf{h}}|^2$ then $(\psi_s)_{\mathbf{h}} = -(\psi_s)_{-\mathbf{h}}$. It should now be clear that $(\psi_s)_{\mathbf{h}}$ is the Fourier coefficient of an *odd* function $P_s(\mathbf{u})$ which will have a peak corresponding to the transform of $\Delta f_1(f_k)_R$ at $(\mathbf{r}_k - \mathbf{r}_1)$ and a peak corresponding to the transform of $-\Delta f_1(f_k)_R$ at $-(\mathbf{r}_k - \mathbf{r}_1)$. The function $P_s(\mathbf{u})$ is given by

$$P_s(\mathbf{u}) = \frac{1}{V}\sum_{\mathbf{h}}(\psi_s)_{\mathbf{h}}\sin(2\pi\mathbf{h}\cdot\mathbf{u}). \quad (8.33)$$

This function will have peaks *from* an anomalous scatterer, say atom $m$, *to* any other atom, say atom $n$, of weight approximately proportional to $\Delta f_m Z_n$ and peaks *from* atom $n$ *to* an anomalous scatterer, atom $m$, of weight $-\Delta f_m Z_n$. The significance of this anti-centrosymmetric function is that it not only gives a pattern of peaks, positive and negative, which can be used to solve the structure but the absolute direction from an anomalous scatterer to other atoms is clearly indicated. Thus if the positive directions of the axes with respect to the crystal are defined and the record of the reflection of index $\mathbf{h}$ is distinguished from that of index $-\mathbf{h}$ then the absolute configuration of the crystal structure can be determined. When a molecular structure exists in enantiomorphic forms then, if it contains an atom for which a suitable wavelength can be found to give anomalous scattering, the absolute configuration of the enantiomers can be determined. Without anomalous scattering the diffraction patterns of the two forms are identical and they cannot be distinguished.

To illustrate the principle of the $P_s(\mathbf{u})$ function there is shown in fig. 8.30 a hypothetical structure in space group $p1$ with the sulphur atom acting as the anomalous scatterer. Anomalous structure factors were calculated using STRUCFAC and the output was used as input for PSCOEFF (appendix X). This program produces an output file with amplitudes and phases which, used as input for FOUR2 gives a $P_s$-function map. The map, reproduced in fig. 8.31, shows positive peaks from the anomalous scatterer to other atoms and negative peaks in the reverse directions. Because two carbon atoms are nearly centrosymmetrically disposed about the sulphur atom the positive and negative peaks related to these two atoms partially cancel each other – although the peaks can still be clearly seen.

Synchrotron sources of X-rays (§ 5.8) can give radiation tuned to a particular wavelength and increasingly this is being used to make anomalous scattering measurements for a given structure at several different wavelengths. The anomalous scatterers have values of $f_1$ and $f_2$ which depend on the X-ray wavelength and if the wavelengths used are tuned to be close to an absorption edge for the anomalous scatterer then there will be large changes in the anomalous components of the scattering factor (*see* fig. 6.17). Many-wavelength anomalous scattering (MAS) overcomes the phase ambiguity of OAS (Woolfson and Fan, 1995).

Anomalous scattering can also be used to resolve the ambiguity of phase

Fig. 8.30.
The synthetic structure
$C_5H_9S$.

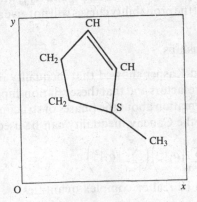

Fig. 8.31.
The $P_s$-function map for $C_5H_9S$. Positive contours are drawn with solid lines and negative contours with dashed lines.

```
FACTOR CONVERTING TO DENSITY/UNIT AREA IS   1.1507
OUTPUT WITH X HORIZONTAL AND ORIGIN BOTTOM LEFT

        0   1    2    3    4    5    6    7    8    9   10   11   12   13   14   15   16   17   18   19   20

20   0  17   4  -16  -28  -34  -18   13   18    2    0   -2  -18  -13   18   34   28   16   -4  -17    0
19 -10      -2 -11  -25  -49  -74  -53    1   18   -2   -5    4   -4   -4    8    7    2    3   -7  -18  -10
18  -4     -11 -13   -1  -15  -59  -61   17   -2  -17  -16    1   -7   -2   17   12    0    3    2   -4   -4
17 -12     -12  -7   32   53  -14  -19  -10  -11  -24  -16   -4  -21  -16   12   15    4    2   -9  -19  -12
16 -51     -25 -15   26   70   51   12    4   -3  -13    2    6  -17  -20    0    3    3   -2  -41  -72  -51
15 -76     -34 -27   -8   21    2  -27  -14   -3   -7   12   21    5   -6   -2   -5    1    2  -50   11  -76
14 -50     -16 -16  -16  -21  -65  -91  -53  -18  -16    3   23   13    2    3   -6   -5    4  -26  -66  -50
13  -5       6  -2   -2  -12  -66  -98  -61   24  -21   -5   17    5   -4    3   -8  -20   -5   -1  -14   -5
12  10      -1 -27  -22   -1  -22  -49  -30   -8   -9    6   17    2   -6    4   -5  -14    6   18   11   10
11   3     -23 -74  -78  -27    0   -7   -2    5    1   13   16    4    0    6   24   57   52   18    3
10   0     -28 -85 ****   48   -6   -1    1    1   -5    0    5   -1   -1    1    6   48  100   85   28    0
 9  -3     -18 -52  -57  -24   -6   -6    0   -4  -16  -13   -1   -5    2    7    0   27   78   74   23   -3
 8 -10     -11 -18   -6   14    5   -4    6   -2  -17   -6    9    8   30   49   22    1   22   27    1  -10
 7   5      14   1    5   20    8   -3    4   -5  -17    5   21   24   64   98   66   12    2    2   -6    5
 6  50      66  26   -4    5    6   -3   -2  -13  -23   -3   16   18   53   91   65   21   16   16   16   50
 5  76     100  50   -2   -1    5    2    6   -5  -21  -12    7    3   14   27   -2  -21    8   27   34   76
 4  51      72  41    2   -3   -3    0   20   17   -6   -2   13    3   -4  -12  -51   70  -26   15   25   51
 3  12      19   9   -2   -4  -15  -12   16   21    4   16   24   11   10   19  -14   53   32    7   12   12
 2   4       4  -2   -3    0  -12  -17    2    7   -1   16   17    2   17   61   59   15    1   13   11    4
 1  10      18   7   -3   -2   -7   -8    4   -4    5    2  -18   -1   53   74   49   25   11    2   10
 0   0      17   4  -16  -28  -34  -18   13   18    2    0   -2  -18  -13   18   34   28   16   -4  -17    0
```

which results from the use of a single pair of isomorphous compounds. If the simplest case of a single $i$-$r$ atom is taken then the phase ambiguity is shown in fig. 8.32. When X-rays are used for which the $i$-$r$ atom $Z$ scatters anomalously, a component $\Delta f_z$ is added to the scattering factor (ignoring the change in the real part) and the new structure factor is shown in fig. 8.32 as $OQ'$ or $OR'$ depending on whether it was originally $OQ$ or $OR$. Since $OQ'$ and $OR'$ will have different magnitudes it will be possible to distinguish which of the phases is the correct one.

Another way of combining SIR and OAS is by the probability curve method illustrated earlier in fig. 8.26. Although the atomic positions for both isomorphous replacement and anomalous scattering are the same, the central phases of the phase ambiguities, $\phi_{\chi h}$ and $\phi_{2,h}$, will differ by $\pi/2$ so that the multiplication of the probability curves will not give a symmetrical result.

## 8.6 Inequality relationships

In 1948, Harker and Kasper showed that inequality relationships existed between the structure factors and that these relationships could occasionally lead to definite information about the phases of structure factors. To derive these relationships, the Cauchy inequality can be used. This is

$$\left| \sum_{j=1}^{N} a_j b_j \right|^2 \leqslant \left( \sum_{j=1}^{N} |a_j|^2 \right) \left( \sum_{j=1}^{N} |b_j|^2 \right) \tag{8.34}$$

where $a_j$ and $b_j$ can be real or complex quantities.

Fig. 8.32.
The resolution of the
phase ambiguity from
isomorphous
replacement with
anomalous scattering.

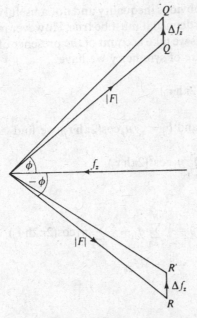

We shall apply this inequality to derive relationships between unitary structure factors. The assumption will be made that the unitary scattering factors, the $n$'s remain constant over reciprocal space. If the structure contains atoms of equal weight, or nearly so, this assumption will be justified; if there are heavy atoms present then, since their scattering factors fall off more slowly with $\sin \theta$ than do those of light atoms, the assumption is not really valid.

Consider the unitary structure factor

$$U_{\mathbf{h}} = \sum_{j=1}^{N} n_j \exp(2\pi i \mathbf{h} \cdot \mathbf{r}_j). \tag{8.35}$$

Taking $a_j = \sqrt{n_j}$ and $b_j = \sqrt{n_j} \exp(2\pi i \mathbf{h} \cdot \mathbf{r}_j)$ then from Cauchy's inequality we find

$$|U_{\mathbf{h}}|^2 \leqslant \sum_{j=1}^{N} n_j \times \sum_{j=1}^{N} n_j |\exp(2\pi i \mathbf{h} \cdot \mathbf{r}_j)|^2. \tag{8.36}$$

Since

$$|\exp(2\pi i \mathbf{h} \cdot \mathbf{r}_j)| = |\cos(2\pi \mathbf{h} \cdot \mathbf{r}_j) + i \sin(2\pi \mathbf{h} \cdot \mathbf{r}_j)| = 1$$

and

$$\sum_{j=1}^{N} n_j = 1$$

inequality (8.36) gives

$$|U_{\mathbf{h}}|^2 \leqslant 1. \tag{8.37}$$

This is a fairly obvious inequality and not a useful one, since by the very way in which $U_h$ is defined it must be true. However more useful results are obtained as soon as we take account of the presence of symmetry elements.

Thus for a centre of symmetry we have

$$U_h = \sum_{j=1}^{N} n_j \cos(2\pi \mathbf{h}\cdot\mathbf{r}_j). \tag{8.38}$$

Putting $a_j = \sqrt{n_j}$ and $b_j = \sqrt{n_j}\cos(2\pi\mathbf{h}\cdot\mathbf{r}_j)$ we find

$$U_h^2 \leqslant \sum_{j=1}^{N} n_j \times \sum_{j=1}^{N} n_j \cos^2(2\pi\mathbf{h}\cdot\mathbf{r}_j). \tag{8.39}$$

Now

$$\sum_{j=1}^{N} n_j \cos^2(2\pi\mathbf{h}\cdot\mathbf{r}_j) = \tfrac{1}{2}\sum_{j=1}^{N} n_j + \tfrac{1}{2}\sum_{j=1}^{N} n_j \cos(2\pi 2\mathbf{h}\cdot\mathbf{r}_j)$$

$$= \tfrac{1}{2} + \tfrac{1}{2}U_{2h}$$

so that

$$U_h^2 \leqslant \tfrac{1}{2}(1 + U_{2h}). \tag{8.40}$$

This inequality can be used to give the sign of $U_{2h}$; if it is written in the form

$$U_{2h} \geqslant 2U_h^2 - 1 \tag{8.41}$$

this can be seen a little more easily. If $|U_h|$ and $|U_{2h}|$ are both sufficiently large then one can show that $U_{2h}$ must be positive. Thus if $|U_h| = 0.6$ and $|U_{2h}| = 0.5$ then, since from inequality (8.41)

$$U_{2h} \geqslant -0.28,$$

it is clear that $U_{2h}$ must be positive since the negative sign for $U_{2h}$ would contravene the inequality.

Let us now see what inequalities result from having a twofold axis. Atoms will occur in pairs with coordinates $(x, y, z)$ and $(-x, y, -z)$ and the unitary-structure-factor equation appears as

$$U_{hkl} = \sum_{j=1}^{N} n_j \exp(2\pi iky_j)\cos\{2\pi(hx_j + lz_j)\}. \tag{8.42}$$

For $a_j = \sqrt{n_j}\exp(2\pi iky_j)$ and $b_j = \sqrt{n_j}\cos\{2\pi(hx_j + lz_j)\}$ this gives

$$|U_{hkl}|^2 \leqslant \tfrac{1}{2}(1 + U_{2h,0,2l}). \tag{8.43}$$

This inequality can be very useful. The $b$-axis projection is centrosymmetric and it is possible to determine the sign of a large structure factor $U_{2h,0,2l}$ if *any* of the structure factors $U_{hkl}$ has a sufficiently high magnitude.

If, together with the twofold axis, there is a centre of symmetry then the unitary-structure-factor equation has the form

$$U_{hkl} = \sum_{j=1}^{N} n_j \cos(2\pi ky_j)\cos\{2\pi(hx_j + lz_j)\}. \tag{8.44}$$

This can lead to two different inequality relationships depending on the way the summation terms are partitioned.

For $a_j = \sqrt{n_j}\cos(2\pi ky_j)$ (and $b_j = \sqrt{n_j}\cos\{2\pi(hx_j + lz_j)\}$) we have

$$U_{hkl}^2 \leqslant \tfrac{1}{4}(1 + U_{0,2k,0})(1 + U_{2h,0,2l}) \tag{8.45}$$

while for $a_j = \sqrt{n_j}$ and $b_j = \sqrt{n_j}\cos(2\pi ky_j)\cos\{2\pi(hx_j + lz_j)\}$ the inequality becomes

$$U_{hkl}^2 \leqslant \tfrac{1}{4}(1 + U_{0,2k,0} + U_{2h,0,2l} + U_{2h,2k,2l}). \tag{8.46}$$

The general rules for formulating inequalities of this type should be now be apparent. The unitary-structure-factor equation is written down for the particular space group with which one is dealing, the summation terms are partitioned in all possible ways and, for each partition, an inequality relationship can be deduced.

The reader should be able to confirm the following inequalities for the space group $P2_1/m$ for which the sets of related atoms have coordinates $\pm(x, y, z)$, $\pm(x, \tfrac{1}{2} - y, z)$:

$$U_{hkl}^2 \leqslant \tfrac{1}{4}\{1 + (-1)^k U_{2h,0,2l}\}\{1 + (-1)^k U_{0,2k,0}\} \tag{8.47}$$

and

$$U_{hkl}^2 \leqslant \tfrac{1}{4}\{1 + (-1)^k U_{2h,0,2l} + (-1)^k U_{0,2k,0} + U_{2h,2k,2l}\}. \tag{8.48}$$

We now consider a number of situations for $P2_1/m$ and study the conclusions therefrom.

(1)   $|U_{hkl}| = 0.5$, $|U_{2h,0,2l}| = 0.3$, $|U_{0,2k,0}| = 0.1$.

The left-hand side of the inequality equals 0.25. If $(-1)^k U_{2h,0,2l}$ is negative then the maximum value of the right-hand side of inequality (8.47) is $0.25 \times 1.1 \times 0.7 = 0.19$. Thus $(-1)^k U_{2h,0,2l}$ must be positive. This proves that $U_{2h,0,2l}$ is positive if $k$ is even and negative if $k$ is odd.

(2)   $|U_{hkl}| = 0.4$, $|U_{2h,0,2l}| = 0.25$, $|U_{0,2k,0}| = 0.25$.

The left-hand side of the inequality equals 0.16. If $(-1)^k U_{2h,0,2l}$ and $(-1)^k U_{0,2k,0}$ are both negative then the maximum value of the right-hand side of inequality (8.47) is $0.25 \times 0.75 \times 0.75 = 0.14$. However if either of them is positive or both then the inequality can be satisfied. Hence one can only show in this case that at least one of $(-1)^k U_{2h,0,2l}$ and $(-1)^k U_{0,2k,0}$ must be positive.

(3)   $|U_{hkl}| = 0.5$, $|U_{2h,0,2l}| = |U_{0,2k,0}| = 0.05$, $|U_{2h,2k,2l}| = 0.3$.

From inequality (8.47) it can be shown that at least one of $(-1)^k U_{2h,0,2l}$ and $(-1)^k U_{0,2k,0}$ must be positive. From inequality (8.48) it can be shown that $U_{2h,2k,2l}$ must be positive.

There is another type of inequality relationship which can be deduced by forming the sum and differences of unitary structure factors. We have

$$U_{\mathbf{h}} + U_{\mathbf{h}'} = \sum_{j=1}^{N} n_j\{\cos(2\pi\mathbf{h}\cdot\mathbf{r}_j) + \cos(2\pi\mathbf{h}'\cdot\mathbf{r}_j)\}$$

$$= \sum_{j=1}^{N} 2n_j \cos\{\pi(\mathbf{h} + \mathbf{h}')\cdot\mathbf{r}_j\}\cos\{\pi(\mathbf{h} - \mathbf{h}')\cdot\mathbf{r}_j\}. \tag{8.49}$$

Taking the partition

$$a_j = \sqrt{2n_j}\cos\{\pi(\mathbf{h} + \mathbf{h}')\cdot\mathbf{r}_j\}$$

and

$$b_j = \sqrt{2n_j}\cos\{\pi(\mathbf{h} - \mathbf{h}')\cdot\mathbf{r}_j\}$$

we find

$$(U_\mathbf{h} + U_{\mathbf{h}'})^2 \leqslant (1 + U_{\mathbf{h}+\mathbf{h}'})(1 + U_{\mathbf{h}-\mathbf{h}'}). \tag{8.50}$$

Similarly

$$U_\mathbf{h} - U_{\mathbf{h}'} = \sum_{j=1}^{N} - 2n_j\sin\{\pi(\mathbf{h} + \mathbf{h}')\cdot\mathbf{r}_j\}\sin\{\pi(\mathbf{h} - \mathbf{h}')\cdot\mathbf{r}_j\} \tag{8.51}$$

and the corresponding inequality becomes

$$(U_\mathbf{h} - U_{\mathbf{h}'})^2 \leqslant (1 - U_{\mathbf{h}+\mathbf{h}'})(1 - U_{\mathbf{h}-\mathbf{h}'}). \tag{8.52}$$

Both the inequalities (8.50) amd (8.52) are valid for a given set of $U_\mathbf{h}$, $U_{\mathbf{h}'}$, $U_{\mathbf{h}+\mathbf{h}'}$ and $U_{\mathbf{h}-\mathbf{h}'}$, and they can, under suitable circumstances, give rise to definite relationships between structure factors. Let $U_\mathbf{h}$ and $U_{\mathbf{h}'}$ be known in both magnitude and sign as 0.45 and $-0.45$, respectively, and $|U_{\mathbf{h}+\mathbf{h}'}| = 0.3$ and $|U_{\mathbf{h}-\mathbf{h}'}| = 0.1$. Then inequality (8.52) gives a left-hand side equal to 0.81 and it can be seen that if $U_{\mathbf{h}+\mathbf{h}'}$ is positive the inequality cannot be satisfied so that $U_{\mathbf{h}+\mathbf{h}'}$ must be negative.

If $U_\mathbf{h}$ and $U_{\mathbf{h}'}$ have the same sign then one uses (8.50) while if they have opposite signs (8.52) is used to get the largest left-hand side. This means that (8.50) and (8.52) can be combined into a single inequality

$$(|U_\mathbf{h}| + |U_{\mathbf{h}'}|)^2 \leqslant \{1 + s(\mathbf{h})s(\mathbf{h}')U_{\mathbf{h}+\mathbf{h}'}\}\{1 + s(\mathbf{h})s(\mathbf{h}')U_{\mathbf{h}-\mathbf{h}'}\} \tag{8.53}$$

where $s(\mathbf{h})$ means 'the sign of $U_\mathbf{h}$'. The reader should confirm that if $s(\mathbf{h})s(\mathbf{h}')$ is positive then (8.53) is equivalent to (8.50) while if $s(\mathbf{h})s(\mathbf{h}')$ is negative then (8.53) is equivalent to (8.52). As a next step one can transform (8.53) into

$$(|U_\mathbf{h}| + |U_{\mathbf{h}'}|)^2 \leqslant \{1 + s(\mathbf{h})s(\mathbf{h}')s(\mathbf{h} + \mathbf{h}')|U_{\mathbf{h}+\mathbf{h}'}|\}$$
$$\{1 + s(\mathbf{h})s(\mathbf{h}')s(\mathbf{h} - \mathbf{h}')|U_{\mathbf{h}-\mathbf{h}'}|\}. \tag{8.54}$$

This relationship is interesting in that what it can show, if the $U$'s are sufficiently large, is that either or both of the triple products of signs $s(\mathbf{h})s(\mathbf{h}')s(\mathbf{h} + \mathbf{h}')$ and $s(\mathbf{h})s(\mathbf{h}')s(\mathbf{h} - \mathbf{h}')$ are positive.

We shall consider a number of examples.

(1)     $|U_\mathbf{h}| = |U_{\mathbf{h}'}| = |U_{\mathbf{h}+\mathbf{h}'}| = |U_{\mathbf{h}-\mathbf{h}'}| = 0.5$.

The left-hand side equals 1.0. It can easily be shown that unless both the triple products of sign are positive then the inequality (8.54) cannot be satisfied. For each of these triple products if two of the $U$'s are known in sign then the sign of the third $U$ will be found.

(2)     $|U_\mathbf{h}| = |U_{\mathbf{h}'}| = |U_{\mathbf{h}+\mathbf{h}'}| = |U_{\mathbf{h}-\mathbf{h}'}| = 0.4$.

The left-hand side equals 0.64. At least one of the triple products must be positive otherwise the inequality cannot be satisfied.

(3)  $|U_{\mathbf{h}}| = |U_{\mathbf{h'}}| = |U_{\mathbf{h+h'}}| = |U_{\mathbf{h-h'}}| = 0.3.$

The left-hand side equals 0.36. The inequality is satisfied even if both the triple products are negative and no conclusions can be drawn.

In fact inequality (8.54) does not completely replace (8.50) and (8.52) as the following example will show. We consider the situation

$|U_{\mathbf{h}}| = 0.6, |U_{\mathbf{h'}}| = 0, |U_{\mathbf{h+h'}}| = |U_{\mathbf{h-h'}}| = 0.45.$

For either (8.50) or (8.52) the left-hand side has value 0.36. Let us assume that $U_{\mathbf{h+h'}}$ and $U_{\mathbf{h-h'}}$ are both positive; from (8.52) we have the right-hand side equal to

$(1 - 0.45)(1 - 0.45) = 0.3025$

and since this is less than the left-hand side the assumption cannot be valid. Similarly if one assumes that $U_{\mathbf{h+h'}}$ and $U_{\mathbf{h-h'}}$ are both negative then the use of inequality (8.50) shows that the assumption is invalid. The only valid assumption is that $U_{\mathbf{h+h'}}$ and $U_{\mathbf{h-h'}}$ have opposite signs when either of the inequalities gives the valid conclusion

$0.36 \leqslant (1 + 0.45)(1 - 0.45).$

We shall demonstrate the application of inequality relationships by the solution of the $c$-axis projection of tetraethyl diphosphine disulphide (Dutta and Woolfson, 1961).

Before we do this, however, we should acquaint ourselves with a useful piece of information – that the signs of a number of structure factors may be chosen arbitrarily. The reason for this is the existence of a number of non-equivalent centres of symmetry which may be chosen as origin. As the origin is moved from one centre of symmetry to another so the signs of the structure factors will change but they change in such a way that all the members of one parity group change together. A parity group is a set of structure factors whose three indices are odd or even in the same way. Suppose we take one centre of symmetry $C$ as the point $(0, 0, 0)$ and one structure factor from each parity group each of which, with respect to this origin, is positive. Then as the origin is moved to other centres of symmetry the signs of the structure factors will change. For example if a structure factor with respect to $C$ as origin can be written as

$$[F_{hkl}]_{000} = \sum_{j=1}^{N} f_j \cos\{2\pi(hx_j + ky_j + lz_j)\} \tag{8.55}$$

then with respect to an origin at $(\frac{1}{2}, 0, 0)$ it will be

$$[F_{hkl}]_{\frac{1}{2}00} = \sum_{j=1}^{N} f_j \cos[2\pi\{(h(x_j - \frac{1}{2}) + ky_j + lz_j\}]$$

or

$$[F_{hkl}]_{\frac{1}{2}00} = (-1)^h [F_{hkl}]_{000}. \tag{8.56}$$

Table 8.7. *hkl parity*

| Origin | eee | oee | eoe | eeo | ooe | oeo | eoo | ooo |
|---|---|---|---|---|---|---|---|---|
| $(0, 0, 0)$ | + | + | + | + | + | + | + | + |
| $(\frac{1}{2}, 0, 0)$ | + | − | + | + | − | − | + | − |
| $(0, \frac{1}{2}, 0)$ | + | + | − | + | − | + | − | − |
| $(0, 0, \frac{1}{2})$ | + | + | + | − | + | − | − | − |
| $(\frac{1}{2}, \frac{1}{2}, 0)$ | + | − | − | + | + | − | − | + |
| $(\frac{1}{2}, 0, \frac{1}{2})$ | + | − | + | − | − | + | − | + |
| $(0, \frac{1}{2}, \frac{1}{2})$ | + | + | − | − | − | − | + | + |
| $(\frac{1}{2}, \frac{1}{2}, \frac{1}{2})$ | + | − | − | − | + | + | + | − |

The complete pattern of changing signs is shown in table 8.7.

The *eee* group does not change its sign with change of origin and these structure factors are therefore called *structure invariants*. If three groups are taken, such as *oee, eoe* and *ooe*, where adding the parity of the indices gives *eee*, then the product of three such structure factors (called linearly dependent) is also a structure invariant. The reader may check that the products of signs in these three columns is always positive. On the other hand, if one takes three linearly independent reflections from three different parity groups and not including *eee* then all eight $(2^3)$ combinations of signs occur at one or other of the origins. With three-dimensional data we are thus enabled to select three such reflections and arbitrarily to fix their signs (as positive, say) and this amounts to fixing the origin of the cell. In two dimensions there are four origins available and one may arbitrarily select the signs of two reflections taken from different parity groups of *oe, eo* and *oo*.

The $|U|$ data for tetraethyl diphosphine disulphide is given in fig. 8.33. The two-dimensional space group is *p*2 implying only a centre of symmetry. First one can look for pairs of large $|U|$'s of indices of form $U_h$ and $U_{2h}$ to see whether inequality (8.41) can be applied. We find

$$|U_{2,\bar{3}}| = 0.56, \quad |U_{4,\bar{6}}| = 0.45.$$

From inequality (8.41) we must have $U_{4,\bar{6}}$ greater than $-0.37$ and hence, from its magnitude, $U_{4,\bar{6}}$ must be positive. Similarly $U_{8,2}$ can also be shown to be positive. The logical nature of these sign determinations should be noted; the two $U$'s can be shown to be positive and they must therefore be structure invariants otherwise their signs would change with change of origin.

At this point two origin-fixing signs may be chosen. It is clearly advisable to choose large $|U|$'s and two which satisfy the rules for selection are $U_{3,\bar{7}}$ and $U_{6,\bar{7}}$ which are both taken as positive.

Further progress can now be made with inequality (8.54). To use this inequality it is necessary to select four indices of form $\mathbf{h}, \mathbf{h}', \mathbf{h} + \mathbf{h}'$ and $\mathbf{h} - \mathbf{h}'$. This may conveniently be done by preparing a transparent replica (*t-r*) of fig. 8.33. If the origin of the *t-r* is placed over $U_\mathbf{h}$ of the figure then under the $U_{\mathbf{h}'}$ and $U_{-\mathbf{h}'}$ of the *t-r* are seen $U_{\mathbf{h}+\mathbf{h}'}$ and $U_{\mathbf{h}-\mathbf{h}'}$. The reader should make a *t-r* of fig. 8.33 and try finding sets of four indices in this way.

$k$

```
            86  0 85 91  0  0  0  0  0
         42 65 42  0 45  0  0  0  0  0  0 51 34 44
    0  0 45  0  0  0  0  0 20 54 26 28 73 58  0 34 61  0
    0  0  0  0  0 63 45 13 70 65  0 42 75 43  0 18  0  0  0
 0  0 62 46  0 33 70 36 15 56 55 14 28  0  0 16 33  0 49 35  0
 0 21 53 35 23 39 41  0 38  0  6  0  8 17  0 21 42  0 69 53  0  0
 0 42 49  0 48 30 20 14 17 16 30 33 14 13 61 52  0 26 49 12  0  0
 0 17 14 10 10  0  6 41 57 42 30  X  30 42 57 41  6  0 10 10 14 17  0   → h
    0  0 12 49 26  0 52 61 13 14 33 30 16 17 14 20 30 48  0 49 42  0
    0  0 53 69  0 42 21  0 17  8  0  6  0 38  0 41 39 23 35 53 21  0
       0 35 49  0 33 16  0  0 28 14 55 56 15 36 70 33  0 46 62  0  0
          0  0  0 18  0 43 75 42  0 65 70 13 45 63  0  0  0  0  0
          0 61 34  0 58 73 28 26 54 20  0  0  0  0  0 45  0  0
            44 34 51  0  0  0  0  0  0 45  0 42 65 42
                0  0  0  0  0 91 85  0 86
```

Fig. 8.33.
The values of $100|U_{hk0}|$ for tetraethyl diphosphine disulphide.

A few sign determinations follow:

| h | h' | h + h' | h − h' | Conclusion |
|---|---|---|---|---|
| $U_{3,\bar{7}} = 0.91$ | $|U_{3,0}| = 0.57$ | $U_{6,\bar{7}} = 0.86$ | $U_{0,\bar{7}} = 0$ | $s(3,0) = +1$ |
| $U_{4,\bar{6}} = 0.45$ | $U_{3,0} = 0.57$ | $|U_{7,\bar{6}}| = 0.65$ | $U_{1,\bar{6}} = 0$ | $s(7,\bar{6}) = +1$ |
| $U_{3,\bar{7}} = 0.91$ | $U_{8,2} = 0.53$ | $|U_{11,\bar{5}}| = ?$ | $|U_{5,5}| = 0.34$ | $s(5,5) = +1$ |

The last relationship is interesting in that one of the $U$'s is outside the observed region. However it is known for this as for all other $U$'s that $|U| \leqslant 1$ and it can be shown that $U_{5,5}$ must be positive no matter what is the value of $U_{11,\bar{5}}$.

It will be noticed that progress so far has only enabled signs to be determined as positive since all the initial signs were positive. When all the signs for a structure are positive then a Fourier synthesis gives a very large peak at the origin. If it is known that there is no atom at the origin then we must expect the progress in determining new signs to break down at some stage. This situation cannot be avoided by choosing one or more origin-fixing signs as negative; this will merely give a sign-development pattern where a large peak will build up at a non-origin centre of symmetry.

Once progress in sign determination is baulked by an inability to relate fresh signs to ones already determined then it is possible to introduce sign symbols. If the sign of $U_{4,\bar{7}}$ is indicated as $p$ then one may find:

| h | h' | h + h' | h − h' | Conclusion |
|---|---|---|---|---|
| $U_{4,\bar{7}} = 0.85p$ | $U_{3,\bar{7}} = 0.91$ | $U_{7,\overline{14}}| = ?$ | $|U_{1,0}| = 0.30$ | $s(1,0) = p$ |
| $U_{4,\bar{7}} = 0.85p$ | $U_{6,\bar{7}} = 0.86$ | $|U_{10,\overline{14}}| = ?$ | $|U_{2,0}| = 0.42$ | $s(2,0) = p$ |

Whenever progress stops then another symbol can be introduced.

At the end of the sign-determining process one should have determined the signs of a number of the largest $U$'s in terms of a few sign symbols. If the number of symbols is $n$ then there are $2^n$ possible sets of signs and, if Fourier syntheses are calculated for all these, one might recognize the correct structure in one of them.

For tetraethyl diphosphine disulphide most of the large unitary structure factors were found in terms of two symbols $p$ and $q$. In fact it was possible to determine both $p$ and $q$ by the use of inequalities (8.50) and (8.52) which, as we have seen, can show that $s(\mathbf{h} + \mathbf{h}')s(\mathbf{h} - \mathbf{h}') = -1$. Thus

| $\mathbf{h}$ | $\mathbf{h}'$ | $\mathbf{h} + \mathbf{h}'$ | $\mathbf{h} - \mathbf{h}'$ | Conclusion |
|---|---|---|---|---|
| $U_{7,\bar{3}} = 0$ | $|U_{2,0}| = 0.42$ | $U_{9,\bar{3}} = 0.62pq$ | $U_{5,\bar{3}} = 0.70q$ | $pq^2 = p = -1$ |
| $U_{2,\bar{2}} = 0$ | $|U_{5,\bar{4}}| = 0.63$ | $U_{7,\bar{6}} = 0.65$ | $U_{3,\bar{2}} = 0.38q$ | $q = -1.$ |

The Fourier synthesis based on the inequality-derived signs is shown in fig. 8.34 and it can be seen that this shows the form of the molecule quite well.

This structure was solved by the use of the simple basic inequalities for $P\bar{1}$ – (8.41), (8.50), (8.52) *and* (8.54). For space groups with more symmetry other inequalities may be used but it is found in practice that it is with the simple inequalities that most progress is made.

In general, inequality relationships can only be used with the simplest structures. If there are $N$ equal atoms per centrosymmetric unit cell then the $U$'s will have a normal distribution with variance $N^{-1}$ and mean zero. Only one in twenty structure factors should have a magnitude greater than two times the standard deviation, or $2N^{-\frac{1}{2}}$, and three in a thousand should be greater in magnitude than $3N^{-\frac{1}{2}}$. For $N = 36$ this means that very few $|U|$'s (about 0.3%) will be greater than 0.5 and one needs to have a fair number of $|U|$'s this large in order for inequalities to be effective. For $N > 40$ or so it is extremely unlikely that the use of inequality relationships alone will enable the structure to be solved.

## 8.7  Sign relationships

From inequality (8.54) it can be shown that if the corresponding $|U|$'s are all sufficiently large then

Fig. 8.34.
The electron-density map with inequality-determined signs for tetraethyl diphosphine disulphide (from Dutta & Woolfson, 1961).

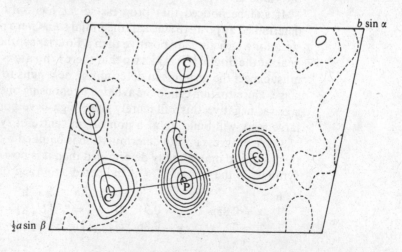

$$s(\mathbf{h})s(\mathbf{h}')s(\mathbf{h} - \mathbf{h}') = + 1. \tag{8.57}$$

In 1952 Sayre, Cochran and Zachariasen separately showed that even when the $|U|$'s were smaller than was necessary to satisfy the inequality relations it could still be shown that equation (8.57) was probably true. Sayre's approach to this problem is the one we shall consider.

Let us consider a structure containing equal resolved atoms as demonstrated by the one-dimensional electron-density distribution in fig. 8.35(a). If this electron density $\rho(x)$ is squared to give $\rho(x)^2$ then this distribution, shown in fig. 8.35(b), also consists of equal resolved peaks. From equation (3.31) we may write for three dimensions

$$F_{\mathbf{h}} = \sum_{j=1}^{N} f_{\mathbf{h}} \exp(2\pi i \mathbf{h} \cdot \mathbf{r}_j) \tag{8.58}$$

where $f_{\mathbf{h}}$ is the scattering factor for each of the atoms. The value of $f_{\mathbf{h}}$ also corresponds to that of the Fourier transform of an atomic peak at the point corresponding to $\mathbf{h}$ in reciprocal space.

We may also write

$$G_{\mathbf{h}} = \sum_{j=1}^{N} g_{\mathbf{h}} \exp(2\pi i \mathbf{h} \cdot \mathbf{r}_j) \tag{8.59}$$

where $G_{\mathbf{h}}$ is the hth Fourier coefficient of $\rho(\mathbf{r})^2$ and $g_{\mathbf{h}}$ is the Fourier transform of a 'squared' peak. From equations (8.58) and (8.59) we may write

$$F_{\mathbf{h}} = \frac{f_{\mathbf{h}}}{g_{\mathbf{h}}} G_{\mathbf{h}} = \theta_{\mathbf{h}} G_{\mathbf{h}}, \text{ say.} \tag{8.60}$$

It follows from the convolution theorem given in §4.6 that the Fourier coefficients of $\rho(\mathbf{r})^2$ will be given by the self-convolution of the Fourier coefficients of $\rho(\mathbf{r})$.

Fig. 8.35.
(a) $\rho(x)$ for equal, resolved atoms in one dimension.
(b) $\rho(x)^2$ for equal, resolved atoms in one dimension.

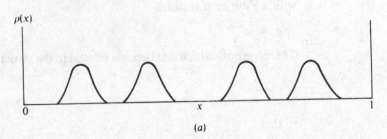

(a)

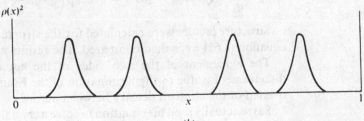

(b)

From equation (4.72) it can be seen that the Fourier coefficients of $\rho(\mathbf{r})$ are $(1/V)F_\mathbf{h}$ and similarly the Fourier coefficients of $\rho_\mathbf{h}(\mathbf{r})^2$ will be $(1/V)G_\mathbf{h}$. Hence, from the convolution theorem

$$\frac{1}{V}G_\mathbf{h} = \sum_{\mathbf{h'}}\frac{1}{V}F_\mathbf{h'}\frac{1}{V}F_{\mathbf{h}-\mathbf{h'}}$$

and, using equation (8.60), we find

$$F_\mathbf{h} = \frac{\theta_\mathbf{h}}{V}\sum_{\mathbf{h'}}F_{\mathbf{h'}}F_{\mathbf{h}-\mathbf{h'}}. \tag{8.61}$$

This equation, known as Sayre's equation, applies to any equal-atom structure whether centrosymmetrical or not. It will apply to one- or two-dimensional data if the atoms are resolved in the appropriate projection and then $l$, the length, or $A$, the area, of projection will replace $V$ in equation (8.61).

Sayre illustrated the application of equation (8.61) with a centrosymmetrical one-dimensional model structure. This had atoms at $\pm 0.113$, $\pm 0.234$, $\pm 0.361$ and $\pm 0.438$ in a cell of length $20\,\text{Å}$. The electron density in the atoms was taken as

$$\rho(u) = e^{-2\pi u^2} \tag{8.62}$$

where $u$ is the distance from the centre of the atom, and this gives a scattering factor

$$f_s = \frac{1}{\sqrt{2}}e^{-\frac{1}{2}\pi s^2}. \tag{8.63}$$

The squared density was

$$\rho(u)^2 = e^{-4\pi u^2} \tag{8.64}$$

with a Fourier transform

$$g_s = \tfrac{1}{2}e^{-\frac{1}{4}\pi s^2}. \tag{8.65}$$

The reciprocal-lattice distance $s$ is related to the structure-factor index $h$ by

$$s = \frac{h}{a} \tag{8.66}$$

so that

$$\theta_h = \frac{f_h}{g_h} = \sqrt{2}\exp(-\tfrac{1}{4}\pi s^2) = \sqrt{2}\exp\left(-\tfrac{1}{4}\pi\frac{h^2}{a^2}\right). \tag{8.67}$$

Structure factors were calculated for the structure and the two sides of equation (8.61) were then compared. The results are shown in table 8.8.

The agreement of the two sides of the equation is excellent; any disagreement is due to the termination of the Fourier series and to slight overlap of the electron density between atoms.

Sayre actually used his equation to solve a crystal structure but generally this is not feasible. What can be seen from these equations is that if $F_\mathbf{h}$ is

Table 8.8.

| $h$ | $F_h$ | $\dfrac{\theta_h}{a}\displaystyle\sum_{h'} F_{h'-h'}$ | $h$ | $F_h$ | $\dfrac{\theta_h}{a}\displaystyle\sum_{h'} F_{h'}F_{h-h'}$ |
|---|---|---|---|---|---|
| 0  | 5.66  | 5.66  | 14 | 0.38  | 0.37  |
| 1  | −1.00 | −1.00 | 15 | −1.81 | −1.78 |
| 2  | −0.41 | −0.41 | 16 | 0.77  | 0.76  |
| 3  | −0.48 | −0.48 | 17 | 0.72  | 0.72  |
| 4  | −1.28 | −1.27 | 18 | 0.38  | 0.38  |
| 5  | 0.35  | 0.35  | 19 | −0.05 | −0.07 |
| 6  | −1.78 | −1.78 | 20 | −0.07 | −0.05 |
| 7  | −0.54 | −0.54 | 21 | −0.10 | −0.10 |
| 8  | 1.41  | 1.41  | 22 | −0.03 | −0.03 |
| 9  | 2.80  | 2.79  | 23 | −0.17 | −0.16 |
| 10 | −1.29 | −1.30 | 24 | −0.37 | −0.36 |
| 11 | 0.48  | 0.47  | 25 | 0.36  | 0.35  |
| 12 | −0.64 | −0.62 | 26 | 0.03  | 0.03  |
| 13 | −0.52 | −0.52 |    |       |       |

large then any large products on the right-hand side of equation (8.61) are likely to have the same sign as $F_h$ or

$$s(\mathbf{h})s(\mathbf{h'})s(\mathbf{h} - \mathbf{h'}) \approx 1 \tag{8.68}$$

where $\approx$ means 'probably equals'. Cochran and Zachariasen also arrived at this probability relationship although by different lines of reasoning.

The probability that equation (8.68) is true was worked out for the equal-atom case by Woolfson (1954). We write the unitary structure factor as the sum of independent terms

$$U_\mathbf{h} = \sum_{j=1}^{N/2} 2n_j \cos(2\pi\mathbf{h}\cdot\mathbf{r}_j)$$

$$= \sum_{j=1}^{N/2} 2n_j \cos(2\pi\mathbf{h'}\cdot\mathbf{r}_j)\cos\{2\pi(\mathbf{h} - \mathbf{h'})\cdot\mathbf{r}_j\}$$

$$- \sum_{j=1}^{N/2} 2n_j \sin(2\pi\mathbf{h'}\cdot\mathbf{r}_j)\sin\{2\pi(\mathbf{h} - \mathbf{h'})\cdot\mathbf{r}_j\} \tag{8.69}$$

This can be written as

$$U_\mathbf{h} = \sum_{j=1}^{N/2} \alpha_j + \sum_{j=1}^{N/2} \beta_j \tag{8.70}$$

where

$$\alpha_j = 2n_j \cos(2\pi\mathbf{h'}\cdot\mathbf{r}_j)\cos\{2\pi(\mathbf{h} - \mathbf{h'})\cdot\mathbf{r}_j\} \tag{8.71a}$$

and

$$\beta_j = 2n_j \sin(2\pi\mathbf{h'}\cdot\mathbf{r}_j)\sin\{2\pi(\mathbf{h} - \mathbf{h'})\cdot\mathbf{r}_j\}. \tag{8.71b}$$

The central-limit theorem tells us that the probability distribution of $U_\mathbf{h}$ is normal with a mean

$$\overline{U_\mathbf{h}} = \sum_{j=1}^{N/2} \overline{\alpha_j} + \sum_{j=1}^{N/2} \overline{\beta_j} \tag{8.72}$$

and a variance from equation (7.18)

$$\sigma_\mathbf{h}^2 = \sum_{j=1}^{N/2} (\overline{\alpha_j^2} - \overline{\alpha_j}^2) + \sum_{j=1}^{N/2} (\overline{\beta_j^2} - \overline{\beta_f}^2). \tag{8.73}$$

Now

$$\overline{\alpha_j} = \overline{2n_j \cos(2\pi\mathbf{h'}\cdot\mathbf{r}_j)\cos\{2\pi(\mathbf{h} - \mathbf{h'})\cdot\mathbf{r}_j\}}$$
$$= 2n_j \overline{\cos(2\pi\mathbf{h'}\cdot\mathbf{r}_j)} \times \overline{\cos\{2\pi(\mathbf{h} - \mathbf{h'})\cdot\mathbf{r}_j\}} \tag{8.74}$$

taking the product of the averages is valid since the two quantities are independent.

When the atoms are equal $n_j = N^{-1}$ and

$$\sum_{j=1}^{N/2} 2n_j(\overline{\cos 2\pi\mathbf{h}\cdot\mathbf{r}_j}) = 2n_j \times \frac{N}{2} \times \overline{(\cos 2\pi\mathbf{h}\cdot\mathbf{r}_j)}$$

$$= \overline{\cos(2\pi\mathbf{h}\cdot\mathbf{r}_j)} = U_\mathbf{h}. \tag{8.75}$$

Hence

$$\overline{\alpha_j} = 2N^{-1}U_{\mathbf{h'}}U_{\mathbf{h}-\mathbf{h'}}. \tag{8.76}$$

We also have

$$\overline{\beta_j} = 2n_j \overline{\sin(2\pi\mathbf{h'}\cdot\mathbf{r}_j)} \times \overline{\sin\{2\pi(\mathbf{h} - \mathbf{h'})\cdot\mathbf{r}_j\}} = 0 \tag{8.77}$$

since the values of $\sin(2\pi\mathbf{h}\cdot\mathbf{r}_j)$ are equally probably positive and negative and not constrained by the value of $U_\mathbf{h}$ – as long as $|U_\mathbf{h}|$ is not exceptionally large.

This gives, from equations (8.72) and (8.76),

$$\overline{U_\mathbf{h}} = \sum_{j=1}^{N/2} 2N^{-1}U_{\mathbf{h'}}U_{\mathbf{h}-\mathbf{h'}} = U_{\mathbf{h'}}U_{\mathbf{h}-\mathbf{h'}}. \tag{8.78}$$

As long as $U_{\mathbf{h'}}U_{\mathbf{h}-\mathbf{h'}}$ is not too large we can write

$$\sigma_\mathbf{h}^2 = \sum_{j=1}^{N/2} \overline{\alpha_j^2} + \sum_{j=1}^{N/2} \overline{\beta_j^2}$$

$$= \sum_{j=1}^{N/2} 4n_j^2 \overline{\cos^2(2\pi\mathbf{h'}\cdot\mathbf{r}_j)} \times \overline{\cos^2\{2\pi(\mathbf{h} - \mathbf{h'})\cdot\mathbf{r}_j\}}$$

$$+ \sum_{j=1}^{N/2} 4n_j^2 \overline{\sin^2(2\pi\mathbf{h}\cdot\mathbf{r}_j)} \times \overline{\sin^2\{2\pi(\mathbf{h} - \mathbf{h'})\cdot\mathbf{r}_j\}}. \tag{8.79}$$

If we take $\overline{\sin^2(2\pi\mathbf{h}\cdot\mathbf{r}_j)} = \overline{\cos^2(2\pi\mathbf{h}\cdot\mathbf{r}_j)} = \frac{1}{2}$ then we find for equal atoms
$$\sigma_\mathbf{h}^2 = N^{-1}. \tag{8.80}$$

Thus for a particular pair of unitary structure factors $U_{h'}$ and $U_{h-h'}$ the distribution of the values of $U_h$ is

$$P(U_h) = (2\pi N^{-1}) \exp\{ -\tfrac{1}{2}N(U_h - U_{h'}U_{h-h'})^2\}. \tag{8.81}$$

However we know the value of $|U_h|$ and the value of $U_h$ is either $s(h')s(h-h')|U_h|$ or $-s(h')s(h-h')|U_h'|$. The probability of it having each of these values is proportional to the ordinate of the distribution (8.81). These two values correspond to $s(h)s(h')s(h-h')$ being positive and negative, respectively; if we denote the probabilities by $P_+(h,h')$ and $P_-(h,h')$ then

$$\frac{P_+(h,h')}{P_-(h,h')} = \frac{\exp\{-\tfrac{1}{2}N(|U_h| - |U_{h'}U_{h-h'}|)^2\}}{\exp\{-\tfrac{1}{2}N(|U_h| + |U_{h'}U_{h-h'}|)^2\}} = \exp(2N|U_hU_{h'}U_{h-h'}|). \tag{8.82}$$

Since we must have $P_+(h,h') + P_-(h,h') = 1$ we can find

$$P_+(h,h') = \tfrac{1}{2} + \tfrac{1}{2}\tanh(N|U_hU_{h'}U_{h-h'}|). \tag{8.83}$$

More accurate probability expressions, and ones which apply to structures with non-equal atoms, have been given by Cochran and Woolfson (1955), Bertaut (1955) and Klug (1958).

In table 8.9 there are shown some typical values of $P_+(h,h')$ for $N = 20$, 40 and 80 and for a range of values of $\bar{U} = (|U_hU_{h'}U_{h-h'}|)^{\frac{1}{3}}$.

The probabilities given by equation (8.83) tend to be somewhat lower than those given by the more precise theories although the expression does become more accurate as $N$ increases. The expression also gives $P_+(h,h') \neq 1$ in some situations where inequalities apply. However, in practice, one is not normally interested in the exact probabilities when one is using expression (8.68), the triple-product sign relationship (t.p.s.r.).

These relationships were first used by Zachariasen (1952) in solving the structure of metaboric acid. By inequality relationships he was able to find the signs of forty structure factors in terms of five sign symbols. Next Zachariasen extended the sign information and found relationships between the symbols by the use of

$$s(h) \approx s\left\{\sum_{h'} s(h')s(h-h')\right\} \tag{8.84}$$

where the summation was taken for all products of structure factors for

Table 8.9.

| $\bar{U}$ | $N = 20$ | 40 | 80 |
|-----|------|------|------|
| 0.0 | 0.500 | 0.500 | 0.500 |
| 0.1 | 0.510 | 0.520 | 0.540 |
| 0.2 | 0.579 | 0.654 | 0.782 |
| 0.3 | 0.757 | 0.896 | 0.987 |
| 0.4 | 0.928 | 0.994 | 1.000 |
| 0.5 | 0.993 | 1.000 | 1.000 |

which signs or sign symbols were known. This relationship is an approximation to the more accurate relationship

$$s(\mathbf{h}) \approx s\left\{\sum_{\mathbf{h'}} U_{\mathbf{h'}} U_{\mathbf{h-h'}}\right\} \tag{8.85}$$

which has an associated probability

$$P_+(\mathbf{h}) = \tfrac{1}{2} + \tfrac{1}{2}\tanh\left(N|U_{\mathbf{h}}|\sum_{\mathbf{h'}} U_{\mathbf{h'}} U_{\mathbf{h+h'}}\right) \tag{8.86}$$

where $P_+(\mathbf{h})$ is the probability that $U_{\mathbf{h}}$ is positive. It will be noticed that if the summation is negative then $P_+(\mathbf{h}) < \tfrac{1}{2}$.

In a typical application of equation (8.84) it might be found that the separate products in the summation were $a\,a\,a\,a\,ab\,ab\,ab$. This would be taken as an indication that $s(\mathbf{h}) = a$ and $b = +1$. If other indications gave $b = +1$ it could soon be accepted with some confidence.

As an example we consider the determination of signs for the $c$-axis projection of Roussin's red ethyl ester (Thomas, Robertson & Cox, 1958). The structure has two molecules of $(NO)_4Fe_2S_2(C_2H_5)_2$ per unit cell and has space group $P2_1/a$ with $a = 7.81$, $b = 12.67$, $c = 7.01$ Å and $\beta = 111° 24'$. The $c$-axis projection has the two-dimensional space group $pgg$ and the table of $|U|$'s is shown in fig. 8.36.

We should note that, for this space group, equivalent reflections in neighbouring quadrants have signs related by

$$s(h, \bar{k}) = (-1)^{h+k} s(h, k). \tag{8.87}$$

The two origin-fixing reflections are chosen by taking

$$s(7, 3) = s(5, 8) = +1.$$

The first few stages of sign determination by inequalities give:

| $\mathbf{h}$ | $\mathbf{h'}$ | $\mathbf{h + h'}$ | $\mathbf{h - h'}$ | Conclusion |
|---|---|---|---|---|
| $U_{7,3}$ $= 0.78$ | $U_{5,8}$ $= 0.58$ | $\lvert U_{12,11}\rvert$ $= ?$ | $\lvert U_{2,3}\rvert$ $= 0.24$ | $s(2, \bar{5}) = -s(2, 5)$ $= +1$ |
| $U_{7,1}$ $= 0.74a$ | $U_{7,\bar{1}}$ $= 0.74a$ | $\lvert U_{14,0}\rvert$ $= ?$ | $\lvert U_{0,2}\rvert$ $= 0.26$ | $s(0, 2) = +1$ |
| $U_{7,3}$ $= 0.78$ | $U_{0,2}$ $= 0.26$ | $\lvert U_{7,5}\rvert$ $= 0.57$ | $U_{7,1}$ $= 0.74a$ | $s(7, 5) = +1$ and $a = +1$ |
| $U_{5,8}$ $= 0.58$ | $\lvert U_{2,3}\rvert$ $= 0.27$ | $U_{7,5}$ $= 0.57$ | $\lvert U_{3,11}\rvert$ $= 0$ | $s(2, \bar{3}) = -s(2, 3)$ $= +1$ |

Inequalities can be applied, with the introduction of more sign symbols, until virtually all the larger $U$'s have their signs determined. In fig. 8.36 there is shown a stage where 23 signs are determined in terms of four symbols. One can easily go further and find explicitly what all the symbols represent but the process has been stopped here to demonstrate the application of the t.p.s.r.

All the signs and symbols in fig. 8.36 should be marked on a transparent

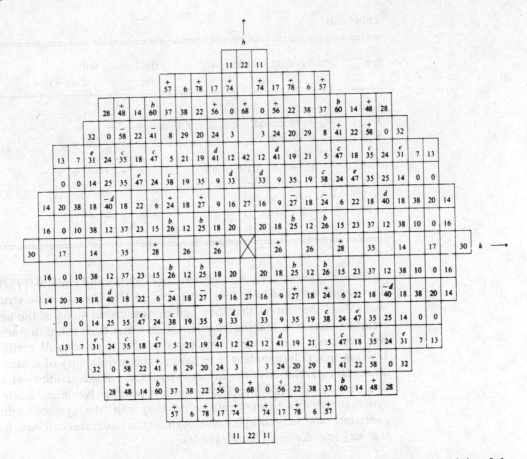

Fig. 8.36.
Values of $100|U_{hko}|$ for Roussin's red ethyl ester showing inequality-derived signs and symbols.

sheet so that they exactly fit over the ones in the figure. The origin of the transparent sheet is then moved in turn to lie over points in the figure whose signs have not been determined and for which $|U| \geqslant 0.35$. The overlap of symbols and signs then give probable sign indications for these new $U$'s. The results of doing this gives the independent indications shown in table 8.10.

An inspection of this table reveals that the almost certain values for the symbols are $b = +1$, $c = d = e = -1$ and when these are taken the final column of table 8.10 gives the sign indications for the new $U$'s. Some points of interest are:

(a) Sign relationships (as distinct from inequality relationships) do occasionally break down and three examples of breakdown can be seen.

(b) The sign of $U_{6,5}$ seems not to be given. If the magnitudes of the $U$'s are taken into account one finds from equation (8.86) that $P_+(6, 5) = 0.89$ which is a correct indication that the sign is positive. In calculating this probability it has been assumed that there are 32 *equal* atoms in the unit cell.

(c) Superimposing the chart on points for which signs or symbols have been found can give symbol relationships directly. For example, superimposing on (3, 5) with sign symbol $c$ gives indications $c \approx bc$, $bd$, $bd$, $be$, $c$, $c$, $c$, $d$, $d$, $e$, all of which confirm the indications in table 8.10.

Table 8.10.

| $h, k$ | Independent indications | Indications with $b = -c = -d = -e = 1$ |
|--------|-------------------------|------------------------------------------|
| 4, 0   | $-, -, -, bc, c, c, d, d$ | $- - - - - - - -$ |
| 3, 3   | $-, -, bc, bc, bd, c, c, c, d, d, d, d, d$ | $- - - - - - - - - - - - -$ |
| 6, 4   | $+, +, b, b, b, cd$ | $+ + + + + +$ |
| 6, 5   | $b, c, -d, e$ | $+ - + -$ |
| 0, 8   | $-, +, +, +, b, cd, ed, ed$ | $- + + + + + + +$ |
| 1, 8   | $-, -, -, -b, -b, c, d$ | $- - - - - - -$ |
| 3, 9   | $-b, -b$ | $- -$ |
| 1, 10  | $-, -b, c, d, e$ | $- - - - -$ |
| 2, 11  | $-, -, -, -, c, c, c, d$ | $- - - - - - - -$ |

In solving the structure of L-glutamine, Cochran and Penfold (1952) used a procedure very similar to that just described. However when the structure was solved they found that they had overestimated some of the unitary structure factors so that the inequalities, on which their solution seemed to rest, were not really valid. This suggests that a reasonable method of tackling sign determination is to assume the inviolability of some of the strongest sign relationships – in fact to treat them as inequalities – and then to use relationship (8.84) or (8.85) as soon as possible. Karle and Karle have systematized this process into what they call 'the symbolic-addition procedure' and very complex centrosymmetric structures can be solved in this way (e.g. Karle & Karle, 1964a).

## 8.8  General phase relationships

Sayre's equation (8.61) applies quite generally and for non-centrosymmetric structure factors can be written as

$$|F_{\mathbf{h}}| \exp(i\phi_{\mathbf{h}}) = \frac{\theta_{\mathbf{h}}}{V} \sum_{\mathbf{h'}} |F_{\mathbf{h'}}| |F_{\mathbf{h}-\mathbf{h'}}| \exp(i\phi_{\mathbf{h'}}) \exp(i\phi_{\mathbf{h}-\mathbf{h'}})$$

$$= \frac{\theta_{\mathbf{h}}}{V} \sum_{\mathbf{h'}} |F_{\mathbf{h'}}| |F_{\mathbf{h}-\mathbf{h'}}| \exp\{i(\phi_{\mathbf{h'}} + \phi_{\mathbf{h}-\mathbf{h'}})\}. \tag{8.88}$$

If equation (8.88) is written with $U$'s or $E$'s instead of $F$'s then the summation must be over an infinite number of terms because the data then corresponds to point atoms for which there is no scattering-factor fall-off. However when only a few terms (perhaps only one) are known on the right-hand side then the most probable value of $\phi_{\mathbf{h}}$, $\langle \phi_{\mathbf{h}} \rangle$, is that given by those terms. Thus with a single term involving reflections with indices $\mathbf{h'}$ and $\mathbf{h} - \mathbf{h'}$

$$\langle \phi_{\mathbf{h}} \rangle = \phi_{\mathbf{h'}} + \phi_{\mathbf{h}-\mathbf{h'}} \tag{8.89a}$$

or, in the form of a three-phase relationship,

$$\phi_{\mathbf{h}} - \phi_{\mathbf{h}'} - \phi_{\mathbf{h}-\mathbf{h}'} \approx 0 \text{ (modulo } 2\pi) \tag{8.89b}$$

which means that the combination of three phases has a distribution about 0 or some multiple of $2\pi$.

If there are a number of pairs of phases which, on the right-hand side of equation (8.89a), all contribute to an estimate of $\phi_{\mathbf{h}}$ then the probable value of $\tan \phi_{\mathbf{h}}$ may be found from

$$\langle \tan \phi_{\mathbf{h}} \rangle = \frac{\sum_{\mathbf{h}'} |U_{\mathbf{h}'}||U_{\mathbf{h}-\mathbf{h}'}|\sin(\phi_{\mathbf{h}'} + \phi_{\mathbf{h}-\mathbf{h}'})}{\sum_{\mathbf{h}'} |U_{\mathbf{h}'}||U_{\mathbf{h}-\mathbf{h}'}|\cos(\phi_{\mathbf{h}'} + \phi_{\mathbf{h}-\mathbf{h}'})}. \tag{8.90a}$$

Equation (8.90a) with $E$'s (see equation (7.44)) instead of $U$'s was first given by Karle and Hauptman (1956). This equation is referred to as the *tangent formula* and it has played a very important role in the development of direct methods of solving crystal structures. Another form of the tangent formula, which has a better theoretical basis, is

$$\langle \tan \phi_{\mathbf{h}} \rangle = \frac{\sum_{\mathbf{h}'} \kappa(\mathbf{h}, \mathbf{h}')\sin(\phi_{\mathbf{h}'} + \phi_{\mathbf{h}-\mathbf{h}'})}{\sum_{\mathbf{h}'} \kappa(\mathbf{h}, \mathbf{h}')\cos(\phi_{\mathbf{h}'} + \phi_{\mathbf{h}-\mathbf{h}'})} = \frac{T_{\mathbf{h}}}{B_{\mathbf{h}}} \tag{8.90b}$$

where

$$\kappa(\mathbf{h}, \mathbf{h}') = 2N^{-\frac{1}{2}}|E_{\mathbf{h}}E_{\mathbf{h}'}E_{\mathbf{h}-\mathbf{h}'}|. \tag{8.90c}$$

If the atoms in the structure are not all equal then $N$, the number of atoms in the unit cell, is replaced by a more complicated expression but normally its value will not be very different from $N$.

The actual value of $\phi_{\mathbf{h}}$ will in general be different from the most probable value given by equation (8.89a) but the larger the value of $\kappa(\mathbf{h}, \mathbf{h}')$ for that relationship the smaller the difference one is likely to get. The general form of the probability distribution of $\phi_{\mathbf{h}}$ has been given by Cochran (1955) and is shown in fig. 8.37. It is of the form

$$P(|\phi - \langle\phi\rangle|) = \frac{1}{2\pi I_0\{\kappa(\mathbf{h}, \mathbf{h}')\}} \exp\{\kappa(\mathbf{h}, \mathbf{h}')\cos(|\phi - \langle\phi\rangle|)\}. \tag{8.91}$$

$I_0$ is the modified Bessel function which also appears in equation (8.17). The larger is $\kappa(\mathbf{h}, \mathbf{h}')$ the smaller is the variance of the distribution and the more likely it is that the estimated value from equation (8.89) will be close to the true value. If equation (8.90) is used to estimate $\phi_{\mathbf{h}}$ then the same probability relationship is valid except that $\kappa(\mathbf{h}, \mathbf{h}')$ is replaced by $\alpha_{\mathbf{h}}$ where

$$\alpha_{\mathbf{h}}^2 = T_{\mathbf{h}}^2 + B_{\mathbf{h}}^2. \tag{8.92}$$

Karle and Karle (1964b) solved the structure of L-arginine dihydrate, the first non-centrosymmetric structure to be solved by direct methods. They used what they called the *symbolic addition* method in which unknown phases are represented by letter symbols, much as was done for signs in the

Fig. 8.37.
The Cochran
distribution.

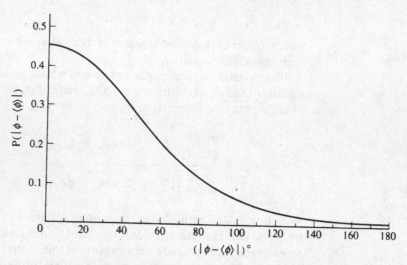

solution of L-glutamine. The so-called $E$-map, the density map obtained with $E$'s as coefficients, obtained by the Karles is shown in fig. 8.38 and clearly shows every atom. Subsequent to this important structure solution other procedures were developed which were amenable to implementation by computer and the vast bulk of small-to-medium-sized structures are now solved by direct-methods computer programs. One very simple program, RANTAN, was developed by Yao (1981). It is a so-called multi-solution method where several sets of phases are generated and are then tested by various figures of merit to see which are the most plausible. In RANTAN a set of reflections of large $|E|$ are given random phases and these are then subjected to an iterative refinement by the tangent formula. However, for RANTAN a weighted form of tangent formula was used

$$\langle \tan \phi_{\mathbf{h}} \rangle = \frac{\sum\limits_{\mathbf{h}'} w(\mathbf{h}')w(\mathbf{h}-\mathbf{h}')\kappa(\mathbf{h},\mathbf{h}')\sin(\phi_{\mathbf{h}'}+\phi_{\mathbf{h}-\mathbf{h}'})}{\sum\limits_{\mathbf{h}'} w(\mathbf{h}')w(\mathbf{h}-\mathbf{h}')\kappa(\mathbf{h},\mathbf{h}')\cos(\phi_{\mathbf{h}'}+\phi_{\mathbf{h}-\mathbf{h}'})} = \frac{T_{\mathbf{h}}^{\mathrm{W}}}{B_{\mathbf{h}}^{\mathrm{W}}} \qquad (8.93)$$

where $w(\mathbf{h})$ is proportional to $\{(T_{\mathbf{h}}^{\mathrm{W}})^2 + (B_{\mathbf{h}}^{\mathrm{W}})^2\}^{\frac{1}{2}}$. For each reflection in turn a new phase estimate is found from the weighted tangent formula using current phase estimates for all other reflections. A new weight is also found but the phase is not changed to the new value unless the new weight is bigger than the previous weight. An initial system of weights is set up; the initial weight for each general reflection is usually 0.25 but other reflections, which define the origin for example, may be given initial weights of 1.0. After several cycles the phases converge to stable values and one phase-set has been obtained. Different initial random phase sets give different final phase-sets and, depending on the size of the structure, anything from 50 to 1000 different sets of phases will be generated. Because of the way the phases are determined the phases of each phase-set give reasonable satisfaction of the three-phase relationships (8.89b).

Many figures of merit have been suggested from time to time but they all depend on how well certain expected conditions are satisfied. They tend to

Fig. 8.38.
The a-axis projection of
a three-dimensional
electron-density map for
L-arginine dihydrate with
400 phases determined
by phase relationships
(from Karle & Karle,
1964b).

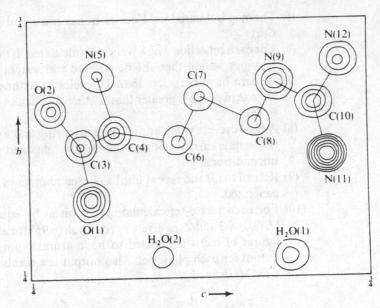

be rather complicated but a very simple figure of merit, not now used, is

$$A = \sum_{h} \alpha_{h} \qquad (8.94)$$

where the **h**'s are for all large $|E_{h}|$ and all large $|E|$'s are used in evaluating the $\alpha$'s – which is just a measure of how well the three-phase relationships are satisfied. Another simple figure of merit is

$$Z = \sum_{k} \alpha_{k} \qquad (8.95)$$

where, in this case, the values of $|E_{k}|$ are small but the large $|E|$'s go into the summations which give the $\alpha$'s. This figure of merit can best be understood in terms of Sayre's equation (8.61) which can also be written in terms of $E$'s. The values of $\alpha$ are just proportional to the magnitude of the right-hand side of equation (8.61) so this figure of merit is simply saying that if $|E|$ is small for normal density then it is expected to be small for squared density.

We can illustrate the application of direct methods with MINDIR (appendix XI), a simple two-dimensional form of RANTAN which can be applied to structures with any two-dimensional space group. From Sayre's equation it is evident that the phase obtained from equation (8.93) is just that from squaring the density of a map with coefficients $w(\mathbf{h})E(\mathbf{h})$. The steps in MINDIR are:

(1) A file is read in giving $h$, $k$ and $|E(h,k)|$ values in a form given by the output of FTOUE.
(2) A subset of large $|E|$'s is found for which $|E| \geq 1.0$.
(3) Random phases are generated for the subset of reflections and weights of 0.25 are allocated.
(4) A map is calculated with magnitudes of coefficients $w(\mathbf{h})|E(\mathbf{h})|$ and current phase estimates.

(5) The map is squared and Fourier-transformed to give Fourier coefficients $G(\mathbf{h})$.

(6) For each reflection a new weight is calculated. If this is greater than the previous weight then the new phase and weight are accepted.

(7) If there have been less than 16 cycles of refinement or if the mean change in phase is greater than 2° then go back to (4). Otherwise go on to (8).

(8) A final cycle of calculating a map, squaring and Fourier-transforming the map is carried out and the final weights and phases are accepted unconditionally.

(9) Return to (3) and repeat until a chosen number of phase-sets has been generated.

(10) For each phase-set calculate $A$, as given by equation (8.94), for all $|E|$'s $\geq 1.3$ and $Z$, as given by equation (8.95), for all $|E|$'s $\leq 0.5$. These figures of merit are scaled to have a maximum value of 1000 and output for each phase-set. Also output is a combined figure of merit, $CFOM$, defined as

$$CFOM = \frac{A - A_{min}}{A_{max} - A_{min}} + \frac{Z_{max} - Z}{Z_{max} - Z_{min}}. \tag{8.96}$$

The subscripts refer to the maximum and minimum values of $A$ and $Z$ for all the phase-sets generated and $CFOM$ would have a value of 2.0 if it had the greatest value of $A$ and the least value of $Z$ and a value of zero if it had the least value of $A$ and the greatest value of $Z$.

(11) MINDIR automatically outputs the map for the largest value of $CFOM$. If this does not reveal the structure then MINDIR can be run again using a switch which simply finds any other designated phase-set from a file and then computes and outputs the map without re-running the complete program.

Structure factors were calculated with STRUCFAC for the artificial structure, with space group *pg*, shown earlier in fig. 8.17 and the output file from this was used with FTOUE to give a set of $|E|$'s. This was used as input for MINDIR which generated 50 phase-sets and gave the list of figures of merit shown in table 8.11. Solution 17 had the highest value of $CFOM$, 1.330, followed by solution 43 with a value 1.321. The program automatically produced a map for solution 17 which is shown in fig. 8.39. The following points will be noted. The sodium atoms appear clearly but overweight compared with other peaks in the map. This is a common feature of solutions produced with the tangent formula because, as we have seen with MINDIR, the application of the tangent formula is tantamount to squaring the electron-density, which tends to increase density differences and makes heavier atoms appear stronger than they should be. The correct positions of the light atoms all fall on positive density although two do not fall within contoured regions; in addition there is a considerable amount of spurious density. All these features are seen in the results from the large and complicated direct-methods computer packages which are commonly used, but to a lesser extent. Finding new phases in series, one at a time, by the tangent formula, rather than by squaring the map and producing large

Table 8.11.  *Figures of merit from MINDIR for the structure* $C_3H_5N_2ONa_2$

The values of $A$ and $Z$ are as given in equations (8.94) and (8.95). Solution 17 has the highest combined figure of merit, *CFOM*.

THE SEED USED IS   9000

| SOLN | A | Z | CFOM | SOLN | A | Z | CFOM |
|---|---|---|---|---|---|---|---|
| 1 | 997.3 | 948.1 | 1.051 | 2 | 311.0 | 198.3 | 1.122 |
| 3 | 970.2 | 872.8 | 1.102 | 4 | 35.3 | 41.5 | 1.000 |
| 5 | 963.7 | 889.4 | 1.078 | 6 | 925.0 | 817.3 | 1.113 |
| 7 | 488.9 | 369.9 | 1.128 | 8 | 300.5 | 177.5 | 1.133 |
| 9 | 1000.0 | 1000.0 | 1.000 | 10 | 991.4 | 971.7 | 1.021 |
| 11 | 258.6 | 152.6 | 1.116 | 12 | 926.5 | 716.8 | 1.219 |
| 13 | 983.7 | 942.1 | 1.043 | 14 | 991.2 | 964.7 | 1.028 |
| 15 | 889.4 | 656.2 | 1.244 | 16 | 516.0 | 336.2 | 1.191 |
| 17 | 628.0 | 314.1 | 1.330 | 18 | 766.5 | 517.3 | 1.262 |
| 19 | 980.7 | 940.8 | 1.042 | 20 | 996.0 | 972.5 | 1.025 |
| 21 | 399.9 | 264.4 | 1.145 | 22 | 988.2 | 955.9 | 1.034 |
| 23 | 263.8 | 238.9 | 1.031 | 24 | 990.9 | 966.5 | 1.025 |
| 25 | 707.6 | 426.7 | 1.295 | 26 | 239.5 | 112.9 | 1.137 |
| 27 | 998.5 | 941.0 | 1.060 | 28 | 404.6 | 152.5 | 1.267 |
| 29 | 996.1 | 984.1 | 1.013 | 30 | 264.5 | 268.7 | 1.001 |
| 31 | 928.6 | 761.7 | 1.175 | 32 | 41.4 | 44.7 | 1.003 |
| 33 | 996.2 | 985.3 | 1.011 | 34 | 80.1 | 64.0 | 1.023 |
| 35 | 734.0 | 725.5 | 1.011 | 36 | 555.3 | 410.1 | 1.155 |
| 37 | 426.8 | 400.9 | 1.031 | 38 | 243.6 | 136.4 | 1.117 |
| 39 | 404.7 | 257.2 | 1.158 | 40 | 408.7 | 285.6 | 1.132 |
| 41 | 950.2 | 854.0 | 1.101 | 42 | 986.4 | 928.6 | 1.060 |
| 43 | 705.8 | 399.6 | 1.321 | 44 | 730.3 | 479.9 | 1.263 |
| 45 | 925.7 | 785.3 | 1.147 | 46 | 931.0 | 796.9 | 1.140 |
| 47 | 499.4 | 349.8 | 1.159 | 48 | 851.7 | 648.5 | 1.213 |
| 49 | 952.6 | 847.0 | 1.111 | 50 | 277.3 | 339.3 | 0.940 |

numbers of new phases in parallel, dampens the squaring effect. Again, in three dimensions the ratio of the number of data employed to the number of atomic positions to be determined is much greater than in two dimensions and gives better resolution of the structure; it should be noted that 400 reflections in three dimensions have gone into the $E$-map given in fig. 8.38. Nevertheless the general principles of direct methods are well illustrated by MINDIR.

A characteristic of direct methods when applied to non-centrosymmetric structures, as with almost all other methods, is that the structure obtained may be the true structure or its enantiomorph. If, for example, the positions of all atoms are changed from $(x, y)$ to $(\bar{x}, \bar{y})$ then the magnitudes of the structure factors are unchanged. For that reason, when solving an unknown non-centrosymmetric crystal structure when only the experimental $|F|$'s are available, either form could be found by the structure-solving method. Organic molecules very often occur in either a right-handed or left-handed form represented by the prefixes D (dextro) or L (laevo). The two forms are related to each other as a left hand is to a right hand and two molecular structures of different chirality may be arranged so as to be

Fig. 8.39.
The direct-methods solution of $C_3H_5N_2ONa_2$ from the program MINDIR. (Solution 17 from table 8.11 with $A = 628.0$, $Z = 314.1$ and $CFOM = 1.330$.)

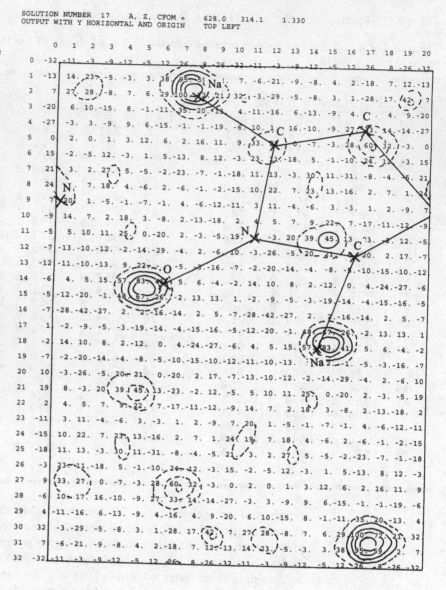

SOLUTION NUMBER 17   A, Z, CFOM =   628.0   314.1   1.330
OUTPUT WITH Y HORIZONTAL AND ORIGIN   TOP LEFT

related by either a centre of symmetry or a mirror plane. In fig. 8.40 two enantiomorphic structures are illustrated; in this case the molecules in the two enantiomorphs are related by a mirror plane. The solution shown in fig. 8.39 has a change both of origin and enantiomorph compared to the structure shown in fig. 8.17.

The exception to the rule that either enantiomorph may be found in the structure solution is when anomalous scattering is used. If for one enantiomorph $|F(\mathbf{h})| > |F(\mathbf{\bar{h}})|$ then for the other $|F(\mathbf{h})| < |F(\mathbf{\bar{h}})|$ and so the solution may be found for the correct form of the crystal.

Fig. 8.40.
An illustration of
enantiomorphic
structures in two
dimensions.

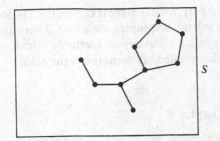

S

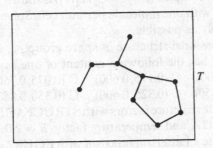

T

Fig. 8.40.
An illustration of
enantiomorphic
structures in two
dimensions.

## 8.9   A general survey of methods

The range of methods available to the crystallographer represents a formidable armoury, even for tackling structures of very great complexity. The examples given in this chapter have tended to be organic, or pseudo-organic, structures. This is partly because the majority of structures solved are of this type and also because, as molecular structures, they are clearer for demonstration purposes. On the other hand, metallic alloys and minerals are often three-dimensional framework-like structures and frequently they involve arrangements of atoms in special positions.

The methods which have had the greatest impact on structural crystallography are the heavy-atom method and that of isomorphous replacement which, applied to proteins and other macromolecular structures, have led to tremendous advances in the subjects of biochemistry and biophysics. However the widescale availability of synchrotron radiation is changing the pattern in macromolecular crystallography so that the use of anomalous scattering should increase, perhaps to the point where it may become the method of choice. Many native proteins contain atoms suitable for anomalous-scattering experiments, for example, iron, calcium or even sulphur, and anomalous differences can be relied upon to the resolution limits of the data – unlike isomorphous replacement where the addition of a heavy-atom-containing group somewhat distorts the structure so that isomorphism breaks down at higher resolutions.

For most small-to-medium structures – up to 150 or so non-hydrogen atoms in the asymmetric unit – direct methods are usually used because they are so automatic. Indeed, where high resolution ($\sim 1$ Å) data of good quality are available it has been shown that small proteins with up to 500 or so atoms in the asymmetric unit can also be solved by direct methods. The oldest of the systematic approaches, the Patterson method, still has an important role to play, for example it can be used to locate heavy atoms as the first stage of solving a complex structure. There are also available very powerful image-seeking Patterson methods which often succeed when direct methods have failed, as sometimes they do.

## Problems to Chapter 8

8.1 The $c$-axis Patterson projection of the structure of cuprous chloride azo-methane complex $(Cu_2Cl_2C_2H_2N_2)$ is shown in fig. 8.41. The space group is $P\bar{1}$ with one molecule per unit cell. Solve the crystal structure as completely as possible.

8.2 A two-dimensional structure of space group $p2$, with $a = 10$, $b = 5\,\text{Å}$ and $\gamma = 98°$ has the following content of one asymmetric unit:
S (0.150, 0.300),   C (0.175, 0.600),   C (0.075, 0.750),
C (0.375, 0.350),   C (0.325, 0.600),   C (0.350, 0.850).
 (i) Calculate structure factors with STRUCFAC for X-ray wavelength $\lambda = 1.542\,\text{Å}$ and temperature factor $B = 2.0\,\text{Å}^{-2}$.
 (ii) Calculate a Patterson function with FOUR2 and find the position of the S atom.
 (iii) Calculate the contribution of the S atoms with STRUCFAC.
 (iv) Use the program HEAVY to produce a data file, SF.DAT with the phases given by S and structure amplitudes with centrosymmetric heavy-atom weights.
 (v) Calculate a Fourier synthesis with program FOUR2 and data file SF.DAT to find the remaining atoms.

8.3 A structure which is isomorphous with that described in Problem 8.2 has sulphur replaced by oxygen.
 (i) Calculate structure factors for this crystal, again for $\lambda = 1.542\,\text{Å}$ and temperature factor $B = 2.0\,\text{Å}^{-2}$, putting the result in data file SF2.DAT.
 (ii) Run program ISOFILE to give coefficients for a difference Patterson in an output file SF.DAT.
 (iii) Run FOUR2 with SF.DAT to give a difference Patterson and locate the $i$-$r$ atom.

8.4 A two-dimensional structure, space group $p1$ with $a = 8$, $b = 5\,\text{Å}$ and $\gamma = 95°$, has atoms with the following coordinates:
S (0.563, 0.300),   S (0.375, 0.200),   S (0.125, 0.600),
C (0.313, 0.500),   C (0.469, 0.750),   C (0.656, 0.650),
C (0.813, 0.800),   C (0.875, 0.500).
An isomorphous compound exists with oxygen replacing sulphur.
 (i) Calculate the structure factors for the sulphur compound for $\lambda = 1.542\,\text{Å}$ with temperature factor $B = 2.0\,\text{Å}^{-2}$ using STRUC-FAC and place the results in file SF1.DAT.
 (ii) Similarly calculate structure factors for the oxygen compound and place in file SF2.DAT.
 (iii) Use ISOFILE to find the coefficients of the difference Patterson and place them in data file SF.DAT.
 (iv) Use FOUR2 to calculate the difference Patterson. Confirm that

Fig. 8.41.
The *c*-axis Patterson
projection of
$Cu_2Cl_2C_2H_2N_2$ (see
Problem 8.1).

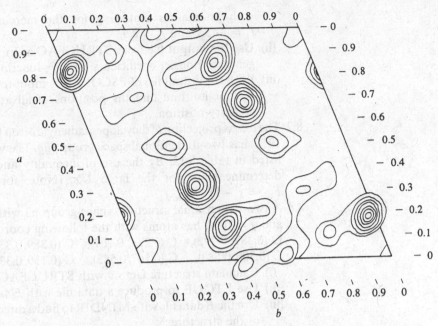

this can be interpreted as vectors from the three *i-r* atoms.

(v) Calculate structure factors for the sulphur atoms alone and place in data file SFIRA.DAT.

(vi) Calculate structure factors for the oxygen atoms alone and place in data file SFIRB.DAT.

(vii) Use the program ISOCOEFF with the data files SF1.DAT, SF2.DAT, SFIRA.DAT and SFIRB.DAT to generate the Fourier coefficients of a map which should show the carbon atoms.

(viii) Use FOUR2 with the output file from ISOCOEFF and find the carbon positions.

8.5 Data is taken for the structure described in Problem 8.4 with the modification that the second sulphur atom is replaced by carbon. Fe $K\alpha_1$ radiation ($\lambda = 1.936$ Å) is used giving anomalous scattering from the remaining two sulphur atoms with $f_1 = 0.385$ and $f_2 = 0.846$.

(i) Calculate structure factors with STRUCFAC including the anomalous contributions from sulphur.

(ii) Use the data file from STRUCFAC as the input file to ANOFILE to generate anomalous difference coefficients.

(iii) Use the output file from ANOFILE as input to FOUR2 to generate an anomalous-difference Patterson map.

(iv) Interpret the map to show that it is consistent with the peak between the sulphur atoms.

8.6 A two-dimensional structure, space group *p*1 with $a = 9.0$, $b = 6.0$ Å and $\gamma = 98°$, has atoms with the following coordinates:
Hg (0.333, 0.167),   O (0.167, 0.250),   O (0.500, 0.292),
O (0.778, 0.500),   N (0.472, 0.833),   C (0.111, 0.500),
C (0.500, 0.500),   C (0.389, 0.667),   C (0.639, 0.625),
C (0.611, 0.833).

(i) Calculate structure factors for $\lambda = 1.542$ Å using STRUCFAC with anomalous scattering for the mercury atom for which $f_1 = -5.1$ and $f_2 = 8.9$.

(ii) Use the output file from STRUCFAC as input to PSCOEFF to generate Fourier coefficients for a $P_s$-function map.

(iii) With the output file of PSCOEFF as input to FOUR2 calculate a $P_s$ function and find the positions of all atoms relative to the mercury position.

8.7 The *a*-axis projection of dicyclopentadienyldi-iron tetracarbonyl (Mills, 1958) has two-dimensional space group *pgg*. The values of $100|U|$ are listed in table 8.12. By the use of inequality and sign relationships determine signs for the large $U$'s. (Note: for this space group $F_{0\bar{k}l} = (-1)^{k+l}F_{0kl}$.)

8.8 A two-dimensional structure, space group *p*2 with $a = 9.0$, $b = 6.0\,\text{Å}$ and $\gamma = 100°$, has atoms with the following coordinates:

O (0.389, 0.792), C (0.278, 0.167), C (0.389, 0.333),
C (0.306, 0.583), C (0.167, 0.583), C (0.139, 0.333).

(i) Calculate structure factors with STRUCFAC.

(ii) Use FTOUE to produce a data file with $E$'s.

(iii) Use the $E$ data file with MINDIR to find a direct-methods solution of the structure.

Table 8.12.

| k | l | 100\|U\| | k | l | 100\|U\| | k | l | 100\|U\| | k | l | 100\|U\| |
|---|---|---|---|---|---|---|---|---|---|---|---|
| 2 | 0 | 43 | 4 | 2 | 13 | 0 | 4 | 40 | 11 | 5 | 11 |
| 4 | 0 | 6 | 5 | 2 | 32 | 1 | 4 | 17 | 0 | 6 | 13 |
| 6 | 0 | 33 | 6 | 2 | 6 | 2 | 4 | 42 | 1 | 6 | 8 |
| 8 | 0 | 40 | 7 | 2 | 17 | 3 | 4 | 2 | 2 | 6 | 10 |
| 10 | 0 | 33 | 8 | 2 | 13 | 4 | 4 | 5 | 3 | 6 | 29 |
| 12 | 0 | 16 | 9 | 2 | 16 | 5 | 4 | 11 | 4 | 6 | 0 |
| 1 | 1 | 33 | 10 | 2 | 6 | 6 | 4 | 28 | 5 | 6 | 26 |
| 2 | 1 | 36 | 11 | 2 | 33 | 7 | 4 | 5 | 6 | 6 | 0 |
| 3 | 1 | 1 | 12 | 2 | 0 | 8 | 4 | 40 | 7 | 6 | 0 |
| 4 | 1 | 6 | 13 | 2 | 21 | 9 | 4 | 11 | 8 | 6 | 7 |
| 5 | 1 | 9 | 1 | 3 | 36 | 10 | 4 | 24 | 9 | 6 | 21 |
| 6 | 1 | 10 | 2 | 3 | 34 | 11 | 4 | 15 | 1 | 7 | 100 |
| 7 | 1 | 44 | 3 | 3 | 20 | 12 | 4 | 0 | 2 | 7 | 24 |
| 8 | 1 | 6 | 4 | 3 | 30 | 1 | 5 | 17 | 3 | 7 | 13 |
| 9 | 1 | 32 | 5 | 3 | 31 | 2 | 5 | 12 | 4 | 7 | 31 |
| 10 | 1 | 21 | 6 | 3 | 20 | 3 | 5 | 13 | 5 | 7 | 4 |
| 11 | 1 | 16 | 7 | 3 | 37 | 4 | 5 | 31 | 6 | 7 | 9 |
| 12 | 1 | 28 | 8 | 3 | 11 | 5 | 5 | 0 | 7 | 7 | 16 |
| 13 | 1 | 14 | 9 | 3 | 27 | 6 | 5 | 12 | 0 | 8 | 38 |
| 0 | 2 | 18 | 10 | 3 | 21 | 7 | 5 | 20 | 1 | 8 | 0 |
| 1 | 2 | 15 | 11 | 3 | 6 | 8 | 5 | 0 | 2 | 8 | 28 |
| 2 | 2 | 27 | 12 | 3 | 21 | 9 | 5 | 36 | 3 | 8 | 0 |
| 3 | 2 | 22 | 13 | 3 | 0 | 10 | 5 | 12 | 4 | 8 | 0 |

# 9 Accuracy and refinement processes

## 9.1 The determination of unit-cell parameters

When a crystal structure is solved and refined the solution appears as a set of fractional coordinates from which can be determined bond lengths and angles, van der Waals distances, etc. However the accuracy with which these quantities can be determined will depend not only on the accuracy of the atomic coordinates but also on the accuracy of determination of the unit-cell parameters.

By the measurement of layer-line spacings or from Weissenberg photographs one can usually measure cell edges to about 1% and angles with an error of about $\frac{1}{2}°$. The order of accuracy of cell dimension required to match that of coordinate determination is about one part in a thousand or perhaps a little better. This would correspond to less than 0.002 Å in a bond of length 1.500 Å and rarely is this order of accuracy really required.

For some other purposes more accurate unit-cell parameters may be required – for example for measurement of thermal expansion coefficients of crystalline materials or for investigating small changes in cell parameters with changes of composition of the material.

There has been a great deal of work in this field and it would be difficult to mention it all. What will be done is to select an example of each of the main types of method to illustrate the ranges of techniques and accuracy which are available.

The basic idea behind all the methods is to measure the Bragg angle for a number of reflections. This is related to the reciprocal-lattice constants as follows. From equation (3.29) we have

$$s^2 = \frac{4\sin^2\theta}{\lambda^2} = h^2 a^{*2} + k^2 b^{*2} + l^2 c^{*2} + 2hka^*b^*\cos\gamma^*$$
$$+ 2hla^*c^*\cos\beta^* + 2klb^*c^*\cos\alpha^*. \tag{9.1}$$

Equation (9.1) takes on special forms for the various crystal systems. Once the reciprocal-space quantities have been found they can be transformed into crystal-space quantities by the relationships in table 3.1.

In a method proposed by Farquhar and Lipson (1946) a single-crystal photograph is taken with a camera arranged as in fig. 9.1(a). The collimator passes through a hole in the film so that reflections for $\theta$ close to 90° are recorded near the centre of the film – not at the edge as is usually the situation. By oscillating an orthorhombic crystal about an $a$ axis in such a way that the centre of the oscillation range has the $c$ axis pointing at the collimator one will record equivalent reflections $(0kl)$ and $(0\bar{k}l)$ on the zero

Fig. 9.1.
(a) The arrangement of the film in the Farquhar and Lipson camera.
(b) A symmetrical oscillation diagram for the Farquhar and Lipson technique.
(c) The appearance of the zero layer of the film. The collimator passes through the central hole and high-angle reflections are near the centre of the film.

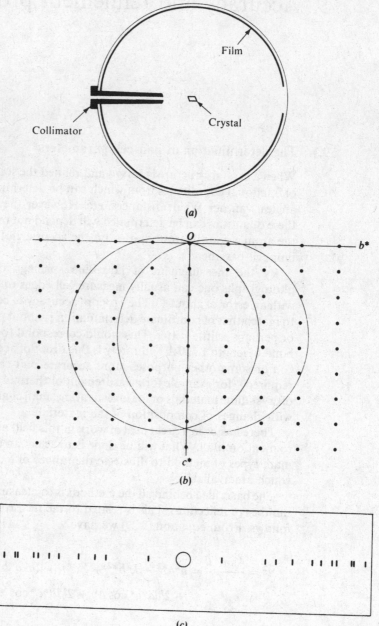

layer symmetrically disposed about the centre of the film. The indexing chart for a photograph produced in this way is shown in fig. 9.1(b) and the zero layer on the straightened-out film is represented in fig. 9.1(c). The value of $\theta$ for a particular reflection is found by measuring the distance $x$ between equivalent reflections. If the camera radius is $R$ then it is clear from fig. 9.1(a) that

$$\theta = \frac{\pi}{2} - \frac{x}{4R}. \tag{9.2}$$

The values of $\theta$ are measured for a number of high-angle reflections. The advantages of this are:

(i) Since $x$ will be comparatively small the effect of film shrinkage is minimized. However the effect of errors in $x$ or $R$ can be seen to diminish and to tend to zero as $\theta \to \pi/2$. This provides the basis of an extrapolation technique to eliminate errors due to this cause.

(ii) At high angles the $\alpha_1 - \alpha_2$ doublet is resolved and the position of the reflection can be measured more precisely. At lower angles the lack of resolution leads to a fuzziness of the spot on the film.

(iii) As will be seen from fig. 9.1(b) the high-angle reflections will be those for large values of $l$ and low values of $k$. They are therefore very suitable for precisely determining the value of $c$ since a small error in $b$ will not matter very much.

(iv) The quantity which appears in equation (9.1) is $\sin \theta$, not $\theta$, and for a given error $d\theta$ the error in $\sin \theta$ reduces as $\theta \to 90°$. This can be seen from

$$d(\sin \theta) = \cos \theta d\theta. \tag{9.3}$$

Farquhar and Lipson illustrated the method by determining the axial lengths for thallium hydrogen tartrate. Optical goniometry (Groth, 1910) gave the axial ratios as

$$a : b : c = 0.6976 : 1 : 0.7275.$$

From equation (9.1), for the $0kl$ reflections of an orthorhombic crystal, one finds

$$c = \frac{1}{2}\lambda \sqrt{\frac{c^2}{b^2}k^2 + l^2} \, \text{cosec} \, \theta. \tag{9.4}$$

With the value of $c/b$ given by Groth the values of $\theta$ were measured for a number of reflections and the values of $c$ calculated from equation (9.4) were plotted as a function of $\sin^2 \theta$. The graph for this is shown in fig. 9.2(a) and the scatter of the points was taken as an indication of an incorrect ratio $c/b$. For a ratio $c/b = 0.7197_5$ however the graph in fig. 9.2(b) was obtained and extrapolation to $\theta = 90°$ gave the value $c = 7.9099 \pm 0.0003$ kX. The unit kX is given by

$$1 \, \text{kX} = 1.00202 \, \text{Å}$$

and was used before X-ray wavelengths had been precisely related to the centimetre.

The $a$ and $b$ axes were similarly determined and Farquhar and Lipson discussed the extension of the method to the measurement of unit-cell parameters for monoclinic and triclinic crystals. The overall accuracy of the method is about 1 part in 20 000 if care is taken in measuring the quantities $x$ but even if the $x$'s are measured more roughly, with a steel scale for example, an accuracy of 1 part in 500 is still attainable.

A new principle for measuring unit-cell parameters was introduced by

Fig. 9.2.
(*a*) A Farquhar and
Lipson graph for an
incorrect axial ratio.
(*b*) The improvement of
the extrapolation when
the axial ratio is
corrected (from
Farquhar & Lipson,
1946).

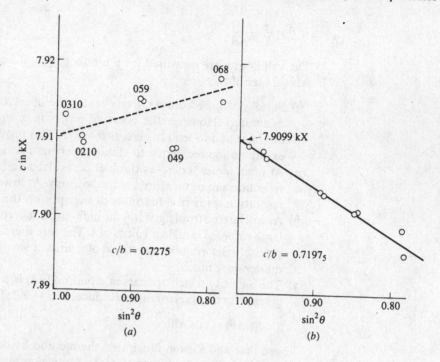

Weiss *et al.* (1948). This involved the measurement of the angle between two
positions of a crystal when it gives the same reflection on opposite sides of
the incident beam. In fig. 9.3 the normal to the reflecting planes is
perpendicular to the rotation axis and, for the two positions in which the
reflection occurs, the normals are along $ON$ and $ON'$. It should be clear
from the figure that the angle turned through by the crystal, $NON'$, equals
$2\theta$. The technique used by Weiss *et al.* was first to locate the crystal position
for the desired reflection to within 5° by taking 5° oscillation pictures. The
crystal was then moved by steps of $\frac{1}{2}$° within this range to locate the position
even more closely and finally intensities for a fixed exposure were recorded
at steps of 3′ with a very finely collimated beam of X-rays to find the
position of maximum reflection. For crystals which reflect over 15′ of arc,
due to the finite size of the crystal and its mosaic structure, the position of
maximum reflection could be found to within ±1′.

By measurement of ($h00$), ($0k0$) and ($00l$) reflections the values of $a^*$, $b^*$
and $c^*$ could be found directly. In fig. 9.4 the position of the reflecting planes
is shown for the two positions of a ($h00$) reflection. It should be clear from
this figure that when the crystal is half-way between these two positions the
$a^*$ axis lies along the direction of the incident beam. Thus if the crystal is
being rotated about the $c$ axis and ($h00$) and ($0k0$) reflections are recorded
then it is possible to deduce $\gamma^*$, the angle between the $a^*$ and $b^*$ axes. One is
thus able to measure all the reciprocal-cell parameters and thereby deduce
the real-cell parameters. As used by Weiss *et al.* with angles measured on the
goniometer circle of the oscillation camera the method gave an accuracy of
about one part in $10^4$. This method was also used by Bond (1960) who used
a counter to record the diffracted beam. The apparatus was accurately
made and a finely collimated incident beam of X-rays was used. Corrections

Fig. 9.3.
The two reflecting positions with normals along *ON* and *ON'*.

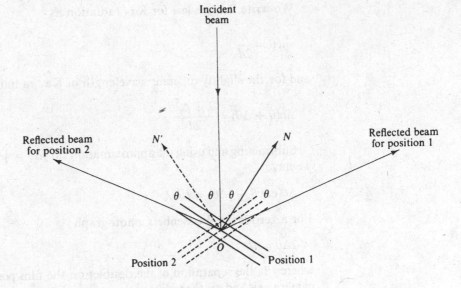

Fig. 9.4.
Location of the *a\** axis from the two positions which give an (*h*00) reflection.

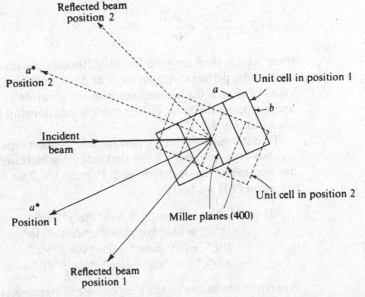

were made for the various errors that could occur and the temperature of the crystal was kept constant. Bond claimed an accuracy of about one part in $10^6$. An automated instrument on similar principles has been built by Baker *et al.* (1966) who have measured lattice parameter changes due to radiation damage and also coefficients of thermal expansion. The accuracy claimed is one part in $10^7$.

A method proposed by Main and Woolfson (1963) is useful in that it uses information from Weissenberg photographs, which are usually available anyway, and gives an accuracy of better than 1 part in $10^3$. The method rests on the fact that $\sin \theta$ can be determined for high-angle reflections from the separation of $\alpha_1 - \alpha_2$ doublets.

We write Bragg's law for $K\alpha_2$ radiation as

$$\sin\theta = \frac{\lambda}{2d} \tag{9.5}$$

and for the slightly different wavelength of $K\alpha_1$ radiation

$$\sin(\theta + \Delta\theta) = \frac{\lambda + \Delta\lambda}{2d}. \tag{9.6}$$

Subtracting and using the approximation $\sin\frac{1}{2}\Delta\theta = \frac{1}{2}\Delta\theta$ and $\cos\frac{1}{2}\Delta\theta = 1$ we have

$$\Delta\theta(\cot\theta - \tfrac{1}{2}\Delta\theta) = \Delta\lambda/\lambda. \tag{9.7}$$

For a zero-layer Weissenberg photograph

$$2\Delta\theta = t/r \tag{9.8}$$

where $t$ is the separation of the doublet on the film perpendicular to the camera axis and $r$ is the radius of the camera. From equations (9.7) and (9.8) we find

$$\cot\theta = \frac{2r}{t}\frac{\Delta\lambda}{\lambda} + \frac{t}{4r} \tag{9.9}$$

from which $\sin\theta$ may be found. Although $t$ cannot be measured very accurately, perhaps with an error of 3% or so for a 1 mm separation, the value of $\sin\theta$ is determined much more accurately. For Cu $K\alpha$ radiation and a camera diameter of 57.3 mm the relationship between $\sin\theta$ and $t$ is shown in fig. 9.5.

To determine the lattice parameters, doublet separations are measured on zero-layer photographs so that more doublets are measured than there are parameters to be determined. Putting $s = 2\sin\theta/\lambda$ and differentiating equation (9.1) we find

$$\begin{aligned}
s\,\mathrm{d}s = {} & (h^2a^* + hlc^*\cos\beta^* + hkb^*\cos\gamma^*)\mathrm{d}a^* \\
& + (k^2b^* + kha^*\cos\gamma^* + klc^*\cos\alpha^*)\mathrm{d}b^* \\
& + (l^2c^* + lkb^*\cos\alpha^* + lha^*\cos\beta^*)\mathrm{d}c^* \\
& - klb^*c^*\sin\alpha^*\mathrm{d}\alpha^* - lhc^*a^*\sin\beta^*\mathrm{d}\beta^* - hka^*b^*\sin\gamma^*\mathrm{d}\gamma^*. 
\end{aligned} \tag{9.10}$$

Approximate values of the reciprocal-cell parameters will be known and they can be used to evaluate the coefficients on the right-hand side. On the left-hand side one makes $s$ equal to the measured value and $\mathrm{d}s = s - s_c$ where $s_c$ is obtained from equation (9.1) by using the approximate lattice parameters. This process yields a set of linear equations in the quantities $\mathrm{d}a^*$, $\mathrm{d}b^*$, $\mathrm{d}c^*$, $\mathrm{d}\alpha^*$, $\mathrm{d}\beta^*$ and $\mathrm{d}\gamma^*$. Since there are more equations than unknowns one can find a least-squares solution (see §9.4). The quantities $a^* + \mathrm{d}a^*$, etc., are better approximations to the reciprocal-lattice parameters and can be used for the next cycle of refinement.

The measurements on the films can be made very quickly with a travelling microscope and the parameter-refinement process is carried out by a standard computer program.

Fig. 9.5.
The relationship
between sin θ and the
separation of the $\alpha_1 - \alpha_2$
reflections in a
Weissenberg
photograph.

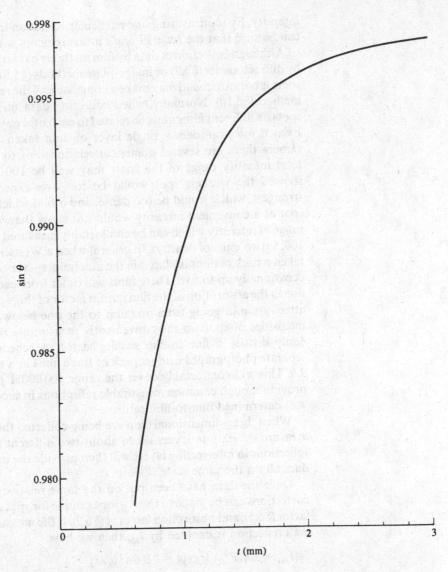

An extension of this technique so that measurements on higher-layer photographs can be utilized has been given by Alcock and Sheldrick (1967).

The methods mentioned in this section are only a small selection of those available. A number of papers from a conference on 'The Precision Determination of Lattice Parameters' at Stockholm in 1959 will be found in *Acta Cryst.* (1960) **13**, 818–50.

## 9.2   The scaling of observed data

The first problem of the crystallographer in collecting diffraction data is to make sure that it is all on the same relative scale. With diffractometers, for example, the running conditions of the X-ray tube can be strictly stabilized and the number of counts in a given time will then be a measure of the intensity. By re-measuring one particular reflection from time to time one can be sure that the scale of one's measurements is remaining steady.

Although collection of data by film methods has largely been superseded by diffractometer (§ 5.7) or image-plate methods (§ 5.9) they are still used in some laboratories and may make a comeback by the resurgence of the Laue method (§ 5.10). Normally, when collecting data on film, the diffraction spots on different films must be related to each other by empirical processes. Even if one considers a single layer of data taken with a Weissenberg camera there are several numerical relationships to be established. The total intensity range of the spots may well be 100 000:1; a film which showed the weakest spot would be too over-exposed to measure the strongest, which would be too dense, and a film which gave the strongest spot at a convenient intensity would not show the weakest. The greatest range of intensity which can be comfortably measured on one film is about 100:1 (two units of density). In general when a Weissenberg photograph is taken a pack of films is placed in the film holder – very often three films but occasionally up to five. These films will differ from each other in intensity due to the absorption of the film itself, a factor of three or so being the usual attenuation in going from one film to the one below. By comparing the intensities of spots in the conveniently measurable range one can quite easily deduce a film-to-film scaling factor. A scheme for taking three separate photographs using a pack of three films at a time is shown in fig. 9.6. This will comfortably cover the range 100 000:1 or even greater and provide enough common measurable reflections in successive films to give well-determined film-to-film ratios.

When three-dimensional data are being collected then it is necessary to measure the data in layers taken about two different axes. The common reflections in intersecting layers will then provide the means of putting the data all on the same scale.

Once the data have been put on the same relative scale then various corrections can be made to them – for example for spot shape (§ 5.6); the $Lp$ factor (§ 6.1); and absorption factor $A$ (§ 6.2). If the measured intensity of the $(hkl)$ reflection is denoted by $I_{hkl}$ then we have

$$I_{hkl} = K|(F_{hkl})_r|^2 \exp(-2B\sin^2\theta/\lambda^2)$$

(9.11)

where $K$ is some scale factor, $(F_{hkl})_r$ is the structure factor corresponding to scattering factors for atoms at rest and the exponential term is the Debye–Waller factor (§ 6.5). It is implicitly assumed in equation (9.11) that the temperature factor $B$ is the same for each atom and is isotropic.

In order to determine the values of $K$ and $B$ we can make use of the result in § 7.5 which is that

Fig. 9.6.
Arranging three packs of three films to cover an intensity range $10^5$:1.

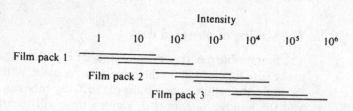

$$\langle |F_r|^2 \rangle = \sum_{j=1}^{N} f_j^2 = \Sigma. \tag{9.12}$$

In equation (9.12) the structure factor has been denoted as that for atoms at rest and the scattering factors will therefore be those tabulated for rest atoms.

If we consider together a number of observed intensities all within a narrow range of $\sin\theta$ then we may write from equations (9.11) and (9.12)

$$\langle I_\theta \rangle = K \langle |F_r|^2 \rangle \exp(-2B \sin^2\theta/\lambda^2)$$
$$= K\Sigma_\theta \exp(-2B \sin^2\theta/\lambda^2) \tag{9.13}$$

where $\langle I_\theta \rangle$ is the average of the intensities in the $\sin\theta$ range and $\Sigma_\theta$ is the value of $\Sigma$ for the particular mean value of $\theta$. From equation (9.13) we find

$$\ln\left\{\frac{\langle I_\theta \rangle}{\Sigma_\theta}\right\} = \ln K - 2B \sin^2\theta/\lambda^2. \tag{9.14}$$

Thus if the data are divided into ranges of $\sin\theta$ and $\ln(\langle I_\theta \rangle/\Sigma_\theta)$ is plotted against $\sin^2\theta/\lambda^2$ one should obtain a straight line whose slope is $-2B$ and whose intercept on one axis is $\ln K$. Such a procedure was first suggested by Wilson (1942) and the graph is known as a Wilson plot.

In fig. 9.7(a) is shown a reciprocal-lattice net for the $b$-axis projection of salicylic acid (Cochran, 1951) with intensities corrected for $Lp$ and absorption factors. The net shows one asymmetric unit of the reciprocal-lattice layer $k = 0$ and arcs of radius $\sin\theta/\lambda = 0.1, 0.2$, etc., are drawn.

Since the space group is $P2_1/a$ the $(h0l)$ reflections only occur for $h$ even. The value of $\langle I_{h0l} \rangle$ for the reflections which are not systematically absent is twice the average for a general reflection; this has been allowed for by taking $\Sigma_\theta' = 2\Sigma_\theta$ while $\langle I_\theta \rangle$ is the average for the observable reflections. In table 9.1 there are derived the values of $\ln(\langle I_\theta \rangle/\Sigma_\theta')$ for four ranges of $\sin^2\theta/\lambda^2$. It is assumed in this table that $\langle \sin^2\theta/\lambda^2 \rangle$ is the average for the extremities of the range and the value of $\Sigma_\theta'$ is that computed for $s = 2\langle \sin^2\theta/\lambda^2 \rangle^{\frac{1}{2}}$.

The values of $\ln(\langle I_\theta \rangle/\Sigma_\theta')$ are plotted in fig. 9.7(b) and a straight line through three of the points is shown. The innermost point has been ignored; the number of reflections in the corresponding annular region is small but in any case Wilson gives reasons for not using data for low values of $\sin\theta/\lambda$.

From the graph one finds $B = 2.3$ and $K = 0.33$. The values should be approximately $B = 4.0$ and $K = 0.40$, and the errors from this method of scaling are seen to be comparatively large. However, initially, one needs only a rough scale to set about the task of structure determination. Once the structure has been solved the observed data can be scaled to those calculated and the differences in the observed and calculated data then give a measure of the accuracy of the atomic parameters.

## 9.3 Fourier refinement

We have seen in chapter 8 that there are a large number of ways to solve a crystal structure. Basically these methods give information of one of two types – either a trial structure or a phase-set for the structure factors.

Let us imagine that a structure has been solved by a method that leads to

Table 9.1. *Calculations for determining the Wilson plot*

| $\sin\theta/\lambda$ range | $\left\langle\dfrac{\sin^2\theta}{\lambda^2}\right\rangle$ | $\left\langle\dfrac{\sin^2\theta}{\lambda^2}\right\rangle^{\frac{1}{2}}$ | $f_o$ | $f_c$ | $f_H$ | $\Sigma'_\theta$ | $\langle I_\theta\rangle$ | $\ln\left\{\dfrac{\langle I_\theta\rangle}{\Sigma'_\theta}\right\}$ |
|---|---|---|---|---|---|---|---|---|
| 0.1–0.2 | 0.025 | 0.158 | 5.6 | 3.5 | 0.5 | 1448 | 206 | −1.95 |
| 0.2–0.3 | 0.065 | 0.255 | 3.9 | 2.4 | 0.2 | 688 | 170 | −1.40 |
| 0.3–0.4 | 0.125 | 0.354 | 2.8 | 1.8 | 0.0 | 370 | 67 | −1.71 |
| 0.4–0.5 | 0.205 | 0.452 | 2.1 | 1.6 | 0.0 | 250 | 33 | −2.04 |

Fig. 9.7.
(a) Reciprocal-lattice net for the *b*-axis projection of salicyclic acid (Cochran, 1951).
(b) Wilson plot for the *b*-axis projection of salicylic acid.

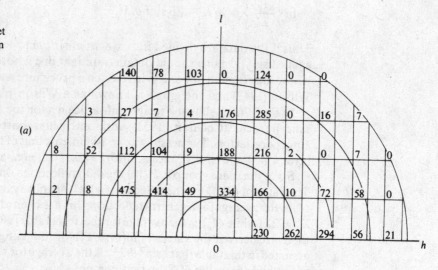

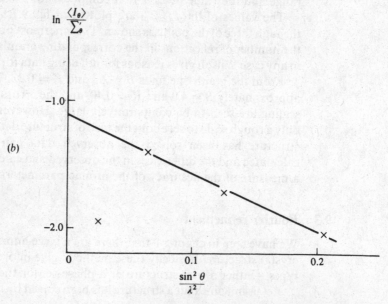

a trial set of atomic positions – for example by an examination of a Patterson function. In order to test the plausibility of the structure one would calculate structure factors and compare them with those deduced from the experimental data. To put the observed structure factors, the $F_o$'s, on an absolute scale they could be multiplied by the factor $K^{-\frac{1}{2}}$ deduced from a Wilson plot (§ 9.2) while the calculated structure factors, the $F_c$'s, should include a temperature-factor term $\exp(-B\sin^2\theta/\lambda^2)$ where $B$ is also given by the Wilson plot. In fact, once a trial structure has been obtained it is customary to determine the multiplying factor which puts the observed structure factors on an absolute scale as

$$\psi = \frac{\Sigma|F_c|}{\Sigma|F_o|}, \tag{9.15}$$

where the summation is over all the structure factors for which $F_c$ is found.

The agreement between the observed and calculated structure factors is expressed by $R$, the reliability index, as given in equation (8.1). The significance of the value of $R$ depends on whether or not the structure has a centre of symmetry. If observed structure factors are compared with structure factors calculated for a structure with the correct number and type of atoms in randomly incorrect positions the most probable values of $R$ would be 0.586 for a non-centrosymmetric structure and 0.828 for a centrosymmetric structure (Wilson, 1950). We can take as a rough rule that a value of $R$ less than 0.40 for a centrosymmetric structure or less than 0.30 for a non-centrosymmetric structure would indicate a satisfactory, or at least plausible, first trial structure.

Once an acceptable set of calculated structure factors is available a cyclic process of refinement can be initiated. An electron-density map is calculated with the magnitudes of the observed structure factors and the phases from the calculated structure factors, i.e.

$$\rho_r = \frac{1}{V}\sum_h |(F_o)_h|\exp\{-2\pi i\mathbf{h\cdot r} + i\phi_h\}. \tag{9.16a}$$

If the structure is centrosymmetric this will appear as

$$\rho_r = \frac{1}{V}\sum_h s(\mathbf{h})|(F_o)_h|\cos(2\pi\mathbf{h\cdot r}) \tag{9.16b}$$

where $s(\mathbf{h})$ is the sign of the calculated structure factor.

This electron-density map will indicate positions for the atomic centres which will be displaced somewhat from those originally inserted and the new atomic positions should be a better estimate of the true atomic positions. From the new positions another set of $F_c$'s can be calculated and this will give a lower value of $R$. After a number of cycles of this process the solution will converge and, typically, one will be left with an $R$ somewhere between 0.02 and 0.10 depending to some extent on the complexity of the structure and also on the accuracy with which the data were collected. The situation with a centrosymmetric structure is that the refinement process is terminated when the signs of the calculated structure factors are the same for two successive cycles; with non-centrosymmetric structure factors, on the other hand, the phases change continuously and when to terminate the refinement process is a matter of judgement. It should be stressed that for effective refinement the problem should be greatly over-determined – that is to say, that there should be many more items of data than there are parameters to be refined.

It has been assumed above that one began with a trial structure but the same process is valid if one starts with a trial phase-set instead. The Fourier synthesis with the trial phases – perhaps only for some subset of the data – may give a rough but adequate picture of the structure from which a trial set of coordinates may be found and refinement initiated.

What are the underlying causes of error in this process? There will of course be errors in the measurements of the $F_o$'s but, in addition to this, there is also an error due to series termination, i.e. the limitation in the amount of data being collected. The errors in the $F_o$'s will lead to an error in the electron density and we can estimate the extent of this.

Let $\rho_r$ be the calculated electron density, $\rho_r'$ be the electron density for error-free data, $(F_o)_\mathbf{h}$ the observed structure factor of index $\mathbf{h}$, and $F_\mathbf{h}'$ the error-free structure factor. We may now write

$$\Delta\rho_r = \rho_r - \rho_r' = \frac{1}{V}\sum_\mathbf{h}\{(F_o)_\mathbf{h} - F_\mathbf{h}'\}\exp(-2\pi i\mathbf{h}\cdot\mathbf{r}). \tag{9.17}$$

In fig. 9.8 $(F_o)_\mathbf{h}$ and $F_\mathbf{h}'$ are depicted on an Argand diagram. If in the completely refined structure the phase angles $\phi_\mathbf{h}'$ and $\phi_\mathbf{h}$ are not very different then it is clear that

$$(F_o)_\mathbf{h} - F_\mathbf{h}' = |\Delta F_\mathbf{h}|\exp(i\xi_\mathbf{h}) \tag{9.18a}$$

where

$$|\Delta F_\mathbf{h}| \simeq \left||(F_o)_\mathbf{h}| - |F_\mathbf{h}'|\right|. \tag{9.18b}$$

Fig. 9.8.
The observed structure factor $(F_o)_\mathbf{h}$ with the error-free structure factor $F_\mathbf{h}'$ on an Argand diagram.

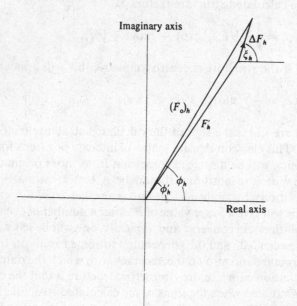

Combining equations (9.17) and (9.18)

$$\Delta\rho_r = \frac{1}{V}\sum_h |\Delta F_h| \exp(-2\pi i h \cdot r + i\xi_h). \tag{9.19}$$

From an application of the central-limit theorem (§ 7.5) we can find that the probability distribution of $\Delta\rho_r$ will be normal with a zero mean and a variance

$$\overline{(\Delta\rho_r)^2} = \frac{1}{V^2}\sum_h |\Delta F_h|^2. \tag{9.20}$$

In deriving equation (9.20) the result was used that

$$|\exp(i\alpha)| = 1.$$

If the standard deviation of $\rho_r$ from $\rho'_r$ averaged over the whole cell is $\sigma(\rho)$ then

$$\{\sigma(\rho)\}^2 = \overline{(\Delta\rho_r)^2} \tag{9.21}$$

or

$$\sigma(\rho) = \frac{M^{\frac{1}{2}}}{V}\sigma(F) \tag{9.22}$$

where $M$ is the total number of terms in the summation and $\sigma(F)$ is the standard deviation of the errors in $F_o$ which are assumed to be uncorrelated with the magnitude of $F_o$. When $\lambda$ is the wavelength of the X-radiation being used then the explorable volume of reciprocal space is $\frac{4}{3}\pi(2/\lambda)^3$ and, since the volume of a reciprocal unit cell is $V^{-1}$, we find $M = (32\pi/3)V\lambda^{-3}$. This gives

$$\sigma(\rho) = \left\{\frac{32\pi}{3V\lambda^3}\right\}^{\frac{1}{2}}\sigma(F). \tag{9.23}$$

Another assumption which could be made about the errors of measurement is that $\overline{|\Delta F|^2}$ is proportional to $|F_o|^2$. From this assumption one would have from equations (9.20) and (9.21) that

$$\sigma(\rho) = \frac{Q}{V^{\frac{1}{2}}}\sqrt{\frac{1}{V}\sum |(F_o)_h|^2}$$

$$= \frac{Q}{V^{\frac{1}{2}}}(P(0))^{\frac{1}{2}} \tag{9.24}$$

where $Q^2$ is the constant of proportionality linking $\overline{|\Delta F|^2}$ with $|F_o|^2$ and $P(0)$ is the value at the origin of the Patterson function. For atoms which are not too heavy and for an average temperature factor one finds that

$$P(0) \simeq 0.8 \sum_{j=1}^{N} Z_j^2. \tag{9.25}$$

We can now make some order-of-magnitude assessment of $\sigma(\rho)$. For example with $V = 500\,\text{Å}^3$, $\lambda = 1.54\,\text{Å}$ and $\sigma(F) \simeq 1$ electron we find from equation (9.23) that $\sigma(\rho) \simeq 0.14e\,\text{Å}^{-3}$. If the structure contains 40 carbon atoms ($Z = 6$) and $Q = 0.10$, that is to say that the average error in measuring a structure factor is of the order of 10% of its magnitude, we find $\sigma(\rho) \simeq 0.15e\,\text{Å}^{-3}$.

The actual errors of electron density would be of interest if one was looking for evidence of a redistribution of electron density in some particular chemical bonding situation. However in normal circumstances what one is concerned with are the atomic coordinates, and a uniform raising or lowering of the electron density in the neighbourhood of an atomic peak would not affect the precision with which this could be determined. What will affect the peak position will be the gradient of $\Delta\rho$ in the vicinity of the atomic centre. This can be seen in the one-dimensional example in fig. 9.9 which shows $\rho$ and $\rho + \Delta\rho$ in the vicinity of a maximum of $\rho$ which represents the atomic position. If the maximum is displaced a distance $\Delta X$ from the true atomic centre, taken as the coordinate origin, then one has for a maximum of $\rho + \Delta\rho$

$$\left[\frac{d}{dx}(\rho + \Delta\rho)\right]_{x=\Delta X} = 0 \tag{9.26}$$

or

$$\left(\frac{d\rho}{dx}\right)_{x=\Delta X} + \left(\frac{d}{dx}(\Delta\rho)\right)_{x=\Delta X} = 0. \tag{9.27}$$

The slope of $\Delta\rho$ may be assumed constant over a small range so that

$$\left(\frac{d}{dx}(\Delta\rho)\right)_{x=\Delta X} \simeq \left(\frac{d}{dx}(\Delta\rho)\right)_{x=0} = G_0, \text{ say.} \tag{9.28}$$

In addition, from Taylor's theorem

$$\left(\frac{d\rho}{dx}\right)_{x=\Delta X} = \left(\frac{d\rho}{dx}\right)_{x=0} + \Delta X\left(\frac{d^2\rho}{dx^2}\right)_{x=0} + \cdots \tag{9.29}$$

Now the curvature of $\rho$, $C_x$, is given by the well-known result

$$C_x = \frac{d^2\rho/dx^2}{\{1 + (d\rho/dx)^2\}^{\frac{3}{2}}} \tag{9.30}$$

and at the atomic centre where $(d\rho/dx)_{x=0} = 0$ this gives

$$C_0 = \left(\frac{d^2\rho}{dx^2}\right)_{x=0}. \tag{9.31}$$

Fig. 9.9.
The shift in a peak maximum due to the gradient of the error in electron density.

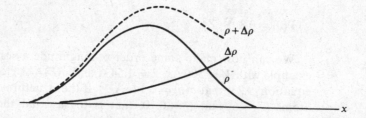

From this we find

$$\Delta X = - \frac{\{d(\Delta\rho)/dx\}_{x=0}}{(d^2\rho/dx^2)_{x=0}} = -\frac{G_0}{C_0}. \qquad (9.32)$$

The general three-dimensional result corresponding to equation (9.32) replaces differentiation by partial differentiation thus

$$\Delta X = - \frac{\{\partial(\Delta\rho)/\partial x\}_{x,y,z=0}}{(\partial^2\rho/\partial x^2)_{x,y,z=0}}. \qquad (9.33)$$

One can find a good estimate of $C_0$ by assuming that the atomic peak changes position but not shape so that $C_0$ is the curvature at the centre of the 'perfect' electron-density distribution. Hence the error in coordinate will depend on the slope produced by the electron-density errors at the atomic centres and one can make an assessment of the magnitude of this on a statistical basis.

If we write equation (9.19) in terms of the individual Cartesian coordinates we have

$$\Delta\rho_{x,y,z} = \frac{1}{V}\sum_h\sum_k\sum_l |\Delta F_{hkl}| \exp\{-2\pi i(hx + ky + lz) + i\xi_{hkl}\} \qquad (9.34)$$

which gives

$$\left(\frac{\partial}{\partial x}(\Delta\rho)\right)_{x,y,z=0} = G_0' = \frac{1}{V}\sum_h\sum_k\sum_l -2\pi ih|\Delta F_{hkl}|\exp(i\xi_{hkl}). \qquad (9.35)$$

The origin of the unit cell has been placed arbitrarily at the centre of the atom of interest. From equation (9.35) we can deduce that $G_0'$ should have a normal probability distribution with zero mean and with standard deviation

$$\sigma(G_0') = \frac{2\pi}{V}(M\overline{h^2})^{\frac{1}{2}}\sigma(F) \qquad (9.36)$$

where $\sigma(F)$, the standard deviation of the errors in the $F_o$'s, is assumed not to be correlated with $|F_o|$ and M is the number of terms in the summation. In fig. 9.10 there is shown the limiting sphere with the various layers of constant h. A few moments study of this should convince the reader that $(\overline{h^2}/h_{max}^2)$ is the mean-square distance in a sphere of unit radius from a diametral plane, where $h_{max}$ is the 'index' (perhaps non-integral) corresponding to $\sin\theta = 1$. This ratio may easily be found to be equal to 0.2 so that

$$\overline{h^2} = 0.2h_{max}^2 \qquad (9.37a)$$

and, since $h_{max} = 2a/\lambda$ then

$$\overline{h^2} = 0.8\frac{a^2}{\lambda^2}. \qquad (9.37b)$$

Thus, with the value of M found previously, we obtain

$$\sigma(G_0') = 2\pi\left\{\frac{25.6\pi}{3V\lambda^5}\right\}^{\frac{1}{2}}a\sigma(F) \qquad (9.38)$$

Fig. 9.10.
Layers of constant $h$
within the limiting
sphere.

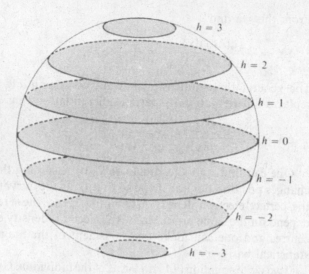

where it should be noted that the units of $G_0'$ are rate of change of electron density per unit increase of *fractional* coordinate. To change this to rate of change of electron density per unit distance along the $x$ direction, $G_0$, we must divide by the cell edge $a$. Hence

$$\sigma(G_0) = 2\pi \left\{ \frac{25.6\pi}{3V\lambda^5} \right\}^{\frac{1}{2}} \sigma(F)$$

$$= 32.6 \left( \frac{1}{V\lambda^5} \right)^{\frac{1}{2}} \sigma(F). \tag{9.39}$$

A representation of the electron density near the centre of an atom which is often found useful is

$$\rho_r = Z \left( \frac{p}{\pi} \right)^{\frac{3}{2}} \exp(-pr^2) \tag{9.40}$$

where $p$ is typically about $4\,\text{Å}^{-2}$ for an atom of moderate weight with an average temperature factor. In equation (9.40), $r$ is the distance from the atomic centre; the curvature at the centre can be found by differentiating twice with respect to $r$ and then making $r = 0$. This gives

$$C_0 = -2Zp \left( \frac{p}{\pi} \right)^{\frac{3}{2}}. \tag{9.41}$$

The standard deviation of the error of coordinate may be obtained from equation (9.32) and is

$$\sigma(\Delta X) = \frac{\sigma(G_0)}{C_0} \tag{9.42}$$

which, from equations (9.39) and (9.41) gives

$$\sigma(\Delta X) = \frac{16.3}{Zp} \left( \frac{\pi}{p} \right)^{\frac{3}{2}} \left( \frac{1}{V\lambda^5} \right)^{\frac{1}{2}} \sigma(F) \tag{9.43a}$$

or, with $p = 4\,\text{Å}^{-2}$,

$$\sigma(\Delta X) = \frac{2.84}{Z}\left(\frac{1}{V\lambda^5}\right)^{\frac{1}{2}}\sigma(F). \tag{9.43b}$$

Thus with $V = 500\,\text{Å}^3$, $\lambda = 1.54\,\text{Å}$, $Z = 6$ and $\sigma(F) = 1$ electron then $\sigma(\Delta X) = 0.007\,\text{Å}$.

It can be shown that for a projection of area $\mathfrak{A}$ the standard deviation of a coordinate error is

$$\sigma(\Delta X) = \frac{2\pi^2}{p^2\lambda^2}\left(\frac{\pi}{\mathfrak{A}}\right)^{\frac{1}{2}}\frac{1}{Z}\sigma(F) \tag{9.44}$$

where the two-dimensional electron density is given by

$$P_r = Z\frac{p}{\pi}\exp(-pr^2).$$

The effect of series termination is rather less easy to deal with. Effectively, it is as though the scattering factors for the atoms sharply terminate at $s = 2/\lambda$ and this means that the computed electron density for an atom, instead of smoothly decreasing with increasing distance from the atomic centre, will have diffraction ripples. These will be particularly disturbing round a heavy or even moderately heavy atom (sulphur for example) and the error induced in the coordinates of a neighbouring atom can be greater than that due to errors in measurement of the $|F_o|$'s.

## 9.4    Least-squares refinement

Let us imagine for now that we have an error-free set of $|F_o|$'s and an almost correct set of atomic coordinates and temperature factors. For simplicity of our discussion we shall assume that the structure is centrosymmetric and the temperature factors are isotropic. The calculated structure factors will be given by

$$(F_c)_{hkl} = \sum_{j=1}^{N/2} 2f_j\exp\left(-B_j\frac{\sin^2\theta}{\lambda^2}\right)\cos\{2\pi(hx_j + ky_j + lz_j)\}. \tag{9.45}$$

The correct parameters for the $j$th atom can be expressed as

$$(B_j + \Delta B_j, x_j + \Delta x_j, y_j + \Delta y_j, z_j + \Delta z_j)$$

and hence we can write

$$-(F_o)_{hkl} = \sum_{j=1}^{N/2} 2f_j\exp\left\{-(B_j + \Delta B_j)\frac{\sin^2\theta}{\lambda^2}\right\}\cos[2\pi\{h(x_j + \Delta x_j) + k(y_j + \Delta y_j)$$

$$+ l(z_j + \Delta z_j)\}]. \tag{9.46}$$

As long as the errors in the parameters are small this enables us to write

$$(F_o)_{hkl} - (F_c)_{hkl} = \Delta F_{hkl} \tag{9.47}$$

and

$$\Delta F_{hkl} = \sum_{j=1}^{N/2} \left( \frac{\partial (F_c)_{hkl}}{\partial B_j} \Delta B_j + \frac{\partial (F_c)_{hkl}}{\partial x_j} \Delta x_j + \frac{\partial (F_c)_{hkl}}{\partial y_j} \Delta y_j + \frac{\partial (F_c)_{hkl}}{\partial z_j} \Delta z_j \right).$$

$$(9.48)$$

An equation of type (9.48) can be produced for each reflection and, in general, there will be many more equations than parameters. Thus for space group $P\bar{1}$ with 40 atoms in the unit cell and hence 160 parameters there could be 2000 or more independent reflections.

The equations are linear in the corrections to be made to the parameters and, ideally, if a subset of these equations was taken such that the number of equations equalled the number of parameters then the correct parameters could be found and recalculated values of the $F_c$'s would then equal the $F_o$'s. However in practice the $|F_o|$'s are not error-free, and so one finds a least-squares solution of the complete set of equations (9.48). This solution is the one such that when the parameters are changed to $B_j + \Delta B_j, x_j + \Delta x_j$, etc., the quantity

$$R_s = \sum_{\mathbf{h}} \{ (F_o)_{\mathbf{h}} - (F'_c)_{\mathbf{h}} \}^2 \tag{9.49}$$

is a minimum, where $(F'_c)_{\mathbf{h}}$ is the revised calculated structure factor. From Taylor's theorem we find, ignoring second-order small quantities, that

$$(F'_c)_{\mathbf{h}} = (F_c)_{\mathbf{h}} + \sum_{j=1}^{N/2} \left( \frac{\partial (F_c)_{\mathbf{h}}}{\partial B_j} \Delta B_j + \dots \right) \tag{9.50}$$

and hence

$$(F_o)_{\mathbf{h}} - (F'_c)_{\mathbf{h}} = \Delta F_{\mathbf{h}} - \sum_{j=1}^{N/2} \left( \frac{\partial (F_c)_{\mathbf{h}}}{\partial B_j} \Delta B_j + \dots \right). \tag{9.51}$$

Hence $R_s$ is the sum for all the equations of the squares of the differences between the left-hand sides and right-hand sides of equations (9.48) when the solution is inserted and it is the minimization of this quantity which is usually implied by the term 'least-squares solution'.

It is possible to weight the equations according to the expected reliability of the quantity $\Delta F$. If the measured $|F_o|$ is expected to have a large random error then the value of $\Delta F$ may well be dominated by this and the $\Delta F$ may not be very useful in indicating the changes to be made in the atomic parameters. Errors of measurement tend to be related to the actual value of $|F_o|$ and many weighting schemes have been proposed based on the value of $|F_o|$. One such scheme proposed by Hughes (1941) takes $\omega = $ constant for $|F_o| < F_{lim}$ and $\omega \propto 1/|F_o|^2$ for $|F_o| > F_{lim}$ where $F_{lim}$ is some chosen upper limit of $|F_o|$ for a constant weight. Another weighting scheme suggested by Cruickshank *et al.* (1961) is $\omega = (a + |F_o| + c|F_o|^2)^{-1}$ where $a \simeq 2F_{min}$ and $c \simeq 2/F_{max}$, where $F_{min}$ and $F_{max}$ are the minimum and maximum observed intensities.

When a reasonable set of weights has been found, each equation is multiplied throughout by the appropriate weight and a least-squares solution of the modified set of equations is then sought in the usual way. The quantity which is being minimized in this case is

$$R_{sw} = \sum_{h} \omega\{(F_o)_h - (F_c)_h\}^2. \tag{9.52}$$

A brief description will be given of the standard process for finding a least-squares solution of a set of equations. In general there will be $m$ equations for $n$ unknowns $x_1\, x_2, \ldots, x_n$ where $m > n$ and these can be written as

$$
\begin{aligned}
a_{11}x_1 + a_{12}x_2 + \cdots + a_{1n}x_n &= b_1 \\
a_{21}x_1 + a_{22}x_2 + \cdots + a_{2n}x_n &= b_2 \\
\vdots \qquad \vdots \qquad \vdots \qquad \vdots & \\
a_{m1}x_1 + a_{m2}x_2 + \cdots + a_{mn}x_n &= b_m.
\end{aligned}
\tag{9.53}
$$

A description of these equations is most neatly given in the terminology of matrix algebra. Thus equation (9.53) can be written

$$A\mathbf{x} = \mathbf{b} \tag{9.54}$$

where $A$ is the $(m \times n)$ matrix

$$
\begin{bmatrix}
a_{11} & a_{12} \cdots a_{1n} \\
a_{21} & a_{22} \cdots a_{2n} \\
\vdots & \\
a_{m1} & a_{m2} \cdots a_{mn}
\end{bmatrix}
$$

and $\mathbf{x}$ and $\mathbf{b}$ are the column vectors

$$
\begin{bmatrix}
x_1 \\
x_2 \\
\vdots \\
x_n
\end{bmatrix}
\quad \text{and} \quad
\begin{bmatrix}
b_1 \\
b_2 \\
\vdots \\
b_m
\end{bmatrix}.
$$

The first step in the solution is to reduce the equations (9.53) to a normal set of equations. The first equation of the normal set is obtained by multiplying each equation by its coefficient of $x_1$ and then adding together the modified equations. This is illustrated below:

$$
\begin{aligned}
a_{11}^2 x_1 + a_{11}a_{12}x_2 + \cdots + a_{11}a_{1n}x_n &= a_{11}b_1 \\
a_{21}^2 x_1 + a_{21}a_{22}x_2 + \cdots + a_{21}a_{2n}x_n &= a_{21}b_2 \\
\vdots \qquad\qquad & \\
a_{m1}^2 x_1 + a_{m1}a_{m2}x_2 + \cdots + a_{m1}a_{mn}x_n &= a_{m1}b_m.
\end{aligned}
\tag{9.55}
$$

$$\overline{\sum_{r=1}^{n} \left\{ \sum_{s=1}^{m} a_{s1}a_{sr} \right\} x_r = \sum_{s=1}^{m} a_{s1}b_s}$$

If this process is repeated by multiplying by the coefficients of $x_2, x_3$, etc., in turn there is obtained a set of $n$ equations, of which equation (9.55) is one member, in the $n$ unknown $x$'s. The solution of this normal set of equations is the required least-squares solution of the original set.

In matrix-algebra terminology the normal set of equations is obtained as

$$\tilde{A}A\mathbf{x} = \tilde{A}\mathbf{b} \tag{9.56}$$

where $\tilde{A}$ is the transpose of $A$ and is the $(n \times m)$ matrix

$$\tilde{A} = \begin{bmatrix} a_{11} & a_{21} \cdots a_{m1} \\ a_{12} & a_{22} \cdots a_{m2} \\ \vdots & \\ a_{1n} & a_{2n} \cdots a_{mn} \end{bmatrix}. \tag{9.57}$$

This gives as the solution vector for the $x$'s

$$\mathbf{x} = (\tilde{A}A)^{-1}\tilde{A}\mathbf{b} \tag{9.58}$$

where $(\tilde{A}A)^{-1}$ is the inverse matrix of $\tilde{A}A$.

In practice the crystallographer need not get deeply involved in the mathematical complexities of refinement by the least-squares method. Excellent standard computer programs are available which make the refinement process quite automatic once the data are provided. A fairly complete account of least-squares refinement methods can be found in *Computing Methods in Crystallography* edited by Rollett (1965).

The technique of least-squares refinement can be used for non-centrosymmetric structures and other types of parameter can also be refined. These latter include a scale factor for the $|F_o|$'s, the coefficients for anisotropic temperature factors and even the scattering factors themselves if it is felt that the quality of the observed data justifies this.

From the final value of $R_{sw}$ the standard errors of the final parameters can be estimated. Further details of this are given in *Computing Methods in Crystallography*.

## 9.5 The parameter-shift method

The methods of refinement so far discussed occasionally run into trouble in unfavourable circumstances. If a projection is being refined and there is some overlap of atoms then Fourier methods are difficult to apply because of the interference of the overlapped atoms with each other. Again, least-squares methods are also difficult to apply in such circumstances and, indeed, even in three dimensions where overlap is no problem the refinement process will not work if the trial coordinates are not very close to the correct positions.

A very successful method of refinement, which works well even under the difficult circumstances described above, is the parameter-shift method of Bhuiya and Stanley (1963). Let us say that the structure can be described in terms of $n$ parameters whose initial values $u_1, u_2, \ldots, u_n$ give a value of residual

$$R(u_1, u_2, \ldots, u_n) = \frac{\sum ||F_o| - |F_c||}{\sum |F_o|}. \tag{9.59}$$

In the parameter-shift method the first parameter is varied in steps of $\Delta u_1$ from $u_1 - k\Delta u_1$ to $u_1 + k\Delta u_1$ and the $2k + 1$ values of $R$ are calculated for each value. The steps can be quite large; typically in the initial stages of refinement one would have $k = 5$ and $\Delta u_1 = 0.02$ Å. The first parameter is shifted to the value which gives the lowest residual and the other parameters are treated in turn in the same way. It should be noted that the $2k + 1$ values of $R$ are worked out by exact calculation and *not* by

$$R(u_1 + k\Delta u_1, u_2, \ldots, u_n) = R(u_1, u_2, \ldots, u_n) + \frac{\partial R}{\partial u_1} k\Delta u_1, \tag{9.60}$$

which would not be valid for the large shifts which might be used.

The advantage of this method is that it can move atoms away from false positions which give a local minimum of $R$. This is shown in fig. 9.11; the parameter-shift method would move the parameter to $N$, at the end of the range of investigation, whereas with the least-squares method it would be unlikely that the parameter would climb out of the local minimum.

Once the complete set of parameters has been shifted in the first cycle one begins again with $u_1$. As the residual is reduced so the steps $\Delta u$ are also reduced since the parameter space must be explored more closely as the correct values are approached.

An example given by Stanley (1964) was the $c$-axis projection of naphthocinnoline. This has space group $pgg$ and contains sixteen carbon atoms and two nitrogen atoms in the asymmetric unit. The initial value of $R$ was 0.37 and it was decided just to refine the coordinates for the first cycle. The values of $k$ and $\Delta u$ were 5 and 0.02 Å for both $x$ and $y$ and the progress of the first refinement cycle is shown in table 9.2. The refinement was very rapid and all the atoms were displaced to the extreme end of their permitted range for at least one of the coordinates. It turned out when the refinement was complete that the root-mean-square displacement of the atoms from their initial positions was 0.5 Å! When least-squares refinement was tried on this projection it reduced $R$ from 0.37 to 0.33 in two cycles and then could refine no further.

The parameter-shift method is very fast in application and the time taken by the computer is proportional to the product of the number of parameters

Fig. 9.11.
The residual $R$ as a function of parameter shift showing the parameter at a false minimum.

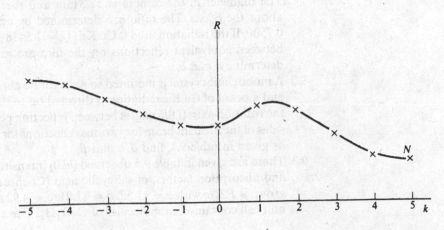

Table 9.2.

| Atom no. | $\lvert \Delta x \rvert$ (Å) | $\lvert \Delta y \rvert$ (Å) | R |
|---|---|---|---|
| 1 | 0.10 | 0.10 | 0.3651 |
| 2 | 0.10 | 0.10 | 0.3593 |
| 3 | 0.06 | 0.10 | 0.3549 |
| 4 | 0.10 | 0.10 | 0.3497 |
| 5 | 0.10 | 0.10 | 0.3449 |
| 6 | 0.10 | 0.02 | 0.3425 |
| 7 | 0.10 | 0.10 | 0.3376 |
| 8 | 0.10 | 0.10 | 0.3323 |
| 9 | 0.08 | 0.10 | 0.3277 |
| 10 | 0.06 | 0.10 | 0.3233 |
| 11 | 0.10 | 0.10 | 0.3185 |
| 12 | 0.10 | 0.10 | 0.3138 |
| 13 | 0.08 | 0.10 | 0.3091 |
| 14 | 0.08 | 0.10 | 0.3045 |
| 15 | 0.10 | 0.10 | 0.2994 |
| 16 | 0.10 | 0.10 | 0.2947 |
| 17 | 0.10 | 0.10 | 0.2898 |
| 18 | 0.10 | 0.10 | 0.2850 |

and the number of reflections. The least-squares method, on the other hand, which involves eventually the inversion of a matrix of order $n$, the number of parameters, takes a time proportional to $n^3$. Because of this, in the early stages of least-squares refinement, where one often uses approximate methods of solving the normal equations, it could be argued that the parameter-shift method should be used instead.

## Problems to Chapter 9

9.1 An oscillation photograph of an orthorhombic crystal is taken on a camera modified in the way described by Farquhar and Lipson (1946). The diameter of the camera is 57.3 mm and the crystal is oscillated about the $c$ axis. The ratio $a:b$ determined by optical goniometry is 0.7500. If the radiation used is Cu $K\alpha_1$ ($\lambda = 1.5418$ Å) and the distances between equivalent reflections on the film are as given in table 9.3, determine $a$ and $b$.

9.2 A monoclinic crystal is mounted so as to rotate about the unique $b$ axis and a beam of Cu $K\alpha$ radiation is directed on to the crystal normal to the rotation axis. If the angles between reflecting positions on opposite sides of the incident beam for various reflections for $K\alpha_1$ radiation were as given in table 9.4, find $a$, $c$ and $\beta$.

9.3 There are given in table 9.5 observed $(hk0)$ intensities, corrected for $Lp$ and absorption factors, of salicyclic acid (Cochran, 1953). The space group is $P2_1/c$ with $a = 11.52$, $b = 11.21$, $c = 4.92$ Å and $\beta = 91°$. The unit cell contains four molecules of $C_7O_3H_6$. Use a Wilson plot to put

Table 9.3.

| h | k | l | Distance (mm) |
|---|---|---|---|
| 7 | 1 | 0 | 51.09 |
| 7 | 2 | 0 | 46.83 |
| 7 | 3 | 0 | 38.83 |
| 7 | 4 | 0 | 24.50 |
| 6 | 5 | 0 | 50.18 |
| 6 | 6 | 0 | 32.20 |

Table 9.4.

| h | k | l | Angle |
|---|---|---|---|
| 3 | 0 | 0 | 64° 31′ |
| 4 | 0 | 0 | 90° 46′ |
| 5 | 0 | 0 | 125° 42′ |
| 0 | 0 | 4 | 61° 9′ |
| 0 | 0 | 5 | 78° 56′ |
| 0 | 0 | 6 | 99° 24′ |
| 0 | 0 | 7 | 125° 41′ |
| 3 | 0 | 4 | 129° 4′ |
| 4 | 0 | $\bar{5}$ | 85° 15′ |

the data on an absolute scale and also determine an overall temperature factor.

9.4 (i) Estimate the root-mean-square error of electron density when data are taken from a crystal with Cu K$\alpha$ radiation ($\lambda = 1.54$ Å), the unit cell has volume 300 Å$^3$ and the root-mean-square error of the observed structure amplitudes is 0.5 electrons and is not correlated with the magnitude of $|F_o|$.

(ii) Under the conditions of part (i) what would be the root-mean-square error in the $x$ coordinate (in Å) for a carbon atom when the electron density near the maximum is given by equation (9.40) with $p = 3.6$ Å$^{-2}$?

9.5 (i) Calculate structure factors with STRUCFAC for the synthetic structure $C_3H_5N_2ONa_2$, the coordinates for which are given in §8.3. Take the temperature factor $B = 2.0$ Å$^{-2}$ and X-ray wavelength $\lambda = 1.542$ Å. This gives simulated observed structure amplitudes.

(ii) Approximate coordinates are available as follows:

Na (0.950, 0.275), Na (0.480, 0.600), O (0.000, 0.650),
N (0.200, 0.400), N (0.200, 0.000), C (0.100, 0.200),
C (0.400, 0.150), C (0.350, 0.400).

Calculate structure factors for these coordinates with the same $B$ and $\lambda$ as in (i). This gives calculated structure amplitudes and phases.

Table 9.5.

| h | k | I | h | k | I | h | k | I | h | k | I | h | k | I |
|---|---|---|---|---|---|---|---|---|---|---|---|---|---|---|
| 0 | 2 | 142 | 2 | 13 | 5 | 5 | 5 | 126 | 7 | 11 | 0 | 10 | 7 | 11 |
|   | 4 | 246 |   | 14 | 0 |   | 6 | 21 |   | 12 | 1 |   | 8 | 4 |
|   | 6 | 63 | 3 | 1 | 1 |   | 7 | 5 | 8 | 0 | 58 |   | 9 | 2 |
|   | 8 | 24 |   | 2 | 407 |   | 8 | 16 |   | 1 | 2 |   | 10 | 1 |
|   | 10 | 1 |   | 3 | 23 |   | 9 | 11 |   | 2 | 43 | 11 | 1 | 1 |
|   | 12 | 1 |   | 4 | 28 |   | 10 | 8 |   | 3 | 11 |   | 2 | 0 |
|   | 14 | 0 |   | 5 | 226 |   | 11 | 2 |   | 4 | 76 |   | 3 | 1 |
| 1 | 1 | 529 |   | 6 | 94 |   | 12 | 0 |   | 5 | 19 |   | 4 | 0 |
|   | 2 | 112 |   | 7 | 31 |   | 13 | 1 |   | 6 | 27 |   | 5 | 19 |
|   | 3 | 14 |   | 8 | 0 | 6 | 0 | 300 |   | 7 | 7 |   | 6 | 13 |
|   | 4 | 83 |   | 9 | 2 |   | 1 | 67 |   | 8 | 11 |   | 7 | 0 |
|   | 5 | 7 |   | 10 | 1 |   | 2 | 124 |   | 9 | 7 |   | 8 | 9 |
|   | 6 | 225 |   | 11 | 9 |   | 3 | 21 |   | 10 | 0 |   | 9 | 1 |
|   | 7 | 67 |   | 12 | 34 |   | 4 | 19 |   | 11 | 0 | 12 | 0 | 1 |
|   | 8 | 0 |   | 13 | 1 |   | 5 | 10 |   | 12 | 0 |   | 1 | 2 |
|   | 9 | 2 |   | 14 | 1 |   | 6 | 23 | 9 | 1 | 0 |   | 2 | 1 |
|   | 10 | 19 | 4 | 0 | 260 |   | 7 | 18 |   | 2 | 6 |   | 3 | 0 |
|   | 11 | 6 |   | 1 | 35 |   | 8 | 4 |   | 3 | 35 |   | 4 | 4 |
|   | 12 | 6 |   | 2 | 63 |   | 9 | 10 |   | 4 | 86 |   | 5 | 1 |
|   | 13 | 0 |   | 3 | 9 |   | 10 | 4 |   | 5 | 9 |   | 6 | 2 |
|   | 14 | 1 |   | 4 | 17 |   | 11 | 1 |   | 6 | 7 |   | 7 | 8 |
| 2 | 0 | 211 |   | 5 | 61 |   | 12 | 0 |   | 7 | 3 |   | 8 | 10 |
|   | 1 | 1140 |   | 6 | 263 |   | 13 | 1 |   | 8 | 11 | 13 | 1 | 4 |
|   | 2 | 4 |   | 7 | 2 | 7 | 1 | 92 |   | 9 | 0 |   | 2 | 7 |
|   | 3 | 13 |   | 8 | 7 |   | 2 | 134 |   | 10 | 0 |   | 3 | 2 |
|   | 4 | 194 |   | 9 | 26 |   | 3 | 40 |   | 11 | 0 |   | 4 | 0 |
|   | 5 | 3 |   | 10 | 1 |   | 4 | 6 | 10 | 0 | 22 |   | 5 | 3 |
|   | 6 | 122 |   | 11 | 69 |   | 5 | 19 |   | 1 | 0 |   | 6 | 0 |
|   | 7 | 47 |   | 12 | 0 |   | 6 | 22 |   | 2 | 1 | 12 | 0 | 5 |
|   | 8 | 42 |   | 13 | 0 |   | 7 | 2 |   | 3 | 18 |   | 1 | 3 |
|   | 9 | 2 | 5 | 1 | 134 |   | 8 | 3 |   | 4 | 98 |   | 2 | 3 |
|   | 10 | 50 |   | 2 | 3 |   | 9 | 0 |   | 5 | 18 |   | 3 | 2 |
|   | 11 | 10 |   | 3 | 22 |   | 10 | 3 |   | 6 | 17 |   | 4 | 1 |
|   | 12 | 38 |   | 4 | 107 |   |   |   |   |   |   |   |   |   |

(iii) A program CALOBS is available (appendix XII) which combines the observed structure factors and calculated phases in an output file suitable for input to FOUR2 and also gives a reliability index. Use this program with the output from (i) and (ii).

(iv) Using the output file from CALOBS as input to FOUR2 calculate a map.

(v) Contour the map and find new estimates of the coordinates.

(vi) Calculate structure factors for the new coordinates.

(vii) Using CALOBS find the new residual.

# Physical constants and tables

Among the basic units used by the SI system are the following:

| length | metre | m |
| mass | kilogramme | kg |
| time | second | s |
| current | ampere | A |

Some other units derived from these are:

| force | newton | $N = kg\,m\,s^{-2}$ |
| energy | joule | $J = Nm$ |
| power | watt | $W = J\,s^{-1}$ |
| electric charge | coulomb | $C = A\,s$ |
| electric potential | volt | $V = W\,A^{-1}$ |
| electric capacitance | farad | $F = A\,s\,V^{-1}$ |

Acceptable multiples or sub-multiples can be used which differ from the basic unit in steps of $10^3$. Thus the millimetre (mm) is an acceptable unit.

Field $E$ due to charge $q$ at distance $r$ is given by

$$E = \frac{q}{4\pi\varepsilon_0 r^2}$$

where $\varepsilon_0$ is the permittivity of free space.

The force on a charge $q$ in a field $E$ is given by

$$F = qE.$$

The electric-field amplitude $E$ in a beam of electromagnetic radiation of intensity $I$ is given by

$$E^2 = \frac{I}{c\varepsilon_0}.$$

Some useful physical constants:

| electron charge | $1.602 \times 10^{-19}\,C$ |
| electron mass | $9.110 \times 10^{-31}\,kg$ |
| atomic mass unit | $1.660 \times 10^{-27}\,kg$ |
| proton mass | $1.675 \times 10^{-27}\,kg$ |
| Avogadro's number | $6.022 \times 10^{23}\,mol^{-1}$ |
| Boltzmann's constant | $1.381 \times 10^{-23}\,J\,K^{-1}$ |
| Planck's constant | $6.626 \times 10^{-34}\,J\,s$ |
| electron volt | $1.602 \times 10^{-19}\,J$ |
| speed of light | $2.998 \times 10^8\,m\,s^{-1}$ |
| permittivity of free space | $8.854 \times 10^{-12}\,F\,m^{-1}$ |

*Scattering-factor table*

| $\dfrac{\sin\theta}{\lambda}$ (Å$^{-1}$) | H | Li | C | N | O | Na | Al | P | S | Cl | Fe | Br | Hg |
|---|---|---|---|---|---|---|---|---|---|---|---|---|---|
| 0.0 | 1.000 | 3.000 | 6.000 | 7.000 | 8.000 | 11.00 | 13.00 | 15.00 | 16.00 | 17.00 | 26.00 | 35.00 | 80.00 |
| 0.1 | 0.811 | 2.215 | 5.126 | 6.203 | 7.250 | 9.76 | 11.23 | 13.17 | 14.33 | 15.33 | 23.68 | 32.43 | 75.48 |
| 0.2 | 0.481 | 1.741 | 3.581 | 4.600 | 5.634 | 8.34 | 9.16 | 10.34 | 11.21 | 12.00 | 20.09 | 27.70 | 67.14 |
| 0.3 | 0.251 | 1.512 | 2.502 | 3.241 | 4.094 | 6.89 | 7.88 | 8.59 | 8.99 | 9.44 | 16.77 | 23.82 | 59.31 |
| 0.4 | 0.130 | 1.269 | 1.950 | 2.397 | 3.010 | 5.47 | 6.77 | 7.54 | 7.83 | 8.07 | 13.84 | 20.84 | 52.65 |
| 0.5 | 0.071 | 1.032 | 1.685 | 1.944 | 2.338 | 4.29 | 5.69 | 6.67 | 7.05 | 7.29 | 11.47 | 18.27 | 47.04 |
| 0.6 | 0.040 | 0.823 | 1.536 | 1.698 | 1.944 | 3.40 | 4.71 | 5.83 | 6.31 | 6.64 | 9.71 | 15.91 | 42.31 |
| 0.7 | 0.024 | 0.650 | 1.426 | 1.550 | 1.714 | 2.76 | 3.88 | 5.02 | 5.56 | 5.96 | 8.47 | 13.78 | 38.22 |
| 0.8 | 0.015 | 0.513 | 1.322 | 1.444 | 1.566 | 2.31 | 3.21 | 4.28 | 4.82 | 5.27 | 7.60 | 11.93 | 34.64 |
| 0.9 | 0.010 | 0.404 | 1.218 | 1.350 | 1.462 | 2.00 | 2.71 | 3.64 | 4.15 | 4.60 | 6.99 | 10.41 | 31.43 |
| 1.0 | 0.007 | 0.320 | 1.114 | 1.263 | 1.374 | 1.78 | 2.32 | 3.11 | 3.56 | 4.00 | 6.51 | 9.19 | 28.59 |

$10 \times$ *Mass absorption coefficients* $\mu/\rho\,(\mathrm{m^2\,kg^{-1}})$

| | λ(Å) | H | Li | C | N | O | Na | Al | P | S | Cl | Fe | Br | Hg |
|---|---|---|---|---|---|---|---|---|---|---|---|---|---|---|
| Ag Kβ | 0.4970 | 0.366 | 0.177 | 0.333 | 0.433 | 0.570 | 1.22 | 1.90 | 2.85 | 3.44 | 4.09 | 14.0 | 30.8 | 47.1 |
| Ag Kα | 0.5608 | 0.371 | 0.187 | 0.400 | 0.544 | 0.740 | 1.67 | 2.65 | 4.01 | 4.84 | 5.77 | 19.7 | 42.7 | 64.7 |
| Mo Kβ | 0.6323 | 0.376 | 0.200 | 0.495 | 0.700 | 0.98 | 2.32 | 3.71 | 5.64 | 6.82 | 8.14 | 27.7 | 58.8 | 87.9 |
| Mo Kα | 0.7107 | 0.380 | 0.217 | 0.625 | 0.916 | 1.31 | 3.21 | 5.16 | 7.89 | 9.55 | 11.4 | 38.5 | 79.8 | 117 |
| Cu Kβ | 1.3922 | 0.421 | 0.571 | 3.44 | 5.60 | 8.52 | 22.3 | 36.2 | 55.2 | 66.5 | 79.0 | 238 | 79.8 | 117 |
| Cu Kα | 1.5418 | 0.435 | 0.716 | 4.60 | 7.52 | 11.5 | 30.1 | 48.6 | 74.1 | 89.1 | 106 | 238 | 74.4 | 166 |
| Co Kβ | 1.6208 | 0.443 | 0.80 | 5.31 | 8.70 | 13.3 | 34.8 | 56.2 | 85.5 | 103 | 122 | 308 | 99.6 | 216 |
| Co Kα | 1.7902 | 0.464 | 1.03 | 7.07 | 11.6 | 17.8 | 46.5 | 74.8 | 114 | 136 | 161 | 349 | 115 | 245 |
| Fe Kβ | 1.7565 | 0.459 | 0.98 | 6.69 | 11.0 | 16.8 | 44.0 | 70.9 | 108 | 129 | 152 | 52.8 | 152 | 312 |
| Fe Kα | 1.9373 | 0.483 | 1.25 | 8.90 | 14.6 | 22.4 | 58.6 | 93.9 | 142 | 170 | 200 | 50.0 | 144 | 298 |
| Cr Kβ | 2.0848 | 0.507 | 1.52 | 11.0 | 18.2 | 27.8 | 72.5 | 116 | 175 | 209 | 246 | 66.4 | 190 | 377 |
| Cr Kα | 2.2909 | 0.545 | 1.96 | 14.5 | 23.9 | 36.6 | 95.3 | 152 | 229 | 272 | 318 | 82.2 | 234 | 446 |
| | | | | | | | | | | | | 108 | 305 | 547 |

The mass absorption coefficients in SI units (m$^2$ kg$^{-1}$) are one-tenth of the tabulated values. The values come from the *International Tables for X-ray Crystallography* and are in units (cm$^2$ g$^{-1}$).

# Appendices

The FORTRAN® listings given in these appendices relate to programs described and illustrated in the text and used for the solutions to examples. They are heavily interrelated, in that the output files from some of them become the input files for others. Readers are advised to examine the listings before use as they are well provided with COMMENT, C, statements which describe the workings of the programs. In addition, when running the programs users are guided by screen output and these should be carefully followed. In particular, it is important that data-file names should be correctly given and in all programs it is possible to designate the names of the input files if the default values are invalid.

These files can be downloaded free of charge from –
http://www.cup.cam.ac.uk/onlinepubs/412714/412714top.html

Appendix I

```
      PROGRAM STRUCFAC
C THIS CALCULATES STRUCTURE FACTORS FOR ALL THE 17 PLANE GROUPS.
C IT WILL DO SO BOTH FOR NORMAL AND FOR ANOMALOUS SCATTERING.
C THE INITIAL PART OF THE PROGRAM REQUIRES THE INPUT OF THE
C FOLLOWING INFORMATION:
C X-RAY WAVELENGTH - XLAM
C CELL DIMENSIONS AND ANGLE - A, B, GAMMA
C   NOTE THAT OTHER PROJECTIONS CAN BE USED IN WHICH CASE
C   A, B, GAMMA MAY, FOR EXAMPLE, BE REPLACED BY A, C, BETA SO
C   THE INSTRUCTIONS GIVEN BY THE PROGRAM WILL HAVE TO BE
C   INTERPRETED ACCORDINGLY.
C PLANE GROUP - PG
C CONTENTS OF THE CELL IN THE FORM OF:
C   TYPES OF ATOMS - ATYPE(I)
C   NUMBER OF ATOMS OF EACH TYPE IN THE ASYMMETRIC UNIT - NATOM(I)
C   THEIR COORDINATES - X, Y
C   WHETHER OR NOT THIS TYPE OF ATOM SCATTERS ANOMALOUSLY
C   IF SO THE VALUES OF F' AND F"
C THE MAXIMUM NUMBER OF ATOMS IN THE UNIT CELL IS LIMITED TO 100 BY
C THE DIMENSION OF THE ARRAYS X & Y.   THE NUMBER OF DIFFERENT TYPES
C OF ATOM IS LIMITED TO 20 BY THE DIMENSIONS OF IBEG, IEND, FP AND
C FPP AND THE MAXIMUM NUMBER OF REFLECTIONS FOR WHICH STRUCTURE
C FACTORS ARE CALCULATED IS LIMITED TO 1000 BY THE DIMENSIONS OF
C THE ARRAYS HT, KT, F & PHI.
      DIMENSION EQP(17,72),X(100),Y(100),F(1000),PHI(1000),FP(20)
     +FPP(20),NAU(17),MTYPE(13),NATOM(20),IBEG(20),IEND(20),
     +HT(1000),KT(1000)
      INTEGER HMAX,H,HT
      CHARACTER PGG*(*),PG*4,ATOM*(*),ASYMB*3,Q*4,FNAME*10
      PARAMETER (PGG='p1   p2   pm   pg   cm   pmm pmg pgg cmm p4   p4m p4g p3
     +  p3m1p31mp6   p6m ')
      PARAMETER (ATOM='H  LI C  N  O  F  NA AL P  S  CL FE BR HG END')
C THE FOLLOWING DATA STATEMENTS GIVE FOR ALL THE 17 PLANE GROUPS
C THE NUMBER OF ASYMMETRIC UNITS IN THE CELL AND CONSTANTS FOR
C CALCULATING THE POSITIONS OF ATOMS EQUIVALENT TO (X, Y) IN THE
C FORM (C*X+D*Y+E,  F*X+G*Y+H).
      DATA NAU(1),(EQP(1,I),I=1,6)/1,1,0,0,0,1,0/
      DATA NAU(2),(EQP(2,I),I=1,12)/2,1,0,0,0,1,0,-1,0,0,0,-1,0/
      DATA NAU(3),(EQP(3,I),I=1,12)/2,1,0,0,0,1,0,-1,0,0,0,1,0/
      DATA NAU(4),(EQP(4,I),I=1,12)/2,1,0,0,0,1,0,-1,0,0,0,1,0.5/
      DATA NAU(5),(EQP(5,I),I=1,24)/4,1,0,0,0,1,0,-1,0,0,0,1,0,
     +1,0,0.5,0,1,0.5,-1,0,0.5,0,1,0.5/
      DATA NAU(6),(EQP(6,I),I=1,24)/4,1,0,0,0,1,0,-1,0,0,0,1,0,
     +-1,0,0,0,-1,0,1,0,0,0,-1,0/
      DATA NAU(7),(EQP(7,I),I=1,24)/4,1,0,0,0,1,0,-1,0,0,0,-1,0,
     +1,0,0.5,0,-1,0,-1,0,0.5,0,1,0/
      DATA NAU(8),(EQP(8,I),I=1,24)/4,1,0,0,0,1,0,-1,0,0,0,-1,0,
     +1,0,0.5,0,-1,0.5,-1,0,0.5,0,1,0.5/
      DATA NAU(9),(EQP(9,I),I=1,48)/8,1,0,0,0,1,0,-1,0,0,0,1,0,
     +-1,0,0,0,-1,0,1,0,0,0,-1,0,1,0,0.5,0,1,0.5,-1,0,0.5,0,1,0.5,
     +-1,0,0.5,0,-1,0.5,1,0,0.5,0,-1,0.5/
      DATA NAU(10),(EQP(10,I),I=1,24)/4,1,0,0,0,1,0,-1,0,0,0,-1,0,
     +0,1,0,-1,0,0,0,-1,0,1,0,0/
      DATA NAU(11),(EQP(11,I),I=1,48)/8,1,0,0,0,1,0,-1,0,0,0,-1,0,
     +0,1,0,-1,0,0,0,-1,0,1,0,0,-1,0,0,0,1,0,1,0,0,0,-1,0,
     +0,-1,0,-1,0,0,0,1,0,1,0,0,0/
      DATA NAU(12),(EQP(12,I),I=1,48)/8,1,0,0,0,1,0,0,1,0,0,-1,0,0,
     +-1,0,0.5,0,1,0.5,0,-1,0.5,-1,0,0.5,-1,0,0,0,-1,0,0,1,0,0,
     +1,0,0.5,0,-1,0.5,0,1,0.5,1,0,0.5/
      DATA NAU(13),(EQP(13,I),I=1,18)/3,1,0,0,0,1,0,0,-1,0,1,-1,0,
     +-1,1,0,-1,0,0/
      DATA NAU(14),(EQP(14,I),I=1,36)/6,1,0,0,0,1,0,0,-1,0,1,-1,0,
     +-1,1,0,-1,0,0,0,-1,0,-1,0,0,1,0,0,1,-1,0,-1,1,0,0,1,0/
```

```
          DATA NAU(15), (EQP(15,I),I=1,36)/6,1,0,0,0,1,0,0,-1,0,1,-1,0,
         +-1,1,0,-1,0,0,0,1,0,1,0,0,-1,0,0,-1,1,0,1,-1,0,0,-1,0/
          DATA NAU(16), (EQP(16,I),I=1,36)/6,1,0,0,0,1,0,0,-1,0,1,-1,0,
         +-1,1,0,-1,0,0,-1,0,0,0,-1,0,0,1,0,-1,1,0,1,-1,0,1,0,0/
          DATA NAU(17), (EQP(17,I),I=1,72)/12,1,0,0,0,1,0,0,-1,0,1,-1,0,
         +-1,1,0,-1,0,0,0,1,0,1,0,0,-1,0,0,-1,1,0,1,-1,0,0,0,-1,0,
         +-1,0,0,0,-1,0,0,1,0,-1,1,0,1,-1,0,1,0,0,0,-1,0,-1,0,0,
         +1,0,0,1,-1,0,-1,1,0,0,1,0/
C
          OPEN (UNIT=9, FILE='LPT1')
C   CHANGE NAME OF OUTPUT FILE IF REQUIRED
          FNAME='SF1.DAT'
       68 WRITE(6,'('' THE OUTPUT FILE, WHICH CONTAINS H K F AND PHI '')')
          WRITE(6,'('' IS NAMED "SF1.DAT". DO YOU WANT TO CHANGE IT Y/N'')')
          READ(5,50)Q
          IF(Q.EQ.'Y'.OR.Q.EQ.'y')GOTO 66
          IF(Q.EQ.'N')GOTO 67
          IF(Q.EQ.'n')GOTO 67
          GOTO 68
       66 WRITE(6,'('' READ IN THE FILENAME [<= 10 CHARACTERS] '')')
          READ(5,58)FNAME
       58 FORMAT(A10)
       67 OPEN (UNIT=10, FILE=FNAME)
          WRITE(6,'('' INPUT X-RAY WAVELENGTH IN ANGSTROMS '')')
          READ(5,*)XLAM
          WRITE(6,'('' INPUT CELL DIMENSIONS A,B [IN ANGSTROMS '')')
          WRITE(6,'('' AND GAMMA IN DEGREES '')')
          READ(5,*)A,B,GAMMA
C   READ IN PLANE GROUP SYMBOL AND TRANSLATE INTO IDENTIFYING INTEGER
        1 WRITE(6,'('' INPUT PLANE GROUP - ONE OF THE FOLLOWING '')')
          WRITE(6,'('' p1, p2, pm, pg, pgg, cm, pmm, pmg, cmm, p4,'')')
          WRITE(6,'('' p4m, p4g, p3, p3m1, p31m, p6, p6m '')')
          WRITE(6,'('' Note: use lower case symbols as shown '')')
          READ(5,50)PG
       50 FORMAT(A4)
C   FORM IDENTIFYING INTEGER 1 TO 17 FOR PLANE GROUP
          NX=INDEX(PGG,PG)+3
          NPG=NX/4
          IF(NPG*4.NE.NX)GOTO 1
C   NOW INITIALISE THE FOLLOWING QUANTITIES: NTOTAT -TOTAL NUMBER OF
C   ATOMS IN THE CELL; NOTYPE - THE NUMBER OF DIFFERENT TYPE OF ATOMS
C   IBEG AND IEND ARE ARRAYS WHICH FLAG THE FIRST AND LAST ATOMS OF
C   THE VARIOUS TYPES IN THE COORDINATE LIST
          NTOTAT=0
          NOTYPE=0
          IBEG(1)=1
        2 WRITE(6,'('' INPUT ATOMIC TYPE-ONE OF H, LI, C, N, O, F, NA,'')')
          WRITE(6,'('' AL, P, S, CL, FE, BR, HG - OR INPUT END IF ALL '')')
          WRITE(6,'('' COORDINATES HAVE BEEN READ IN '')')
          READ(5,60)ASYMB
       60 FORMAT(A3)
C   GENERATE A NUMBER FROM 1 TO 14 TO IDENTIFY THE TYPE OF ATOM WHERE
C   1 = H, 2 = LI, ......... 14 = HG, OR 15 TO END INPUT.
          NX=INDEX(ATOM,ASYMB)+2
          NASYMB=NX/3
          IF(NASYMB*3.NE.NX)GOTO 2
          IF(NASYMB.EQ.15)GOTO 33
C   A NEW TYPE OF ATOM HAS BEEN IDENTIFIED.
          NOTYPE=NOTYPE+1
          MTYPE(NOTYPE)=NASYMB
          WRITE(6,'('' INPUT THE NUMBER OF ATOMS OF THIS TYPE IN ONE '')')
          WRITE(6,'('' ASYMMETRIC UNIT '')')
          READ(5,*)NATOM(NOTYPE)
C   INPUT COORDINATES
```

```
      WRITE(6,'('' INPUT COORDINATES X, Y - ONE ATOM PER LINE '')')
      WRITE(6,'('' IF YOU ENTER A COORDINATE INCORRECTLY THERE '')')
      WRITE(6,'('' IS AN OPPORTUNITY LATER TO CORRECT IT '')')
      DO 5 I=1,NATOM(NOTYPE)
      READ(5,*)XX,YY  .
C  GENERATE COORDINATES OF ALL SYMMETRY RELATED ATOMS
      DO 5 J=1,NAU(NPG)
      NTOTAT=NTOTAT+1
      X(NTOTAT)=EQP(NPG,6*J-5)*XX+EQP(NPG,6*J-4)*YY+EQP(NPG,6*J-3)
      Y(NTOTAT)=EQP(NPG,6*J-2)*XX+EQP(NPG,6*J-1)*YY+EQP(NPG,6*J)
    5 CONTINUE
      IEND(NOTYPE)=NTOTAT
      IBEG(NOTYPE+1)=NTOTAT+1
C  IF THIS ATOMIC TYPE IS AN ANOMALOUS SCATTERER THEN THE REAL AND
C  IMAGINARY PARTS OF THE ANOMALOUS COMPONENT OF THE SCATTERING
C  FACTOR MUST BE GIVEN
   16 WRITE(6,'('' IS THIS TYPE AN ANOMALOUS SCATTERER? Y/N'')')
      READ(5,50)Q
      IF(Q.EQ.'Y   '.OR.Q.EQ.'y   ')GOTO 6
      IF(Q.EQ.'N   ')GOTO 2
      IF(Q.EQ.'n   ')GOTO 2
      GOTO 16
    6 WRITE(6,'('' INPUT REAL AND IMAGINARY PART OF A.S FACTOR'')')
      READ(5,*)FP(NOTYPE),FPP(NOTYPE)
      GOTO 2
C  AT THIS STAGE ALL ATOMIC COORDINATES HAVE BEEN INPUT
C  IN CASE OF ERROR THEY MAY NOW BE DISPAYED WITH A POSSIBILITY
C  OF CHANGING ANY INCORRECT ONE
   33 WRITE(6,'('' DO YOU WANT TO INSPECT THE COORDINATES? Y/N '')')
      READ(5,50)Q
      IF(Q.EQ.'Y'.OR.Q.EQ.'y')GOTO 80
      IF(Q.EQ.'N')GOTO 3
      IF(Q.EQ.'n')GOTO 3
      GOTO 33
   80 DO 90 I=1,NTOTAT,NAU(NPG)
      WRITE(6,200)X(I),Y(I)
  200 FORMAT(2F8.4)
   70 WRITE(6,'('' DO YOU WANT TO CHANGE THIS? Y/N'')')
      READ(5,50)Q
      IF(Q.EQ.'Y'.OR.Q.EQ.'y')GOTO 71
      IF(Q.EQ.'N')GOTO 90
      IF(Q.EQ.'n')GOTO 90
      GOTO 70
   71 WRITE(6,'('' INPUT REPLACEMENT X,Y '')')
      READ(5,*)XX,YY
      DO 72 J=1,NAU(NPG)
      X(I+J-1)=EQP(NPG,6*J-5)*XX+EQP(NPG,6*J-4)*YY+EQP(NPG,6*J-3)
   72 Y(I+J-1)=EQP(NPG,6*J-2)*XX+EQP(NPG,6*J-1)*YY+EQP(NPG,6*J)
   90 CONTINUE
    3 WRITE(6,'('' INPUT TEMPERATURE FACTOR.   IF ONE IS NOT '')')
      WRITE(6,'('' REQUIRED THEN INPUT 0 '')')
      READ(5,*)BTEMP
C  PROVIDE A TEST QUANTITY FOR THE PRESENCE OF ANOMALOUS SCATTERING
      TESTAS=0
      DO 7 I=1,NOTYPE
    7 TESTAS=TESTAS+FP(I)**2+FPP(I)**2
C  ALL INPUT IS NOW COMPLETE.   NOW STRUCTURE FACTORS ARE CALCULATED
C  TO THE LIMIT OF RESOLUTION GIVEN BY THE X-RAY WAVELENGTH.
C  NORMALLY A COMPLETE SEMICIRCLE OF STRUCTURE FACTORS WILL BE
C  CALCULATED ALTHOUGH THE ASYMMETRIC UNIT OF RECIPROCAL SPACE
C  MAY BE SMALLER THAN THIS.   CALCULATIONS ARE FOR ALL INDEPENDENT
C  REFLECTIONS  WITH k >= 0.   IF ANOMALOUS SCATTERING IS PRESENT THEN
C  THE COMPLETE CIRCLE IS CALCULATED.  FIRST CALCULATE MAXIMUM INDICES.
      HMAX=INT(2.0*A/XLAM)
```

```
          KMAX=INT(2.0*B/XLAM)
C  CO1, CO2 AND CO3 ARE USED FOR CALCULATING SIN(THETA)/LAMBDA
          CO1=0.25/A/A
          CO2=0.25/B/B
          CO3=0.5/A/B*COS(GAMMA*0.0174533)
C  SLIM IS THE LIMITING VALUE OF (SIN(THETA)/LAMBDA)**2 AND NOREF
C  RECORDS THE TOTAL NUMBER OF REFLECTIONS.
          SLIM=1/XLAM/XLAM
          NOREF=0
          DO 10 H=-HMAX,HMAX
          DO 10 K=0,KMAX
          IF(K.EQ.0.AND.H.LT.0)GOTO 10
C  NOW CALCULATE (SIN(THETA)/LAMBDA)**2 AND CHECK THAT < SLIM
          SL2=CO1*H*H+CO2*K*K-CO3*H*K
          IF(SL2.GT.SLIM)GOTO 10
          SL=SQRT(SL2)
C  NOW THE STRUCTURE FACTOR IS CALCULATED.   IF THERE IS ANOMALOUS
C  SCATTERING THEN BOTH F(h k l) AND F(-h -k -l) ARE CALCULATED.
          SUMA=0
          SUMB=0
          SMAASP=0
          SMAASM=0
          SMBASP=0
          SMBASM=0
          DO 15 I=1,NOTYPE
C  CALL SUBROUTINE TO DERIVE SCATTERING FACTOR [SCF] FOR ATOM
C  OF TYPE I FOR SIN(THETA)/LAMBDA = SL
          CALL SCAT(MTYPE(I),SL,SCF)
          PSUMA=0
          PSUMB=0
C  CALCULATE CONTRIBUTIONS TO REAL AND IMAGINARY PARTS OF THE
C  STRUCTURE FACTOR OF ATOMS OF TYPE I
          DO 26 J=IBEG(I),IEND(I)
          Z=6.2831853*(H*X(J)+K*Y(J))
          PSUMA=PSUMA+COS(Z)
     26   PSUMB=PSUMB+SIN(Z)
          SUMA=SUMA+SCF*PSUMA
          SUMB=SUMB+SCF*PSUMB
C  TEST FOR PRESENCE OF ANOMALOUS SCATTERING
          IF(TESTAS.LT.1.0E-12)GOTO 15
C  TEST FOR ATOM TYPE I AS ANOMALOUS SCATTERER
          IF(FP(I)**2+FPP(I)**2.LT.1.0E-12)GOTO 15
          SMAASP=SMAASP+PSUMA*FP(I)-PSUMB*FPP(I)
          SMAASM=SMAASM+PSUMA*FP(I)+PSUMB*FPP(I)
          SMBASP=SMBASP+PSUMB*FP(I)+PSUMA*FPP(I)
          SMBASM=SMBASM-PSUMB*FP(I)+PSUMA*FPP(I)
     15   CONTINUE
          NOREF=NOREF+1
          HT(NOREF)=H
          KT(NOREF)=K
          BTFAC=1.0
          IF(BTEMP.GT.1.0E-12)BTFAC=EXP(-BTEMP*SL2)
          F(NOREF)=SQRT((SUMA+SMAASP)**2+(SUMB+SMBASP)**2)*BTFAC
          PHI(NOREF)=ATAN2(SUMB+SMBASP,SUMA+SMAASP)*57.2958
          IF(F(NOREF).LT.1.0E-4)PHI(NOREF)=0
          IF(TESTAS.LT.1.0E-12)GOTO 10
          NOREF=NOREF+1
          HT(NOREF)=-H
          KT(NOREF)=-K
          F(NOREF)=SQRT((SUMA+SMAASM)**2+(-SUMB+SMBASM)**2)*BTFAC
          PHI(NOREF)=ATAN2(-SUMB+SMBASM,SUMA+SMAASM)*57.2958
          IF(F(NOREF).LT.1.0E-4)PHI(NOREF)=0
     10   CONTINUE
          WRITE(10,*)NOREF,(HT(I),KT(I),F(I),PHI(I),I=1,NOREF)
```

```
      WRITE(9,700)
  700 FORMAT(3(24H      h    k    F      PHI ))
      WRITE(9,100)(HT(I),KT(I),F(I),PHI(I),I=1,NOREF)
  100 FORMAT(3(I6,I4,F8.2,F6.0))
      STOP
      END



      SUBROUTINE SCAT(I,SL,SCF)
C   THERE NOW FOLLOW THE SCATTERING FACTORS FROM INTERNATIONAL TABLES
C   FOR ATOMS H, LI, C, N, O, NA, AL, P, S, CL, FE, BR AND HG AT
C   INTERVALS OF 0.1 FOR SIN(THETA)/LAMBDA. PARABOLIC INTERPOLATION IS
C   USED TO OBTAIN THE SCATTERING FACTOR FOR SIN(THETA)/LAMBDA = SL.
      DIMENSION SF(14,0:10)
      DATA (SF(1,I),I=0,10)/1.000,0.811,0.481,0.251,0.130,0.071,
     +0.040,0.024,0.015,0.010,0.007/
      DATA (SF(2,I),I=0,10)/3.000,2.215,1.741,1.521,1.269,1.032,
     +0.823,0.650,0.513,0.404,0.320/
      DATA (SF(3,I),I=0,10)/6.000,5.126,3.581,2.502,1.950,1.685,
     +1.536,1.426,1.322,1.218,1.114/
      DATA (SF(4,I),I=0,10)/7.000,6.203,4.600,3.241,2.397,1.944,
     +1.698,1.550,1.444,1.350,1.263/
      DATA (SF(5,I),I=0,10)/8.000,7.250,5.634,4.094,3.010,2.338,
     +1.944,1.714,1.566,1.462,1.374/
      DATA (SF(6,I),I=0,10)/9.000,8.293,6.691,5.044,3.760,2.878,
     +2.312,1.958,1.735,1.587,1.481/
      DATA (SF(7,I),I=0,10)/11.00,9.76,8.34,6.89,5.47,4.29,3.40,
     +2.76,2.31,2.00,1.78/
      DATA (SF(8,I),I=0,10)/13.00,11.23,9.16,7.88,6.72,5.69,4.71,
     +3.88,3.21,2.71,2.32/
      DATA (SF(9,I),I=0,10)/15.00,13.17,10.34,8.59,7.54,6.67,5.83,
     +5.02,4.28,3.64,3.11/
      DATA (SF(10,I),I=0,10)/16.00,14.33,11.21,8.99,7.83,7.05,6.31,
     +5.56,4.82,4.15,3.56/
      DATA (SF(11,I),I=0,10)/17.00,15.33,12.00,9.44,8.07,7.29,6.64,
     +5.96,5.27,4.60,4.00/
      DATA (SF(12,I),I=0,10)/26.00,23.68,20.09,16.77,13.84,11.47,
     +9.71,8.47,7.60,6.99,6.51/
      DATA (SF(13,I),I=0,10)/35.00,32.43,27.70,23.82,20.84,18.27,
     +15.91,13.78,11.93,10.41,9.19/
      DATA (SF(14,I),I=0,10)/80.00,75.48,67.14,59.31,52.65,47.04,
     +42.31,38.22,34.64,31.43,28.59/
      J=INT(SL/0.1)
      IF(J.EQ.10)J=9
      IF(J.EQ.0)J=1
      A=(SF(I,J+1)+SF(I,J-1)-2*SF(I,J))/0.02
      B=(SF(I,J+1)-SF(I,J-1))/0.2
      SCF=A*(SL-0.1*J)**2+B*(SL-0.1*J)+SF(I,J)
      RETURN
      END
```

## Appendix II

```
      PROGRAM FOUR1
C   THIS PROGRAM COMPUTES A ONE-DIMENSIONAL FOURIER SERIES.
C   THE INPUT COEFFICIENTS CAN BE IN THE FORM EITHER A(h), B(h)
C   OR |F(h)|, PHI(h).
C   THE OUTPUT RANGE FOR X/A CAN BE EITHER (-1/2,1/2) OR (0,1).
      DIMENSION X(0:100),Y(0:100),IC(0:200),A(0:50),B(0:50),Z(0:200)
      INTEGER H,HMAX
      CHARACTER*4 NC,FX
      NC='n'
      FX='f(x)'
      OPEN(UNIT=9,FILE='LPT1')
      TPI=6.2831853
C   CONSTANT TO CONVERT DEGREES TO RADIANS
      DTR=0.0174533
      WRITE(6,'('' IF DATA IS IN FORM A & B THEN INPUT 0 '')')
      WRITE(6,'('' IF IN THE FORM F & PHI THEN INPUT 1 '')')
   13 READ(5,*)KEY
      IF(KEY.EQ.0.OR.KEY.EQ.1)GOTO 12
      GOTO 13
   12 WRITE(6,'('' INDICATE INTERVAL OF X/A AT WHICH FUNCTION IS'')')
      WRITE(6,'('' CALCULATED,1/N, BY INPUT OF EVEN INTEGER N <=200'')')
      READ(5,*)N
      WRITE(6,'('' FOR OUTPUT OF X/A IN RANGE (-1/2,1/2) INPUT 0'')')
      WRITE(6,'('' FOR THE RANGE (0,1) INPUT 1'')')
    7 READ(5,*)IOUT
      IF(IOUT.EQ.0.OR.IOUT.EQ.1)GOTO 11
      GOTO 7
   11 WRITE(6,'('' INPUT h(+ve) FOLLOWED BY EITHER A(h),B(h) OR '')')
      WRITE(6,'('' F(h),PHI(h) [in degrees].   VALUES OF A, B'')')
      WRITE(6,'('' AND F MUST ALL BE < 1000.  FOR THE TERM OF'')')
      WRITE(6,'('' INDEX 0 INPUT  0, A [OR F], 0.  THE MAXIMUM'')')
      WRITE(6,'('' VALUE OF h IS 50.  TO TERMINATE DATA INPUT'')')
      WRITE(6,'('' H = 100. '')')
C   CLEAR A, B ARRAYS
      DO 50 I=0,50
      A(I)=0
   50 B(I)=0
C   READ IN DATA. CONVERT TO A,B FORM IF NECESSARY. ALSO FIND HMAX.
      HMAX=0
    3 READ(5,*)H
      IF(H.EQ.100)GOTO 4
      READ(5,*)XX,YY
      IF(KEY.EQ.0)THEN
      A(H)=XX
      B(H)=YY
      GOTO 2
      ENDIF
      A(H)=XX*COS(DTR*YY)
      B(H)=XX*SIN(DTR*YY)
    2 IF(H.GT.HMAX)HMAX=H
      GOTO 3
C   START THE SUMMATION.   SYMMETRY ABOUT THE ORIGIN IS USED BY DOING
C   SEPARATE SUMMATIONS FOR THE COSINE AND SINE TERMS FOR X/A FROM
C   0 TO 1/2 AND THEN COMBINING THEM.
    4 DO 5 I=0,N/2
      X(I)=A(0)
      Y(I)=0
      DO 6 J=1,HMAX
      ARG=TPI*FLOAT(J*I)/N
      X(I)=X(I)+2*A(J)*COS(ARG)
    6 Y(I)=Y(I)+2*B(J)*SIN(ARG)
    5 CONTINUE
      DO 8 I=0,N/2
```

```
      IF(IOUT.EQ.0)THEN
      IC(I)=I-N/2
      Z(I)=X(N/2-I)-Y(N/2-I)
      IC(N-I)=-IC(I)
      Z(N-1)=X(N/2-I)+Y(N/2-I)
      GOTO 8
      ENDIF
      IC(I)=I
      Z(I)=X(I)+Y(I)
      IC(N-I)=N-I
      Z(N-I)=X(I)-Y(I)
    8 CONTINUE
C CALCULATION COMPLETE.    OUTPUT STAGE BEGINS.
      WRITE(9,100)N
  100 FORMAT(41H THE COORDINATES ARE X/A = n/N WHERE N = ,I3)
      WRITE(9,'('' '')')
      WRITE(9,150)NC,FX,NC,FX,NC,FX,NC,FX
  150 FORMAT(4(6X,A4,1X,A4))
      WRITE(9,200)(IC(I),Z(I),I=0,N)
  200 FORMAT(4(I8,F7.2))
      STOP
      END
```

Appendix III

```
      PROGRAM SIMP1
C  THIS EVALUATES THE INTEGRALS FOR FINDING A(h) AND B(h) BY
C  SIMPSON'S RULE.    BECAUSE OF THE PERIODIC NATURE OF f(x) THE
C  SUMMATION IS DONE OVER THE POINTS 0 TO N-1 (NOT N) AND THE
C  POINT 0 IS GIVEN WEIGHT 2 RATHER THAN 1 AS USUAL.
      REAL F(0:199),A(0:50),B(0:50)
      INTEGER H,HMAX
      CHARACTER*4 HC,AC,BC,ANS*1
      HC='h'
      AC='A(h)'
      BC='B(h)'
      OPEN(UNIT=9,FILE='LPT1')
      TPI=6.2831853
      WRITE(6,'('' INPUT THE INTERVAL, 1/N, FOR QUADRATURE AS AN'')')
      WRITE(6,'('' EVEN INTEGER N <= 200 '')')
      READ(5,*)NN
      WRITE(6,'('' INPUT N VALUES OF f(x) FOR x = 0 TO 1-1/N '')')
      WRITE(6,'('' IN STEPS OF 1/N. IF YOU MAKE AN ERROR THERE  '')')
      WRITE(6,'('' WILL BE AN OPPORTUNITY TO CORRECT IT LATER. '')')
      DO 1 I=0,NN-1
      WRITE(6,100)I
  100 FORMAT(11H INPUT f{X[,I3,1H])
      READ(5,*)F(I)
    1 CONTINUE
    3 WRITE(6,'('' DO YOU WANT TO CHECK THE VALUES OF f(x)?  Y/N '')')
      READ(5,25)ANS
   25 FORMAT(A1)
      IF(ANS.EQ.'N'.OR.ANS.EQ.'n')GOTO 2
      IF(ANS.EQ.'Y'.OR.ANS.EQ.'y')GOTO 4
      GOTO 3
    4 DO 5 I=0,NN-1
      WRITE(6,200)I,NN,F(I)
  200 FORMAT(4H X =,I4,1H/,I3,7H f(x) =,F6.2)
    7 WRITE(6,'('' DO YOU WANT TO CHANGE THIS? Y/N '')')
      READ(5,25)ANS
      IF(ANS.EQ.'N'.OR.ANS.EQ.'n')GOTO 5
      IF(ANS.EQ.'Y'.OR.ANS.EQ.'y')GOTO 6
      GOTO 7
    6 WRITE(6,'('' INPUT REPLACEMENT f(x) '')')
      READ(5,*)F(I)
    5 CONTINUE
    2 WRITE(6,'('' INPUT HMAX FOR THE CALCULATION [<=50] '')')
      READ(5,*)HMAX
C  THE VALUES OF A(h) AND B(h) ARE NOW TO BE CALCULATED
      DO 10 H=0,HMAX
      AA=0
      BB=0
      W=4
C  SUMMATIONS FOR A(h) AND B(h)
      DO 12 J=0,NN-1
      W=6-W
      ARG=TPI*H*J/NN
      AA=AA+W*F(J)*COS(ARG)
      BB=BB+W*F(J)*SIN(ARG)
   12 CONTINUE
      A(H)=AA/3.0/NN
      B(H)=BB/3.0/NN
   10 CONTINUE
C  OUTPUT RESULTS
      WRITE(9,150)HC,AC,BC,HC,AC,BC,HC,AC,BC,HC,AC,BC
  150 FORMAT(4(4X,A4,A4,3X,A4))
      WRITE(9,250)(I,A(I),B(I),I=0,HMAX)
  250 FORMAT(4(I5,2F7.2))
      STOP
      END
```

Appendix IV

```
      PROGRAM FOUR2
C  FOUR2 CALCULATES A TWO-DIMENSIONAL PROJECTED ELECTRON DENSITY MAP
C  FROM STRUCTURE FACTORS INPUT FROM THE OUTPUT FILE OF STRUCFAC.
C  THE DENSITY IS SCALED TO A MAXIMUM VALUE OF 100 AND THE SCALING
C  FACTOR TO CONVERT TO ELECTRONS/ANGSTROM**2 IS GIVEN.    TO TAKE
C  ACCOUNT OF THE LIMITED WIDTH OF PRINTER PAPER THE OUTPUT IS IN
C  SECTIONS WHICH HAVE TO BE JOINED SIDEWAYS.    FOR ECONOMY THE
C  OUTPUT IS ON AN ORTHOGONAL GRID AND THE SCALING WILL NOT BE
C  ACCURATE ALONG THE AXES.    THE VALUES OF HMAX AND HMIN MUST BE
C  <= 15 AND THE MINIMUM GRID SPACING IS 1/60.
C  THERE IS A PATTERSON FUNCTION OPTION IN WHICH CASE THE SCALING
C  FACTOR IS ELECTRONS**2/ANGSTROM**2.
      DIMENSION A(-15:15,0:15),B(-15:15,0:15),AC(0:15,0:30),
     +AS(0:15,0:30),BC(0:15,0:30),BS(0:15,0:30),CSX(0:60),
     +CSY(0:60),RO(0:60,0:60),LIST(0:60),HH(500),KK(500),
     +F(500),PHI(500)
      INTEGER H,HH
      CHARACTER ANS*1,FNAME*10,DIR*1,BT*6
      DATA NPAGE/15/
      OPEN(UNIT=9,FILE='LPT1')
      TPI=6.2831853
C  CONSTANT FOR CONVERTING DEGREES TO RADIANS
      DTR=0.0174533
C  CLEAR A AND B ARRAYS
      DO 1 H=-15,15
      DO 1 K=0,15
      A(H,K)=0
      B(H,K)=0
    1 CONTINUE
      FNAME='SF1.DAT'
    4 WRITE(6,'('' IT IS ASSUMED THAT THE INPUT DATA FILE IS NAMED'')')
      WRITE(6,'('' SF1.DAT.    DO YOU WISH TO CHANGE THE NAME?  Y/N'')')
      READ(5,50)ANS
   50 FORMAT(A1)
      IF(ANS.EQ.'N'.OR.ANS.EQ.'n')GOTO 2
      IF(ANS.EQ.'Y'.OR.ANS.EQ.'y')GOTO 3
      GOTO 4
    3 WRITE(6,'('' READ IN FILENAME [<= 10 CHARACTERS] '')')
      READ(5,55)FNAME
   55 FORMAT(A10)
    2 OPEN(UNIT=10,FILE=FNAME)
      HMAX=0
      KMAX=0
C  INPUT THE CELL DIMENSIONS A & B AND THE CELL ANGLE GAMMA.
      WRITE(6,'('' INPUT CELL DIMENSIONS A,B AND GAMMA (degrees)'')')
      WRITE(6,'('' MAKE SURE THAT THEY ARE CONSISTENT WITH THE '')')
      WRITE(6,'('' X,Y COORDINATES USED IN PROGRAM STRUCFAC. '')')
      READ(5,*)AA,BB,GAMMA
      AREA=AA*BB*SIN(DTR*GAMMA)
      READ(10,*)NOREF,(HH(I),KK(I),F(I),PHI(I),I=1,NOREF)
   95 WRITE(6,'('' DO YOU WANT THE PATTERSON OPTION?  Y/N '')')
      WRITE(6,'('' REMEMBER - IF THE INPUT TAPE IS FROM PROGRAM'')')
      WRITE(6,'('' ISOFILE THEN YOU SHOULD INPUT N '')')
      READ(5,50)ANS
      IF(ANS.EQ.'Y'.OR.ANS.EQ.'y')GOTO 93
      IF(ANS.EQ.'N'.OR.ANS.EQ.'n')GOTO 94
      GOTO 95
   93 DO 96 I=1,NOREF
      F(I)=F(I)*F(I)
      PHI(I)=0
   96 CONTINUE
   94 DO 77 I=1,NOREF
      IF(ABS(HH(I)).GT.HMAX)HMAX=ABS(HH(I))
```

```
        IF(KK(I).GT.KMAX)KMAX=KK(I)
        A(HH(I),KK(I))=F(I)*COS(DTR*PHI(I))
        B(HH(I),KK(I))=F(I)*SIN(DTR*PHI(I))
     77 CONTINUE
C  INPUT OF DATA COMPLETE
        WRITE(6,'('' INPUT NX AND NY, THE NUMBER OF DIVISIONS IN THE'')')
        WRITE(6,'('' X & Y DIRECTIONS. THEY SHOULD BE DIVISIBLE BY 4.'')')
        READ(5,*)NX,NY
        NX4=NX/4
        NY4=NY/4
        NX16=16*NX
        NY16=16*NY
C  TABLES ARE NOW SET UP OF COS(2*PI*N/NX) AND COS(2*PI*M/NY) WITH N
C  FROM 0 TO NX AND M FROM 0 TO NY. THESE TABLES ARE USED IN THE
C  SUMMATIONS WHICH FOLLOW AND THEIR USE IS FAR MORE EFFICIENT THAN
C  CALCULATING EACH SINE AND COSINE AS IT IS REQUIRED.
        DO 10 I=0,NX
     10 CSX(I)=COS(TPI*I/NX)
        DO 11 I=0,NY
     11 CSY(I)=COS(TPI*I/NY)
C  NOW DO THE SUMMATIONS AC, AS, BC AND BS AFTER CLEARING ARRAYS
        DO 12 I=0,15
        DO 12 J=0,30
        AC(I,J)=0
        AS(I,J)=0
        BC(I,J)=0
        BS(I,J)=0
     12 CONTINUE
        DO 13 K=0,KMAX
        DO 13 IX=0,NX/2
        DO 13 H=-HMAX,HMAX
        IF(H.EQ.0.AND.K.EQ.0)GOTO 13
        COZ=CSX(MOD(H*IX+NX16,NX))
        ZIN=CSX(MOD(H*IX+NX16-NX4,NX))
        AC(K,IX)=AC(K,IX)+A(H,K)*COZ
        AS(K,IX)=AS(K,IX)+A(H,K)*ZIN
        BC(K,IX)=BC(K,IX)+B(H,K)*COZ
        BS(K,IX)=BS(K,IX)+B(H,K)*ZIN
     13 CONTINUE
C  THE SUMMATIONS OVER X FOR AC,AS,BC AND BS ARE FOR X = 0 TO 1/2.
C  THE REMAINING REGION FROM X = 1/2 TO 1 WILL BE GIVEN BY SYMMETRY.
C  THE F(0,0) TERM HAS BEEN EXCLUDED.   IT HAS HALF THE WEIGHT OF THE
C  OTHER TERMS AND ITS CONTRIBUTION IS NOW LOADED INTO ARRAY RO WHICH
C  WILL CONTAIN THE SCALED DENSITY.
        DO 15 IX=0,NX
        DO 15 IY=0,NY
     15 RO(IX,IY)=A(0,0)/2.0
        DO 16 IX=0,NX/2
        W=1.0
        IF(IX.EQ.NX/2)W=0.5
        DO 16 IY=0,NY
        DO 16 K=0,KMAX
        COZ=CSY(MOD(K*IY+NY16,NY))
        ZIN=CSY(MOD(K*IY+NY16-NY4,NY))
        RO(IX,IY)=RO(IX,IY)+W*(AC(K,IX)+BS(K,IX))*COZ
       ++(BC(K,IX)-AS(K,IX))*ZIN
        RO(NX-IX,IY)=RO(NX-IX,IY)+W*(AC(K,IX)-BS(K,IX))*COZ
       ++W*(BC(K,IX)+AS(K,IX))*ZIN
     16 CONTINUE
C  THE DENSITY TABLE IS NOW COMPLETE AND IT IS SCALED TO A MAXIMUM
C  MAGNITUDE OF 100.
        RMAX=0
        DO 20 I=0,NX-1
        DO 20 J=0,NY-1
```

```
        IF(ABS(RO(I,J)).GT.RMAX)RMAX=ABS(RO(I,J))
   20 CONTINUE
        FAC=100/RMAX
        DO 21 I=0,NX
        DO 21 J=0,NY
        RO(I,J)=FAC*RO(I,J)
   21 CONTINUE
C  CALCULATE FACTOR WHICH GIVES PROJECTED DENSITY FROM PRINTED VALUES
        CAF=1/FAC/AREA
C  OUTPUT THE FACTOR
        WRITE(9,700)CAF
  700 FORMAT(42H FACTOR CONVERTING TO DENSITY/UNIT AREA IS,F8.4)
        DO 66 I=0,60
        LIST(I)=I
   66 CONTINUE
C  FIND A CONVENIENT ORIENTATION FOR THE OUTPUT WHICH IS THAT WITH
C  THE LEAST NUMBER OF DIVISIONS ACROSS THE PAGE.
        DIR='X'
        BT='BOTTOM'
        M=NX/NPAGE+1
        L=0
        IF(NY.LT.NX)THEN
        DIR='Y'
        BT='   TOP'
        M=NY/NPAGE+1
        L=1
        ENDIF
        WRITE(9,100)DIR,BT
  100 FORMAT(13H OUTPUT WITH ,A1,23H HORIZONTAL AND ORIGIN ,A6,5H LEFT)
        IF(L.EQ.1)GOTO 30
C  THE MAP IS PRINTED IN SECTIONS WHICH MUST BE JOINED AT SIDE EDGES.
C  THE NUMBER OF GRIDPOINTS ACROSS THE PAGE IS NPAGE WHICH IS SET AT
C  15.    FOR OTHER AVAILABLE PAGE WIDTHS THIS CAN BE ALTERED BY THE
C  DATA STATEMENT NEAR THE BEGINNING OF THE PROGRAM.
        DO 25 N0=1,M
        WRITE(9,'('' '')')
        WRITE(9,'('' '')')
        N1=(N0-1)*NPAGE
        N2=NPAGE*N0
        IF(N0.EQ.M)N2=NX
        WRITE(9,150)(LIST(I),I=N1,N2)
  150 FORMAT(5X,21I4)
        WRITE(9,'('' '')')
        DO 25 IY=NY,0,-1
        WRITE(9,200)IY,(RO(IX,IY),IX=N1,N2)
  200 FORMAT(I4,2X,21F4.0)
        WRITE(9,'('' '')')
   25 CONTINUE
        GOTO 45
   30 DO 26 N0=1,M
        WRITE(9,'('' '')')
        WRITE(9,'('' '')')
        N1=(N0-1)*NPAGE
        N2=NPAGE*N0
        IF(N0.EQ.M)N2=NY
        WRITE(9,150)(LIST(I),I=N1,N2)
        WRITE(9,'('' '')')
        DO 26 IX=0,NX
        WRITE(9,200)IX,(RO(IX,IY),IY=N1,N2)
        WRITE(9,'('' '')')
   26 CONTINUE
   45 STOP
        END
```

## Appendix V

```
      PROGRAM FTOUE
C  THIS PROGRAM ACCEPTS THE OUTPUT FROM STRUCFAC AND CONVERTS THE
C  F's INTO EITHER U's OR E's.    THE F's ARE FIRST SORTED INTO GROUPS
C  OF ROUGHLY 40 IN ORDER OF INCREASING VALUE OF SIN(THETA).    FOR
C  EACH GROUP A CONVERSION FACTOR IS FOUND WHICH WILL, FOR U's, MAKE
C  <|U**2|>= SIGMA(f**2)/(SIGMA(f))**2 OR, FOR E's MAKE <|E**2|> = 1.
C  THESE FACTORS ARE THEORETICALLY WHAT IS REQUIRED TO CONVERT F**2 TO
C  U**2 OR E**2 AT <SIN(THETA)> FOR EACH GROUP.    A BEST FIT OF THE
C  KIND     " FACTOR = C(1)+C(2)*s+C(3)*s**2 " [s = SIN(THETA)]   IS FOUND
C  TO FIT THE CONVERSION FACTORS AND <SIN(THETA)> FOR EACH GROUP.
C  THE QUADRATIC FORMULA IS THEN USED WITH THE S FOR EACH REFLECTION
C  TO CONVERT THE VALUE OF F TO U OR E.    THESE ARE THEN PLACED IN A
C  FILE UE.DAT FOR SUBSEQUENT USE.
      REAL F(1000),THETLIST(100),F2LIST(100),AVSUMF2(30),AVTHET(30),
     +EPS(30),C(3),UE(1000)
      INTEGER ISYMB(13),NATOM(13),HT(1000),KT(1000),NUMLIST(100),
     +IEND(0:30)
      CHARACTER Q*4,FNAME*10,ASYMB*3,CU*1,ATOM*(*)
      PARAMETER (ATOM='H  LI C  N  O  NA AL P  S  CL FE BR HG END')
      OPEN(UNIT=9,FILE='LPT1')
      FNAME='UE.DAT'
      WRITE(6,'('' THE OUTPUT FILE WHICH CONTAINS H,K,U (or E) IS'')')
      WRITE(6,'('' NAMED UE.DAT.   DO YOU WANT TO CHANGE IT? (Y/N)'')')
   3  READ(5,50)Q
      IF(Q.EQ.'Y'.OR.Q.EQ.'y')GOTO 1
      IF(Q.EQ.'N'.OR.Q.EQ.'n')GOTO 2
      GOTO 3
   1  WRITE(6,'('' READ IN THE FILENAME [<= 10 CHARACTERS] '')')
      READ(5,75)FNAME
  50  FORMAT(A4)
  75  FORMAT(A10)
   2  OPEN(UNIT=10,FILE=FNAME)
C  INPUT THE UNIT CELL DATA
      WRITE(6,'('' INPUT CELL DIMENSIONS A, B [IN ANGSTROMS] AND'')')
      WRITE(6,'('' GAMMA IN DEGREES. '')')
      READ(5,*)A,B,GAMMA
C  NOW CALCULATE FACTORS WHICH WILL BE USED LATER TO FIND VALUES
C  OF SIN[THETA]/LAMDA
      CO1=0.25/A/A
      CO2=0.25/B/B
      CO3=0.5/A/B*COS(GAMMA*0.0174533)
      WRITE(6,'('' INPUT THE RADIATION WAVELENGTH IN ANGSTROMS'')')
      READ(5,*)XLAM
C  DECIDE WHETHER TO OUTPUT U's OR E's.
      WRITE(6,'('' DO YOU WANT U OR E?    (U/E) '')')
      READ(5,50)Q
      IF(Q.EQ.'U'.OR.Q.EQ.'u')THEN
      IUE=1
      GOTO 4
      ENDIF
      IF(Q.EQ.'E'.OR.Q.EQ.'e')THEN
      IUE=2
      GOTO 4
      ENDIF
C  INPUT THE ATOMIC CONTENTS OF THE WHOLE UNIT CELL
   4  WRITE(6,'('' THE CONTENTS OF THE WHOLE UNIT CELL [NOT JUST'')')
      WRITE(6,'('' ONE ASYMMETRIC UNIT] IS INPUT IN THE FORM X1,N1'')')
      WRITE(6,'('' X2,N2,ETC WHERE X1 IS THE TYPE OF ATOM, WHICH '')')
      WRITE(6,'('' MUST BE ONE OF H,LI,C,N,O,NA,AL,P,S,CL,FE,BR,HG,'')')
      WRITE(6,'('' AND N1 IS THE TOTAL NUMBER OF THAT TYPE. '')')
      WRITE(6,'(''   '')')
      ITOT=0
   6  WRITE(6,'('' INPUT ATOMIC TYPE'')')
```

```
      WRITE(6,'('' IF ALL ATOMS HAVE BEEN INPUT THEN INPUT END'')')
      READ(5,100)ASYMB
  100 FORMAT(A3)
      NX=INDEX(ATOM,ASYMB)+2
      NASYMB=NX/3
      IF(NASYMB*3.NE.NX)GOTO 6
      IF(NASYMB.EQ.14)GOTO 7
      ITOT=ITOT+1
      ISYMB(ITOT)=NASYMB
      WRITE(6,'('' INPUT NUMBER OF ATOMS OF THIS TYPE IN WHOLE CELL'')')
      READ(5,*)NATOM(ITOT)
      GOTO 6
C THE DATA FROM STRUCFAC CAN NOW BE READ IN
    7 FNAME='SF1.DAT'
   10 WRITE(6,'('' THE OUTPUT FILE FROM STRUCFAC IS ASSUMED TO BE'')')
      WRITE(6,'('' NAMED SF1.DAT. DO YOU WANT TO CHANGE THIS? (Y/N)'')')
      READ(5,50)Q
      IF(Q.EQ.'Y'.OR.Q.EQ.'y')GOTO 8
      IF(Q.EQ.'N'.OR.Q.EQ.'n')GOTO 9
      GOTO 10
    8 WRITE(6,'('' INPUT THE FILE NAME YOU WANT [<= 10 CHARACTERS]'')')
      READ(5,75)FNAME
    9 OPEN(UNIT=11,FILE=FNAME)
      READ(11,*)NOREF,(HT(I),KT(I),F(I),DUMMY,I=1,NOREF)
C EACH REFLECTION IS READ IN AND VALUES OF NUMBER OF REFLECTIONS
C SIGMA(F**2) AND SIGMA(SIN[THETA]) ARE FOUND FOR SIN[THETA] RANGES
C OF 0.01. REFLECTIONS WITH SIN[THETA] LESS THAN 0.05 ARE EXCLUDED
C SINCE THEY DISTURB THE STATISTICS.  FIRST ALL ARRAYS ARE CLEARED
      DO 66 I=1,100
      THETLIST(I)=0
      F2LIST(I)=0
      NUMLIST(I)=0
   66 CONTINUE
      DO 20 I=1,NOREF
      STHET=XLAM*SQRT(CO1*HT(I)**2+CO2*KT(I)**2-CO3*HT(I)*KT(I))
      IF(STHET.LT.0.05)GOTO 20
      ITG=INT(100*STHET)+1
      IF(ITG.GT.100)ITG=100
      THETLIST(ITG)=THETLIST(ITG)+STHET
      F2LIST(ITG)=F2LIST(ITG)+F(I)**2
      NUMLIST(ITG)=NUMLIST(ITG)+1
   20 CONTINUE
C NOW FIND GROUPS OF 40 OR SO FOR FINDING AVERAGES
C THESE GROUPS WILL SLIGHTLY OVERLAP
      DO 21 I=1,30
      IEND(I)=0
   21 CONTINUE
      IEND(0)=1
      IX=1
      IN=1
   31 SUM=0
      DO 30 I=IN,100
      SUM=SUM+NUMLIST(I)
      IF(SUM.GE.40)THEN
      IEND(IX)=I
      IX=IX+1
      IN=I
      IF(IN.EQ.100)GOTO 30
      GOTO 31
      ENDIF
   30 CONTINUE
C IF THE FINAL GROUP HAS MORE THAN 20 MEMBERS THEN IT IS ACCEPTED
C OTHERWISE IT IS ADDED TO THE PREVIOUS GROUP
      IF(SUM.LT.20)IX=IX-1
```

```
      IEND(IX)=100
C  THE TOTAL NUMBER OF GROUPS IS IX.    THE BEGINNINGS AND ENDS OF THE
C  GROUPS ARE IEND(0) TO IEND(1), IEND(1) TO IEND(2) ..........
C  ...., IEND(IX-1) TO IEND(IX).   NOW THERE ARE CALCULATED THE AVERAGE
C  F**2 IN EACH GROUP AND <SIN[THETA]>.
      DO 40 I=1,IX
      SUMF2=0
      SUMTHET=0
      NSUM=0
      DO 41 J=IEND(I-1),IEND(I)
      SUMF2=SUMF2+F2LIST(J)
      SUMTHET=SUMTHET+THETLIST(J)
      NSUM=NSUM+NUMLIST(J)
   41 CONTINUE
      AVSUMF2(I)=SUMF2/NSUM
      AVTHET(I)=SUMTHET/NSUM
   40 CONTINUE
C  NOW CALCULATE CONVERSION FACTORS FOR CALCULATION OF U OR E
      DO 60 I=1,IX
      STLAM=AVTHET(I)/XLAM
      IF(IUE.EQ.2)THEN
      FACTOR=SQRT(1.0/AVSUMF2(I))
      GOTO 62
      ENDIF
      SUM1=0
      SUM2=0
      DO 61 J=1,ITOT
      IA=ISYMB(ITOT)
      CALL SCAT(IA,STLAM,SCF)
      SUM1=SUM1+NATOM(J)*SCF
      SUM2=SUM2+NATOM(J)*SCF*SCF
   61 CONTINUE
      FACTOR=SQRT(SUM2/AVSUMF2(I))/SUM1
   62 EPS(I)=FACTOR
   60 CONTINUE
C  A BEST PARABOLIC FIT IS NOW TO BE FOUND TO GIVE EPS AS A FUNCTION
C  OF SIN[THETA]/LAMDA [=s].
      CALL PARAFIT(AVTHET,EPS,IX,C)
C  C(1)+C(2)*s+C(3)*s*s GIVES THE CONVERSION FACTOR FROM F TO U OR E
C  THE PROPER VALUE OF E(0,0) OR U(0,0) IS KNOWN AND IS NOW CALCULATED
   94 WRITE(6,'('' DO YOU WANT TO OUTPUT THE VALUES OF a, b AND c '')')
      WRITE(6,'('' GIVEN IN EQUATION 2.48?     Y/N       '')')
      READ(5,50)Q
      IF(Q.EQ.'Y'.OR.Q.EQ.'y')THEN
      IABC=1
      GOTO 93
      ENDIF
      IF(Q.EQ.'N'.OR.Q.EQ.'n')THEN
      IABC=0
      GOTO 93
      ENDIF
      GOTO 94
   93 IF(IUE.EQ.1)THEN
      Z=1.0
      GOTO 89
      ENDIF
      SUM1=0
      SUM2=0
      DO 87 I=1,ITOT
      NN=ISYMB(I)
      CALL SCATFAC(NN,0.0,SCF)
      SUM1=SUM1+NATOM(I)*SCF
      SUM2=SUM2+NATOM(I)*SCF**2
   87 CONTINUE
```

```
          Z=SUM1/SQRT(SUM2)
C   FIND E(0)
       89 DO 90 I=1,NOREF
          IF(HT(I).EQ.0.AND.KT(I).EQ.0)THEN
          UE(I)=Z
          GOTO 90
          ENDIF
       90 CONTINUE
          DO 91 I=1,NOREF
          IF(HT(I).EQ.0.AND.KT(I).EQ.0)GOTO 91
          STHET=XLAM*SQRT(CO1*HT(I)**2+CO2*KT(I)**2-CO3*HT(I)*KT(I))
          CONV=C(1)+C(2)*STHET+C(3)*STHET**2
          UE(I)=CONV*F(I)
C   CHECK THAT THE CALCULATED E IS NOT BIGGER THAN E(0) WHICH CAN
C   HAPPEN FOR A LARGE STRUCTURE FACTOR TAKING INTO ACCOUNT THE
C   INACCURACIES IN SCALING.
          IF(UE(I).GT.0.95*Z)UE(I)=0.95*Z
       91 CONTINUE
C   ALL F's HAVE NOW BEEN CONVERTED TO U OR E.   THEY ARE NOW WRITTEN ON
C   A FILE FOR FUTURE USE.
          WRITE(10,*)IUE,NOREF,(HT(I),KT(I),UE(I),I=1,NOREF)
C   THE VALUES OF a, b AND c ARE WRITTEN IF REQUESTED
          IF(IABC.EQ.0)GOTO 95
          WRITE(9,500)C
      500 FORMAT(5H a = ,E9.3,5H b = ,E9.3,5H c = ,E9.3)
       95 CU='U'
          IF(IUE.EQ.2)CU='E'
          WRITE(9,200)CU
      200 FORMAT(11H VALUES OF ,A1)
          WRITE(9,300)CU,CU,CU
      300 FORMAT(3(16H        H     K     ,A1,3X))
          WRITE(9,400)(HT(I),KT(I),UE(I),I=1,NOREF)
      400 FORMAT(3(5X,I3,I5,F7.3))
          STOP
          END


          SUBROUTINE SCATFAC(I,SL,SCF)
C   THERE NOW FOLLOW THE SCATTERING FACTORS FROM INTERNATIONAL TABLES
C   FOR ATOMS H, LI, C, N, O, NA, AL, P, S, CL, FE, BR AND HG AT
C   INTERVALS OF 0.1 FOR SIN(THETA)/LAMBDA. PARABOLIC INTERPOLATION IS
C   USED TO OBTAIN THE SCATTERING FACTOR FOR SIN(THETA)/LAMBDA = SL.
          DIMENSION SF(13,0:10)
          DATA (SF(1,I),I=0,10)/1.000,0.811,0.481,0.251,0.130,0.071,
         +0.040,0.024,0.015,0.010,0.007/
          DATA (SF(2,I),I=0,10)/3.000,2.215,1.741,1.521,1.269,1.032,
         +0.823,0.650,0.513,0.404,0.320/
          DATA (SF(3,I),I=0,10)/6.000,5.126,3.581,2.502,1.950,1.685,
         +1.536,1.426,1.322,1.218,1.114/
          DATA (SF(4,I),I=0,10)/7.000,6.203,4.600,3.241,2.397,1.944,
         +1.698,1.550,1.444,1.350,1.263/
          DATA (SF(5,I),I=0,10)/8.000,7.250,5.634,4.094,3.010,2.338,
         +1.944,1.714,1.566,1.462,1.374/
          DATA (SF(6,I),I=0,10)/11.00,9.76,8.34,6.89,5.47,4.29,3.40,
         +2.76,2.31,2.00,1.78/
          DATA (SF(7,I),I=0,10)/13.00,11.23,9.16,7.88,6.72,5.69,4.71,
         +3.88,3.21,2.71,2.32/
          DATA (SF(8,I),I=0,10)/15.00,13.17,10.34,8.59,7.54,6.67,5.83,
         +5.02,4.28,3.64,3.11/
          DATA (SF(9,I),I=0,10)/16.00,14.33,11.21,8.99,7.83,7.05,6.31,
         +5.56,4.82,4.15,3.56/
          DATA (SF(10,I),I=0,10)/17.00,15.33,12.00,9.44,8.07,7.29,6.64,
         +5.96,5.27,4.60,4.00/
          DATA (SF(11,I),I=0,10)/26.00,23.68,20.09,16.77,13.84,11.47,
         +9.71,8.47,7.60,6.99,6.51/
```

```
      DATA (SF(12,I),I=0,10)/35.00,32.43,27.70,23.82,20.84,18.27,
     +15.91,13.78,11.93,10.41,9.19/
      DATA (SF(13,I),I=0,10)/80.00,75.48,67.14,59.31,52.65,47.04,
     +42.31,38.22,34.64,31.43,28.59/
      J=INT(SL/0.1)
      IF(J.EQ.10)J=9
      IF(J.EQ.0)J=1
      A=(SF(I,J+1)+SF(I,J-1)-2*SF(I,J))/0.02
      B=(SF(I,J+1)-SF(I,J-1))/0.2
      SCF=A*(SL-0.1*J)**2+B*(SL-0.1*J)+SF(I,J)
      RETURN
      END

      SUBROUTINE PARAFIT(X,Y,NDATA,C)
C THIS SUBROUTINE FINDS THE COEFFICIENTS C(1), C(2), C(3) TO GIVE
C THE BEST FIT OF Y(I)=C(1)+C(2)*X(I)+C(3)*X(I)**2 TO A SET OF
C POINTS X(I),Y(I), I=1 TO NDATA.  IT USES THE SUBROUTINE GAUSSJ
C WHICH IS DERIVED FROM "NUMERICAL RECIPES" BY W.H.Press,
C B.P.Flannery, S.A.Teukolsky & W.T.Vetterling, PUBLISHED BY
C Cambridge University Press, 1986.
      DIMENSION X(NDATA),Y(NDATA),C(3),A(3,3),B(3),FUN(3)
      DO 1 J=1,3
      DO 2 K=1,3
      A(J,K)=0.
    2 CONTINUE
      B(J)=0.
    1 CONTINUE
      DO 3 I=1,NDATA
      FUN(1)=1
      FUN(2)=X(I)
      FUN(3)=X(I)*X(I)
      DO 4 J=1,3
      DO 5 K=1,J
      A(J,K)=A(J,K)+FUN(K)*FUN(J)
    5 CONTINUE
      B(J)=B(J)+Y(I)*FUN(J)
    4 CONTINUE
    3 CONTINUE
      DO 6 J=1,3
      DO 7 K=1,J
      A(K,J)=A(J,K)
    7 CONTINUE
    6 CONTINUE
      CALL GAUSSJ(A,3,3,B,1,1)
      DO 8 I=1,3
      C(I)=B(I)
    8 CONTINUE
      RETURN
      END

      SUBROUTINE GAUSSJ(A,N,NP,B,M,MP)
      PARAMETER (NMAX=50)
      DIMENSION A(NP,NP),B(NP,MP),IPIV(NMAX),INDXR(NMAX),INDXC(NMAX)
      DO 11 J=1,N
      IPIV(J)=0
   11    CONTINUE
      DO 22 I=1,N
      BIG=0.
      DO 13 J=1,N
      IF(IPIV(J).NE.1)THEN
      DO 12 K=1,N
      IF (IPIV(K).EQ.0) THEN
      IF (ABS(A(J,K)).GE.BIG)THEN
      BIG=ABS(A(J,K))
```

```
                IROW=J
                ICOL=K
                ENDIF
                ELSE IF (IPIV(K).GT.1) THEN
                PAUSE 'Singular matrix'
                ENDIF
12              CONTINUE
                ENDIF
13              CONTINUE
                IPIV(ICOL)=IPIV(ICOL)+1
                IF (IROW.NE.ICOL) THEN
                DO 14 L=1,N
                DUM=A(IROW,L)
                A(IROW,L)=A(ICOL,L)
                A(ICOL,L)=DUM
14              CONTINUE
                DO 15 L=1,M
                DUM=B(IROW,L)
                B(IROW,L)=B(ICOL,L)
                B(ICOL,L)=DUM
15              CONTINUE
                ENDIF
                INDXR(I)=IROW
                INDXC(I)=ICOL
                IF (A(ICOL,ICOL).EQ.0.) PAUSE 'Singular matrix.'
                PIVINV=1./A(ICOL,ICOL)
                A(ICOL,ICOL)=1.
                DO 16 L=1,N
                A(ICOL,L)=A(ICOL,L)*PIVINV
16              CONTINUE
                DO 17 L=1,M
                B(ICOL,L)=B(ICOL,L)*PIVINV
17              CONTINUE
                DO 21 LL=1,N
                IF(LL.NE.ICOL)THEN
                DUM=A(LL,ICOL)
                A(LL,ICOL)=0.
                DO 18 L=1,N
                A(LL,L)=A(LL,L)-A(ICOL,L)*DUM
18              CONTINUE
                DO 19 L=1,M
                B(LL,L)=B(LL,L)-B(ICOL,L)*DUM
19              CONTINUE
                ENDIF
21              CONTINUE
22              CONTINUE
                DO 24 L=N,1,-1
                IF(INDXR(L).NE.INDXC(L))THEN
                DO 23 K=1,N
                DUM=A(K,INDXR(L))
                A(K,INDXR(L))=A(K,INDXC(L))
                A(K,INDXC(L))=DUM
23              CONTINUE
                ENDIF
24              CONTINUE
                RETURN
                END


                SUBROUTINE SCAT(I,SL,SCF)
C       THERE NOW FOLLOW THE SCATTERING FACTORS FROM INTERNATIONAL TABLES
C       FOR ATOMS H, LI, C, N, O, NA, AL, P, S, CL, FE, BR AND HG AT
C       INTERVALS OF 0.1 FOR SIN(THETA)/LAMBDA. PARABOLIC INTERPOLATION IS
C       USED TO OBTAIN THE SCATTERING FACTOR FOR SIN(THETA)/LAMBDA = SL.
                DIMENSION SF(13,0:10)
```

```
 DATA (SF(1,I),I=0,10)/1.000,0.811,0.481,0.251,0.130,0.071,
+0.040,0.024,0.015,0.010,0.007/
 DATA (SF(2,I),I=0,10)/3.000,2.215,1.741,1.521,1.269,1.032,
+0.823,0.650,0.513,0.404,0.320/
 DATA (SF(3,I),I=0,10)/6.000,5.126,3.581,2.502,1.950,1.685,
+1.536,1.426,1.322,1.218,1.114/
 DATA (SF(4,I),I=0,10)/7.000,6.203,4.600,3.241,2.397,1.944,
+1.698,1.550,1.444,1.350,1.263/
 DATA (SF(5,I),I=0,10)/8.000,7.250,5.634,4.094,3.010,2.338,
+1.944,1.714,1.566,1.462,1.374/
 DATA (SF(6,I),I=0,10)/11.00,9.76,8.34,6.89,5.47,4.29,3.40,
+2.76,2.31,2.00,1.78/
 DATA (SF(7,I),I=0,10)/13.00,11.23,9.16,7.88,6.72,5.69,4.71,
+3.88,3.21,2.71,2.32/
 DATA (SF(8,I),I=0,10)/15.00,13.17,10.34,8.59,7.54,6.67,5.83,
+5.02,4.28,3.64,3.11/
 DATA (SF(9,I),I=0,10)/16.00,14.33,11.21,8.99,7.83,7.05,6.31,
+5.56,4.82,4.15,3.56/
 DATA (SF(10,I),I=0,10)/17.00,15.33,12.00,9.44,8.07,7.29,6.64,
+5.96,5.27,4.60,4.00/
 DATA (SF(11,I),I=0,10)/26.00,23.68,20.09,16.77,13.84,11.47,
+9.71,8.47,7.60,6.99,6.51/
 DATA (SF(12,I),I=0,10)/35.00,32.43,27.70,23.82,20.84,18.27,
+15.91,13.78,11.93,10.41,9.19/
 DATA (SF(13,I),I=0,10)/80.00,75.48,67.14,59.31,52.65,47.04,
+42.31,38.22,34.64,31.43,28.59/
 J=INT(SL/0.1)
 IF(J.EQ.10)J=9
 IF(J.EQ.0)J=1
 A=(SF(I,J+1)+SF(I,J-1)-2*SF(I,J))/0.02
 B=(SF(I,J+1)-SF(I,J-1))/0.2
 SCF=A*(SL-0.1*J)**2+B*(SL-0.1*J)+SF(I,J)
 RETURN
 END
```

Appendix VI

```
      PROGRAM HEAVY
C THIS PROGRAM ALLOWS A SIMULATION OF A HEAVY-ATOM-METHOD SOLUTION
C OF EITHER A CENTROSYMMETRIC OR NON-CENTROSYMMETRIC STRUCTURE.
C DATA FILES ARE READ IN [SF1.DAT & SF2.DAT OR OTHER CHOSEN NAMES]
C CONTAINING THE OUTPUT FROM "STRUCFAC" FOR THE WHOLE STRUCTURE AND
C FOR THE HEAVY ATOMS RESPECTIVELY.  THE OUTPUT FILE, SUITABLE FOR
C INPUT TO "FOUR2", HAS THE PHASE OF THE HEAVY ATOM CONTRIBUTION AND
C MAGNITUDE OF THE WHOLE STRUCTURE MODIFIED BY EITHER THE SIM WEIGHT
C FOR A NON-CENTROSYMMETRIC STRUCTURE OR THAT GIVEN BY WOOLFSON FOR
C A CENTROSYMMETRIC STRUCTURE.
C THERE IS ALSO AN OPTION FOR ACCEPTING THE HEAVY ATOM PHASE WITHOUT
C WEIGHTING THE MAGNITUDE.
      REAL F1(1000),F2(1000),PHI(1000)
      INTEGER H(1000),K(1000),NATOM(14),MTYPE(14)
      CHARACTER FNAME*10,ANS*1,ATYPE*3,ATOM*(*)
      PARAMETER (ATOM='H  LI C  N  O  F  NA AL P  S  CL FE BR HG END')
      FNAME='SF1.DAT'
    3 WRITE(6,'('' IT IS ASSUMED THAT THE FILE CONTAINING THE DATA'')')
      WRITE(6,'('' FOR THE WHOLE STRUCTURE IS CALLED SF1.DAT.  DO '')')
      WRITE(6,'('' YOU WANT TO CHANGE IT?   Y/N '')')
      READ(5,50)ANS
   50 FORMAT(A1)
      IF(ANS.EQ.'N'.OR.ANS.EQ.'n')GOTO 1
      IF(ANS.EQ.'Y'.OR.ANS.EQ.'y')GOTO 2
      GOTO 3
    2 WRITE(6,'('' READ IN THE REQUIRED FILE NAME '')')
      READ(5,100)FNAME
    1 OPEN(UNIT=10,FILE=FNAME)
      FNAME='SF2.DAT'
    6 WRITE(6,'('' IT IS ASSUMED THAT THE FILE CONTAINING THE DATA'')')
      WRITE(6,'('' FOR THE HEAVY ATOM STRUCTURE IS CALLED SF2.DAT.'')')
      WRITE(6,'('' DO YOU WANT TO CHANGE IT?   Y/N '')')
      READ(5,50)ANS
      IF(ANS.EQ.'N'.OR.ANS.EQ.'n')GOTO 4
      IF(ANS.EQ.'Y'.OR.ANS.EQ.'y')GOTO 5
      GOTO 6
    5 WRITE(6,'('' READ IN THE REQUIRED FILE NAME '')')
      READ(5,100)FNAME
    4 OPEN(UNIT=11,FILE=FNAME)
      FNAME='SF.DAT'
    9 WRITE(6,'('' IT IS ASSUMED THAT THE FILE CONTAINING OUTPUT'')')
      WRITE(6,'('' DATA FOR INPUT TO FOUR2 IS CALLED SF.DAT.  DO '')')
      WRITE(6,'('' YOU WANT TO CHANGE IT?   Y/N '')')
      READ(5,50)ANS
      IF(ANS.EQ.'N'.OR.ANS.EQ.'n')GOTO 7
      IF(ANS.EQ.'Y'.OR.ANS.EQ.'y')GOTO 8
      GOTO 9
    8 WRITE(6,'('' READ IN THE REQUIRED FILE NAME '')')
      READ(5,100)FNAME
  100 FORMAT(A10)
    7 OPEN(UNIT=12,FILE=FNAME)
   46 WRITE(6,'('' DO YOU WANT TO TAKE THE STRUCTURE AMPLITUDE'')')
      WRITE(6,'('' WITHOUT WEIGHT?  Y/N '')')
      READ(5,50)ANS
      IF(ANS.EQ.'N'.OR.ANS.EQ.'n')GOTO 12
      IF(ANS.EQ.'Y'.OR.ANS.EQ.'y')THEN
      KEY=1
      GOTO 13
      ENDIF
      GOTO 46
   12 WRITE(6,'('' INPUT C FOR A CENTROSYMMETRIC STRUCTURE '')')
      WRITE(6,'('' OR N FOR A NON-CENTROSYMMETRIC STRUCTURE '')')
      READ(5,50)ANS
```

```
      IF(ANS.EQ.'C'.OR.ANS.EQ.'c')GOTO 10
      IF(ANS.EQ.'N'.OR.ANS.EQ.'n')GOTO 11
      GOTO 12
   10 KEY=2
      GOTO 13
   11 KEY=3
   13 READ(10,*)NOREF,(H(I),K(I),F1(I),PHI(I),I=1,NOREF)
      READ(11,*)NOREF,(H(I),K(I),F2(I),PHI(I),I=1,NOREF)
      WRITE(6,'('' INPUT CELL DIMENSIONS A, B IN ANGSTROMS '')')
      WRITE(6,'('' AND GAMMA IN DEGREES '')')
      READ(5,*)A,B,GAMMA
C  CO1, CO2 AND CO3 ARE USED FOR CALCULATING SIN(THETA)/LAMBDA
      CO1=0.25/A/A
      CO2=0.25/B/B
      CO3=0.5/A/B*COS(GAMMA*0.0174533)
      WRITE(6,'(''THERE ARE NOW INPUT THE TYPES AND NUMBERS '')')
      WRITE(6,'(''OF ATOMS WHOSE COORDINATES ARE NOT KNOWN. '')')
      WRITE(6,'(''NOTE-THE NUMBERS ARE FOR THE WHOLE UNIT CELL!'')')
      NOTYPE=0
   30 WRITE(6,'('' INPUT ATOMIC TYPE - ONE OF H, LI, C, N, O, F,'')')
      WRITE(6,'('' NA, AL, P, S, CL, FE, BR, HG OR END IF ALL '')')
      WRITE(6,'('' ATOMS HAVE BEEN READ IN '')')
      READ(5,150)ATYPE
  150 FORMAT(A3)
C  GENERATE A NUMBER FROM 1 TO 14 TO IDENTIFY THE TYPE OF ATOM
C  OR 15 IF ALL ATOMS HAVE BEEN READ IN
      NX=INDEX(ATOM,ATYPE)+2
      NASYMB=NX/3
      IF(NASYMB*3.NE.NX)GOTO 30
      IF(NASYMB.EQ.15)GOTO 40
      NOTYPE=NOTYPE+1
      MTYPE(NOTYPE)=NASYMB
      WRITE(6,'('' INPUT THE NUMBER OF THIS TYPE OF ATOM IN THE'')')
      WRITE(6,'('' WHOLE UNIT CELL '')')
      READ(5,*)NATOM(NOTYPE)
      GOTO 30
   40 DO 31 I=1,NOREF
C  CALCULATE THE VALUE OF (SIN(THETA)/LAMBDA)
      SL=SQRT(H(I)*H(I)*CO1+K(I)*K(I)*CO2-H(I)*K(I)*CO3)
      SIG=0
      DO 32 J=1,NOTYPE
      CALL SCAT(MTYPE(J),SL,SCF)
      SIG=SIG+NATOM(J)*SCF*SCF
   32 CONTINUE
      X=2*F1(I)*F2(I)/SIG
      GOTO(33,34,35)KEY
C  THIS GIVES UNWEIGHTED STRUCTURE AMPLITUDE UNLESS THE HEAVY ATOM
C  CONTRIBUTION IS ZERO WHEN THE TERM IS REMOVED FROM THE SYNTHESIS.
   33 FACTOR=1
      IF(F2(I).LT.1.0E-6)FACTOR=0
      GOTO 36
C  THIS GIVES THE WEIGHT GIVEN BY WOOLFSON
   34 FACTOR=EXP(X)
      FACTOR=FACTOR/(1.0+FACTOR)
      GOTO 36
C  THERE FOLLOWS A CALCULATION OF THE SIM WEIGHT.  IF X>6 THEN IT
C  IS EFFECTIVELY 1.  FOR SMALLER VALUES OF X A POLYNOMIAL
C  APPROXIMATION IS USED.
   35 IF(X.GT.6.0)THEN
      FACTOR=1
      GOTO 36
      ENDIF
      FACTOR=0.5658*X-0.1304*X*X+0.0106*X*X*X
   36 F1(I)=F1(I)*FACTOR
```

```
   31 CONTINUE
      WRITE(12,*)NOREF,(H(I),K(I),F1(I),PHI(I),I=1,NOREF)
      STOP
      END




      SUBROUTINE SCAT(I,SL,SCF)
C  THERE NOW FOLLOW THE SCATTERING FACTORS FROM INTERNATIONAL TABLES
C  FOR ATOMS H, LI, C, N, O, NA, AL, P, S, CL, FE, BR AND HG AT
C  INTERVALS OF 0.1 FOR SIN(THETA)/LAMBDA. PARABOLIC INTERPOLATION IS
C  USED TO OBTAIN THE SCATTERING FACTOR FOR SIN(THETA)/LAMBDA = SL.
      DIMENSION SF(14,0:10)
      DATA (SF(1,I),I=0,10)/1.000,0.811,0.481,0.251,0.130,0.071,
     +0.040,0.024,0.015,0.010,0.007/
      DATA (SF(2,I),I=0,10)/3.000,2.215,1.741,1.521,1.269,1.032,
     +0.823,0.650,0.513,0.404,0.320/
      DATA (SF(3,I),I=0,10)/6.000,5.126,3.581,2.502,1.950,1.685,
     +1.536,1.426,1.322,1.218,1.114/
      DATA (SF(4,I),I=0,10)/7.000,6.203,4.600,3.241,2.397,1.944,
     +1.698,1.550,1.444,1.350,1.263/
      DATA (SF(5,I),I=0,10)/8.000,7.250,5.634,4.094,3.010,2.338,
     +1.944,1.714,1.566,1.462,1.374/
      DATA (SF(6,I),I=0,10)/9.000,8.293,6.691,5.044,3.760,2.878,
     +2.312,1.958,1.735,1.587,1.481/
      DATA (SF(7,I),I=0,10)/11.00,9.76,8.34,6.89,5.47,4.29,3.40,
     +2.76,2.31,2.00,1.78/
      DATA (SF(8,I),I=0,10)/13.00,11.23,9.16,7.88,6.72,5.69,4.71,
     +3.88,3.21,2.71,2.32/
      DATA (SF(9,I),I=0,10)/15.00,13.17,10.34,8.59,7.54,6.67,5.83,
     +5.02,4.28,3.64,3.11/
      DATA (SF(10,I),I=0,10)/16.00,14.33,11.21,8.99,7.83,7.05,6.31,
     +5.56,4.82,4.15,3.56/
      DATA (SF(11,I),I=0,10)/17.00,15.33,12.00,9.44,8.07,7.29,6.64,
     +5.96,5.27,4.60,4.00/
      DATA (SF(12,I),I=0,10)/26.00,23.68,20.09,16.77,13.84,11.47,
     +9.71,8.47,7.60,6.99,6.51/
      DATA (SF(13,I),I=0,10)/35.00,32.43,27.70,23.82,20.84,18.27,
     +15.91,13.78,11.93,10.41,9.19/
      DATA (SF(14,I),I=0,10)/80.00,75.48,67.14,59.31,52.65,47.04,
     +42.31,38.22,34.64,31.43,28.59/
      J=INT(SL/0.1)
      IF(J.EQ.10)J=9
      IF(J.EQ.0)J=1
      A=(SF(I,J+1)+SF(I,J-1)-2*SF(I,J))/0.02
      B=(SF(I,J+1)-SF(I,J-1))/0.2
      SCF=A*(SL-0.1*J)**2+B*(SL-0.1*J)+SF(I,J)
      RETURN
      END
```

Appendix VII

```
      PROGRAM ISOFILE
C THIS PROGRAM TAKES DATA FILES FROM STRUCFAC FOR TWO ISOMORPHOUS
C COMPOUNDS.    THEY ARE COMBINED TO GIVE AN INPUT TAPE FOR FOUR2
C TO GIVE A DIFFERENCE PATTERSON MAP.    IF THE STRUCTURE IS
C CENTROSYMMETRIC THEN THE COEFFICIENTS ARE (|FA| - |FB|)**2;
C IF NON-CENTROSYMMETRIC THEN THEY ARE |FA|**2 - |FB|**2.    IN BOTH
C CASES THE PHASE IS PUT AT ZERO.    BECAUSE THE COEFFICIENTS ARE
C ALREADY IN PATTERSON FORM FOUR2 SHOULD BE USED IN NORMAL MODE -
C NOT IN PATTERSON MODE.
      REAL FTAB(1000,2),PHI(1000)
      INTEGER HTAB(1000,2),KTAB(1000,2)
      CHARACTER FNAME*10,FNAME1*10,FNAME2*10,ANS*1
      OPEN(UNIT=9,FILE='LPT1')
      WRITE(6,'('' READ IN THE NAME OF THE FIRST DATA FILE '')')
      WRITE(6,'('' [<= 10 CHARACTERS] '')')
      READ(5,50)FNAME1
      OPEN(UNIT=10,FILE=FNAME1)
      WRITE(6,'('' READ IN THE NAME OF THE SECOND DATA FILE '')')
      WRITE(6,'('' [,= 10 CHARACTERS] '')')
      READ(5,50)FNAME2
      OPEN(UNIT=11,FILE=FNAME2)
   50 FORMAT(A10)
      FNAME='SF1.DAT'
    1 WRITE(6,'(''IT IS ASSUMED THAT THE OUTPUT DATA FILE IS NAMED'')')
      WRITE(6,'(''SF1.DAT.  DO YOU WANT TO CHANGE THIS?  Y/N '')')
      READ(5,100)ANS
  100 FORMAT(A1)
      IF(ANS.EQ.'N'.OR.ANS.EQ.'n')GOTO 2
      IF(ANS.EQ.'Y'.OR.ANS.EQ.'y')GOTO 3
      GOTO 1
    3 WRITE(6,'('' READ IN FILENAME [<= 10 CHARACTERS]'')')
      READ(5,50)FNAME
    2 OPEN(UNIT=12,FILE=FNAME)
      READ(10,*)NOREF,(HTAB(I,1),KTAB(I,1),FTAB(I,1),DUMMY,I=1,NOREF)
      READ(11,*)NOREF,(HTAB(I,2),KTAB(I,2),FTAB(I,2),DUMMY,I=1,NOREF)
    7 WRITE(6,'('' INPUT C FOR A CENTROSYMMETRIC STRUCTURE OR N '')')
      WRITE(6,'('' FOR A NON-CENTROSYMMETRIC STRUCTURE '')')
      READ(5,100)ANS
      IF(ANS.EQ.'C'.OR.ANS.EQ.'c')GOTO 5
      IF(ANS.EQ.'N'.OR.ANS.EQ.'n')GOTO 6
      GOTO 7
    5 KEY=1
      GOTO 8
    6 KEY=2
C THE OUTPUT TAPE WILL NOW BE PRODUCED.    HOWEVER, A CHECK WILL BE
C MADE THAT THE INDICES ON THE TWO INPUT TAPES CORRESPOND.    IF NOT
C THEN THE PROGRAM IS TERMINATED WITH AN ERROR MESSAGE.
    8 DO 9 I=1,NOREF
      GOTO(10,11)KEY
   10 FTAB(I,1)=(FTAB(I,1)-FTAB(I,2))**2
      GOTO 12
   11 FTAB(I,1)=FTAB(I,1)**2-FTAB(I,2)**2
   12 IF(HTAB(I,1).NE.HTAB(I,2))GOTO 99
      IF(KTAB(I,1).NE.KTAB(I,2))GOTO 99
      PHI(I)=0
    9 CONTINUE
      WRITE(12,*)NOREF,(HTAB(I,1),KTAB(I,1),FTAB(I,1),PHI(I),I=1,NOREF)
      GOTO 101
   99 WRITE(6,'('' THE INDICES ON THE TWO INPUT TAPES DO NOT AGREE'')')
      WRITE(6,'('' THE PROGRAM IS ABORTED.  CHECK PRINTOUTS OF '')')
      WRITE(6,'('' INPUT FILES '')')
  101 STOP
      END
```

Appendix VIII

```
      PROGRAM ISOCOEFF
C  THIS PROGRAM CALCULATES THE FOURIER COEFFICIENTS AND PHASES FOR
C  A DOUBLE FOURIER SUMMATION GIVEN THAT THE POSITIONS OF I-R ATOMS
C  AND THE STRUCTURE AMPLITUDES FOR THE TWO ISOMORPHOUS STRUCTURES
C  ARE KNOWN.    THE SIMULATION REQUIRES:
C      (i)    STRUCFAC OUTPUT FILE FOR STRUCTURE A
C      (ii)   STRUCFAC OUTPUT FILE FOR STRUCTURE B
C      (iii)  STRUCFAC OUTPUT FILE FOR I-R ATOMS A
C      (iv)   STRUCFAC OUTPUT FILE FOR I-R ATOMS B
C  THE OUTPUT FILE WILL BE SUITABLE AS AN INPUT FILE FOR FOUR2, THE
C  PRINT-OUT OF WHICH WILL BE THE DOUBLE FOURIER SUMMATION.
      REAL FA(1000),FB(1000),FIRA(1000),FIRB(1000),PHI(1000)
      INTEGER H(1000),K(1000)
      CHARACTER ANS*3,FNAME*10
C  SET NAME OF INPUT FILE FOR STRUCTURE A
      FNAME='SF1.DAT'
    3 WRITE(6,'('' IT IS ASSUMED THAT THE STRUCTURE FACTOR '')')
      WRITE(6,'('' INFORMATION FOR STRUCTURE A IS IN FILE SF1.DAT.'')')
      WRITE(6,'('' DO YOU WANT TO CHANGE THIS?   Y/N '')')
      READ(5,50)ANS
   50 FORMAT(A3)
      IF(ANS.EQ.'N'.OR.ANS.EQ.'n')GOTO 1
      IF(ANS.EQ.'Y'.OR.ANS.EQ.'y')GOTO 2
      GOTO 3
    2 WRITE(6,'(''READ IN NAME OF FILE [<= 10 CHARACTERS] '')')
      READ(5,100)FNAME
  100 FORMAT(A10)
    1 OPEN(UNIT=10,FILE=FNAME)
C  SET NAME OF INPUT FILE FOR STRUCTURE B
      FNAME='SF2.DAT'
   13 WRITE(6,'('' IT IS ASSUMED THAT THE STRUCTURE FACTOR '')')
      WRITE(6,'('' INFORMATION FOR STRUCTURE B IS IN FILE SF2.DAT.'')')
      WRITE(6,'('' DO YOU WANT TO CHANGE THIS?   Y/N '')')
      READ(5,50)ANS
      IF(ANS.EQ.'N'.OR.ANS.EQ.'n')GOTO 11
      IF(ANS.EQ.'Y'.OR.ANS.EQ.'y')GOTO 12
      GOTO 13
   12 WRITE(6,'(''READ IN NAME OF FILE [<= 10 CHARACTERS] '')')
      READ(5,100)FNAME
   11 OPEN(UNIT=11,FILE=FNAME)
C  SET NAME OF INPUT FILE FOR I-R ATOMS A
      FNAME='SFIRA.DAT'
   23 WRITE(6,'('' IT IS ASSUMED THAT THE STRUCTURE FACTOR '')')
      WRITE(6,'('' INFORMATION FOR I-R ATOMS A IS IN FILE '')')
      WRITE(6,'('' SFIRA.DAT. DO YOU WANT TO CHANGE THIS?   Y/N '')')
      READ(5,50)ANS
      IF(ANS.EQ.'N'.OR.ANS.EQ.'n')GOTO 21
      IF(ANS.EQ.'Y'.OR.ANS.EQ.'y')GOTO 22
      GOTO 23
   22 WRITE(6,'(''READ IN NAME OF FILE [<= 10 CHARACTERS] '')')
      READ(5,100)FNAME
   21 OPEN(UNIT=12,FILE=FNAME)
C  SET NAMES OF INPUT FILE FOR I-R ATOMS B
      FNAME='SFIRB.DAT'
   33 WRITE(6,'('' IT IS ASSUMED THAT THE STRUCTURE FACTOR '')')
      WRITE(6,'('' INFORMATION FOR I-R ATOMS B IS IN FILE '')')
      WRITE(6,'('' SFIRB.DAT. DO YOU WANT TO CHANGE THIS?   Y/N '')')
      READ(5,50)ANS
      IF(ANS.EQ.'N'.OR.ANS.EQ.'n')GOTO 31
      IF(ANS.EQ.'Y'.OR.ANS.EQ.'y')GOTO 32
      GOTO 33
   32 WRITE(6,'(''READ IN NAME OF FILE [<= 10 CHARACTERS] '')')
      READ(5,100)FNAME
```

```
   31 OPEN(UNIT=13,FILE=FNAME)
C  SET NAME OF OUTPUT FILE
      FNAME='SDF.DAT'
   43 WRITE(6,'('' IT IS ASSUMED THAT THE OUTPUT FILE FOR THE  '')')
      WRITE(6,'('' DOUBLE FOURIER SUMMATION WILL BE IN SDF.DAT. '')')
      WRITE(6,'('' DO YOU WANT TO CHANGE THIS?   Y/N  '')')
      READ(5,50)ANS
      IF(ANS.EQ.'N'.OR.ANS.EQ.'n')GOTO 41
      IF(ANS.EQ.'Y'.OR.ANS.EQ.'y')GOTO 42
      GOTO 43
   42 WRITE(6,'(''READ IN NAME OF FILE [<= 10 CHARACTERS] '')')
      READ(5,100)FNAME
   41 OPEN(UNIT=14,FILE=FNAME)
C  READ IN ALL DATA FROM INPUT FILES
      READ(10,*)NOREF,(H(I),K(I),FA(I),PHI(I),I=1,NOREF)
      READ(11,*)NOREF,(H(I),K(I),FB(I),PHI(I),I=1,NOREF)
      READ(12,*)NOREF,(H(I),K(I),FIRA(I),PHI(I),I=1,NOREF)
      READ(13,*)NOREF,(H(I),K(I),FIRB(I),PHI(I),I=1,NOREF)
C  CALCULATE VALUES OF K*COS(DELTA PHI)
      DO 10 I=1,NOREF
C  APPLY EQUATION (8.22)
      CPSI=((FIRB(I)-FIRA(I))**2+FB(I)**2-FA(I)**2)/2.0/FB(I)
     +/(FIRB(I)-FIRA(I))
C  CHECK FOR COS(PSI) WITHIN PROPER LIMITS.  IF NOT BRING TO LIMIT
      IF(ABS(CPSI).GT.1.0)CPSI=SIGN(1.0,CPSI)
C  APPLY EQUATION (8.23)
      XK=SQRT(FB(I)**2+FIRB(I)**2-2.0*FB(I)*FIRB(I)*CPSI)
C  APPLY EQUATION (8.24) TO FIND COS(DELTA PHI)
      CDPHI=(FB(I)**2-XK**2-FIRB(I)**2)/2.0/XK/FIRB(I)
      FA(I)=XK*CDPHI
   10 CONTINUE
C  WRITE OUTPUT FILE
      WRITE(14,*)NOREF,(H(I),K(I),FA(I),PHI(I),I=1,NOREF)
      STOP
      END
```

## Appendix IX

```fortran
      PROGRAM ANOFILE
C  THIS PROGRAM TAKES THE OUTPUT FILE FROM STRUCFAC FOR ANOMALOUS
C  SCATTERING AND GIVES AN OUTPUT FILE WITH FOURIER AMPLITUDES
C  (|F(H)|-|F(-H)|).    THIS CAN BE USED AS INPUT TO FOUR2 TO
C  CALCULATE A PATTERSON FOR LOCATING THE ANOMALOUS SCATTERERS.
      REAL F(1000),PHI(1000)
      INTEGER H(1000),K(1000)
      CHARACTER ANS*1, FNAME*10
C  GIVE THE NAME OF THE INPUT DATA FILE.
      FNAME='SF1.DAT'
    3 WRITE(6,'('' THE INPUT FILE IS ASSUMED TO BE SF1.DAT '')')
      WRITE(6,'('' DO YOU WANT TO CHANGE THIS?   Y/N   '')')
      READ(5,50)ANS
   50 FORMAT(A1)
      IF(ANS.EQ.'N'.OR.ANS.EQ.'n')GOTO 1
      IF(ANS.EQ.'Y'.OR.ANS.EQ.'y')GOTO 2
      GOTO 3
    2 WRITE(6,'('' READ IN THE FILE NAME [<= 10 CHARACTERS] '')')
      READ(5,100)FNAME
  100 FORMAT(A10)
    1 OPEN(UNIT=10,FILE=FNAME)
      FNAME='SANO.DAT'
   13 WRITE(6,'('' THE OUPUT FILE IS ASSUMED TO BE SANO.DAT '')')
      WRITE(6,'('' DO YOU WANT TO CHANGE THIS?   Y/N   '')')
      READ(5,50)ANS
      IF(ANS.EQ.'N'.OR.ANS.EQ.'n')GOTO 11
      IF(ANS.EQ.'Y'.OR.ANS.EQ.'y')GOTO 12
      GOTO 13
   12 WRITE(6,'('' READ IN THE FILE NAME [<= 10 CHARACTERS] '')')
      READ(5,100)FNAME
   11 OPEN(UNIT=11,FILE=FNAME)
C  MODIFY ARRAY F TO CONTAIN ANOMALOUS DIFFERENCE.   SINCE THE
C  REFLECTIONS ARE IN ADJACENT PAIRS WITH INDICES h AND -h THE
C  OUTPUT FILE HAS ONLY ONE HALF THE NUMBER OF REFLECTIONS AND
C  THE STRUCTURE AMPLITUDES ARE THE DIFFERENCES OF NEIGHBOURING
C  F VALUES ON THE INPUT FILE.
      READ(10,*)NOREF,(H(I),K(I),F(I),PHI(I),I=1,NOREF)
      DO 10 I=1,NOREF-1,2
      F(I)=ABS(F(I)-F(I+1))
   10 CONTINUE
C IDENTIFY ABOUT 25% OF THE LARGEST ANOMALOUS DIFFERENCES AND
C PUT THE REMAINDER EQUAL TO ZERO IN OUTPUT FILE
      DIFMAX=0
      DO 20 I=1,NOREF-1,2
      IF(F(I).GT.DIFMAX)DIFMAX=F(I)
   20 CONTINUE
      DO 30 I=1,19
      XLIM=(1.0-0.05*I)*DIFMAX
      NUM=0
      DO 31 J=1,NOREF-1,2
      IF(F(J).GT.XLIM)NUM=NUM+1
   31 CONTINUE
      IF(FLOAT(NUM)/FLOAT(NOREF).GT.0.125)GOTO 35
   30 CONTINUE
   35 DO 36 I=1,NOREF-1,2
      IF(F(I).LT.XLIM)F(I)=0
   36 CONTINUE
C  READ IN TO FILE FOR INPUT TO FOUR2
      NOREF2=NOREF/2
      WRITE(11,*)NOREF2,(H(I),K(I),F(I),PHI(I),I=1,NOREF-1,2)
      STOP
      END
```

## Appendix X

```
      PROGRAM PSCOEFF
C  THIS PROGRAM TAKES THE OUTPUT FILE FROM STRUCFAC FOR ANOMALOUS
C  SCATTERING AND GIVES AN OUTPUT FILE WHICH, USED AS AN INPUT FILE
C  FOR FOUR2, GIVES A PS-FUNCTION MAP.
      REAL F(1000),PHI(1000)
      INTEGER H(1000),K(1000)
      CHARACTER ANS*1, FNAME*10
C  GIVE THE NAME OF THE INPUT DATA FILE.
      FNAME='SF1.DAT'
    3 WRITE(6,'('' THE INPUT FILE IS ASSUMED TO BE SF1.DAT '')')
      WRITE(6,'('' DO YOU WANT TO CHANGE THIS?   Y/N   '')')
      READ(5,50)ANS
   50 FORMAT(A1)
      IF(ANS.EQ.'N'.OR.ANS.EQ.'n')GOTO 1
      IF(ANS.EQ.'Y'.OR.ANS.EQ.'y')GOTO 2
      GOTO 3
    2 WRITE(6,'('' READ IN THE FILE NAME [<= 10 CHARACTERS] '')')
      READ(5,100)FNAME
  100 FORMAT(A10)
    1 OPEN(UNIT=10,FILE=FNAME)
      FNAME='PSF.DAT'
   13 WRITE(6,'('' THE OUPUT FILE IS ASSUMED TO BE PSF.DAT '')')
      WRITE(6,'('' DO YOU WANT TO CHANGE THIS?   Y/N   '')')
      READ(5,50)ANS
      IF(ANS.EQ.'N'.OR.ANS.EQ.'n')GOTO 11
      IF(ANS.EQ.'Y'.OR.ANS.EQ.'y')GOTO 12
      GOTO 13
   12 WRITE(6,'('' READ IN THE FILE NAME [<= 10 CHARACTERS] '')')
      READ(5,100)FNAME
   11 OPEN(UNIT=11,FILE=FNAME)
C  MODIFY ARRAY F TO CONTAIN ANOMALOUS INTENSITY DIFFERENCE
C  MAGNITUDES AND ARRAY PHI TO CONTAIN PI/2 OR -PI/2 DEPENDING
C  ON WHETHER |F(h)|**2-|F(-h)|**2 IS POSITIVE OR NEGATIVE.
C  THIS WILL GIVE THE PS-FUNCTION.   SINCE THE REFLECTIONS ARE
C  IN ADJACENT PAIRS WITH INDICES h AND -h THE OUTPUT FILE HAS
C  ONLY ONE HALF THE NUMBER OF REFLECTIONS ON THE INPUT FILE.
      READ(10,*)NOREF,(H(I),K(I),F(I),PHI(I),I=1,NOREF)
      DO 10 I=1,NOREF-1,2
      X=(F(I)**2-F(I+1)**2)
      F(I)=X*X
      PHI(I)=90.0*SIGN(1.0,X)
   10 CONTINUE
C  READ IN TO FILE FOR INPUT TO FOUR2
      NOREF2=NOREF/2
      WRITE(11,*)NOREF2,(H(I),K(I),F(I),PHI(I),I=1,NOREF-1,2)
      STOP
      END
```

Appendix XI

```
      PROGRAM MINDIR
C ------------------------------------------------------------------
C
C                     USER PLEASE NOTE
C  THIS PROGRAM CONTAINS A RANDOM NUMBER GENERATOR WHICH USES
C  INTEGERS.   SOME COMPILERS HAVE DEFAULT INTEGERS WITH ONLY
C  16 BITS WHICH IS NOT SUFFICIENT.   HOWEVER, IN THAT CASE THERE
C  WILL ALWAYS BE A COMPILER OPTION WHICH GIVES LONG INTEGERS
C  AND THIS SHOULD BE USED.
C ------------------------------------------------------------------
C  THIS PROGRAM GIVES A SIMULATION OF A DIRECT-METHODS SOLUTION
C  FOR EITHER A CENTROSYMMETRIC OR A NON-CENTROSYMMETRIC STRUCTURE.
C  IT FINDS A NUMBER OF SETS OF PHASES WHICH SATISFY THE TANGENT
C  FORMULA, THEN APPLIES SIMPLE FIGURE-OF-MERIT TESTS AND THEN
C  COMPUTES THE E-MAP WITH THE HIGHEST COMBINED FIGURE OF MERIT.
C  IF THIS DOES NOT GIVE A SOLUTION THEN IT SUBSEQUENTLY COMPUTES
C  AND PRINTS E-MAPS FOR ANY PHASE SET SELECTED BY THE USER.
C  THE STEPS IN THE PROCESS ARE:
C      (1) GENERATE RANDOM PHASES FOR ALL REFLECTIONS AND GIVE
C          INITIAL WEIGHTS 0.25.
C      (2) CALCULATE A DENSITY MAP, WITH WEIGHTS AND SQUARE.
C      (3) FOURIER TRANSFORM THE SQUARED MAP AND FIND NEW PHASES
C          AND WEIGHTS. ACCEPT NEW PHASE AND WEIGHT ONLY IF NEW
C          WEIGHT IS GREATER THAN PREVIOUS WEIGHT.
C      (4) COMPARE WITH PREVIOUS PHASES.   IF MEAN CHANGE IS LESS
C          THAN 2 DEGREES OR IF THIS IS SIXTEENTH REFINEMENT CYCLE
C          THEN GOTO (5).    OTHERWISE RETURN TO (2).
C      (5) RETURN TO (1) UNLESS NUMBER OF TRIALS EQUALS NTRIAL.
C      (6) DO ONE LAST CYCLE ACCEPTING NEW PHASES REGARDLESS OF
C          WEIGHTS.
C      (7) OUTPUT FIGURES OF MERIT INCLUDING COMBINED FOM.
C      (8) THE MAP FOR THE BEST FOM IS OUTPUT.
C      (9) THEREAFTER USER REQUESTS E-MAP FOR DESIGNATED PHASE SETS.
C  THE INITIAL INPUT IS AN OUTPUT FILE FROM FTOUE WITH E'S.
      REAL RO(0:60,0:60),E(1000),PHI(1000),FP(1000),PHIP(1000),
     +CFOM(500),A(500),Z(500),WAIT(1000),EP(1000),WATE(1000)
      INTEGER H(1000),K(1000),HMAX,LIST(0:60)
      CHARACTER ANS*3,FNAME*10,DIR*1,BT*6
      CHARACTER PGG*(*),PG*4
      PARAMETER (PGG='p1  p2  pm  pg  cm  pmm pmg pgg cmm p4  p4m p4g p3
     +  p3m1p31mp6  p6m ')
      COMMON RO,E,PHI,H,K,NX,NY,HMAX,KMAX,NOREF,WAIT
      DATA NPAGE,NTRIAL/15,50/
      OPEN(UNIT=9,FILE='LPT1')
      PI=4.0*ATAN(1.0)
      CV=PI/180.0
      VC=1/CV
      MARK=0
   80 WRITE(6,'('' IF YOU ARE RUNNING THE PROGRAM FROM THE'')')
      WRITE(6,'('' BEGINNING THEN INPUT "B"'')')
      WRITE(6,'('' IF YOU WANT TO PRINT A MAP FROM A PREVIOUS'')')
      WRITE(6,'('' RUN THEN INPUT "P" '')')
      READ(5,50)ANS
      IF(ANS.EQ.'B'.OR.ANS.EQ.'b')GOTO 79
      IF(ANS.EQ.'P'.OR.ANS.EQ.'p')THEN
      MARK=1
      GOTO 78
      ENDIF
      GOTO 80
   78 WRITE(6,'('' READ IN THE NUMBER OF THE PHASE SET FROM THE'')')
      WRITE(6,'('' PREVIOUS RUN YOU WISH TO PRINT '')')
      READ(5,*)IBEST
      GOTO 76
C ENTER SPACE GROUP AND CELL DATA
```

```
   79 WRITE(6,'('' INPUT X-RAY WAVELENGTH IN ANGSTROMS '')')
      READ(5,*)XLAM
      WRITE(6,'('' INPUT CELL DIMENSIONS A,B [IN ANGSTROMS '')')
      WRITE(6,'('' AND GAMMA IN DEGREES '')')
      READ(5,*)AC,BC,GAMMA
C  READ IN PLANE GROUP SYMBOL AND TRANSLATE INTO IDENTIFYING INTEGER
  107 WRITE(6,'('' INPUT PLANE GROUP - ONE OF THE FOLLOWING '')')
      WRITE(6,'('' p1, p2, pm, pg, pgg, cm, pmm, pmg, cmm, p4,'')')
      WRITE(6,'('' p4m, p4g, p3, p3m1, p31m, p6, p6m '')')
      WRITE(6,'('' Note: use lower case symbols as shown '')')
      READ(5,51)PG
   51 FORMAT(A4)
C  FORM IDENTIFYING INTEGER 1 TO 17 FOR PLANE GROUP
      NSG=INDEX(PGG,PG)+3
      NPG=NSG/4
      IF(NPG*4.NE.NSG)GOTO 107
      WRITE(6,'('' READ IN NUMBER OF ATOMS IN THE UNIT CELL'')')
      READ(5,*)NATOM
      FAC=2.0/SQRT(FLOAT(NATOM))
      FNAME='UE.DAT'
    3 WRITE(6,'(''IT IS ASSUMED THAT THE INPUT FILE FROM FTOUE'')')
      WRITE(6,'(''IS NAMED UE.DAT. DO YOU WANT TO CHANGE IT? Y/N'')')
      READ(5,50)ANS
   50 FORMAT(A3)
      IF(ANS.EQ.'N'.OR.ANS.EQ.'n')GOTO 1
      IF(ANS.EQ.'Y'.OR.ANS.EQ.'y')GOTO 2
      GOTO 3
    2 WRITE(6,'(''INPUT THE REQUIRED NAME.'')')
      READ(5,100)FNAME
  100 FORMAT(A10)
    1 OPEN(UNIT=10,FILE=FNAME)
   76 FNAME='DIRECT.DAT'
  103 WRITE(6,'(''IT IS ASSUMED THAT THE PROGRAM OUTPUT FILE IS'')')
      WRITE(6,'(''NAMED DIRECT.DAT. DO YOU WANT TO CHANGE IT? Y/N'')')
      READ(5,50)ANS
      IF(ANS.EQ.'N'.OR.ANS.EQ.'n')GOTO 101
      IF(ANS.EQ.'Y'.OR.ANS.EQ.'y')GOTO 102
      GOTO 103
  102 WRITE(6,'(''INPUT THE REQUIRED NAME.'')')
      READ(5,100)FNAME
  101 OPEN(UNIT=11,FILE=FNAME)
      REWIND(11)
      IF(MARK.EQ.1)GOTO 177
      READ(10,*)IUE,NOREF,(H(I),K(I),E(I),I=1,NOREF)
      IF(IUE.EQ.1)THEN
      WRITE(6,'(''U VALUES HAVE BEEN INPUT INSTEAD OF E'')')
      STOP
      ENDIF
C  END OF DATA INPUT.
C  STORE ALL VALUES OF E IN EP AND SELECT SUBSET OF LARGE E'S
C  ALSO LOCATE POSITION OF E(0)
      EMAX=0
      DO 129 I=1,NOREF
      EP(I)=E(I)
      IF(E(I).GT.EMAX)THEN
      EMAX=E(I)
      M0=I
      ENDIF
      IF(E(I).LT.1.0)E(I)=0
  129 CONTINUE
      E(M0)=0.5*E(M0)
C  CALCULATE THE INTERVALS FOR E-MAP
      HMAX=0
      KMAX=0
```

```
         DO 5 I=1,NOREF
         IF(ABS(H(I)).GT.HMAX)HMAX=ABS(H(I))
         IF(ABS(K(I)).GT.KMAX)KMAX=ABS(K(I))
      5 CONTINUE
         IF(HMAX.GT.15)HMAX=15
         IF(KMAX.GT.15)KMAX=15
         NX=4*HMAX
         NY=4*KMAX
C  ALLOCATE RANDOM PHASES. A SEED IS REQUIRED FOR THE RANDOM
C  NUMBER GENERATOR.
         WRITE(6,'('' INPUT A SEED FOR THE RANDOM NUMBER GENERATOR'')')
         WRITE(6,'('' AS AN INTEGER BETWEEN 0 AND 10,000'')')
         READ(5,*)IDUM
         ISEED=IDUM
         IDUM=-IDUM
C  NTRIAL SETS OF PHASES WILL BE GENERATED
         DO 10 ITRIAL=1,NTRIAL
         IEND=0
C  GENERATE INITIAL SET OF RANDOM PHASES
         CALL RANFAZ(NPG,H,K,NOREF,PHI,IDUM,M0)
C  INITIALISE WEIGHTS TO 0.25
         DO 555 I=1,NOREF
         WAIT(I)=0.25
    555 CONTINUE
         ICYCLE=0
     70 CALL EMAP
C  THE E-MAP HAS BEEN CALCULATED.  IT IS NOW SQUARED.
         DO 12 I=0,NX
         DO 12 J=0,NY
         RO(I,J)=RO(I,J)**2
     12 CONTINUE
         CALL FTMAP(FP,PHIP)
C  THE STRUCTURE AMPLITUDES OF THE SQUARED MAP ARE IN ARRAY FP
C  AND THE PHASES IN ARRAY PHIP.  CALCULATE WEIGHTS FOR THE
C  NEW PHASE ESTIMATES.   FIRST A FACTOR MUST BE FOUND WHICH
C  CONVERTS COEFFICIENTS FP TO THOSE OF SQUARED DENSITY.
C  NOW CALCULATE NEXT PHASES UNLESS IT IS FINAL CYCLE
         SUM=0
         DO 536 I=1,NOREF
         SUM=SUM+(WAIT(I)*E(I))**2
    536 CONTINUE
         SCALE=SUM/FP(M0)
         DO 537 I=1,NOREF
         FP(I)=SCALE*FP(I)
    537 CONTINUE
         RF=4.0
C  ADJUST THE CONSTANT RF UNTIL AT LEAST ONE PHASE IS CHANGED.
C  HOWEVER, RF HAS A MINIMUM VALUE OF 2.5
    571 INEW=0
         DO 538 I=1,NOREF
         WATE(I)=FAC*E(I)*FP(I)/RF
         IF(WATE(I).GT.1.0)WATE(I)=1.0
         IF(WATE(I).GT.WAIT(I))INEW=INEW+1
    538 CONTINUE
         IF(INEW.EQ.0.AND.RF.GT.2.51)THEN
         RF=RF-0.5
         GOTO 571
         ENDIF
C  PHASES CHANGED IF WATE > WAIT. FOR FINAL CYCLE ALL WEIGHTS AND
C  PHASES ARE CHANGED.
         DO 667 I=1,NOREF
         IF(IEND.EQ.1)THEN
         WAIT(I)=WATE(I)
         GOTO 667
```

```
        ENDIF
        IF(WATE(I).GT.WAIT(I).OR.WATE(I).GT.0.999999)THEN
        WAIT(I)=WATE(I)
        ELSE
        PHIP(I)=PHI(I)
        ENDIF
   667 CONTINUE
        IF(IEND.EQ.1)GOTO 115
C   CHECK THAT THERE HAVE BEEN AT LEAST FIVE CYCLES
        IF(ICYCLE.LE.5)THEN
        DO 159 I=1,NOREF
        PHI(I)=PHIP(I)
   159 CONTINUE
        GOTO 370
        ENDIF
C   CHECK FOR CONDITION THAT EITHER THERE HAVE BEEN 16 CYCLES
C   OF REFINEMENT OR THAT MEAN PHASE SHIFT IS LESS THAN 2 DEGREE
        SUM=0.0
        DO 13 I=1,NOREF
        DEL=ABS(PHI(I)-PHIP(I))
        IF(DEL.GT.180.0)DEL=360.0-DEL
        SUM=SUM+DEL
        PHI(I)=PHIP(I)
    13 CONTINUE
        IF(SUM/NOREF.LE.2.0)GOTO 60
   370 ICYCLE=ICYCLE+1
        IF(ICYCLE.EQ.16)GOTO 60
        GOTO 70
C   NOW DO A FINAL CYCLE WHERE ALL THE WEIGHTS AND PHASES ARE CHANGED.
    60 IEND=1
        GOTO 70
C   CALCULATE THE FIGURES OF MERIT FOR THIS TRIAL.
   115 A(ITRIAL)=0.0
        Z(ITRIAL)=0
        DO 637 I=1,NOREF
        IF(E(I).GT.1.3)A(ITRIAL)=A(ITRIAL)+FP(I)*WAIT(I)
        IF(EP(I).LT.0.5)Z(ITRIAL)=Z(ITRIAL)+FP(I)
   637 CONTINUE
C   WRITE THE SOLUTION ON AN OUTPUT TAPE.
        WRITE(11,*)NOREF,(H(I),K(I),E(I),WAIT(I),PHI(I),I=1,NOREF)
    10 CONTINUE
C   ALL SETS OF PHASES HAVE NOW BEEN GENERATED TOGETHER WITH FOMS.
C   NORMALIZE THE FOMS TO MAXIMUM OF 1000
        AMAX=0
        ZMAX=0
        AMIN=1.0E8
        ZMIN=1.0E8
        DO 737 I=1,NTRIAL
        IF(A(I).GT.AMAX)AMAX=A(I)
        IF(Z(I).GT.ZMAX)ZMAX=Z(I)
        IF(A(I).LT.AMIN)AMIN=A(I)
        IF(Z(I).LT.ZMIN)ZMIN=Z(I)
   737 CONTINUE
        DO 837 I=1,NTRIAL
        A(I)=1000.0*A(I)/AMAX
        Z(I)=1000.0*Z(I)/ZMAX
   837 CONTINUE
        AMIN=1000.0*AMIN/AMAX
        ZMIN=1000.0*ZMIN/ZMAX
C   CALCULATE COMBINED FIGURE OF MERIT
        DO 937 I=1,NTRIAL
        CFOM(I)=(A(I)-AMIN)/(1000.0-AMIN)+(1000.0-Z(I))/(1000.0-ZMIN)
   937 CONTINUE
        WRITE(9,231)ISEED
```

```
  231 FORMAT(17H THE SEED USED IS,I7)
      WRITE(9,200)
  200 FORMAT(2(28H  SOLN    A        Z      CFOM,4X))
      WRITE(9,300)(I,A(I),Z(I),CFOM(I),I=1,NTRIAL)
  300 FORMAT(2(I6,2F8.1,F6.3,4X))
      WRITE(9,'('' '')')
      WRITE(9,'('' '')')
      CBEST=0
      DO 65 I=1,NTRIAL
      IF(CFOM(I).GT.CBEST)THEN
      CBEST=CFOM(I)
      IBEST=I
      ENDIF
   65 CONTINUE
  177 REWIND(11)
      DO 66 I=1,IBEST
      READ(11,*)NOREF,(H(II),K(II),E(II),WAIT(II),PHI(II),II=1,NOREF)
   66 CONTINUE
      DO 715 I=1,NOREF
      IF(ABS(H(I)).GT.HMAX)HMAX=ABS(H(I))
      IF(ABS(K(I)).GT.KMAX)KMAX=ABS(K(I))
  715 CONTINUE
      IF(HMAX.GT.15)HMAX=15
      IF(KMAX.GT.15)KMAX=15
      NX=4*INT(0.75*HMAX+0.5)
      NY=4*INT(0.75*KMAX+0.5)
      CALL EMAP
C   SCALE THE E-MAP READY FOR OUTPUT
      ROMAX=0
      DO 67 I=0,NX
      DO 67 J=0,NY
      IF(ABS(RO(I,J)).GT.ROMAX)ROMAX=ABS(RO(I,J))
   67 CONTINUE
      DO 68 I=0,NX
      DO 68 J=0,NY
      RO(I,J)=RO(I,J)*100.0/ROMAX
   68 CONTINUE
      DO 166 I=0,60
      LIST(I)=I
  166 CONTINUE
C   OUTPUT BEST OR CHOSEN MAP
C   FIND A CONVENIENT ORIENTATION FOR THE OUTPUT WHICH IS THAT WITH
C   THE LEAST NUMBER OF DIVISIONS ACROSS THE PAGE.
      DIR='X'
      BT='BOTTOM'
      M=NX/NPAGE+1
      L=0
      IF(NY.LT.NX)THEN
      DIR='Y'
      BT='   TOP'
      M=NY/NPAGE+1
      L=1
      ENDIF
      DO 444 IL=1,6
  444 WRITE(9,'('' '')')
      IF(MARK.EQ.1)THEN
      WRITE(9,651)IBEST
      GOTO 652
      ENDIF
  651 FORMAT(16H SOLUTION NUMBER,I4)
      WRITE(9,650)IBEST,A(IBEST),Z(IBEST),CFOM(IBEST)
  650 FORMAT(16H SOLUTION NUMBER,I4,3X,14H A,  Z,  CFOM = ,2F8.1,F9.3)
  652 WRITE(9,500)DIR,BT
  500 FORMAT(13H OUTPUT WITH ,A1,23H HORIZONTAL AND ORIGIN ,A6,5H LEFT)
```

```
      IF(L.EQ.1)GOTO 30
C  THE MAP IS PRINTED IN SECTIONS WHICH MUST BE JOINED AT SIDE EDGES.
C  THE NUMBER OF GRIDPOINTS ACROSS THE PAGE IS NPAGE WHICH IS SET AT
C  15.    FOR OTHER AVAILABLE PAGE WIDTHS THIS CAN BE ALTERED BY THE
C  DATA STATEMENT NEAR THE BEGINNING OF THE PROGRAM.
      DO 25 N0=1,M
      WRITE(9,'('' '')')
      WRITE(9,'('' '')')
      N1=(N0-1)*NPAGE
      N2=NPAGE*N0
      IF(N0.EQ.M)N2=NX
      WRITE(9,150)(LIST(I),I=N1,N2)
  150 FORMAT(5X,21I4)
      WRITE(9,'('' '')')
      DO 25 IY=NY,0,-1
      WRITE(9,200)IY,(RO(IX,IY),IX=N1,N2)
  600 FORMAT(I4,2X,21F4.0)
      WRITE(9,'('' '')')
   25 CONTINUE
      GOTO 45
   30 DO 26 N0=1,M
      WRITE(9,'('' '')')
      WRITE(9,'('' '')')
      N1=(N0-1)*NPAGE
      N2=NPAGE*N0
      IF(N0.EQ.M)N2=NY
      WRITE(9,150)(LIST(I),I=N1,N2)
      WRITE(9,'('' '')')
      DO 26 IX=0,NX
      WRITE(9,600)IX,(RO(IX,IY),IY=N1,N2)
      WRITE(9,'('' '')')
   26 CONTINUE
   45 STOP
      END



      SUBROUTINE EMAP
C  EMAP CALCULATES A TWO-DIMENSIONAL E-MAP FROM THE OUTPUT OF
C  DIRECT WITH PHASE CHANGES CONTROLLED BY WEIGHTS. VALUES OF
C  HMAX AND HMIN MUST BE <= 15 AND THE MINIMUM GRID SPACING IS 1/60.
      DIMENSION A(-15:15,0:15),B(-15:15,0:15),AC(0:15,0:30),
     +AS(0:15,0:30),BC(0:15,0:30),BS(0:15,0:30),CSX(0:60),
     +CSY(0:60),RO(0:60,0:60),H(1000),K(1000),WAIT(1000),
     +E(1000),PHI(1000)
      INTEGER H,HMAX,HH
      COMMON RO,E,PHI,H,K,NX,NY,HMAX,KMAX,NOREF,WAIT
      TPI=6.2831853
C  CONSTANT FOR CONVERTING DEGREES TO RADIANS
      DTR=0.0174533
C  CLEAR A AND B ARRAYS
      DO 1 HH=-15,15
      DO 1 KK=0,15
      A(HH,KK)=0
      B(HH,KK)=0
    1 CONTINUE
      DO 77 I=1,NOREF
      A(H(I),K(I))=E(I)*WAIT(I)*COS(DTR*PHI(I))
      B(H(I),K(I))=E(I)*WAIT(I)*SIN(DTR*PHI(I))
   77 CONTINUE
      NX4=NX/4
```

```
      NY4=NY/4
      NX16=16*NX
      NY16=16*NY
C  TABLES ARE NOW SET UP OF COS(2*PI*N/NX) AND COS(2*PI*M/NY) WITH N
C  FROM 0 TO NX AND M FROM 0 TO NY. THESE TABLES ARE USED IN THE
C  SUMMATIONS WHICH FOLLOW AND THEIR USE IS FAR MORE EFFICIENT THAN
C  CALCULATING EACH SINE AND COSINE AS IT IS REQUIRED.
      DO 10 I=0,NX
   10 CSX(I)=COS(TPI*I/NX)
      DO 11 I=0,NY
   11 CSY(I)=COS(TPI*I/NY)
C  NOW DO THE SUMMATIONS AC, AS, BC AND BS AFTER CLEARING ARRAYS
      DO 12 I=0,15
      DO 12 J=0,30
      AC(I,J)=0
      AS(I,J)=0
      BC(I,J)=0
      BS(I,J)=0
   12 CONTINUE
      DO 13 KK=0,KMAX
      DO 13 IX=0,NX/2
      DO 13 HH=-HMAX,HMAX
      IF(HH.EQ.0.AND.KK.EQ.0)GOTO 13
      COZ=CSX(MOD(HH*IX+NX16,NX))
      ZIN=CSX(MOD(HH*IX+NX16-NX4,NX))
      AC(KK,IX)=AC(KK,IX)+A(HH,KK)*COZ
      AS(KK,IX)=AS(KK,IX)+A(HH,KK)*ZIN
      BC(KK,IX)=BC(KK,IX)+B(HH,KK)*COZ
      BS(KK,IX)=BS(KK,IX)+B(HH,KK)*ZIN
   13 CONTINUE
C  THE SUMMATIONS OVER X FOR AC,AS,BC AND BS ARE FOR X = 0 TO 1/2.
C  THE REMAINING REGION FROM X = 1/2 TO 1 WILL BE GIVEN BY SYMMETRY.
C  THE E(0,0) TERM HAS BEEN EXCLUDED.   IT HAS HALF THE WEIGHT OF THE
C  OTHER TERMS AND ITS CONTRIBUTION IS NOW LOADED INTO ARRAY RO WHICH
C  WILL CONTAIN THE E-MAP.
      DO 15 IX=0,NX
      DO 15 IY=0,NY
   15 RO(IX,IY)=A(0,0)/2.0
      DO 16 IX=0,NX/2
      W=1.0
      IF(IX.EQ.NX/2)W=0.5
      DO 16 IY=0,NY
      DO 16 KK=0,KMAX
      COZ=CSY(MOD(KK*IY+NY16,NY))
      ZIN=CSY(MOD(KK*IY+NY16-NY4,NY))
      RO(IX,IY)=RO(IX,IY)+W*(AC(KK,IX)+BS(KK,IX))*COZ
     ++W*(BC(KK,IX)-AS(KK,IX))*ZIN
      RO(NX-IX,IY)=RO(NX-IX,IY)+W*(AC(KK,IX)-BS(KK,IX))*COZ
     ++W*(BC(KK,IX)+AS(KK,IX))*ZIN
   16 CONTINUE
C  THE DENSITY TABLE IS NOW COMPLETE.
      RETURN
      END


      SUBROUTINE FTMAP(FP,PHIP)
C  THIS CALCULATES THE FOURIER TRANSFORM OF THE SQUARED E-MAP - G(h).
C  NO ATTEMPT IS MADE TO SCALE THE AMPLITUDES
      REAL RO(0:60,0:60),E(1000),PHI(1000),PHIP(1000),FP(1000),
     +CSX(0:60),CSY(0:60),A(1:60,0:15),B(1:60,0:15),AF(-15:15,0:15),
     +BF(-15:15,0:15),WAIT(1000)
      INTEGER H(1000),HH,HMAX,K(1000)
      COMMON RO,E,PHI,H,K,NX,NY,HMAX,KMAX,NOREF,WAIT
      NX16=16*NX
```

```
         NX4=NX/4
         NY16=16*NY
         NY4=NY/4
         TPI=8.0*ATAN(1.0)
         CONV=360.0/TPI
C  SET UP COSINE TABLES FOR USE IN SUMMATIONS
         DO 1 I=0,NX
       1 CSX(I)=COS(TPI*I/NX)
         DO 2 I=0,NY
       2 CSY(I)=COS(TPI*I/NY)
C  SUMMATION OVER THE Y DIRECTION FOR PARTICULAR K AND X VALUES
C  FIRST CLEAR ARRAYS
         DO 11 IX=1,NX
         DO 11 KK=0,KMAX
         A(IX,KK)=0
         B(IX,KK)=0
      11 CONTINUE
         DO 12 IX=1,NX
         DO 12 KK=0,KMAX
         DO 12 IY=1,NY
         COZ=CSY(MOD(KK*IY+NY16,NY))
         ZIN=CSY(MOD(KK*IY+NY16-NY4,NY))
         A(IX,KK)=A(IX,KK)+RO(IX,IY)*COZ
         B(IX,KK)=B(IX,KK)+RO(IX,IY)*ZIN
      12 CONTINUE
C  NOW FOR EACH H AND K DO THE SUMMATIONS OVER X
C  FIRST CLEAR ARRAYS
         DO 14 HH=-HMAX,HMAX
         DO 14 KK=0,KMAX
         AF(HH,KK)=0
         BF(HH,KK)=0
      14 CONTINUE
         DO 15 HH=-HMAX,HMAX
         DO 15 KK=0,KMAX
         DO 16 IX=1,NX
         COZ=CSX(MOD(HH*IX+NX16,NX))
         ZIN=CSX(MOD(HH*IX+NX16-NX4,NX))
         AF(HH,KK)=AF(HH,KK)+A(IX,KK)*COZ-B(IX,KK)*ZIN
         BF(HH,KK)=BF(HH,KK)+A(IX,KK)*ZIN+B(IX,KK)*COZ
      16 CONTINUE
         FF=SQRT(AF(HH,KK)**2+BF(HH,KK)**2)
         PHIF=ATAN2(BF(HH,KK),AF(HH,KK))
         AF(HH,KK)=FF
         BF(HH,KK)=PHIF*CONV
      15 CONTINUE
C  TRANSFER MAGNITUDES AND PHASES TO ARRAYS FP AND PHIP
         DO 20 I=1,NOREF
         FP(I)=AF(H(I),K(I))
         PHIP(I)=BF(H(I),K(I))
      20 CONTINUE
         RETURN
         END




         SUBROUTINE RANFAZ(NPG,H,K,NOREF,PHI,IDUM,M0)
         REAL PHI(1000)
         INTEGER H(1000),K(1000),KEY(1000)
C  RANDOM PHASES ARE ALLOCATED TO ALL REFLECTIONS.  SYMMETRY-RELATED
```

```
C  REFLECTIONS ARE ALLOCATED PHASES WITH THE PROPER RELATIONSHIPS.
C  THE PROGRAM CAN BE APPLIED TO ALL 17 TWO-DIMENSIONAL SPACE
C  GROUPS.
       PI=4.0*ATAN(1.0)
       GOTO (1,2,3,4,6,6,7,8,6,10,11,11,13,14,14,13,14)NPG
    1 DO 101 I=1,NOREF
      XX=RAN2(IDUM)
      PHI(I)=360.00*XX
  101 CONTINUE
      GOTO 500
    2 DO 102 I=1,NOREF
      XX=RAN2(IDUM)
      PHI(I)=0
      IF(XX.GT.0.5)PHI(I)=180
  102 CONTINUE
      GOTO 500
    3 DO 103 I=1,NOREF
      KEY(I)=0
  103 CONTINUE
      DO 203 I=1,NOREF
      IF(KEY(I).EQ.1)GOTO 203
      XX=RAN2(IDUM)
      IF(K(I).EQ.0)GOTO 303
      PHI(I)=360.0*XX
      GOTO 403
  303 PHI(I)=0
      IF(XX.GT.0.5)PHI(I)=180
  403 DO 503 J=I+1,NOREF
      IF(KEY(J).EQ.1)GOTO 503
      IF(H(J).EQ.-H(I).AND.K(J).EQ.K(I))GOTO 603
      GOTO 503
  603 KEY(J)=1
      PHI(J)=PHI(I)
  503 CONTINUE
  203 CONTINUE
      GOTO 500
    4 DO 104 I=1,NOREF
      KEY(I)=0
  104 CONTINUE
      DO 204 I=1,NOREF
      IF(KEY(I).EQ.1)GOTO 204
      XX=RAN2(IDUM)
      IF(K(I).EQ.0)GOTO 304
      PHI(I)=360.0*XX
      GOTO 404
  304 PHI(I)=0
      IF(XX.GT.0.5)PHI(I)=180
  404 DO 504 J=I+1,NOREF
      IF(KEY(J).EQ.1)GOTO 504
      IF(H(J).EQ.-H(I).AND.K(J).EQ.K(I))GOTO 604
      GOTO 504
  604 KEY(J)=1
      PHI(J)=PHI(I)+(1.0-1.0*(-1.0)**K(I))*90.0
  504 CONTINUE
  204 CONTINUE
      GOTO 500
    6 DO 106 I=1,NOREF
      KEY(I)=0
  106 CONTINUE
      DO 206 I=1,NOREF
      IF(KEY(I).EQ.1)GOTO 206
      XX=RAN2(IDUM)
      PHI(I)=0
      IF(XX.GT.0.5)PHI(I)=180.0
```

```
      DO 306 J=I+1,NOREF
      IF(KEY(J).EQ.1)GOTO 306
      IF(H(J).EQ.-H(I).AND.K(J).EQ.K(I))GOTO 406
      GOTO 306
406   KEY(J)=1
      PHI(J)=PHI(I)
306   CONTINUE
206   CONTINUE
      GOTO 500
  7   DO 107 I=1,NOREF
      KEY(I)=0
107   CONTINUE
      DO 207 I=1,NOREF
      IF(KEY(I).EQ.1)GOTO 207
      XX=RAN2(IDUM)
      PHI(I)=0
      IF(XX.GT.0.5)PHI(I)=180.0
      DO 307 J=I+1,NOREF
      IF(KEY(J).EQ.1)GOTO 307
      IF(H(J).EQ.-H(I).AND.K(J).EQ.K(I))GOTO 407
      GOTO 307
407   KEY(J)=1
      PHI(J)=PHI(I)+(1.0-1.0*(-1.0)**H(I))*90.0
307   CONTINUE
207   CONTINUE
      GOTO 500
  8   DO 108 I=1,NOREF
      KEY(I)=0
108   CONTINUE
      DO 208 I=1,208
      IF(KEY(I).EQ.1)GOTO 208
      XX=RAN2(IDUM)
      PHI(I)=0
      IF(XX.GT.0.5)PHI(I)=180.0
      DO 308 J=I+1,NOREF
      IF(KEY(J).EQ.1)GOTO 308
      IF(H(J).EQ.-H(I).AND.K(J).EQ.K(I))GOTO 408
      GOTO 308
408   KEY(J)=1
      PHI(J)=PHI(I)+(1.0-1.0*(-1.0)**(H(I)+K(I)))*90.0
308   CONTINUE
208   CONTINUE
      GOTO 500
 10   DO 110 I=1,NOREF
      KEY(I)=0
110   CONTINUE
      DO 210 I=1,210
      IF(KEY(I).EQ.1)GOTO 210
      XX=RAN2(IDUM)
      PHI(I)=0
      IF(XX.GT.0.5)PHI(I)=180.0
      DO 310 J=I+1,NOREF
      IF(KEY(J).EQ.1)GOTO 310
      IF(H(J).EQ.-K(I).AND.K(J).EQ.H(I))GOTO 410
      IF(H(J).EQ.K(I).AND.K(J).EQ.-H(I))GOTO 410
      GOTO 310
410   KEY(J)=1
      PHI(J)=PHI(I)
310   CONTINUE
210   CONTINUE
      GOTO 500
 11   DO 111 I=1,NOREF
      KEY(I)=0
111   CONTINUE
```

```
      DO 211 I=1,NOREF
      IF(KEY(I).EQ.1)GOTO 211
      XX=RAN2(IDUM)
      PHI(I)=0
      IF(XX.GT.0.5)PHI(I)=180.0
      DO 311 J=I+1,NOREF
      IF(KEY(J).EQ.1)GOTO 311
      IF(H(J).EQ.K(I).AND.K(J).EQ.H(I))GOTO 511
      IF(H(J).EQ.-K(I).AND.K(J).EQ.-H(I))GOTO 511
      IF(H(J).EQ.-H(I).AND.K(J).EQ.K(I))GOTO 411
      IF(H(J).EQ.-K(I).AND.K(J).EQ.H(I))GOTO 411
      IF(H(J).EQ.K(I).AND.K(J).EQ.-H(I))GOTO 411
      GOTO 311
  511 KEY(J)=1
      PHI(J)=PHI(I)
      GOTO 311
  411 KEY(J)=1
      PHI(J)=PHI(I)+(1.0-1.0*(-1.0)**(H(I)+K(I)))*90.0
  311 CONTINUE
  211 CONTINUE
      GOTO 500
   13 DO 113 I=1,NOREF
      KEY(I)=0
  113 CONTINUE
      DO 213 I=1,NOREF
      IF(KEY(I).EQ.1)GOTO 213
      XX=RAN2(IDUM)
      IF(NPG.EQ.13)PHI(I)=XX*360.0
      IF(NPG.EQ.16)PHI(I)=0
      IF(NPG.EQ.16.AND.XX.GT.0.5)PHI(I)=180.0
      DO 313 J=I+1,NOREF
      IF(KEY(J).EQ.1)GOTO 313
      IF(H(J).EQ.K(I).AND.K(J).EQ.-H(I)-K(I))GOTO 413
      IF(H(J).EQ.-K(I).AND.K(J).EQ.H(I)+K(I))GOTO 513
      IF(H(J).EQ.-H(I)-K(I).AND.K(J).EQ.H(I))GOTO 413
      IF(H(J).EQ.H(I)+K(I).AND.K(J).EQ.-H(I))GOTO 513
      GOTO 313
  413 PHI(J)=PHI(I)
      GOTO 313
  513 PHI(J)=360.0-PHI(I)
  313 CONTINUE
  213 CONTINUE
      GOTO 500
   14 DO 114 I=1,NOREF
      KEY(I)=0
  114 CONTINUE
      DO 214 I=1,NOREF
      IF(KEY(I).EQ.1)GOTO 214
      XX=RAN2(IDUM)
      PHI(I)=XX*360.0
      IF(NPG.EQ.19)PHI(I)=0
      IF(NPG.EQ.19.AND.XX.GT.0.5)PHI(I)=180.0
      DO 314 J=I+1,NOREF
      IF(H(J).EQ.-K(I).AND.K(J).EQ.-H(I))GOTO 414
      IF(H(J).EQ.K(I).AND.K(J).EQ.H(I))GOTO 514
      IF(H(J).EQ.K(I).AND.K(J).EQ.-H(I)-K(I))GOTO 614
      IF(H(J).EQ.-K(I).AND.K(J).EQ.H(I)+K(I))GOTO 714
      IF(H(J).EQ.H(I)+K(I).AND.K(J).EQ.-K(I))GOTO 414
      IF(H(J).EQ.-H(I)-K(I).AND.K(J).EQ.K(I))GOTO 514
      IF(H(J).EQ.H(I).AND.K(J).EQ.-H(I)-K(I))GOTO 514
      IF(H(J).EQ.-H(I).AND.K(J).EQ.H(I)+K(I))GOTO 414
      IF(H(J).EQ.H(I)+K(I).AND.K(J).EQ.-H(I))GOTO 714
      IF(H(J).EQ.-H(I)-K(I).AND.K(J).EQ.H(I))GOTO 614
      GOTO 314
```

```
 414 PHI(J)=PHI(I)
     IF(NPG.EQ.15)PHI(J)=360-PHI(I)
     GOTO 314
 514 PHI(J)=360.0-PHI(I)
     IF(NPG.EQ.15)PHI(J)=PHI(I)
     GOTO 314
 614 PHI(J)=PHI(I)
     GOTO 314
 714 PHI(J)=PHI(I)
 314 CONTINUE
 214 CONTINUE
 500 PHI(M0)=0
     RETURN
     END
C  THE FOLLOWING FUNCTION ROUTINE WHICH GENERATES RANDOM NUMBERS IN
C  THE RANGE 0.0 TO 1.0 COMES FROM "NUMERICAL RECIPES" BY W.H.PRESS,
C  B.P.FLANNERY, S.A.TEUKOLSKY AND W.T.VETTERLING, CAMBRIDGE
C  UNIVERSITY PRESS, 1986.
     FUNCTION RAN2(IDUM)
     PARAMETER (M=714025,IA=1366,IC=150889,RM=1.0/M)
     DIMENSION IR(97)
     DATA IFF/0/
     IF(IDUM.LT.0.OR.IFF.EQ.0)THEN
     IFF=1
     IDUM=MOD(IC-IDUM,M)
     DO 1 J=1,97
     IDUM=MOD(IA*IDUM+IC,M)
     IR(J)=IDUM
   1 CONTINUE
     IDUM=MOD(IA*IDUM+IC,M)
     ENDIF
     IY=IDUM
     J=1+(97*IY)/M
     IF(J.GT.97.OR.J.LT.1)PAUSE
     IY=IR(J)
     RAN2=IY*RM
     IDUM=MOD(IA*IDUM+IC,M)
     IR(J)=IDUM
     RETURN
     END
```

Appendix XII

```
      PROGRAM CALOBS
C  THIS PROGRAM COMBINES OUTPUT FILES THE FIRST OF WHICH IS IN
C  THE FORM GIVEN BY STRUCFAC AND HAS "OBSERVED" F's AND THE SECOND
C  OF WHICH HAS "CALCULATED" F's AND PHI's FROM AN APPROXIMATE
C  STRUCTURE.  THE OUTPUT FILE COMBINES THE OBSERVED F's WITH THE
C  CALCULATED PHI'S AND IS SUITABLE FOR INPUT TO FOUR2.   THIS
C  PROGRAM ENABLES THE FOURIER REFINEMENT OF STRUCTURES.
C  THE PROGRAM ALSO OUTPUTS THE RESIDUAL FOR THE CALCULATED F's.
      REAL F1(1000),PHI2(1000),F2(1000)
      INTEGER  H(1000),K(1000)
      CHARACTER ANS*1,FNAME*10
      OPEN(UNIT=9,FILE='LPT1')
      FNAME='SF1.DAT'
    3 WRITE(6,'('' IT IS ASSUMED THAT THE FIRST FILE CONTAINING'')')
      WRITE(6,'(''THE F[OBS] IS SF1.DAT.   DO YOU WANT TO CHANGE '')')
      WRITE(6,'(''THIS?   Y/N '')')
      READ(5,50)ANS
   50 FORMAT(A1)
      IF(ANS.EQ.'N'.OR.ANS.EQ.'n')GOTO 1
      IF(ANS.EQ.'Y'.OR.ANS.EQ.'y')GOTO 2
      GOTO 3
    2 WRITE(6,'(''READ IN THE FILE NAME [<= 10 CHARACTERS]. '')')
      READ(5,100)FNAME
  100 FORMAT(A10)
    1 OPEN(UNIT=10,FILE=FNAME)
      READ(10,*)NOREF,(H(I),K(I),F1(I),FI,I=1,NOREF)
      FNAME='SF2.DAT'
   13 WRITE(6,'('' IT IS ASSUMED THAT THE SECOND FILE CONTAINING'')')
      WRITE(6,'(''F[CALC] AND  PHI[CALC] IS SF2.DAT. DO YOU WANT '')')
      WRITE(6,'(''TO CHANGE THIS?   Y/N '')')
      READ(5,50)ANS
      IF(ANS.EQ.'N'.OR.ANS.EQ.'n')GOTO 11
      IF(ANS.EQ.'Y'.OR.ANS.EQ.'y')GOTO 12
      GOTO 13
   12 WRITE(6,'(''READ IN THE FILE NAME [<= 10 CHARACTERS]. '')')
      READ(5,100)FNAME
   11 OPEN(UNIT=11,FILE=FNAME)
      READ(11,*)NOREF,(H(I),K(I),F2(I),PHI2(I),I=1,NOREF)
      FNAME='COMB.DAT'
   23 WRITE(6,'('' IT IS ASSUMED THAT THE OUTPUT FILE CONTAINING'')')
      WRITE(6,'('' F[OBS] AND PHI[CALC] IS COMB.DAT.   DO YOU '')')
      WRITE(6,'('' WANT TO CHANGE THIS?   Y/N '')')
      READ(5,50)ANS
      IF(ANS.EQ.'N'.OR.ANS.EQ.'n')GOTO 21
      IF(ANS.EQ.'Y'.OR.ANS.EQ.'y')GOTO 22
      GOTO 23
   22 WRITE(6,'(''READ IN THE FILE NAME [<= 10 CHARACTERS]. '')')
      READ(5,100)FNAME
   21 OPEN(UNIT=12,FILE=FNAME)
      WRITE(12,*)NOREF,(H(I),K(I),F1(I),PHI2(I),I=1,NOREF)
C  NOW CALCULATE AND OUTPUT THE RESIDUAL
      SUMDIF=0
      SUM=0
      DO 40 I=1,NOREF
      SUMDIF=SUMDIF+ABS(F1(I)-F2(I))
      SUM=SUM+F1(I)
   40 CONTINUE
      RESID=SUMDIF/SUM
      WRITE(9,300)RESID
  300 FORMAT(39H THE RESIDUAL FOR THE INPUT F[CALC] IS ,F6.3)
      STOP
      END
```

# Solutions to Problems

The solutions are given with outline derivations where appropriate. If graphical processes are used then precise numerical agreement should not be expected.

## Chapter 1

1.1 If unit vectors $\hat{\mathbf{i}}, \hat{\mathbf{j}}, \hat{\mathbf{k}}$ are taken along the edges of the cube then the unit normal to the plane with Miller indices $(hkl)$ is

$$\hat{\mathbf{n}} = (h\hat{\mathbf{i}} + k\hat{\mathbf{j}} + l\hat{\mathbf{k}})/(h^2 + k^2 + l^2)^{\frac{1}{2}}.$$

The angle $\alpha$ between two planes is given by

$$\cos\alpha = \hat{\mathbf{n}}_1 \cdot \hat{\mathbf{n}}_2.$$

(a) 90°. (b) 26° 34′. (c) 54° 44′. (d) 19° 28′.

1.2 (a) Orthorhombic $Pmm2$.
(b) Monoclinic, $P2_1/c$.
(c) Tetragonal, $P\bar{4}$.

1.3 The diagrams, as shown in the *International Tables for X-ray Crystallography*, are given in fig. I-1, overleaf.

## Chapter 2

2.1 (a) With the centre of the group as origin the atomic coordinates are $\pm 3\lambda/2$ and $\pm 9\lambda/2$. The amplitude, as a fraction of that from four scatterers at the origin, is

$$\tfrac{1}{2}\{\cos(2\pi\mathbf{r}_1\cdot\mathbf{s}) + \cos(2\pi\mathbf{r}_2\cdot\mathbf{s})\}$$

where $r_1 = 3\lambda/2$, $r_2 = 9\lambda/2$, $s = 2\sin\theta/\lambda$ and the angle between $\mathbf{r}$ and $\mathbf{s}$ is $\theta$. This gives the amplitudes and intensities listed in table II-1.

Table II-1.

| $2\theta$ | Amplitude | Intensity |
|---|---|---|
| 0 or 180° | 1.000 | 1.000 |
| 20 or 160° | 0.983 | 0.966 |
| 40 or 140° | 0.876 | 0.767 |
| 60 or 120° | 0.248 | 0.062 |
| 80 or 100° | 0.948 | 0.899 |

Fig. I-1.
Solution to Problem 1.3.

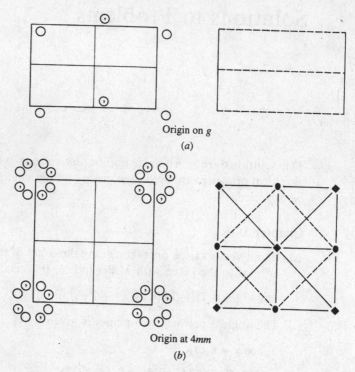

Origin on g

(a)

Origin at 4mm

(b)

(b) With one end of the group as origin the amplitude of scattering compared with that of a single scatterer at the origin is

$$1 + \exp(2\pi i r_1 \cdot s) + \exp(2\pi i r_2 \cdot s) + \exp(2\pi i r_3 \cdot s) = A + iB,$$

say, where $r_1 = 3\lambda$, $r_2 = 6\lambda$ and $r_3 = 9\lambda$. The phase angle required, $\phi$, is given by $\tan \phi = B/A$, the quadrant of $\phi$ being such that the signs of $B$ and $A$ are the signs of $\sin \phi$ and $\cos \phi$. This gives

| $2\theta$ | 0 or 180° | 20° or 160° | 40° or 140° | 60° or 120° | 80° or 100° |
|-----------|-----------|-------------|-------------|-------------|-------------|
| $\phi$    | 0°        | 14°         | 321°        | 323°        | 335°        |

2.2 For an origin at the centre of the cube the scatterers have coordinates $\pm(\frac{1}{2}\lambda, \frac{1}{2}\lambda, \frac{1}{2}\lambda)$, $\pm(\frac{1}{2}\lambda, \frac{1}{2}\lambda, -\frac{1}{2}\lambda)$, $\pm(\frac{1}{2}\lambda, -\frac{1}{2}\lambda, \frac{1}{2}\lambda)$, and $\pm(\frac{1}{2}\lambda, -\frac{1}{2}\lambda, -\frac{1}{2}\lambda)$. The unit vectors in the direction of the incident and scattered beams are $S_0 = 3^{-\frac{1}{2}}(1,1,1)$, $S_1 = (1,0,0)$, $S_2 = (-1,0,0)$, so that

$$s_1 = \frac{S_1 - S_0}{\lambda} = \frac{1}{\lambda}(1 - 3^{-\frac{1}{2}}, -3^{-\frac{1}{2}}, -3^{-\frac{1}{2}})$$

and

$$s_2 = \frac{S_2 - S_0}{\lambda} = -\frac{1}{\lambda}(1 + 3^{-\frac{1}{2}}, 3^{-\frac{1}{2}}, 3^{-\frac{1}{2}}).$$

From this it may be found that the ratio of the scattered intensity to that for a single scatterer is the same in both directions and equals 0.012.

2.3 A 2 kV electron has energy $3.2 \times 10^{-16}$ joule. Equating this to $\frac{1}{2}mv^2$ the electron velocity is found to be approximately $v = 2.7 \times 10^7 \, \text{m s}^{-1}$. If

Fig. I-1. (*cont.*)

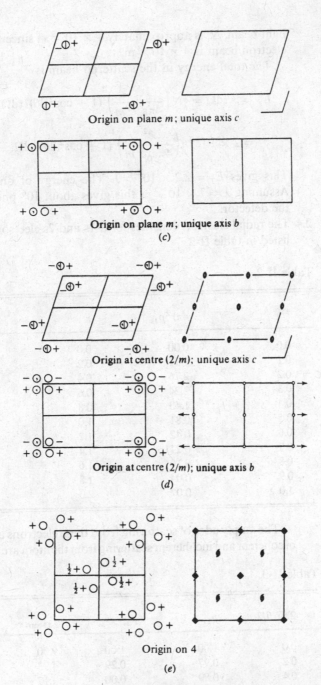

Origin on plane *m*; unique axis *c*

Origin on plane *m*; unique axis *b*

(*c*)

Origin at centre (2/*m*); unique axis *c*

Origin at centre (2/*m*); unique axis *b*

(*d*)

Origin on 4

(*e*)

the electron beam has $n$ electrons per unit volume and its cross-sectional area is $\alpha$ then the current is

$$i = n\alpha ve.$$

This gives $n \simeq 9 \times 10^{19}$ electrons $m^{-3}$. If the laser pulse has intensity $I$, duration $\tau$ and cross-sectional area $s$ then the energy of the pulse $E = Is\tau$. This gives $I = E/s\tau$. The number of electrons traversed by the light beam, $N$, is approximately $4 \times 10^{-3} sn$ since the diameter of the electron beam is $4 \times 10^{-3}$ m.

The total energy in the scattered beam is

$$E_T = \mathscr{I}\tau d\Omega = \tfrac{1}{2}N\left(\frac{e^2}{4\pi\varepsilon_0 c^2 m}\right)^2 (1 + \cos^2 2\theta)I\tau d\Omega$$

$$= 2 \times 10^{-3}n\left(\frac{e^2}{4\pi\varepsilon_0 c^2 m}\right)^2 (1 + \cos^2 2\theta)Ed\Omega.$$

This gives $E_T = 2.2 \times 10^{-13}$ J. The energy of one photon is $hc/\lambda$. Assuming $\lambda \simeq 7 \times 10^{-7}$ m this gives about $10^6$ photons collected by the detector.

2.4 The radial electron densities for a 1s and 2s electron of a Li atom are listed in table II-2.

Table II-2.

| $r$ | $4\pi r^2 \rho_{1s}$ | $r$ | $4\pi r_2 \rho_{2s}$ |
|-----|------|-----|------|
| 0.0 | 0.00 | 0.0 | 0.00 |
| 0.1 | 1.91 | 0.2 | 0.00 |
| 0.2 | 2.76 | 0.4 | 0.04 |
| 0.3 | 2.24 | 0.6 | 0.11 |
| 0.4 | 1.43 | 0.8 | 0.21 |
| 0.5 | 0.81 | 1.0 | 0.32 |
| 0.6 | 0.42 | 1.2 | 0.41 |
| 0.7 | 0.21 | 1.4 | 0.46 |
| 0.8 | 0.10 | 1.6 | 0.48 |
| 0.9 | 0.04 | 1.8 | 0.47 |
| 1.0 | 0.02 | 2.0 | 0.44 |

The amplitudes of scattering from these electrons and the intensities of coherent and incoherent scattering from the atom are given in table II-3.

Table II-3.

| $2\sin\theta/\lambda$ | $p_{1s}$ | $p_{2s}$ | $I_{\text{Thomson}}$ | $I_{\text{Compton}}$ |
|-----|------|------|------|------|
| 0.0 | 1.00 | 1.00 | 9.00 | 0.00 |
| 0.2 | 0.97 | 0.29 | 4.93 | 1.03 |
| 0.4 | 0.89 | 0.00 | 3.17 | 1.42 |
| 0.6 | 0.77 | −0.01 | 2.34 | 1.82 |
| 0.8 | 0.65 | 0.00 | 1.69 | 2.16 |
| 1.0 | 0.52 | | 1.08 | 2.46 |
| 1.2 | 0.42 | | 0.71 | 2.65 |
| 1.4 | 0.33 | | 0.44 | 2.78 |
| 1.6 | 0.26 | | 0.27 | 2.86 |
| 1.8 | 0.20 | | 0.16 | 2.92 |
| 2.0 | 0.16 | | 0.10 | 2.95 |

## Chapter 3

3.1 The values of $K_6$ and $K_6^2$ are given in table III-1.

Table III-1.

| a·s | $K_6$ | $K_6^2$ | a·s | $K_6$ | $K_6^2$ |
|------|-------|---------|------|--------|---------|
| 0.00 | 6.00 | 36.00 | 0.50 | 0.00 | 0.00 |
| 0.05 | 5.17 | 26.76 | 0.55 | −0.82 | 0.67 |
| 0.10 | 3.08 | 9.47 | 0.60 | −1.00 | 1.00 |
| 0.15 | 0.68 | 0.46 | 0.65 | −0.35 | 0.12 |
| 0.20 | −1.00 | 1.00 | 0.70 | 0.73 | 0.53 |
| 0.25 | −1.41 | 2.00 | 0.75 | 1.41 | 2.00 |
| 0.30 | −0.73 | 0.53 | 0.80 | 1.00 | 1.00 |
| 0.35 | 0.35 | 0.12 | 0.85 | −0.68 | 0.46 |
| 0.40 | 1.00 | 1.00 | 0.90 | −3.08 | 9.47 |
| 0.45 | 0.82 | 0.67 | 0.95 | −5.17 | 26.76 |
| 0.50 | 0.00 | 0.00 | 1.00 | −6.00 | 36.00 |

3.2 The orders of diffraction $h$ and angles $\psi_a$ are given in table III-2.

Table III-2.

| $h$ | $\psi_a$ | $h$ | $\psi_a$ |
|------|----------|------|----------|
| 0 | 30° 00′ | −5 | 97° 42′ |
| −1 | 48° 14′ | −6 | 109° 31′ |
| −2 | 62° 13′ | −7 | 122° 16′ |
| −3 | 74° 34′ | −8 | 137° 14′ |
| −4 | 86° 13′ | −9 | 159° 04′ |

3.3 (i) $a^* = 0.231, b^* = 0.115, c^* = 0.067 \text{Å}^{-1}, \alpha^* = \beta^* = 90°, \gamma^* = 60°$.

(ii) $V = 650 \text{Å}^3, V^* = 1.54 \times 10^{-3} \text{Å}^{-3}$.

(iii) $d_{321} = 1.20 \text{Å}$.

(iv) $2\theta = 80° 4′$.

3.4 In table III-3 there is presented a selection of the output of STRUCFAC. If your solution agrees with this then the remainder will certainly be correct.

3.5 In table III-4 there is presented the complete output of STRUCFAC. Notice the relationships between the structure factor magnitudes and phases of reflections with indices $(h\,k\,0)$ and $(h\,\bar{k}\,0)$.

3.6 From $V = abc(1 - \cos^2\alpha - \cos^2\beta - \cos^2\gamma + 2\cos\alpha\cos\beta\cos\gamma)^{\frac{1}{2}}$ the volume of the cell is $2.636 \times 10^{-28} \text{m}^3$. The mass of one molecule is $1.3974 \times 10^{-25}$ kg and for a density of $1.2 \times 10^3$ kg m$^{-3}$ the number of molecules would be $N = 2.26$. The probable value of $N$ is 2 and this gives a density for the crystal of $1.061 \times 10^3$ kg m$^{-3}$.

Table III-3.

| h | k | F | PHI | h | k | F | PHI | h | k | F | PHI |
|---|---|---|---|---|---|---|---|---|---|---|---|
| -14 | 1 | 21.295 | 180.0 | -14 | 2 | 4.754 | -180.0 | -13 | 1 | 9.160 | -180.0 |
| -13 | 2 | 0.415 | 180.0 | -13 | 3 | 1.838 | -180.0 | -13 | 4 | 12.208 | 180.0 |
| -12 | 1 | 7.894 | 0.0 | -12 | 2 | 13.132 | -180.0 | -12 | 3 | 3.796 | 0.0 |
| -12 | 4 | 3.432 | 180.0 | -12 | 5 | 4.386 | 0.0 | -11 | 1 | 9.673 | 0.0 |
| -11 | 2 | 4.383 | 180.0 | -11 | 3 | 1.512 | 0.0 | -11 | 4 | 18.388 | 0.0 |
| -11 | 5 | 23.349 | 0.0 | -11 | 6 | 13.471 | 180.0 | -11 | 7 | 5.241 | 180.0 |
| -10 | 1 | 1.688 | 0.0 | -10 | 2 | 0.947 | -180.0 | -10 | 3 | 6.977 | 180.0 |
| -10 | 4 | 11.091 | 0.0 | -10 | 5 | 8.641 | 0.0 | -10 | 6 | 1.817 | -180.0 |
| -10 | 7 | 3.540 | 180.0 | -10 | 8 | 1.438 | 0.0 | -9 | 1 | 4.693 | 0.0 |
| -9 | 2 | 10.456 | 180.0 | -9 | 3 | 15.519 | 0.0 | -9 | 4 | 5.459 | 0.0 |
| -9 | 5 | 6.089 | 180.0 | -9 | 6 | 11.487 | -180.0 | -9 | 7 | 4.378 | -180.0 |
| -9 | 8 | 5.816 | -180.0 | -8 | 1 | 19.716 | -180.0 | -8 | 2 | 3.534 | 0.0 |
| -8 | 3 | 4.694 | 0.0 | -8 | 4 | 17.292 | 0.0 | -8 | 5 | 13.280 | -180.0 |
| -8 | 6 | 14.910 | 180.0 | -8 | 7 | 22.114 | 0.0 | -8 | 8 | 13.713 | 180.0 |
| -8 | 9 | 1.269 | 0.0 | -7 | 1 | 6.264 | -180.0 | -7 | 2 | 6.490 | 0.0 |
| -7 | 3 | 8.143 | -180.0 | -7 | 4 | 5.985 | 0.0 | -7 | 5 | 33.755 | -180.0 |
| -7 | 6 | 5.373 | 180.0 | -7 | 7 | 4.868 | 0.0 | -7 | 8 | 24.185 | 180.0 |
| -7 | 9 | 6.194 | 0.0 | -7 | 10 | 29.213 | 0.0 | -6 | 1 | 21.644 | 180.0 |
| -6 | 2 | 27.730 | 0.0 | -6 | 3 | 22.614 | 0.0 | -6 | 4 | 19.039 | -180.0 |
| -6 | 5 | 11.490 | 180.0 | -6 | 6 | 16.188 | 0.0 | -6 | 7 | 10.211 | 0.0 |
| -6 | 8 | 21.228 | 180.0 | -6 | 9 | 15.023 | -180.0 | -6 | 10 | 22.119 | 0.0 |
| -6 | 11 | 2.692 | -180.0 | -5 | 1 | 27.541 | -180.0 | -5 | 2 | 28.372 | |
| -5 | 3 | 27.294 | 180.0 | -5 | 4 | 7.910 | 180.0 | -5 | 5 | 12.392 | 0.0 |
| -5 | 6 | 3.735 | 0.0 | -5 | 7 | 0.987 | -180.0 | -5 | 8 | 16.729 | 0.0 |
| -5 | 9 | 39.952 | 0.0 | -5 | 10 | 3.746 | -180.0 | -5 | 11 | 18.870 | 180.0 |
| -4 | 1 | 67.992 | 0.0 | -4 | 2 | 5.672 | 0.0 | -4 | 3 | 32.241 | -180.0 |
| -4 | 4 | 7.022 | -180.0 | -4 | 5 | 8.491 | 0.0 | -4 | 6 | 12.106 | 0.0 |
| -4 | 7 | 35.181 | -180.0 | -4 | 8 | 8.684 | 0.0 | -4 | 9 | 28.160 | 0.0 |
| -4 | 10 | 4.194 | -180.0 | -4 | 11 | 3.080 | -180.0 | -4 | 12 | 6.228 | 180.0 |
| -3 | 1 | 38.858 | 0.0 | -3 | 2 | 9.027 | 0.0 | -3 | 3 | 13.861 | 0.0 |
| -3 | 4 | 11.430 | 180.0 | -3 | 5 | 26.347 | 0.0 | -3 | 6 | 2.834 | 180.0 |
| -3 | 7 | 12.733 | -180.0 | -3 | 8 | 18.989 | 0.0 | -3 | 9 | 2.507 | -180.0 |
| -3 | 10 | 7.850 | 180.0 | -3 | 11 | 22.081 | 180.0 | -3 | 12 | 3.352 | -180.0 |
| -2 | 1 | 43.943 | 0.0 | -2 | 2 | 56.764 | 0.0 | -2 | 3 | 30.501 | -180.0 |
| -2 | 4 | 8.039 | 0.0 | -2 | 5 | 40.944 | 0.0 | -2 | 6 | 13.548 | 180.0 |
| -2 | 7 | 6.167 | 180.0 | -2 | 8 | 20.639 | 0.0 | -2 | 9 | 1.070 | -180.0 |
| -2 | 10 | 20.769 | -180.0 | -2 | 11 | 2.861 | -180.0 | -2 | 12 | 9.499 | 0.0 |
| -1 | 1 | 8.623 | 0.0 | -1 | 2 | 57.283 | 180.0 | -1 | 3 | 15.724 | 0.0 |
| -1 | 4 | 0.584 | 0.0 | -1 | 5 | 8.158 | -180.0 | -1 | 6 | 19.067 | 180.0 |
| -1 | 7 | 2.981 | 0.0 | -1 | 8 | 42.352 | 0.0 | -1 | 9 | 27.479 | 180.0 |
| -1 | 10 | 27.836 | -180.0 | -1 | 11 | 12.692 | 0.0 | -1 | 12 | 7.914 | 180.0 |
| -1 | 13 | 2.005 | 0.0 | 0 | 0 | 256.000 | 0.0 | 0 | 1 | 6.619 | 180.0 |
| 0 | 2 | 65.986 | -180.0 | 0 | 3 | 40.756 | 0.0 | 0 | 4 | 7.510 | -180.0 |
| 0 | 5 | 9.825 | 180.0 | 0 | 6 | 44.845 | -180.0 | 0 | 7 | 8.932 | 0.0 |

Table III-4.

| h | k | F | PHI | h | k | F | PHI | h | k | F | PHI |
|---|---|---|---|---|---|---|---|---|---|---|---|
| -14 | 1 | 8.346 | 176.8 | -14 | 2 | 8.490 | -36.9 | -13 | 1 | 9.786 | 165.9 |
| -13 | 2 | 20.798 | 14.2 | -13 | 3 | 13.324 | 35.7 | -12 | 1 | 18.706 | -40.1 |
| -12 | 2 | 23.391 | -21.1 | -12 | 3 | 11.713 | -155.0 | -12 | 4 | 11.097 | -139.8 |
| -11 | 1 | 7.637 | 96.6 | -11 | 2 | 2.091 | -149.8 | -11 | 3 | 8.868 | 179.6 |
| -11 | 4 | 1.827 | -148.4 | -11 | 5 | 18.392 | -24.7 | -10 | 1 | 19.458 | -154.9 |
| -10 | 2 | 12.396 | 149.4 | -10 | 3 | 18.524 | -138.0 | -10 | 4 | 6.763 | -40.3 |
| -10 | 5 | 14.056 | -64.7 | -9 | 1 | 17.533 | -37.4 | -9 | 2 | 8.181 | 130.9 |
| -9 | 3 | 13.173 | -153.9 | -9 | 4 | 12.658 | -31.1 | -9 | 5 | 6.136 | 23.5 |
| -9 | 6 | 12.231 | 144.9 | -8 | 1 | 7.344 | 113.4 | -8 | 2 | 41.677 | 170.5 |
| -8 | 3 | 3.908 | 34.6 | -8 | 4 | 14.641 | -8.8 | -8 | 5 | 10.325 | -16.4 |
| -8 | 6 | 16.496 | 155.9 | -7 | 1 | 22.071 | -177.6 | -7 | 2 | 12.299 | -162.3 |
| -7 | 3 | 16.864 | -33.6 | -7 | 4 | 20.837 | -7.9 | -7 | 5 | 7.357 | -117.8 |
| -7 | 6 | 22.874 | 160.3 | -7 | 7 | 12.379 | 43.0 | -6 | 1 | 12.094 | 116.3 |
| -6 | 2 | 26.051 | -57.6 | -6 | 3 | 9.729 | 12.6 | -6 | 4 | 2.822 | -139.7 |
| -6 | 5 | 16.262 | -175.2 | -6 | 6 | 9.732 | 117.7 | -6 | 7 | 18.300 | 20.0 |
| -5 | 1 | 3.741 | -8.6 | -5 | 2 | 30.278 | -15.4 | -5 | 3 | 14.162 | 47.4 |
| -5 | 4 | 22.221 | 171.7 | -5 | 5 | 9.933 | 127.7 | -5 | 6 | 2.032 | -30.0 |
| -5 | 7 | 2.623 | 23.5 | -4 | 1 | 35.422 | -3.2 | -4 | 2 | 22.562 | 11.9 |
| -4 | 3 | 11.902 | 121.4 | -4 | 4 | 19.099 | 145.1 | -4 | 5 | 21.011 | 125.1 |
| -4 | 6 | 7.417 | -41.1 | -4 | 7 | 6.762 | -86.4 | -3 | 1 | 36.146 | -29.9 |
| -3 | 2 | 12.182 | -70.5 | -3 | 3 | 10.644 | 44.4 | -3 | 4 | 10.889 | 134.2 |
| -3 | 5 | 20.126 | 120.4 | -3 | 6 | 16.045 | -17.3 | -3 | 7 | 7.102 | 124.9 |
| -2 | 1 | 46.287 | 77.4 | -2 | 2 | 17.364 | 11.4 | -2 | 3 | 25.002 | 66.9 |
| -2 | 4 | 1.958 | 126.9 | -2 | 5 | 10.499 | 19.1 | -2 | 6 | 9.941 | 30.7 |
| -2 | 7 | 11.642 | -99.8 | -1 | 1 | 70.661 | -9.5 | -1 | 2 | 32.605 | 150.1 |
| -1 | 3 | 20.323 | -173.0 | -1 | 4 | 11.046 | -43.3 | -1 | 5 | 10.934 | 73.0 |
| -1 | 6 | 2.869 | -52.1 | -1 | 7 | 12.394 | -125.3 | 0 | 0 | 256.000 | 0.0 |
| 0 | 1 | 0.000 | 15.2 | 0 | 2 | 41.643 | 150.2 | 0 | 3 | 0.000 | 104.6 |
| 0 | 4 | 17.205 | -30.9 | 0 | 5 | 0.000 | 59.7 | 0 | 6 | 24.655 | -118.8 |
| 0 | 7 | 0.000 | 80.9 | 1 | 0 | 30.799 | 0.0 | 1 | 1 | 70.661 | 170.5 |
| 1 | 2 | 32.605 | 150.1 | 1 | 3 | 20.323 | 7.0 | 1 | 4 | 11.046 | -43.3 |
| 1 | 5 | 10.934 | -107.0 | 1 | 6 | 2.869 | -52.1 | 1 | 7 | 12.393 | 54.7 |
| 2 | 0 | 8.745 | -180.0 | 2 | 1 | 46.287 | -102.6 | 2 | 2 | 17.364 | 11.4 |
| 2 | 3 | 25.002 | -113.1 | 2 | 4 | 1.958 | 126.9 | 2 | 5 | 10.499 | -160.9 |
| 2 | 6 | 9.941 | 30.7 | 2 | 7 | 11.642 | 80.2 | 3 | 0 | 13.299 | 0.0 |
| 3 | 1 | 36.146 | 150.1 | 3 | 2 | 12.182 | -70.5 | 3 | 3 | 10.644 | -135.6 |
| 3 | 4 | 10.889 | 134.2 | 3 | 5 | 20.126 | -59.6 | 3 | 6 | 16.045 | -17.3 |
| 3 | 7 | 7.102 | -55.1 | 4 | 0 | 62.284 | 180.0 | 4 | 1 | 35.422 | 176.8 |
| 4 | 2 | 22.562 | 11.9 | 4 | 3 | 11.902 | -58.6 | 4 | 4 | 19.099 | 145.1 |
| 4 | 5 | 21.011 | -54.9 | 4 | 6 | 7.417 | -41.1 | 4 | 7 | 6.762 | 93.6 |
| 5 | 0 | 42.964 | 180.0 | 5 | 1 | 3.741 | 171.4 | 5 | 2 | 30.278 | -15.4 |
| 5 | 3 | 14.162 | -132.6 | 5 | 4 | 22.221 | 171.7 | 5 | 5 | 9.933 | -52.3 |
| 5 | 6 | 2.032 | -30.0 | 5 | 7 | 2.623 | -156.5 | 6 | 0 | 34.322 | -180.0 |
| 6 | 1 | 12.094 | -63.7 | 6 | 2 | 26.051 | -57.6 | 6 | 3 | 9.729 | -167.4 |
| 6 | 4 | 2.822 | -139.7 | 6 | 5 | 16.262 | 4.8 | 6 | 6 | 9.732 | 117.7 |

Table III-4. (*continued*)

| h | k | F | PHI | h | k | F | PHI | h | k | F | PHI |
|---|---|---|---|---|---|---|---|---|---|---|---|
| 6 | 7 | 18.300 | -160.0 | 7 | 0 | 25.766 | 180.0 | 7 | 1 | 22.071 | 2.4 |
| 7 | 2 | 12.299 | -162.3 | 7 | 3 | 16.864 | 146.4 | 7 | 4 | 20.837 | -7.9 |
| 7 | 5 | 7.357 | 62.2 | 7 | 6 | 22.874 | 160.3 | 7 | 7 | 12.379 | -137.0 |
| 8 | 0 | 14.187 | 0.0 | 8 | 1 | 7.344 | -66.6 | 8 | 2 | 41.677 | 170.5 |
| 8 | 3 | 3.908 | -145.4 | 8 | 4 | 14.641 | -8.8 | 8 | 5 | 10.325 | 163.6 |
| 8 | 6 | 16.496 | 155.9 | 9 | 0 | 16.822 | 0.0 | 9 | 1 | 17.533 | 142.6 |
| 9 | 2 | 8.181 | 130.9 | 9 | 3 | 13.173 | 26.1 | 9 | 4 | 12.658 | -31.1 |
| 9 | 5 | 6.136 | -156.5 | 9 | 6 | 12.231 | 144.9 | 10 | 0 | 14.968 | -180.0 |
| 10 | 1 | 19.458 | 25.1 | 10 | 2 | 12.396 | 149.4 | 10 | 3 | 18.524 | 42.0 |
| 10 | 4 | 6.763 | -40.3 | 10 | 5 | 14.056 | 115.3 | 11 | 0 | 25.387 | 0.0 |
| 11 | 1 | 7.637 | -83.4 | 11 | 2 | 2.091 | -149.8 | 11 | 3 | 8.868 | -0.4 |
| 11 | 4 | 1.827 | -148.4 | 11 | 5 | 18.392 | 155.3 | 12 | 0 | 8.978 | 180.0 |
| 12 | 1 | 18.706 | 139.9 | 12 | 2 | 23.391 | -21.1 | 12 | 3 | 11.713 | 25.0 |
| 12 | 4 | 11.097 | -139.8 | 13 | 0 | 22.036 | 180.0 | 13 | 1 | 9.786 | -14.1 |
| 13 | 2 | 20.798 | 14.2 | 13 | 3 | 13.324 | -144.3 | 14 | 0 | 10.056 | 0.0 |
| 14 | 1 | 8.346 | -3.2 | 14 | 2 | 8.490 | -36.9 | | | | |

# Chapter 4

4.1 The output from SIMP1 should be as follows:

| h | A(h) | B(h) | h | A(h) | B(h) | h | A(h) | B(h) | h | A(h) | B(h) |
|---|---|---|---|---|---|---|---|---|---|---|---|
| 0 | 20.05 | 0.00 | 1 | 0.27 | 6.97 | 2 | -0.76 | -1.17 | 3 | 0.62 | 0.17 |
| 4 | -0.15 | 0.31 | 5 | -0.02 | -0.20 | | | | | | |

4.2 The output from FOUR1 is as follows. This may be compared with the table given in fig. 4.21.

```
THE COORDINATES ARE X/A = n/N WHERE N =  20
```

| n | f(x) | n | f(x) | n | f(x) | n | f(x) |
|---|---|---|---|---|---|---|---|
| 0 | 19.97 | 1 | 23.37 | 2 | 26.57 | 3 | 29.09 |
| 4 | 31.40 | 5 | 34.47 | 6 | 37.09 | 7 | 36.00 |
| 8 | 30.11 | 9 | 22.45 | 10 | 16.49 | 11 | 12.52 |
| 12 | 9.35 | 13 | 7.25 | 14 | 7.04 | 15 | 8.07 |
| 16 | 9.22 | 17 | 10.71 | 18 | 13.26 | 19 | 16.57 |
| 20 | 19.97 | | | | | | |

4.3 (i) $F(s) = \dfrac{2}{4\pi^2 s^2 + k^2} \{e^{-ka}[2\pi s \sin(2\pi sa) - k\cos(2\pi sa)] + k\}.$

(ii) $F(s) = \dfrac{1}{2\pi^2 s^2} \{1 - \cos(2\pi s)\}.$

4.4  $-\frac{1}{4} > x$     $c(x) = 0$

$-\frac{1}{4} \leqslant x \leqslant \frac{1}{4}$     $c(x) = \frac{1}{2\pi}\{1 + \sin(2\pi x)\}$

$\frac{1}{4} \leqslant x \leqslant \frac{3}{4}$     $c(x) = \frac{\sin(2\pi x)}{\pi}$

$\frac{3}{4} \leqslant x \leqslant 1\frac{1}{4}$     $c(x) = \frac{1}{2\pi}\{\sin(2\pi x) - 1\}.$

4.5  $(2\pi)^{-\frac{1}{2}}\int_{-\infty}^{\infty} \exp\{-\frac{1}{2}(x-2)^2\}\exp(2\pi ixs)dx$

$= (2\pi)^{-\frac{1}{2}}\exp(-2\pi^2 s^2 + 4\pi is)\int_{-\infty}^{\infty}\exp\{-\frac{1}{2}(x-2-2\pi is)^2\}dx$

$= \exp(-2\pi^2 s^2 + 4\pi is).$

Fourier transform of $g(x) = \exp(-2\pi^2 s^2)$.
Fourier transform of $h(x) = \exp(4\pi is)$.
Hence Fourier transform of $g(x)*h(x) = \exp(-2\pi^2 s^2 + 4\pi is)$.

4.6  $\rho(x,y,z) = \frac{1}{V}\sum_{h=0}^{\infty}\sum_{k=0}^{\infty}\sum_{l=0}^{\infty} \omega_{hkl}F_{hkl}\cos(2\pi hx)\cos(2\pi ky)\cos(2\pi lz)$

where

$\omega_{hkl} = 8$ if none of $h, k, l$ is zero
$= 4$ if one of $h, k, l$ is zero
$= 2$ if two of $h, k, l$ are zero
$= 1$ for $h = k = l = 0$.

The range of summation is $0 - \frac{1}{2}$ for $x, y$ and $z$.

4.7  Part of the projected density is shown in fig. 4.20. If this region is found in the map then the map is almost certainly correct.

## Chapter 5

5.1  This problem can be solved graphically. It will be found that if a reciprocal-lattice net is rotated over a diametral section of the sphere of reflection then one must rotate the net 54° between the (907) and (702) reflections.

5.2  (a) 6.3, 13.5, 24.0 mm.
(b) The layer lines indicate $a = 9.91, 9.97, 10.01$ Å, respectively.

5.3  Fig. V-1 shows the fit of the reciprocal lattice on to the arcs within the allowed regions of reciprocal space. The reflections which are observed are indicated. The reflections too weak to be observed have indices (420), (340) and ($\bar{4}$10).

5.4  Fig. V-2 shows two reciprocal-lattice points, ($hk$) inside circle and ($h'k'$) outside.

For a point inside the circle, the angle through which the crystal must be rotated is $\psi = (\pi/2) - \theta + \phi$ and the intersection will be with the lower section of the circle.

For a point outside the circle, the intersection can be with either the upper section or the lower section – the angle of rotation in the two cases being, respectively,

$$\psi = \frac{\pi}{2} + \theta - \phi' \text{ or } \phi + \theta - \frac{\pi}{2}$$

and

$$\psi = \frac{3\pi}{2} - \theta - \phi' \text{ or } \frac{\pi}{2} - \theta + \phi.$$

A reflection will be produced if $0 \leqslant \psi \leqslant 180°$. With the $y$ coordinate on the upper section taken as positive and on the lower section taken as negative the $(x, y)$ coordinates on the film are as listed in table V-1.

Table V-1.

| $h$ | $k$ | $x$(mm) | $y$(mm) | $h$ | $k$ | $x$(mm) | $y$(mm) |
|-----|-----|---------|---------|-----|-----|---------|---------|
| 1 | 7 | 61 | 65 | 1 | 7 | 35 | −65 |
| 1 | 6 | 68 | 51 | 1 | 6 | 29 | −51 |
| 1 | 5 | 74 | 40 | 1 | 5 | 25 | −40 |
| 1 | 4 | 80 | 31 | 1 | 4 | 21 | −31 |
| 1 | 3 | 85 | 23 | 1 | 3 | 19 | −23 |
| $\bar{1}$ | $\bar{2}$ | 6 | 16 | 1 | 2 | 18 | −16 |
| $\bar{1}$ | $\bar{1}$ | 16 | 9 | 1 | 1 | 23 | −9 |
| $\bar{1}$ | 0 | 42 | 5 | 1 | 0 | 53 | −5 |
| $\bar{1}$ | 1 | 67 | 9 | 1 | $\bar{1}$ | 76 | −9 |
| $\bar{1}$ | 2 | 72 | 16 | 1 | $\bar{2}$ | 88 | −16 |
| $\bar{1}$ | 3 | 71 | 23 | $\bar{1}$ | 3 | 5 | −23 |
| $\bar{1}$ | 4 | 69 | 31 | $\bar{1}$ | 4 | 10 | −31 |
| $\bar{1}$ | 5 | 66 | 40 | $\bar{1}$ | 5 | 16 | −40 |
| $\bar{1}$ | 6 | 61 | 51 | $\bar{1}$ | 6 | 22 | −51 |
| $\bar{1}$ | 7 | 55 | 65 | $\bar{1}$ | 7 | 29 | −65 |

5.5 From equation (5.24) the critical wavelength in ångstroms is given by

$$\lambda_c = \frac{18.64}{BE^2}$$

where $B$ is in tesla and $E$ in GeV. Inserting the given values $\lambda_c = 1.883 \text{ Å}$.
From equation (5.23) the vertical divergence of the beam is given by

$$\delta\phi = \frac{mc^2}{E}$$

where $E$ is the energy of the electron beam. This gives

$$\delta\phi = \frac{9.108 \times 10^{-31} \times (3 \times 10^8)^2}{3 \times 10^9 \times 1.602 \times 10^{-19}} = 1.706 \times 10^{-4} \text{ radians}$$

$$= 35''.$$

Fig. V-1.
Diagram for solution
5.3.

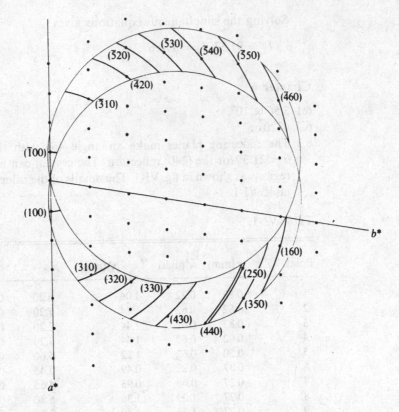

Fig. V-2.
Diagram for solution
5.4.

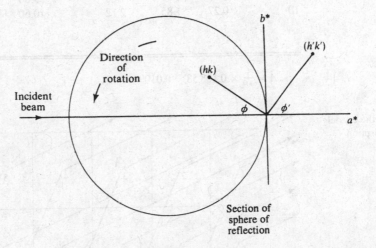

5.6 Let the intensities of the (231) and (462) reflections be $I_1$ and $I_2$, respectively. The (462) reflection corresponds to the shorter wavelength so that

$$I_1 + I_2 = 262$$

and

$$0.6I_1 + 0.3I_2 = 106.$$

Solving the simultaneous equations gives

$$I_1 = 91.33; I_2 = 170.67.$$

## Chapter 6

6.1 $3.56 \times 10^8$.

6.2 $4720 \, \text{m}^{-1}$.

6.3 The reflecting planes make an angle 45° with the cell edges and $\theta = 21.3°$ for the (240) reflection. The crystal can be divided into ten regions as shown in fig. VI-1. The details of the calculation are given in table VI-1.

Table VI-1.

| Point | $x$ (mm) | $x'$ (mm) | $x + x'$ (mm) | $\mu(x + x')$ | $\exp\{-\mu(x + x')\}$ |
|-------|----------|-----------|----------------|----------------|------------------------|
| 1 | 0.82 | 0.22 | 1.04 | 5.20 | 0.005 52 |
| 2 | 0.82 | 0.62 | 1.44 | 7.20 | 0.000 75 |
| 3 | 0.82 | 0.62 | 1.44 | 7.20 | 0.000 75 |
| 4 | 0.82 | 0.62 | 1.44 | 7.20 | 0.000 75 |
| 5 | 0.50 | 0.62 | 1.12 | 5.60 | 0.003 70 |
| 6 | 0.27 | 0.22 | 0.49 | 2.45 | 0.086 30 |
| 7 | 0.27 | 0.66 | 0.93 | 4.65 | 0.009 56 |
| 8 | 0.27 | 1.09 | 1.36 | 6.80 | 0.001 11 |
| 9 | 0.27 | 1.53 | 1.80 | 9.00 | 0.000 12 |
| 10 | 0.27 | 1.85 | 2.12 | 10.60 | 0.000 02 |
| | | | | | 0.108 58 |

$A = \frac{1}{10} \times 0.108\,58 = 0.0109.$

Fig. VI-1.
Division of the crystal
for the absorption
calculation.

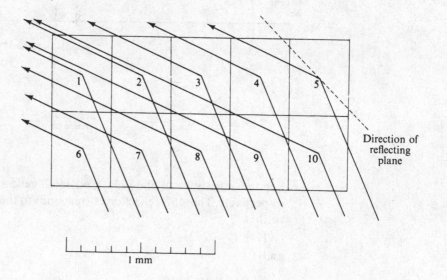

Direction of
reflecting
plane

1 mm

6.4 In table VI-2 there is presented a selection of the output of STRUCFAC. If your solution agrees with this then the remainder will certainly be correct. Compare this solution with that of Problem 3.4 and notice that the effect of the temperature factor becomes more pronounced for reflections with higher Bragg angles.

Table VI-2.

| h | k | F | PHI | h | k | F | PHI | h | k | F | PHI |
|---|---|---|---|---|---|---|---|---|---|---|---|
| -14 | 1 | 6.745 | 180.0 | -14 | 2 | 1.418 | -180.0 | -13 | 1 | 3.387 | -180.0 |
| -13 | 2 | 0.145 | 180.0 | -13 | 3 | 0.597 | -180.0 | -13 | 4 | 3.640 | 180.0 |
| -12 | 1 | 3.370 | 0.0 | -12 | 2 | 5.308 | -180.0 | -12 | 3 | 1.432 | 0.0 |
| -12 | 4 | 1.191 | 180.0 | -12 | 5 | 1.380 | 0.0 | -11 | 1 | 4.713 | 0.0 |
| -11 | 2 | 2.028 | 180.0 | -11 | 3 | 0.654 | 0.0 | -11 | 4 | 7.344 | 0.0 |
| -11 | 5 | 8.480 | 0.0 | -11 | 6 | 4.385 | 180.0 | -11 | 7 | 1.507 | 180.0 |
| -10 | 1 | 0.928 | 0.0 | -10 | 2 | 0.496 | -180.0 | -10 | 3 | 3.428 | 180.0 |
| -10 | 4 | 5.041 | 0.0 | -10 | 5 | 3.581 | 0.0 | -10 | 6 | 0.677 | -180.0 |
| -10 | 7 | 1.168 | 180.0 | -10 | 8 | 0.415 | 0.0 | -9 | 1 | 2.880 | 0.0 |
| -9 | 2 | 6.127 | 180.0 | -9 | 3 | 8.557 | 0.0 | -9 | 4 | 2.792 | 0.0 |
| -9 | 5 | 2.847 | 180.0 | -9 | 6 | 4.842 | -180.0 | -9 | 7 | 1.639 | -180.0 |
| -9 | 8 | 1.907 | -180.0 | -8 | 1 | 13.353 | -180.0 | -8 | 2 | 2.291 | 0.0 |
| -8 | 3 | 2.872 | 0.0 | -8 | 4 | 9.840 | 0.0 | -8 | 5 | 6.929 | -180.0 |
| -8 | 6 | 7.031 | 180.0 | -8 | 7 | 9.290 | 0.0 | -8 | 8 | 5.058 | 180.0 |
| -8 | 9 | 0.405 | 0.0 | -7 | 1 | 4.629 | -180.0 | -7 | 2 | 4.604 | 0.0 |
| -7 | 3 | 5.466 | -180.0 | -7 | 4 | 3.748 | 0.0 | -7 | 5 | 19.431 | -180.0 |
| -7 | 6 | 2.803 | 180.0 | -7 | 7 | 2.269 | 0.0 | -7 | 8 | 9.925 | 180.0 |
| -7 | 9 | 2.206 | 0.0 | -7 | 10 | 8.902 | 0.0 | -6 | 1 | 17.257 | 180.0 |
| -6 | 2 | 21.283 | 0.0 | -6 | 3 | 16.469 | 0.0 | -6 | 4 | 12.968 | -180.0 |
| -6 | 5 | 7.216 | 180.0 | -6 | 6 | 9.238 | 0.0 | -6 | 7 | 5.220 | 0.0 |
| -6 | 8 | 9.582 | 180.0 | -6 | 9 | 5.902 | -180.0 | -6 | 10 | 7.455 | 0.0 |
| -6 | 11 | 0.767 | -180.0 | -5 | 1 | 23.427 | -180.0 | -5 | 2 | 23.296 | 0.0 |
| -5 | 3 | 21.323 | 180.0 | -5 | 4 | 5.796 | 180.0 | -5 | 5 | 8.394 | 0.0 |
| -5 | 6 | 2.306 | 0.0 | -5 | 7 | 0.547 | -180.0 | -5 | 8 | 8.213 | 0.0 |
| -5 | 9 | 17.118 | 0.0 | -5 | 10 | 1.381 | -180.0 | -5 | 11 | 5.898 | 180.0 |
| -4 | 1 | 61.012 | 0.0 | -4 | 2 | 4.926 | 0.0 | -4 | 3 | 26.717 | -180.0 |
| -4 | 4 | 5.472 | -180.0 | -4 | 5 | 6.135 | 0.0 | -4 | 6 | 7.992 | 0.0 |
| -4 | 7 | 20.920 | -180.0 | -4 | 8 | 4.585 | 0.0 | -4 | 9 | 13.011 | 0.0 |
| -4 | 10 | 1.672 | -180.0 | -4 | 11 | 1.044 | -180.0 | -4 | 12 | 1.769 | 180.0 |
| -3 | 1 | 36.371 | 0.0 | -3 | 2 | 8.201 | 0.0 | -3 | 3 | 12.047 | 0.0 |
| -3 | 4 | 9.369 | 180.0 | -3 | 5 | 20.075 | 0.0 | -3 | 6 | 1.978 | 180.0 |
| -3 | 7 | 8.029 | -180.0 | -3 | 8 | 10.661 | 0.0 | -3 | 9 | 1.235 | -180.0 |
| -3 | 10 | 3.346 | 180.0 | -3 | 11 | 8.025 | 180.0 | -3 | 12 | 1.024 | -180.0 |
| -2 | 1 | 42.422 | 0.0 | -2 | 2 | 53.334 | 0.0 | -2 | 3 | 27.494 | -180.0 |
| -2 | 4 | 6.852 | 0.0 | -2 | 5 | 32.533 | 0.0 | -2 | 6 | 9.891 | 180.0 |
| -2 | 7 | 4.078 | 180.0 | -2 | 8 | 12.184 | 0.0 | -2 | 9 | 0.556 | -180.0 |
| -2 | 10 | 9.359 | -180.0 | -2 | 11 | 1.102 | -180.0 | -2 | 12 | 3.085 | 0.0 |
| -1 | 1 | 8.490 | 0.0 | -1 | 2 | 55.041 | 180.0 | -1 | 3 | 14.535 | 0.0 |
| -1 | 4 | 0.512 | 0.0 | -1 | 5 | 6.684 | -180.0 | -1 | 6 | 14.393 | 180.0 |
| -1 | 7 | 2.044 | 0.0 | -1 | 8 | 25.995 | 0.0 | -1 | 9 | 14.883 | 180.0 |
| -1 | 10 | 13.114 | -180.0 | -1 | 11 | 5.127 | 0.0 | -1 | 12 | 2.702 | 180.0 |
| -1 | 13 | 0.570 | 0.0 | 0 | 0 | 256.000 | 0.0 | 0 | 1 | 6.572 | 180.0 |
| 0 | 2 | 64.114 | -180.0 | 0 | 3 | 38.201 | 0.0 | 0 | 4 | 6.694 | -180.0 |
| 0 | 5 | 8.207 | 180.0 | 0 | 6 | 34.612 | -180.0 | 0 | 7 | 6.278 | 0.0 |

6.5 Part of the output of STRUCFAC is shown in table VI-3. Neighbouring pairs of structure factors form Friedel pairs $(hk0)$ and $(\bar{h}\bar{k}0)$. Without anomalous scattering these would have equal amplitudes and opposite phases. Note the breakdown of this condition.

Table VI-3.

| h | k | F | PHI | h | k | F | PHI | h | k | F | PHI |
|---|---|---|---|---|---|---|---|---|---|---|---|
| -12 | 1 | 13.921 | -81.0 | 12 | -1 | 13.767 | 93.0 | -11 | 1 | 13.529 | 136.7 |
| 11 | -1 | 13.754 | -124.5 | -11 | 2 | 13.524 | -125.4 | 11 | -2 | 13.352 | 137.9 |
| -10 | 1 | 13.480 | 23.7 | 10 | -1 | 12.594 | -11.4 | -10 | 2 | 15.743 | 96.9 |
| 10 | -2 | 15.660 | -86.3 | -10 | 3 | 11.997 | -164.8 | 10 | -3 | 11.577 | 178.8 |
| -9 | 1 | 23.473 | -112.9 | 9 | -1 | 22.538 | 119.8 | -9 | 2 | 16.367 | -39.2 |
| 9 | -2 | 16.334 | 49.5 | -9 | 3 | 18.538 | 63.1 | 9 | -3 | 17.849 | -54.1 |
| -9 | 4 | 11.278 | 145.4 | 9 | -4 | 11.128 | -130.5 | -8 | 1 | 27.336 | 88.6 |
| 8 | -1 | 27.564 | -82.5 | -8 | 2 | 17.722 | -172.5 | 8 | -2 | 17.582 | -178.0 |
| -8 | 3 | 21.777 | -90.4 | 8 | -3 | 21.991 | 98.0 | -8 | 4 | 17.338 | 21.2 |
| 8 | -4 | 16.508 | -11.7 | -8 | 5 | 11.545 | 93.2 | 8 | -5 | 11.751 | -78.8 |
| -7 | 1 | 20.582 | -57.3 | 7 | -1 | 21.396 | 65.0 | -7 | 2 | 19.155 | 52.4 |
| 7 | -2 | 18.999 | -43.7 | -7 | 3 | 17.082 | 124.7 | 7 | -3 | 17.837 | -115.4 |
| -7 | 4 | 20.553 | -131.3 | 7 | -4 | 20.573 | 139.4 | -7 | 5 | 14.651 | -30.3 |
| 7 | -5 | 14.193 | 41.7 | -7 | 6 | 10.670 | 43.9 | 7 | -6 | 11.116 | -28.7 |
| -6 | 1 | 20.872 | -175.2 | 6 | -1 | 20.834 | -176.7 | -6 | 2 | 21.248 | -79.3 |
| 6 | -2 | 20.904 | 87.2 | -6 | 3 | 15.853 | 8.5 | 6 | -3 | 15.691 | 2.1 |
| -6 | 4 | 16.945 | 92.5 | 6 | -4 | 17.070 | -82.6 | -6 | 5 | 15.554 | -173.6 |
| 6 | -5 | 15.505 | -175.6 | -6 | 6 | 13.228 | -78.4 | 6 | -6 | 12.962 | 91.2 |
| -5 | 1 | 26.212 | 40.0 | 5 | -1 | 26.625 | -33.7 | -5 | 2 | 26.728 | 142.5 |
| 5 | -2 | 26.504 | -136.3 | -5 | 3 | 21.100 | -125.0 | 5 | -3 | 20.793 | 133.0 |
| -5 | 4 | 21.920 | -33.3 | 5 | -4 | 21.521 | 41.0 | -5 | 5 | 15.582 | 67.6 |
| 5 | -5 | 14.709 | -57.0 | -5 | 6 | 14.166 | 141.1 | 5 | -6 | 14.159 | -129.2 |
| -5 | 7 | 11.889 | -122.3 | 5 | -7 | 11.603 | 136.5 | -4 | 1 | 20.160 | -101.5 |
| 4 | -1 | 20.937 | 109.3 | -4 | 2 | 29.009 | -7.6 | 4 | -2 | 29.536 | 13.2 |
| -4 | 3 | 22.606 | 94.7 | 4 | -3 | 22.555 | -87.3 | -4 | 4 | 25.581 | 172.6 |
| 4 | -4 | 26.116 | -166.3 | -4 | 5 | 25.811 | -71.0 | 4 | -5 | 25.014 | 77.4 |
| -4 | 6 | 14.083 | 18.2 | 4 | -6 | 13.459 | -6.3 | -4 | 7 | 15.587 | 100.6 |
| 4 | -7 | 15.320 | -89.8 | -3 | 1 | 23.092 | 146.7 | 3 | -1 | 22.681 | -139.5 |
| -3 | 2 | 21.858 | -151.5 | 3 | -2 | 22.859 | 158.5 | -3 | 3 | 24.695 | -41.8 |
| 3 | -3 | 24.705 | 48.6 | -3 | 4 | 16.853 | 27.3 | 3 | -4 | 17.956 | -18.4 |
| -3 | 5 | 28.657 | 127.2 | 3 | -5 | 29.196 | -121.5 | -3 | 6 | 20.332 | -115.5 |
| 3 | -6 | 19.552 | 123.6 | -3 | 7 | 15.933 | -43.8 | 3 | -7 | 16.137 | 54.2 |
| -3 | 8 | 15.045 | 59.8 | 3 | -8 | 14.576 | -48.6 | -2 | 1 | 28.797 | -1.8 |
| 2 | -1 | 29.034 | 7.5 | -2 | 2 | 19.534 | 100.1 | 2 | -2 | 19.239 | -91.5 |
| -2 | 3 | 24.690 | 174.1 | 2 | -3 | 25.159 | -167.5 | -2 | 4 | 17.592 | -73.5 |
| 2 | -4 | 16.996 | 83.0 | -2 | 5 | 18.118 | -22.0 | 2 | -5 | 19.402 | 30.0 |
| -2 | 6 | 22.459 | 94.6 | 2 | -6 | 22.416 | -87.1 | -2 | 7 | 13.682 | -176.5 |
| 2 | -7 | 13.815 | -171.3 | -2 | 8 | 16.812 | -90.6 | 2 | -8 | 17.095 | 100.5 |
| -1 | 1 | 18.179 | -123.8 | 1 | -1 | 17.843 | 133.0 | -1 | 2 | 29.910 | -32.3 |
| 1 | -2 | 29.406 | 37.9 | -1 | 3 | 18.820 | 46.9 | 1 | -3 | 18.951 | -38.0 |
| -1 | 4 | 25.297 | 140.2 | 1 | -4 | 25.201 | -133.6 | -1 | 5 | 17.041 | -124.9 |
| 1 | -5 | 16.776 | 134.8 | -1 | 6 | 19.901 | -42.5 | 1 | -6 | 19.988 | 50.9 |
| -1 | 7 | 14.987 | 63.0 | 1 | -7 | 14.354 | -51.9 | -1 | 8 | 12.636 | 132.8 |
| 1 | -8 | 13.079 | -119.9 | 0 | 0 | 71.055 | 1.2 | 0 | 0 | 71.055 | 1.2 |
| 0 | 1 | 37.808 | 125.7 | 0 | -1 | 36.184 | -121.9 | 0 | 2 | 32.420 | -179.3 |
| 0 | -2 | 32.514 | -175.6 | 0 | 3 | 26.740 | -69.0 | 0 | -3 | 25.838 | 75.0 |
| 0 | 4 | 20.091 | -3.4 | 0 | -4 | 20.471 | 11.6 | 0 | 5 | 23.215 | 90.1 |
| 0 | -5 | 23.393 | -82.9 | 0 | 6 | 23.011 | -175.6 | 0 | -6 | 22.974 | -177.1 |
| 0 | 7 | 20.013 | -71.8 | 0 | -7 | 19.299 | 80.0 | 0 | 8 | 14.634 | 16.1 |
| 0 | -8 | 14.104 | -4.7 | 0 | 9 | 12.524 | 105.2 | 0 | -9 | 12.093 | -91.7 |
| 1 | 0 | 47.659 | -169.7 | -1 | 0 | 49.379 | 172.5 | 1 | 1 | 49.924 | -36.9 |

## Chapter 7

7.1 (a) $F_{hkl} = 0$ for $h + k + l = 2n + 1$.
   (b) $F_{h0l} = 0$ for $h + l = 2n + 1$.
   (c) For $3_1$ axis along $b$, $F_{0k0} = 0$ for $h \neq 3n$.

7.2 (a) $P2_1/c$. (b) $Pna2_1$.

7.3 For each projection the reflections are taken in groups of twenty and the intensities are compared with the average within the group. This gives the results shown in table VII-1.

   This indicates that the ($h0l$) projection is centrosymmetric and the ($0kl$) projection is non-centrosymmetric.

Table VII-1.

| $Z$ | ($h0l$) data $N(Z)$ | ($0kl$) data $N(Z)$ |
|---|---|---|
| 0.1 | 0.20 | 0.13 |
| 0.2 | 0.33 | 0.20 |
| 0.3 | 0.40 | 0.28 |
| 0.4 | 0.47 | 0.35 |
| 0.5 | 0.50 | 0.40 |
| 0.6 | 0.60 | 0.43 |
| 0.7 | 0.63 | 0.48 |
| 0.8 | 0.65 | 0.50 |
| 0.9 | 0.68 | 0.56 |
| 1.0 | 0.70 | 0.62 |

7.4 The first part of the output is given as table VII-2. If your results agree with this then you may assume that the remainder is correct.

Table VII-2.

a = 0.423E-01 b = 0.701E-02 c = 0.648E-02
VALUES OF E

| H | K | E | H | K | E | H | K | E |
|---|---|---|---|---|---|---|---|---|
| -14 | 1 | 1.243 | -14 | 2 | 0.303 | -14 | 3 | 1.123 |
| -14 | 4 | 1.028 | -14 | 5 | 0.323 | -14 | 6 | 1.730 |
| -14 | 7 | 0.803 | -14 | 8 | 0.313 | -13 | 1 | 0.523 |
| -13 | 2 | 0.033 | -13 | 3 | 0.124 | -13 | 4 | 0.845 |
| -13 | 5 | 0.118 | -13 | 6 | 2.367 | -13 | 7 | 0.458 |
| -13 | 8 | 0.793 | -13 | 9 | 1.099 | -12 | 1 | 0.439 |
| -12 | 2 | 0.812 | -12 | 3 | 0.274 | -12 | 4 | 0.234 |
| -12 | 5 | 0.367 | -12 | 6 | 0.494 | -12 | 7 | 0.652 |
| -12 | 8 | 0.988 | -12 | 9 | 0.035 | -12 | 10 | 0.505 |
| -11 | 1 | 0.511 | -11 | 2 | 0.259 | -11 | 3 | 0.140 |
| -11 | 4 | 1.219 | -11 | 5 | 1.717 | -11 | 6 | 1.011 |
| -11 | 7 | 0.431 | -11 | 8 | 0.073 | -11 | 9 | 0.760 |
| -11 | 10 | 0.003 | -11 | 11 | 0.221 | -10 | 1 | 0.078 |
| -10 | 2 | 0.038 | -10 | 3 | 0.390 | -10 | 4 | 0.766 |
| -10 | 5 | 0.598 | -10 | 6 | 0.151 | -10 | 7 | 0.272 |
| -10 | 8 | 0.064 | -10 | 9 | 0.675 | -10 | 10 | 1.810 |
| -10 | 11 | 0.823 | -9 | 1 | 0.238 | -9 | 2 | 0.563 |
| -9 | 3 | 0.998 | -9 | 4 | 0.321 | -9 | 5 | 0.414 |
| -9 | 6 | 0.851 | -9 | 7 | 0.365 | -9 | 8 | 0.511 |
| -9 | 9 | 0.837 | -9 | 10 | 0.506 | -9 | 11 | 0.105 |

## Chapter 8

8.1 Let the coordinates of Cu atom be $x, y$, and of Cl atom be $x', y'$. Atomic no. of Cu = 29 and of Cl = 17. Thus there are the following peaks:

| single weight | Cu—Cu at $2x, 2y$ of weight 841 |
| single weight | Cl—Cl at $2x', 2y'$ of weight 289 |
| double weight | Cu—Cl at $x + x', y + y'$ of weight 986 |
| double weight | Cu—Cl at $x - x', y - y'$ of weight 986. |

Examination of the Patterson function gives

$$x = 0.074 \qquad y = 0.193$$
$$x' = 0.249 \qquad y' = 0.146.$$

Superposition methods can be used to locate the two other atoms. They are at

$$x = 0.05 \qquad y = 0.60 \text{ for N}$$

and

$$x = 0.27 \qquad y = 0.67 \text{ for C.}$$

Check that these give vectors with the other atoms which are consistent with the Patterson.

8.2 The Patterson map is shown in fig. VIII-1 and it will be seen that there are two major peaks other than the origin peak. Peak B gives the S—S vector $(2x, 2y)$ at $(0.3, 0.6)$. Where there are alternative peaks it may be necessary to try both to see which one gives a plausible solution. The final Sim map is shown in fig. VIII-2.

8.3 The difference Patterson map is shown in fig. VIII-3 and clearly shows the S—S (or O—O) vector.

8.4 The vectors between the sulphur atoms are at $\pm(0.188, 0.100)$, $\pm(0.438, 0.700)$ and $\pm(0.250, 0.600)$. These can be clearly seen in the difference Patterson map which is given in fig. VIII-4. The final map giving the carbon positions is shown in fig. VIII-5.

8.5 The anomalous-difference Patterson map is shown in fig. VIII-6. The highest non-origin peak corresponds to the S—S vector.

8.6 The $P_s$-function map is given in fig. VIII-7. It would be difficult to interpret it completely for the positions of the light atoms but several could be approximately located. The two atoms not on peaks are approximately centrosymmetrically arranged about the mercury atom so their densities partially cancel each other.

8.7 The complete set of signs for this projection is given in table VIII-1.

Other solutions are possible which represent transformation to other centres of symmetry as origin. If the reflections are divided into four groups ($k$ even, $l$ even); ($k$ even, $l$ odd); ($k$ odd, $l$ even); and ($k$ odd, $l$ odd), then multiplying complete groups by the signs given below corresponds to transforming to the three other origins:

| | | | |
|---|---|---|---|
| (even, even) | + | + | + |
| (odd, even) | − | + | − |
| (even, odd) | + | − | − |
| (odd, odd) | − | − | + |

8.8 The solution obtained will depend on the seed for the random-number generator chosen for MINDIR. With seed = 3000 the data in table VIII-2 was found and the automatic solution corresponding to set 33 (fig. VIII-8) showed the complete structure.

Fig. VIII-1.
The Patterson map showing S–S vectors.

```
FACTOR CONVERTING TO DENSITY/UNIT AREA IS  3.4333
OUTPUT WITH Y HORIZONTAL AND ORIGIN     TOP LEFT
```

|    | 0 | 1 | 2 | 3 | 4 | 5 | 6 | 7 | 8 | 9 | 10 | 11 | 12 | 13 | 14 | 15 | 16 |
|----|---|---|---|---|---|---|---|---|---|---|----|----|----|----|----|----|----|
| 0  | 100 | 73 | 36 | 29 | 32 | 31 | 24 | 23 | 26 | 23 | 24 | 31 | 32 | 29 | 36 | 73 | 100 |
| 1  | 66 | 54. | 31. | 27. | 31. | 32. | 27. | 23. | 28. | 27. | 22. | 22. | 24. | 25. | 28. | 46. | 66 |
| 2  | 27 | 26. | 22. | 21. | 22. | 22. | 21. | 21. | 28. | 31. | 23. | 19. | 19. | 22. | 23. | 24. | 27 |
| 3  | 22 | 21. | 19. | 17. | 17. | 18. | 17. | 18. | 23. | 25. | 20. | 15. | 16. | 17. | 19. | 21. | 22 |
| 4  | 21 | 20. | 18. | 20. | 25. | 26. | 21. | 19. | 21. | 19. | 17. | 15. | 14. | 15. | 16. | 19. | 21 |
| 5  | 30 | 31. | 25. | 4. | 32. | 33. | 25. | 22. | 27. | 28. | 22. | 18. | 16. | 17. | 18. | 22. | 30 |
| 6  | 42 | 48. | 35. | 24. | 26. | 28. | 25. | 22. | 28. | 31. | 25. | 20. | 21. | 21. | 21. | 27. | 42 |
| 7  | 39 | 46. | 36. | 24. | 22. | 22. | 22. | 21. | 24. | 31. | 28. | 23. | 25. | 26. | 28. | 39 |
| 8  | 28 | 32. | 27. | 20. | 19. | 19. | 19. | 19. | 28. | 44. | 44. | 30. | 23. | 27. | 30. | 28. | 28 |
| 9  | 25 | 23. | 21. | 17. | 15. | 14. | 15. | 16. | 24. | 40. | 43. | 29. | 22. | 26. | 32. | 31. | 25 |
| 10 | 21 | 17. | 16. | 16. | 15. | 14. | 14. | 15. | 17. | 21. | 23. | 19. | 19. | 22. | 25. | 25. | 21 |
| 11 | 19 | 18. | 17. | 16. | 15. | 17. | 16. | 15. | 15. | 17. | 17. | 17. | 17. | 18. | 19. | 19. | 19 |
| 12 | 26 | 26. | 25. | 21. | 20. | 22. | 21. | 17. | 17. | 19. | 19. | 18. | 19. | 22. | 26. | 26. | 26 |
| 13 | 33 | 36. | 38. | 32. | 28. | 30. | 29. | 23. | 20. | 22. | 22. | 22. | 25. | 36. | 43. | 38. | 33 |
| 14 | 37 | 44. | 51. | 42. | 31. | 29. | 30. | 25. | 22. | 25. | 30. | 29. | 31. | 42. | 51. | 44. | 37 |
| 15 | 33 | 38. | 43. | 36. | 25. | 22. | 22. | 22. | 20. | 23. | 29. | 30. | 28. | 32. | 38. | 36. | 33 |
| 16 | 26 | 26. | 26. | 22. | 19. | 18. | 19. | 19. | 17. | 17. | 21. | 22. | 20. | 21. | 25. | 26. | 26 |
| 17 | 19 | 19. | 19. | 18. | 17. | 17. | 17. | 17. | 15. | 15. | 16. | 17. | 15. | 16. | 17. | 18. | 19 |
| 18 | 21 | 25. | 25. | 22. | 19. | 19. | 23. | 21. | 17. | 15. | 14. | 14. | 15. | 16. | 16. | 17. | 21 |
| 19 | 25 | 31. | 32. | 26. | 22. | 29. | 43. | 40. | 24. | 16. | 15. | 14. | 15. | 17. | 21. | 23. | 25 |
| 20 | 28 | 28. | 30. | 27. | 23. | 30. | 44. | 44. | 28. | 19. | 19. | 19. | 19. | 20. | 27. | 32. | 28 |
| 21 | 39 | 28. | 26. | 25. | 23. | 22. | 28. | 31. | 24. | 21. | 22. | 22. | 22. | 24. | 36. | 46. | 39 |
| 22 | 42 | 27. | 21. | 21. | 21. | 20. | 25. | 31. | 28. | 22. | 25. | 28. | 26. | 24. | 35. | 48. | 42 |
| 23 | 30 | 22. | 18. | 17. | 16. | 18. | 22. | 28. | 27. | 22. | 25. | 33. | 32. | 24. | 25. | 31. | 30 |
| 24 | 21 | 19. | 16. | 15. | 14. | 15. | 17. | 19. | 21. | 19. | 21. | 26. | 25. | 20. | 18. | 20. | 21 |
| 25 | 22 | 21. | 19. | 17. | 16. | 15. | 20. | 25. | 23. | 18. | 17. | 18. | 17. | 17. | 19. | 21. | 22 |
| 26 | 27 | 24. | 23. | 22. | 19. | 19. | 23. | 31. | 28. | 21. | 21. | 22. | 22. | 21. | 22. | 26. | 27 |
| 27 | 66 | 46. | 28. | 25. | 24. | 22. | 22. | 27. | 28. | 23. | 27. | 32. | 31. | 27. | 31. | 54. | 66 |
| 28 | 100 | 73. | 36. | 29. | 32. | 31. | 24. | 23. | 26. | 23. | 24. | 31. | 32. | 29. | 36. | 73. | 100 |

A  2 B

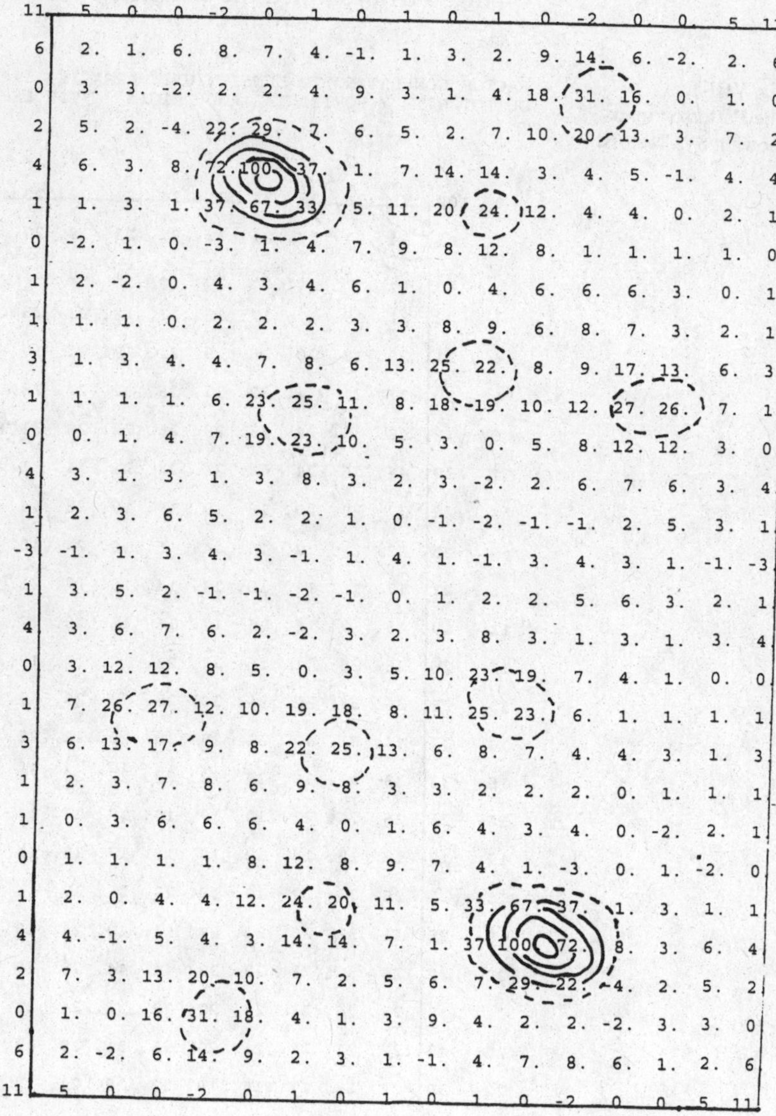

Fig. VIII-2.
A Sim-weighted map based on the sulphur positions.

FACTOR CONVERTING TO DENSITY/UNIT AREA IS 0.1371
OUTPUT WITH Y HORIZONTAL AND ORIGIN TOP LEFT

| | 0 | 1 | 2 | 3 | 4 | 5 | 6 | 7 | 8 | 9 | 10 | 11 | 12 | 13 | 14 | 15 | 16 |
|---|---|---|---|---|---|---|---|---|---|---|---|---|---|---|---|---|---|
| 0 | 11 | 5 | 0 | 0 | -2 | 0 | 1 | 0 | 1 | 0 | 1 | 0 | -2 | 0 | 0 | 5 | 11 |
| 1 | 6 | 2 | 1 | 6 | 8 | 7 | 4 | -1 | 1 | 3 | 2 | 9 | 14 | 6 | -2 | 2 | 6 |
| 2 | 0 | 3 | 3 | -2 | 2 | 2 | 4 | 9 | 3 | 1 | 4 | 18 | 31 | 16 | 0 | 1 | 0 |
| 3 | 2 | 5 | 2 | -4 | 22 | 29 | 2 | 6 | 5 | 2 | 7 | 10 | 20 | 13 | 3 | 7 | 2 |
| 4 | 4 | 6 | 3 | 8 | 72 | 100 | 37 | 1 | 7 | 14 | 14 | 3 | 4 | 5 | -1 | 4 | 4 |
| 5 | 1 | 1 | 3 | 1 | 37 | 67 | 33 | 5 | 11 | 20 | 24 | 12 | 4 | 4 | 0 | 2 | 1 |
| 6 | 0 | -2 | 1 | 0 | -3 | 1 | 4 | 7 | 9 | 8 | 12 | 8 | 1 | 1 | 1 | 1 | 0 |
| 7 | 1 | 2 | -2 | 0 | 4 | 3 | 4 | 6 | 1 | 0 | 4 | 6 | 6 | 6 | 3 | 0 | 1 |
| 8 | 1 | 1 | 1 | 0 | 2 | 2 | 2 | 3 | 3 | 8 | 9 | 6 | 8 | 7 | 3 | 2 | 1 |
| 9 | 3 | 1 | 3 | 4 | 4 | 7 | 8 | 6 | 13 | 25 | 22 | 8 | 9 | 17 | 13 | 6 | 3 |
| 10 | 1 | 1 | 1 | 1 | 6 | 23 | 25 | 11 | 8 | 18 | -19 | 10 | 12 | 27 | 26 | 7 | 1 |
| 11 | 0 | 0 | 1 | 4 | 7 | 19 | 23 | 10 | 5 | 3 | 0 | 5 | 8 | 12 | 12 | 3 | 0 |
| 12 | 4 | 3 | 1 | 3 | 1 | 3 | 8 | 3 | 2 | 3 | -2 | 2 | 6 | 7 | 6 | 3 | 4 |
| 13 | 1 | 2 | 3 | 6 | 5 | 2 | 2 | 1 | 0 | -1 | -2 | -1 | -1 | 2 | 5 | 3 | 1 |
| 14 | -3 | -1 | 1 | 3 | 4 | 3 | -1 | 1 | 4 | 1 | -1 | 3 | 4 | 3 | 1 | -1 | -3 |
| 15 | 1 | 1 | 3 | 5 | 2 | -1 | -1 | -2 | -1 | 0 | 1 | 2 | 5 | 6 | 3 | 2 | 1 |
| 16 | 4 | 3 | 6 | 7 | 6 | 2 | -2 | 3 | 2 | 3 | 8 | 3 | 1 | 3 | 1 | 3 | 4 |
| 17 | 0 | 3 | 12 | 12 | 8 | 5 | 0 | 3 | 5 | 10 | 23 | 19 | 7 | 4 | 1 | 0 | 0 |
| 18 | 1 | 7 | 26 | 27 | 12 | 10 | 19 | 18 | 8 | 11 | 25 | 23 | 6 | 1 | 1 | 1 | 1 |
| 19 | 3 | 6 | 13 | 17 | 9 | 8 | 22 | 25 | 13 | 6 | 8 | 7 | 4 | 4 | 3 | 1 | 3 |
| 20 | 1 | 2 | 3 | 7 | 8 | 6 | 9 | -8 | 3 | 3 | 2 | 2 | 2 | 0 | 1 | 1 | 1 |
| 21 | 1 | 0 | 3 | 6 | 6 | 6 | 4 | 0 | 1 | 6 | 4 | 3 | 4 | 0 | -2 | 2 | 1 |
| 22 | 0 | 1 | 1 | 1 | 1 | 8 | 12 | 8 | 9 | 7 | 4 | 1 | -3 | 0 | 1 | -2 | 0 |
| 23 | 1 | 2 | 0 | 4 | 4 | 12 | 24 | 20 | 11 | 5 | 33 | 67 | 37 | 1 | 3 | 1 | 1 |
| 24 | 4 | 4 | -1 | 5 | 4 | 3 | 14 | 14 | 7 | 1 | 37 | 100 | 72 | 8 | 3 | 6 | 4 |
| 25 | 2 | 7 | 3 | 13 | 20 | 10 | 7 | 2 | 5 | 6 | 7 | 29 | 22 | 4 | 2 | 5 | 2 |
| 26 | 0 | 1 | 0 | 16 | 31 | 18 | 4 | 1 | 3 | 9 | 4 | 2 | 2 | -2 | 3 | 3 | 0 |
| 27 | 6 | 2 | -2 | 6 | 14 | 9 | 2 | 3 | 1 | -1 | 4 | 7 | 8 | 6 | 1 | 2 | 6 |
| 28 | 11 | 5 | 0 | 0 | -2 | 0 | 1 | 0 | 1 | 0 | 1 | 0 | -2 | 0 | 0 | 5 | 11 |

## Solutions to Problems

Fig. VIII-3.
The difference
Patterson map clearly
showing the
isomorphously replaced
atom.

```
FACTOR CONVERTING TO DENSITY/UNIT AREA IS   0.5607
OUTPUT WITH Y HORIZONTAL AND ORIGIN    TOP LEFT
```

|    | 0   | 1   | 2   | 3   | 4   | 5   | 6   | 7   | 8   | 9   | 10  | 11  | 12  | 13  | 14  | 15  | 16  |
|----|-----|-----|-----|-----|-----|-----|-----|-----|-----|-----|-----|-----|-----|-----|-----|-----|-----|
| 0  | 100 | 58  | 7   | 1   | 0   | 0   | 2   | 0   | 1   | 0   | 2   | 0   | 0   | 1   | 7   | 58  | 100 |
| 1  | 51  | 38  | 11  | 7   | 5   | 4   | 6   | 4   | 3   | 4   | 5   | 1   | -4  | -2  | 0   | 20  | 51  |
| 2  | 5   | 9   | 9   | 9   | 6   | 6   | 8   | 7   | 8   | 7   | 5   | 2   | -3  | -1  | 4   | 3   | 5   |
| 3  | 7   | 9   | 8   | 5   | 7   | 11  | 11  | 9   | 10  | 6   | -2  | -3  | 0   | -2  | 0   | 4   | 7   |
| 4  | 1   | 4   | 4   | 4   | 10  | 13  | 8   | 5   | 6   | 2   | -2  | -2  | 1   | 1   | -2  | -2  | 1   |
| 5  | 2   | 8   | 8   | 9   | 14  | 14  | 5   | 4   | 8   | 7   | 4   | 2   | 1   | 2   | 0   | -2  | 2   |
| 6  | 8   | 14  | 10  | 7   | 9   | 9   | 7   | 6   | 8   | 8   | 3   | 4   | 4   | 1   | 1   | 3   | 8   |
| 7  | 5   | 6   | 2   | 2   | 1   | -1  | 2   | 3   | 5   | 13  | 10  | 3   | 3   | 0   | 2   | 5   | 5   |
| 8  | -2  | -3  | -4  | -3  | -1  | -3  | -1  | 0   | 11  | 39  | 39  | 12  | -1  | -2  | 0   | 2   | -2  |
| 9  | 0   | -2  | -1  | 0   | 0   | -4  | -4  | -3  | 3   | 27  | 37  | 18  | 3   | 1   | 3   | 4   | 0   |
| 10 | 2   | 0   | 0   | 2   | 1   | -5  | -7  | -2  | -1  | 0   | 7   | 10  | 6   | 1   | 1   | 4   | 2   |
| 11 | 3   | 2   | 2   | -2  | -5  | -6  | -6  | -3  | 2   | 1   | 1   | 5   | 4   | 1   | 2   | 4   | 3   |
| 12 | 2   | 1   | 1   | -2  | -5  | -4  | -2  | -2  | 0   | 1   | 1   | 2   | 0   | 1   | 5   | 5   | 2   |
| 13 | 2   | 2   | 4   | 2   | 2   | 5   | 5   | 2   | 1   | 0   | 2   | 2   | 1   | 6   | 9   | 5   | 2   |
| 14 | 5   | 8   | 12  | 8   | 5   | 8   | 6   | 1   | 0   | 1   | 6   | 8   | 5   | 8   | 12  | 8   | 5   |
| 15 | 2   | 5   | 9   | 6   | 1   | 2   | 2   | 0   | 1   | 5   | 5   | 2   | 2   | 4   | 2   | 5   | 2   |
| 16 | 2   | 5   | 5   | 1   | 0   | 2   | 1   | 1   | 0   | -2  | -2  | -4  | -5  | -2  | 1   | 1   | 2   |
| 17 | 3   | 4   | 2   | 1   | 4   | 5   | 1   | 1   | 2   | -3  | -6  | -6  | -5  | -2  | 2   | 2   | 3   |
| 18 | 2   | 4   | 1   | 1   | 6   | 10  | 7   | 0   | -1  | -2  | -7  | -5  | 1   | 2   | 0   | 0   | 2   |
| 19 | 0   | 4   | 3   | 1   | 3   | 18  | 37  | 27  | 3   | -3  | -4  | -4  | 0   | 0   | -1  | -2  | 0   |
| 20 | -2  | 2   | 0   | -2  | -1  | 12  | 39  | 39  | 11  | 0   | -1  | -3  | -1  | -3  | -4  | -3  | -2  |
| 21 | 5   | 5   | 2   | 0   | 3   | 3   | 10  | 13  | 5   | 3   | 2   | -1  | 1   | 2   | 2   | 6   | 5   |
| 22 | 8   | 3   | 1   | 1   | 4   | 4   | 3   | 8   | 8   | 6   | 7   | 9   | 9   | 7   | 10  | 14  | 8   |
| 23 | 2   | -2  | 0   | 2   | 1   | 2   | 4   | 7   | 8   | 4   | 5   | 14  | 14  | 9   | 8   | 8   | 2   |
| 24 | 1   | -2  | -2  | 1   | 1   | -2  | -2  | 2   | 6   | 5   | 8   | 13  | 10  | 4   | 4   | 4   | 1   |
| 25 | 7   | 4   | 0   | -2  | 0   | -3  | -2  | 6   | 10  | 9   | 11  | 11  | 7   | 5   | 8   | 9   | 7   |
| 26 | 5   | 3   | 4   | -1  | -3  | 2   | 5   | 7   | 8   | 7   | 8   | 6   | 6   | 9   | 9   | 9   | 5   |
| 27 | 51  | 20  | 0   | -2  | -4  | 1   | 5   | 4   | 3   | 4   | 6   | 4   | 5   | 7   | 11  | 38  | 51  |
| 28 | 100 | 58  | 7   | 1   | 0   | 0   | 2   | 0   | 1   | 0   | 2   | 0   | 0   | 1   | 7   | 58  | 100 |

Fig. VIII-4.
A difference Patterson map showing the six vectors between the sulphur (or oxygen) atoms.

FACTOR CONVERTING TO DENSITY/UNIT AREA IS   4.5557
OUTPUT WITH Y HORIZONTAL AND ORIGIN    TOP LEFT

|      | 0 | 1 | 2 | 3 | 4 | 5 | 6 | 7 | 8 | 9 | 10 | 11 | 12 | 13 | 14 | 15 | 16 |
|------|---|---|---|---|---|---|---|---|---|---|----|----|----|----|----|----|----|
| 0  | 100 | 51 | 5 | 9 | 3 | 3 | 3 | 1 | 6 | 1 | 3 | 3 | 3 | 9 | 5 | 51 | 100 |
| 1  | 59 | 32. | 4. | 10. | 2. | 3. | 6. | 3. | 7. | 4. | 6. | 10. | 7. | 10. | 3. | 24. | 59 |
| 2  | 10 | 5. | 5. | 12. | 5. | 8. | 10. | 3. | 10. | 11. | 7. | 8. | 4. | 8. | 7. | 4. | 10 |
| 3  | 7 | 9. | 10. | 7. | 3. | 7. | 7. | 2. | 12. | 16. | 8. | 7. | 5. | 3. | 11. | 11. | 7 |
| 4  | 8 | 18. | 15. | 3. | 4. | 6. | 6. | 6. | 8. | 10. | 5. | 6. | 8. | 1. | 6. | 9. | 8 |
| 5  | 6 | 27. | 29. | 8. | 7. | 5. | 3. | 7. | 2. | 4. | 5. | 2. | 7. | 5. | 9. | 9. | 6 |
| 6  | 6 | 26. | 31. | 9. | 5. | 3. | 2. | 8. | 8. | 20. | 23. | 7. | 6. | 9. | 11. | 11. | 6 |
| 7  | 3 | 10. | 14. | 3. | 4. | 6. | 7. | 12. | 15. | 32. | 33. | 10. | 9. | 10. | 4. | 6. | 3 |
| 8  | 7 | 5. | 8. | 9. | 8. | 5. | 6. | 11. | 10. | 19. | 21. | 7. | 10. | 13. | 3. | 6. | 7 |
| 9  | 6 | 6. | 12. | 15. | 9. | 1. | 3. | 8. | 6. | 6. | 9. | 6. | 9. | 12. | 6. | 4. | 6 |
| 10 | 1 | 3. | 8. | 9. | 4. | 3. | 5. | 6. | 7. | 5. | 5. | 7. | 6. | 6. | 6. | 2. | 1 |
| 11 | 5 | 5. | 4. | 4. | 2. | 5. | 4. | 0. | 6. | 8. | 10. | 17. | 11. | 4. | 5. | 5. | 5 |
| 12 | 4 | 3. | 3. | 6. | 5. | 4. | 4. | 2. | 6. | 9. | 19. | 34. | 23. | 6. | 5. | 6. | 4 |
| 13 | 5 | 3. | 3. | 5. | 6. | 5. | 6. | 7. | 7. | 4. | 15. | 31. | 21. | 4. | 6. | 8. | 5 |
| 14 | 8 | 9. | 7. | 3. | 9. | 13. | 6. | 3. | 5. | 3. | 6. | 13. | 9. | 3. | 7. | 9. | 8 |
| 15 | 5 | 8. | 6. | 4. | 21. | 31. | 15. | 4. | 7. | 7. | 6. | 5. | 6. | 5. | 3. | 3. | 5 |
| 16 | 4 | 6. | 5. | 6. | 23. | 34. | 19. | 9. | 6. | 2. | 4. | 4. | 5. | 6. | 3. | 3. | 4 |
| 17 | 5 | 5. | 5. | 4. | 11. | 17. | 10. | 8. | 6. | 0. | 4. | 5. | 2. | 4. | 4. | 5. | 5 |
| 18 | 1 | 2. | 6. | 6. | 6. | 7. | 5. | 5. | 7. | 6. | 5. | 3. | 4. | 9. | 8. | 3. | 1 |
| 19 | 6 | 4. | 6. | 12. | 9. | 6. | 9. | 6. | 6. | 8. | 3. | 1. | 9. | 15. | 12. | 6. | 6 |
| 20 | 7 | 6. | 3. | 13. | 10. | 7. | 21. | 19. | 10. | 11. | 6. | 5. | 8. | 9. | 8. | 5. | 7 |
| 21 | 3 | 6. | 4. | 10. | 9. | 10. | 33. | 32. | 15. | 12. | 7. | 6. | 4. | 3. | 14. | 10. | 3 |
| 22 | 6 | 11. | 11. | 9. | 6. | 7. | 23. | 20. | 8. | 8. | 2. | 3. | 5. | 9. | 31. | 26. | 6 |
| 23 | 6 | 9. | 9. | 5. | 7. | 2. | 5. | 4. | 2. | 7. | 3. | 5. | 7. | 8. | 29. | 27. | 6 |
| 24 | 8 | 9. | 6. | 1. | 8. | 6. | 5. | 10. | 8. | 6. | 6. | 6. | 4. | 3. | 15. | 18. | 8 |
| 25 | 7 | 11. | 11. | 3. | 5. | 7. | 8. | 16. | 12. | 2. | 7. | 7. | 3. | 7. | 10. | 9. | 7 |
| 26 | 10 | 4. | 7. | 8. | 4. | 8. | 7. | 11. | 10. | 3. | 10. | 8. | 5. | 12. | 5. | 5. | 10 |
| 27 | 59 | 24. | 3. | 10. | 7. | 10. | 6. | 4. | 7. | 3. | 6. | 3. | 2. | 10. | 4. | 32. | 59 |
| 28 | 100 | 51 | 5 | 9 | 3 | 3 | 3 | 1 | 6 | 1 | 3 | 3 | 3 | 9 | 5 | 51 | 100 |

Fig. VIII-5.
Map showing the
carbon atoms with
phases from sulphur
atom positions as
deduced from fig.
VIII-4.

FACTOR CONVERTING TO DENSITY/UNIT AREA IS  0.0392
OUTPUT WITH Y HORIZONTAL AND ORIGIN    TOP LEFT

| | 0 | 1 | 2 | 3 | 4 | 5 | 6 | 7 | 8 | 9 | 10 | 11 | 12 | 13 | 14 | 15 | 16 |
|---|---|---|---|---|---|---|---|---|---|---|---|---|---|---|---|---|---|
| 0 | 21 | 32. | 13. | 0. | 13. | 6. | 0. | 6. | -9. | -7. | 6. | -4. | -2. | 4. | -4. | 1. | 21 |
| 1 | 39 | 37. | 13. | 6. | 21. | 9. | -7. | -1. | 0. | 3. | -2. | -14. | 0. | 3. | -6. | 15. | 39 |
| 2 | 30 | 24. | 12. | 10. | 9. | -3. | -3. | 2. | -3. | -2. | -3. | -2. | 12. | 9. | 2. | 18. | 30 |
| 3 | 14 | 19. | 12. | 10. | 7. | -6. | -1. | 4. | -7. | -6. | 11. | 23. | 9. | -11. | -5. | 6. | 14 |
| 4 | 4 | 15. | 9. | 8. | 21. | 15. | 0. | -2. | 1. | -2. | 16. | 40. | 13. | -20. | -6. | 3. | 4 |
| 5 | 2 | 9. | 13. | 11. | 22. | 28. | 11. | 1. | 8. | -2. | -2. | 17. | 8. | -8. | 3. | 8. | 2 |
| 6 | 6 | 16. | 34. | 22. | 4. | 13. | 13. | 4. | 3. | -4. | -1. | -3. | -9. | 0. | 4. | 3. | 6 |
| 7 | 6 | 16. | 34. | 18. | -9. | 1. | 4. | 12. | 15. | 2. | 19. | 12. | -5. | 20. | 19. | 1. | 6 |
| 8 | 9 | 4. | 13. | 7. | -2. | 9. | 2. | 33. | 58. | 15. | 9. | 16. | 6. | 37. | 41. | 16. | 9 |
| 9 | 11 | 2. | 8. | 10. | 6. | 14. | 6. | 38. | 73. | 23. | -3. | 14. | 13. | 30. | 35. | 18. | 11 |
| 10 | 2 | 3. | 7. | 10. | -3. | 1. | 14. | 22. | 31. | 15. | 9. | 25. | 16. | 12. | 19. | 7. | 2 |
| 11 | -3 | 2. | -3. | -4. | -13. | -10. | 23. | 13. | -8. | 5. | 12. | 21. | 13. | -2. | 17. | 12. | -3 |
| 12 | 3 | 4. | -7. | -12. | -9. | -7. | 25. | 12. | -8. | 13. | 3. | 25. | 51. | 15. | 20. | 28. | 3 |
| 13 | 4 | 2. | -4. | -11. | -5. | -3. | 19. | 6. | -5. | 18. | -3. | 36. | 100. | 43. | 10. | 25. | 4 |
| 14 | -2 | -2. | 5. | -13. | -22. | -11. | 9. | -7. | -14. | 13. | -8. | 18. | 83. | 33. | -10. | 10. | -2 |
| 15 | -7 | 2. | 8. | -1. | -1. | -9. | -15. | -13. | -11. | 6. | 6. | -6. | -4. | -10. | -10. | -2. | -7 |
| 16 | 0 | 4. | 6. | 0. | 3. | -1. | -25. | -20. | 2. | 7. | 15. | 5. | -17. | -10. | 1. | 0. | 0 |
| 17 | 21 | 8. | 3. | 7. | 8. | 14. | -9. | -18. | 0. | 9. | 24. | 21. | 0. | 7. | 8. | 10. | 21 |
| 18 | 34 | 8. | -5. | 9. | 11. | 17. | 13. | -5. | -11. | 14. | 56. | 43. | 8. | 17. | 19. | 22. | 34 |
| 19 | 26 | 9. | -7. | 0. | 6. | 2. | 8. | 10. | -13. | 1. | 59. | 50. | 7. | 16. | 26. | 23. | 26 |
| 20 | 16 | 16. | 6. | -4. | 6. | -4. | -11. | 13. | -4. | -23. | 5. | 14. | 2. | 7. | 18. | 21. | 16 |
| 21 | 16 | 12. | 13. | -1. | 10. | 12. | 0. | 14. | 6. | -16. | -21. | -15. | 12. | 19. | 12. | 24. | 16 |
| 22 | 16 | 13. | 22. | 8. | 5. | 19. | 28. | 29. | 10. | 8. | 3. | -7. | 35. | 53. | 18. | 16. | 16 |
| 23 | 9 | 31. | 45. | 25. | 4. | 9. | 29. | 42. | 24. | 14. | 13. | 3. | 41. | 69. | 26. | -2. | 9 |
| 24 | 3 | 32. | 48. | 28. | 13. | 10. | 8. | 36. | 52. | 22. | -3. | 0. | 22. | 39. | 25. | 2. | 3 |
| 25 | 0 | 9. | 26. | 16. | 15. | 18. | -2. | 14. | 52. | 30. | -3. | 9. | 10. | -2. | 13. | 18. | 0 |
| 26 | -8 | -5. | 14. | 7. | 2. | 10. | 3. | -1. | 10. | 11. | 12. | 29. | 24. | 1. | 9. | 14. | -8 |
| 27 | -5 | 7. | 12. | 2. | -1. | 0. | 8. | 4. | -17. | -12. | 14. | 24. | 23. | 17. | 9. | 0. | -5 |
| 28 | 21 | 32. | 13. | 0. | 13. | 6. | 0. | 6. | -9. | -7. | 6. | -4. | -2. | 4. | -4. | 1. | 21 |

Fig. VIII-6.
The
anomalous-difference
Patterson map.

FACTOR CONVERTING TO DENSITY/UNIT AREA IS   0.0065
OUTPUT WITH Y HORIZONTAL AND ORIGIN     TOP LEFT

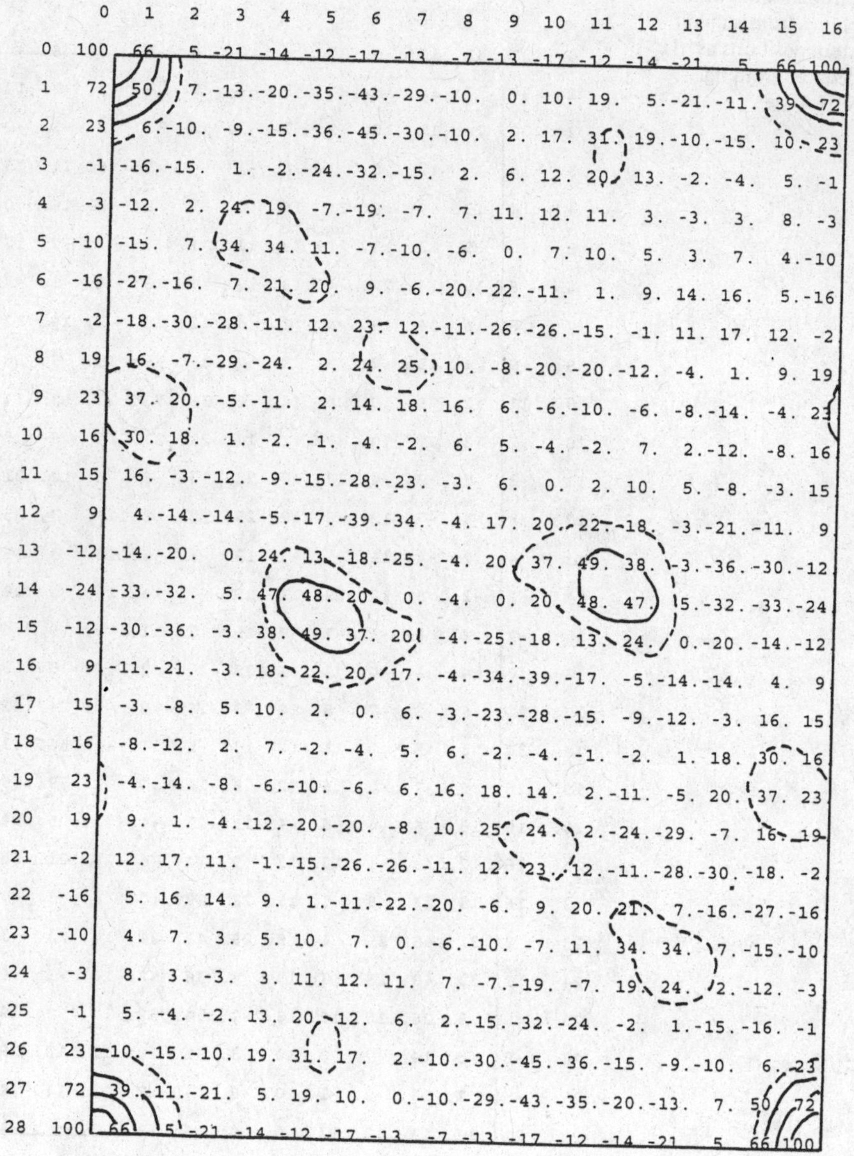

|      | 0 | 1 | 2 | 3 | 4 | 5 | 6 | 7 | 8 | 9 | 10 | 11 | 12 | 13 | 14 | 15 | 16 |
|------|---|---|---|---|---|---|---|---|---|---|----|----|----|----|----|----|----|
| 0  | 100 | 66 | 5 | -21 | -14 | -12 | -17 | -13 | -7 | -13 | -17 | -12 | -14 | -21 | 5 | 66 | 100 |
| 1  | 72 | 50 | 7 | -13 | -20 | -35 | -43 | -29 | -10 | 0 | 10 | 19 | 5 | -21 | -11 | 39 | 72 |
| 2  | 23 | 6 | -10 | -9 | -15 | -36 | -45 | -30 | -10 | 2 | 17 | 31 | 19 | -10 | -15 | 10 | 23 |
| 3  | -1 | -16 | -15 | 1 | -2 | -24 | -32 | -15 | 2 | 6 | 12 | 20 | 13 | -2 | -4 | 5 | -1 |
| 4  | -3 | -12 | 2 | 24 | 19 | -7 | -19 | -7 | 7 | 11 | 12 | 11 | 3 | -3 | 3 | 8 | -3 |
| 5  | -10 | -15 | 7 | 34 | 34 | 11 | -7 | -10 | -6 | 0 | 7 | 10 | 5 | 3 | 7 | 4 | -10 |
| 6  | -16 | -27 | -16 | 7 | 21 | 20 | 9 | -6 | -20 | -22 | -11 | 1 | 9 | 14 | 16 | 5 | -16 |
| 7  | -2 | -18 | -30 | -28 | -11 | 12 | 23 | 12 | -11 | -26 | -26 | -15 | -1 | 11 | 17 | 12 | -2 |
| 8  | 19 | 16 | -7 | -29 | -24 | 2 | 24 | 25 | 10 | -8 | -20 | -20 | -12 | -4 | 1 | 9 | 19 |
| 9  | 23 | 37 | 20 | -5 | -11 | 2 | 14 | 18 | 16 | 6 | -6 | -10 | -6 | -8 | -14 | -4 | 23 |
| 10 | 16 | 30 | 18 | 1 | -2 | -1 | -4 | -2 | 6 | 5 | -4 | -2 | 7 | 2 | -12 | -8 | 16 |
| 11 | 15 | 16 | -3 | -12 | -9 | -15 | -28 | -23 | -3 | 6 | 0 | 2 | 10 | 5 | -8 | -3 | 15 |
| 12 | 9 | 4 | -14 | -14 | -5 | -17 | -39 | -34 | -4 | 17 | 20 | -22 | -18 | -3 | -21 | -11 | 9 |
| 13 | -12 | -14 | -20 | 0 | 24 | 13 | -18 | -25 | -4 | 20 | 37 | 49 | 38 | -3 | -36 | -30 | -12 |
| 14 | -24 | -33 | -32 | 5 | 47 | 48 | 20 | 0 | -4 | 0 | 20 | 48 | 47 | 5 | -32 | -33 | -24 |
| 15 | -12 | -30 | -36 | -3 | 38 | 49 | 37 | 20 | -4 | -25 | -18 | 13 | 24 | 0 | -20 | -14 | -12 |
| 16 | 9 | -11 | -21 | -3 | 18 | 22 | 20 | 17 | -4 | -34 | -39 | -17 | -5 | -14 | -14 | 4 | 9 |
| 17 | 15 | -3 | -8 | 5 | 10 | 2 | 0 | 6 | -3 | -23 | -28 | -15 | -9 | -12 | -3 | 16 | 15 |
| 18 | 16 | -8 | -12 | 2 | 7 | -2 | -4 | 5 | 6 | -2 | -4 | -1 | -2 | 1 | 18 | 30 | 16 |
| 19 | 23 | -4 | -14 | -8 | -6 | -10 | -6 | 6 | 16 | 18 | 14 | 2 | -11 | -5 | 20 | 37 | 23 |
| 20 | 19 | 9 | 1 | -4 | -12 | -20 | -20 | -8 | 10 | 25 | 24 | 2 | -24 | -29 | -7 | 16 | 19 |
| 21 | -2 | 12 | 17 | 11 | -1 | -15 | -26 | -26 | -11 | 12 | 23 | 12 | -11 | -28 | -30 | -18 | -2 |
| 22 | -16 | 5 | 16 | 14 | 9 | 1 | -11 | -22 | -20 | -6 | 9 | 20 | 21 | 7 | -16 | -27 | -16 |
| 23 | -10 | 4 | 7 | 3 | 5 | 10 | 7 | 0 | -6 | -10 | -7 | 11 | 34 | 34 | 7 | -15 | -10 |
| 24 | -3 | 8 | 3 | -3 | 3 | 11 | 12 | 11 | 7 | -7 | -19 | -7 | 19 | 24 | 2 | -12 | -3 |
| 25 | -1 | 5 | -4 | -2 | 13 | 20 | 12 | 6 | 2 | -15 | -32 | -24 | -2 | 1 | -15 | -16 | -1 |
| 26 | 23 | -10 | -15 | -10 | 19 | 31 | 17 | 2 | -10 | -30 | -45 | -36 | -15 | -9 | -10 | 6 | 23 |
| 27 | 72 | 39 | -11 | -21 | 5 | 19 | 10 | 0 | -10 | -29 | -43 | -35 | -20 | -13 | 7 | 50 | 72 |
| 28 | 100 | 66 | 5 | -21 | -14 | -12 | -17 | -13 | -7 | -13 | -17 | -12 | -14 | -21 | 5 | 66 | 100 |

Fig. VIII-7.
The $P_s$-function map showing its antisymmetric form.

FACTOR CONVERTING TO DENSITY/UNIT AREA IS 243.8131
OUTPUT WITH Y HORIZONTAL AND ORIGIN     TOP LEFT

|    |     | 0 | 1 | 2 | 3 | 4 | 5 | 6 | 7 | 8 | 9 | 10 | 11 | 12 | 13 | 14 | 15 | 16 | 17 | 18 | 19 | 20 |
|----|-----|---|---|---|---|---|---|---|---|---|---|----|----|----|----|----|----|----|----|----|----|----|
| 0  | 0   | 1 | 2 | 1 | 0 | -1 | -1 | 1 | 4 | 4 | 0 | -4 | -4 | -1 | 1 | 1 | 0 | -1 | -2 | -1 | 0 |
| 1  | -1  | 2 | 5 | 6 | 7 | 7 | 8 | 14 | 24 | 36 | 41 | 35 | 25 | 18 | 11 | 3 | -3 | -6 | -7 | -5 | -1 |
| 2  | -2  | 10 | 19 | 24 | 25 | 25 | 25 | 27 | 35 | 48 | 58 | 58 | 52 | 43 | 28 | 9 | -9 | -19 | -20 | -13 | -2 |
| 3  | -5  | 23 | 44 | 52 | 52 | 51 | 50 | 47 | 44 | 46 | 52 | 58 | 64 | 64 | 49 | 21 | -10 | -34 | -42 | -30 | -5 |
| 4  | -14 | 32 | 67 | 81 | 79 | 75 | 74 | 68 | 56 | 44 | 39 | 43 | 52 | 59 | 63 | 28 | -7 | -43 | -62 | -50 | -14 |
| 5  | -19 | 28 | 69 | 89 | 89 | 84 | 83 | 77 | 62 | 45 | 32 | 26 | 25 | 30 | 31 | 23 | 0 | -32 | -56 | -52 | -19 |
| 6  | -12 | 16 | 46 | 67 | 74 | 75 | 75 | 71 | 61 | 47 | 33 | 20 | 11 | 8 | 11 | 14 | 9 | -7 | -25 | -28 | -12 |
| 7  | 3   | 10 | 23 | 37 | 48 | 55 | 60 | 63 | 62 | 56 | 44 | 31 | 21 | 16 | 15 | 17 | 17 | 12 | 4 | 1 | 3 |
| 8  | 15  | 15 | 16 | 21 | 28 | 37 | 46 | 58 | 67 | 68 | 58 | 44 | 37 | 35 | 31 | 25 | 20 | 16 | 14 | 15 | 15 |
| 9  | 16  | 18 | 19 | 20 | 22 | 27 | 37 | 51 | 65 | 71 | 63 | 50 | 43 | 41 | 36 | 27 | 18 | 11 | 9 | 12 | 16 |
| 10 | 5   | 11 | 18 | 23 | 28 | 34 | 43 | 52 | 60 | 62 | 57 | 47 | 39 | 35 | 29 | 21 | 11 | 4 | -1 | 0 | 5 |
| 11 | -8  | -1 | 9 | 21 | 37 | 55 | 69 | 72 | 65 | 54 | 44 | 35 | 28 | 24 | 20 | 12 | 4 | -3 | -9 | -11 | -8 |
| 12 | -12 | -8 | 1 | 16 | 40 | 70 | 94 | 97 | 78 | 51 | 28 | 13 | 6 | 5 | 6 | 4 | -2 | -7 | -11 | -13 | -12 |
| 13 | -8  | -5 | 1 | 12 | 33 | 62 | 89 | 97 | 79 | 46 | 13 | -13 | -26 | -26 | -18 | -11 | -8 | -9 | -10 | -10 | -8 |
| 14 | 0   | 5 | 9 | 15 | 28 | 53 | 83 | 100 | 89 | 51 | 0 | -51 | -89 | **** | -83 | -53 | -28 | -15 | -9 | -5 | 0 |
| 15 | 8   | 10 | 10 | 9 | 8 | 11 | 18 | 26 | 26 | 13 | -13 | -46 | -79 | -97 | -89 | -62 | -33 | -12 | -1 | 5 | 8 |
| 16 | 12  | 13 | 11 | 7 | 2 | -4 | -6 | -5 | -6 | -13 | -28 | -51 | -78 | -97 | -94 | -70 | -40 | -16 | -1 | 8 | 12 |
| 17 | 8   | 11 | 9 | 3 | -4 | -12 | -20 | -24 | -28 | -35 | -44 | -54 | -65 | -72 | -69 | -55 | -37 | -21 | -9 | 1 | 8 |
| 18 | -5  | 0 | 1 | -4 | -11 | -21 | -29 | -35 | -39 | -47 | -57 | -62 | -60 | -52 | -43 | -34 | -28 | -23 | -18 | -11 | -5 |
| 19 | -16 | -12 | -9 | -11 | -18 | -27 | -36 | -41 | -43 | -50 | -63 | -71 | -65 | -51 | -37 | -27 | -22 | -20 | -19 | -18 | -16 |
| 20 | -15 | -15 | -14 | -16 | -20 | -25 | -31 | -35 | -37 | -44 | -58 | -68 | -67 | -58 | -46 | -37 | -28 | -21 | -16 | -15 | -15 |
| 21 | -3  | -1 | -4 | -12 | -17 | -17 | -15 | -16 | -21 | -31 | -44 | -56 | -62 | -63 | -60 | -55 | -48 | -37 | -23 | -10 | -3 |
| 22 | 12  | 28 | 25 | 7 | -9 | -14 | -11 | -8 | -11 | -20 | -33 | -47 | -61 | -71 | -75 | -75 | -74 | -67 | -46 | -16 | 12 |
| 23 | 19  | 52 | 56 | 32 | 0 | -23 | -31 | -30 | -25 | -26 | -32 | -45 | -62 | -77 | -83 | -84 | -89 | -89 | -69 | -28 | 19 |
| 24 | 14  | 50 | 62 | 43 | 7 | -28 | -53 | -59 | -52 | -43 | -39 | -44 | -56 | -68 | -74 | -75 | -79 | -81 | -67 | -32 | 14 |
| 25 | 5   | 30 | 42 | 34 | 10 | -21 | -49 | -64 | -64 | -58 | -52 | -46 | -44 | -47 | -50 | -51 | -52 | -52 | -44 | -23 | 5 |
| 26 | 2   | 13 | 20 | 19 | 9 | -9 | -28 | -43 | -52 | -58 | -58 | -48 | -35 | -27 | -25 | -25 | -25 | -24 | -19 | -10 | 2 |
| 27 | 1   | 5 | 7 | 6 | 3 | -3 | -11 | -18 | -25 | -35 | -41 | -36 | -24 | -14 | -8 | -7 | -7 | -6 | -5 | -2 | 1 |
| 28 | 0   | 1 | 2 | 1 | 0 | -1 | -1 | 1 | 4 | 4 | 0 | -4 | -4 | -1 | 1 | 1 | 0 | -1 | -2 | -1 | 0 |

Fig. VIII-8.
An automatic structure
solution from
MINDIR.

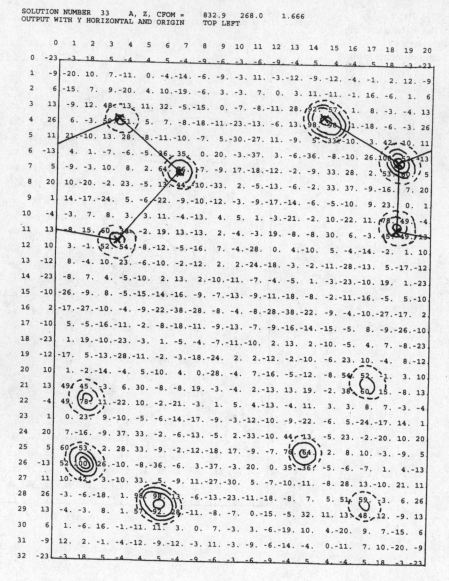

```
SOLUTION NUMBER  33    A, Z, CFOM =    832.9   268.0   1.666
OUTPUT WITH Y HORIZONTAL AND ORIGIN     TOP LEFT
```

Table VIII-1.

| k | l | sign | k | l | sign | k | l | sign | k | l | sign |
|---|---|------|---|---|------|---|---|------|---|---|------|
| 2 | 0 | + | 4 | 2 | + | 0 | 4 | − | 11 | 5 | + |
| 4 | 0 | + | 5 | 2 | − | 1 | 4 | − | 0 | 6 | − |
| 6 | 0 | − | 6 | 2 | − | 2 | 4 | − | 1 | 6 | + |
| 8 | 0 | − | 7 | 2 | − | 3 | 4 | − | 2 | 6 | − |
| 10 | 0 | − | 8 | 2 | + | 4 | 4 | + | 3 | 6 | + |
| 12 | 0 | + | 9 | 2 | + | 5 | 4 | − | 4 | 6 | u |
| 1 | 1 | + | 10 | 2 | − | 6 | 4 | + | 5 | 6 | + |
| 2 | 1 | − | 11 | 2 | + | 7 | 4 | + | 6 | 6 | u |
| 3 | 1 | + | 12 | 2 | u | 8 | 4 | + | 7 | 6 | u |
| 4 | 1 | − | 13 | 2 | + | 9 | 4 | + | 8 | 6 | + |
| 5 | 1 | − | 1 | 3 | − | 10 | 4 | + | 9 | 6 | − |
| 6 | 1 | − | 2 | 3 | − | 11 | 4 | + | 1 | 7 | + |
| 7 | 1 | − | 3 | 3 | − | 12 | 4 | u | 2 | 7 | + |
| 8 | 1 | − | 4 | 3 | − | 1 | 5 | − | 3 | 7 | + |
| 9 | 1 | − | 5 | 3 | + | 2 | 5 | + | 4 | 7 | + |
| 10 | 1 | + | 6 | 3 | − | 3 | 5 | − | 5 | 7 | − |
| 11 | 1 | − | 7 | 3 | + | 4 | 5 | + | 6 | 7 | + |
| 12 | 1 | + | 8 | 3 | + | 5 | 5 | u | 7 | 7 | − |
| 13 | 1 | − | 9 | 3 | + | 6 | 5 | + | 0 | 8 | + |
| 0 | 2 | − | 10 | 3 | + | 7 | 5 | + | 1 | 8 | u |
| 1 | 2 | − | 11 | 3 | + | 8 | 5 | u | 2 | 8 | + |
| 2 | 2 | − | 12 | 3 | + | 9 | 5 | + | 3 | 8 | u |
| 3 | 2 | − | 13 | 3 | u | 10 | 5 | − | 4 | 8 | u |

u = unobserved.

Table VII-2.

THE SEED USED IS 3000

| SOLN | A | Z | CFOM | SOLN | A | Z | CFOM |
|------|---|---|------|------|---|---|------|
| 1 | 670.3 | 296.7 | 1.315 | 2 | 646.9 | 370.3 | 1.171 |
| 3 | 617.7 | 544.2 | 0.880 | 4 | 748.8 | 348.6 | 1.396 |
| 5 | 563.7 | 358.9 | 1.026 | 6 | 608.7 | 428.2 | 1.020 |
| 7 | 779.4 | 491.1 | 1.263 | 8 | 693.2 | 450.5 | 1.152 |
| 9 | 573.6 | 286.2 | 1.144 | 10 | 759.8 | 543.2 | 1.155 |
| 11 | 889.6 | 650.4 | 1.259 | 12 | 758.9 | 601.1 | 1.075 |
| 13 | 722.1 | 454.8 | 1.201 | 14 | 754.8 | 440.1 | 1.284 |
| 15 | 538.6 | 316.6 | 1.035 | 16 | 730.9 | 367.7 | 1.336 |
| 17 | 698.0 | 416.8 | 1.206 | 18 | 640.0 | 513.5 | 0.964 |
| 19 | 527.2 | 284.3 | 1.057 | 20 | 800.3 | 597.2 | 1.160 |
| 21 | 1000.0 | 1000.0 | 1.000 | 22 | 543.3 | 319.1 | 1.041 |
| 23 | 916.2 | 821.2 | 1.080 | 24 | 479.7 | 394.2 | 0.817 |
| 25 | 588.6 | 355.3 | 1.079 | 26 | 1000.0 | 1000.0 | 1.000 |
| 27 | 887.8 | 659.0 | 1.245 | 28 | 565.3 | 411.5 | 0.959 |
| 29 | 598.1 | 464.4 | 0.950 | 30 | 770.3 | 608.2 | 1.087 |
| 31 | 614.2 | 393.4 | 1.077 | 32 | 698.5 | 557.7 | 1.017 |
| 33 | 832.9 | 268.0 | 1.666 | 34 | 810.1 | 513.1 | 1.292 |
| 35 | 707.8 | 349.1 | 1.317 | 36 | 659.9 | 411.6 | 1.140 |
| 37 | 846.7 | 651.3 | 1.176 | 38 | 784.5 | 571.9 | 1.163 |
| 39 | 610.2 | 409.3 | 1.048 | 40 | 711.3 | 356.1 | 1.314 |
| 41 | 715.1 | 443.3 | 1.204 | 42 | 800.8 | 505.2 | 1.285 |
| 43 | 805.9 | 528.6 | 1.263 | 44 | 634.9 | 421.3 | 1.079 |
| 45 | 1000.0 | 1000.0 | 1.000 | 46 | 664.3 | 258.9 | 1.355 |
| 47 | 655.3 | 475.9 | 1.045 | 48 | 524.3 | 481.8 | 0.785 |
| 49 | 706.7 | 381.6 | 1.271 | 50 | 668.3 | 549.7 | 0.970 |

## Chapter 9

9.1 The calculation using equations (9.2) and (9.4) gives the results presented in table IX-1.

Table IX-1.

| h k | $\theta$ | $\sin^2 \theta$ | $a(\text{Å})$ |
|-----|----------|-----------------|---------------|
| 7 1 | 64° 27′ | 0.8140 | 6.013 |
| 7 2 | 66° 35′ | 0.8421 | 6.012 |
| 7 3 | 70° 35′ | 0.8895 | 6.008 |
| 7 4 | 77° 45′ | 0.9560 | 6.005 |
| 6 5 | 64° 55′ | 0.8203 | 6.020 |
| 6 6 | 73° 54′ | 0.9231 | 6.016 |

A plot of the determined values of $a$ against $\sin^2 \theta$ shows considerable scatter, in particular the last two points give values which are far too high. A best straight line for the other points gives $a = 6.002 \pm 0.001$ Å. Try the calculation again with $a/b = 0.747$.

9.2 In the $(h0l)$ layer of the reciprocal lattice we have

$$s^2 = \frac{4 \sin^2 \theta}{\lambda^2} = h^2 a^{*2} + l^2 c^{*2} + 2hla^* c^* \cos \beta^*.$$

From the (300), (400) and (500) reflections one finds an average value of $a^* = 0.23091\,\text{Å}^{-1}$ and the next four reflections similarly give $c^* = 0.16496\,\text{Å}^{-1}$. The final reflections give $\cos \beta^*$ from

$$\cos \beta^* = \frac{(4 \sin^2 \theta / \lambda^2) - h^2 a^{*2} - l^2 c^{*2}}{2hla^* c^*}$$

and the mean value of $\cos \beta^*$ is 0.499 97 which gives $\beta^* = 60° 0'$. For a monoclinic lattice $\beta = 180° - \beta^* = 120° 0'$. We now find

$$a = (a^* \sin \beta)^{-1} = 5.001\,\text{Å}$$

and

$$c = (c^* \sin \beta)^{-1} = 7.000\,\text{Å}.$$

9.3 The data is plotted on a reciprocal-lattice net and is divided into regions of $\sin \theta / \lambda$ enabling table IX-2 to be constructed.
Plotting $\ln(\langle I \rangle / \Sigma)$ indicates that the innermost point does not fit the general trend. A straight line as close as possible to the other points gives

$$K = 0.30 \text{ and } B = 2.6\,\text{Å}^{-2}.$$

9.4 (i) $0.0874\,e\,\text{Å}^{-3}$. (ii) $0.0060\,\text{Å}$.

Table IX-2.

| $\dfrac{\sin\theta}{\lambda}$ | $\left\langle\dfrac{\sin^2\theta}{\lambda^2}\right\rangle$ | $\left\langle\dfrac{\sin^2\theta}{\lambda^2}\right\rangle^{\frac{1}{2}}$ | $28f_c^2$ | $12f_o^2$ | $\Sigma$ | $\langle I\rangle$ | $\ln\dfrac{I}{\Sigma}$ |
|---|---|---|---|---|---|---|---|
| 0.1–0.2 | 0.025 | 0.158 | 481 | 483 | 964 | 99.1 | −2.28 |
| 0.2–0.3 | 0.065 | 0.255 | 235 | 265 | 500 | 80.6 | −1.83 |
| 0.3–0.4 | 0.125 | 0.353 | 129 | 143 | 272 | 59.3 | −1.52 |
| 0.4–0.5 | 0.205 | 0.452 | 85 | 81 | 166 | 14.7 | −2.46 |
| 0.5–0.6 | 0.305 | 0.552 | 72 | 50 | 122 | 7.0 | −2.84 |

9.5 CALOBS gives a reliability index of 0.401 for the original estimated coordinates and the map shown in fig. IX-1. Revised coordinates estimated from the map are:

Na (0.943, 0.255),  Na (0.475, 0.605),  O (0.014, 0.690),

N (0.196, 0.390),  C (0.114, 0.200),  C (0.391, 0.155),

C (0.333, 0.390).

Fig. IX-1.
The map obtained as the first stage in a Fourier refinement procedure.

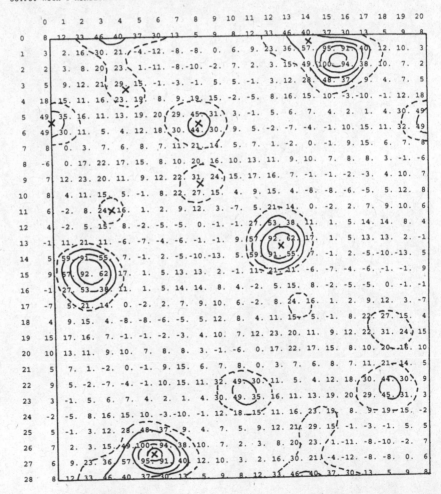

Structure factors calculated with these new estimates gave a reliability index of 0.360. It should be noted that the origin is free to move along the $y$ direction in this space group and that as the refinement proceeds there may be a gradual drift of $y$ coordinates towards higher or lower values. However, this does not prevent the refinement from taking place.

# References

Abrahams, S. C. & Robertson, J. M. (1948), *Acta Cryst.* **1**, 252.

Alcock, N. W. & Sheldrick, G. M. (1967), *Acta Cryst.* **23**, 35.

Archer, E. M. (1948), *Acta Cryst.* **1**, 64.

Baker, T. W., George, J. D., Bellamy, B. A. & Causer, R. (1966), *Nature* **210**, 720.

Beevers, C. A. & Lipson, H. S. (1934), *Phil. Mag.* **17**, 825–6.

Bertaut, E. F. (1955), *Acta Cryst.* **8**, 544.

Bhuïya, A. K. & Stanley, E. (1963), *Acta Cryst.* **16**, 981.

Blow, D. M. & Crick, F. H. C. (1959), *Acta Cryst.* **12**, 794.

Bokhoven, C., Schoone, J. C. & Bijvoet, J. M. (1951), *Acta Cryst.* **4**, 275.

Bond, W. L. (1960), *Acta Cryst.* **13**, 814.

Bragg, W. L. (1913), *Proc. Camb. Phil. Soc.* **17**, 43.

Bragg, W. L., James, R. W. & Bosanquet, C. H. (1921), *Phil. Mag.* **41**, 309.

Brown, G. M. & Bortner, M. H. (1954), *Acta Cryst.* **7**, 139.

Buerger, M. J. (1950a), *Acta Cryst.* **3**, 465.

Buerger, M. J. (1950b), *Proc. Nat. Acad. Sci.* **36**, 376.

Clews, C. J. B. & Cochran, W. (1948), *Acta Cryst.* **1**, 4.

Cochran, W. (1951), *Acta Cryst.* **4**, 376.

Cochran, W. (1952), *Acta Cryst.* **5**, 65.

Cochran, W. (1953), *Acta Cryst.* **6**, 260.

Cochran, W. (1955), *Acta Cryst.* **8**, 473.

Cochran, W. & Penfold, B. R. (1952), *Acta Cryst.* **5**, 644.

Cochran, W. & Woolfson, M. M. (1955), *Acta Cryst.* **8**, 1.

Cruickshank, D. W. J., Pilling, D. E., Bujosa, A., Lovell, F. M. & Truter, M. R. (1961), *Computing Methods and the Phase Problem in X-ray Crystal Analysis.* Oxford: Pergamon.

de Jong, W. F. & Bouman, J. (1938), *Z. Krist.* **98**, 456.

Dunitz, J. D. (1949), *Acta Cryst.* **2**, 1.

Dutta, S. N. & Woolfson, M. M. (1961), *Acta Cryst.* **14**, 178.

Farquhar, M. C. M. & Lipson, H. (1946), *Proc. Phys. Soc. Lond.* **58**, 200.

Fiocco, G. & Thompson, E. (1963), *Phys. Rev. Letters* **10**, 89.

Foster, F. & Hargreaves, A. (1963a), *Acta Cryst.* **16**, 1124.

Foster, F. & Hargreaves, A. (1963b), *Acta Cryst.* **16**, 1133.

Fowweather, F. & Hargreaves, A. (1950), *Acta Cryst.* **3**, 81.

Friedrich, W., Knipping, P. & von Laue, M. (1912) *Interferenz-Erscheinungen bei Röntgenstrahlen.* Sitsungsberichte der Kgl. Bayerischen Akademie der Wissenschaften zu München, pp. 303–22.

Groth, P. (1910), *Chemische Kristallographie.* Leipzig: Engelmann.

Harker, D. (1936), *J. Chem. Phys.* **4**, 381.

Harker, D. & Kasper, J. S. (1948), *Acta Cryst.* **1**, 70.

Hönl, H. (1933), *Ann. Phys.* (Leipzig) **18**, 625–57.

Howells, E. R., Phillips, D. C. & Rogers, D. (1950), *Acta Cryst.* **3**, 210.

Hughes, E. W. (1941), *J. Am. Chem. Soc.* **63**, 1737.

*International Tables for X-ray Crystallography*, Vols. I, II, III. Birmingham: The Kynoch Press.

*International Tables for Crystallography*, Vol. A, Space-Group Symmetry. Dordrecht: Reidel.

*International Tables for Crystallography*, Vol. B, Reciprocal Space. Dordrecht: Reidel.

*International Tables for Crystallography*, Vol. C, Mathematical, Physical and Chemical Tables. Dordrecht: Reidel.

Karle, I. L. & Karle, J. (1964a), *Acta Cryst.* **17**, 1356.

Karle, I. L. & Karle, J. (1964b), *Acta Cryst.* **17**, 835.

Karle, J. & Hauptmann, H. (1956), *Acta Cryst.* **9**, 635.

Klug, A. (1958), *Acta Cryst.* **11**, 515.

Lipson, H. & Woolfson, M. M. (1952), *Acta Cryst.* **6**, 439.

Main, P. & Woolfson, M. M. (1963), *Acta Cryst.* **16**, 731.

Milledge, H. J. (1962), *Proc. Roy. Soc.* A, **267**, 566.

Mills, O. S. (1958), *Acta Cryst.* **11**, 620.

Nakanishi, H. & Sarada, Y. (1978), *Acta Cryst.* **B34**, 332.

Okaya, Y., Saito, Y. & Pepinsky, R. (1955), *Phys. Rev.* **98**, 1857.

Patterson, A. L. (1934), *Phys. Rev.* **46**, 372.

Perales, A. & Garcia-Blanco, S. (1978), *Acta Cryst.* **B34**, 238.

Pitt, G. J. (1948), *Acta Cryst.* **1**, 168.

*Powder Diffraction File*, (Ed. W. L. Berry) Joint Committee on Powder Diffraction Standards (JCPDS), 1601 Park Lane, Swarthmore, PA 19081, U.S.A.

Pradhan, D., Ghosh, S. & Nigam, G. D. (1985), *Structure and Statistics in Crystallography*, (Ed. A. J. C. Wilson). Guilderland, New York: Adenine Press.

Reitveld, H. M. (1967), *Acta Cryst.* **22**, 151.

Renninger, M. (1937), *Z. Krist.* **97**, 107.

Rogers, D. & Wilson, A. J. C. (1953), *Acta Cryst.* **6**, 439.

Rollett, J. S. (1965) *Computing Methods in Crystallography*. Oxford: Pergamon.

Sayre, D. (1952), *Acta Cryst.* **5**, 60.

Sim, G. A. (1957), *Acta Cryst.* **10**, 536.

Sim, G. A. (1959), *Acta Cryst.* **12**, 813.

Sim, G. A. (1960), *Acta Cryst.* **13**, 511.

Stanley, E. (1964), *Acta Cryst.* **17**, 1028.

Thomas, J. T., Robertson, J. H. & Cox, E. G. (1958), *Acta Cryst.* **11**, 599.

Weiss, O., Cochran, W. & Cole, W. F. (1948), *Acta Cryst.* **1**, 83.

Wilson, A. J. C. (1942), *Nature*, **150**, 152.

Wilson, A. J. C. (1949), *Acta Cryst.* **2**, 318.

Wilson, A. J. C. (1950), *Acta Cryst.* **3**, 397.

Woolfson, M. M. (1954), *Acta Cryst.* **7**, 61.

Woolfson, M. M. (1956), *Acta Cryst.* **9**, 804.

Woolfson, M. M. (1958), *Acta Cryst.* **11**, 393.

Woolfson, M. M. & Fan, H.-F. (1995), *Physical and Non-Physical Methods of Solving Crystal Structures*. Cambridge: Cambridge University Press.

Wooster, W. A. (1938), *Crystal Physics*. Cambridge: Cambridge University Press.

Wrinch, D. M. (1939), *Phil. Mag.* **27**, 98.

Yao, J.-X. (1981), *Acta Cryst.* **A37**, 642.

Zachariasen, W. H. (1952), *Acta Cryst.* **5**, 68.

Zachariasen, W. H. (1967), *Acta Cryst.* **23**, 558.

# Bibliography

The books given in the list below, many of them specialist texts, between them cover a much wider range of material and go to much greater depth than does this introductory textbook. They represent a selection of a very large number of books in the field. A comprehensive list is given in the *Crystallographic Book List* edited by Helen D. Megaw and published in 1965 by the International Union of Crystallography Commission on Crystallographic Teaching. An update, produced by Dr J. H. Robertson, appears in the *Journal of Applied Crystallography* (1982) **15**, 640–76.

Arndt, U.W. & Willis, B.T.M. *Single Crystal Diffractometry*. Cambridge: University Press, 1966.

Azároff, L.V., Kaplow, R., Kato, N., Weiss, R.J., Wilson, A.J.C. & Young, R.A. *X-ray Diffraction*. New York: McGraw-Hill, 1974.

Bacon, G.E. *Neutron Diffraction*. Oxford: University Press, 1962.

Brown, P.J. & Forsyth, J.B. *The Crystal Structure of Solids*. London: Edward Arnold, 1973.

Buerger, M.J. *X-ray Crystallography*. New York: Wiley, 1958.

Cracknell, A.P. *Crystals and their Structures*. Oxford: Pergamon, 1969.

Glusker, J.P. & Trueblood, K.N. *Crystal Structure Analysis: A Primer*. Oxford: University Press, 1985.

Guinier, A. (translated by Lorrain, P. & Lorrain, D.S.) *X-ray Diffraction in Crystals, Imperfect Crystals and Amorphous Bodies*. San Francisco: Freeman, 1963.

Henry, N.F.M., Lipson, H. & Wooster, W.A. *The Interpretation of X-ray Diffraction Photographs*. London: Macmillan, 1960.

James, R.W. *X-ray Crystallography*. London: Methuen, 1948.

James, R.W. *The Optical Principles of the Diffraction of X-rays* (*The Crystalline State*, Vol. 2). London: Bell, 1965.

Ladd, M.F.C. & Palmer, R.A. *Structure Determination by X-ray Crystallography*. New York: Plenum, 1978.

Lipson, H. & Cochran, W. *The Determination of Crystal Structures* (*The Crystalline State*, Vol. 3). London: Bell, 1966.

Lonsdale, K. *Crystals and X-rays*. London: Bell, 1948.

McKie, D. & McKie, C. *Essentials of Crystallography*. Oxford: Blackwell, 1986.

Pepinsky, R., Robertson, J.M. & Speakman, J.C. (Eds.) *Computing Methods and the Phase Problem in X-ray Crystal Analysis: Glasgow*. Oxford: Pergamon, 1961.

Phillips, F.C. *An Introduction to Crystallography*. London: Longmans, 1971.

Ramachandran, G.N. & Srinivasan, R. *Fourier Methods in Crystallography*. New York: Wiley, 1970.

Ramaseshan, S. & Abrahams, S.C. (Eds.) *Anomalous Scattering*. Copenhagen: Munksgaard, 1975.

Rollet, J.S. *Computing Methods in Crystallography*. Oxford: Pergamon, 1965

Schenk, H., Wilson, A.J.C. & Parthasarathy, S. (Eds.) *Direct Methods, Macromolecular Crystallography and Crystallographic Statistics*. Singapore: World Scientific, 1987.

Stout, G.H. & Jensen, L.H. *X-ray Structure Determination*. New York: Macmillan, 1968.

Taylor, C.A. & Lipson, H. *Optical Transformations*. London: Bell, 1964.

Wilson, A.J.C. *X-ray Optics*. London: Methuen, 1962.

Wilson, A.J.C. (Ed.) *Structure & Statistics in Crystallography*. Guilderland, New York: Adenine, 1985.

Woolfson, M.M. *Direct Methods in Crystallography*. Oxford: University Press, 1961.

Woolfson, M.M. & Fan Hai-fu. *Physical and Non-Physical Methods of Solving Crystal Structures*. Cambridge: University Press, 1995.

# Index